INTERNATIONAL UNION OF CRYSTALLOGRAPHY

T0177530

Early Days of
X-ray Crystallography

ANDRÉ AUTHIER

Institut de Minéralogie et de Physique des Milieux Condensés,
Université P. et M. Curie, Paris, France

INTERNATIONAL UNION OF CRYSTALLOGRAPHY

UNIVERSITY PRESS

OXFORD

UNIVERSITY PRESS

Great Clarendon Street, Oxford, OX2 6DP,
United Kingdom

Oxford University Press is a department of the University of Oxford.
It furthers the University's objective of excellence in research, scholarship,
and education by publishing worldwide. Oxford is a registered trade mark of
Oxford University Press in the UK and in certain other countries

First published 2013
First published in paperback 2015

Impression: 1

Published in the United States of America by Oxford University Press
198 Madison Avenue, New York, NY 10016, United States of America

British Library Cataloguing in Publication Data
Data available

Library of Congress Cataloging in Publication Data
Data available

ISBN 978–0–19–965984–5 (Hbk.)
ISBN 978–0–19–875405–3 (Pbk.)

Printed and bound in Great Britain by
Clays Ltd, St Ives plc

Front cover, upper figure: Laue diagram of zinc-blende, W. Friedrich, P. Knipping, and M. Laue
(1912) © Photo Deutsches Museum, Munich.

Front cover, lower figures (left-right): Sir W. H. Bragg © The Nobel Foundation; P. P. Ewald,
Polytechnic Institute of Brooklyn, courtesy AIP Emlio Segre Visual Archives; M. von Laue
© The Nobel Foundation; Sir W. L. Bragg © The Nobel Foundation.

Back cover figures: Stamps from the author's private collection. *1st and 3rd stamps from left*: from a
series issued by the Swedish Post Office to commemorate Nobel Prize winners. *2nd from left*: stamp
issued by the German Post Office to commemorate Nobel Prize winners. *2nd from left*: stamp issued
by the German Post Office to commemorate M. von Laue's Nobel Prize. *4th from left*: stamp issued
by the British Post Office to commemorate W. H. Bragg and W. L. Bragg's Nobel Prize.

Dedication

To my wife Irena

PREFACE

The 2012 Nobel Prize for Chemistry was awarded to R. Lefkowitz and B. K. Kobila for studies related to crystal structural insights into 'G-protein-coupled receptors' signalling, just one hundred years after one of the most significant discoveries of the early twentieth century, the discovery of X-ray diffraction by M. Laue, W. Friedrich, and P. Knipping (April 1912), and the birth of X-ray analysis (W. L. Bragg, November 1912). This award is the latest in a long list of Nobel Prizes awarded for work related to X-ray crystallography. Crystallography has come a very long way since the first crystal structure determinations by W. H. and W. L. Bragg, father and son, in the spring of 1913. The momentous impact of Laue's discovery in the fields of chemistry, physics, mineralogy, material science, biochemistry, and biotechnology has been recognized by the General Assembly of the United Nations by establishing 2014 as the International Year of Crystallography.

The immediate result of Friedrich, Knipping, and Laue's discovery was to confirm the wave nature of X-rays and the regular three-dimensional arrangement of atoms in crystals. It had two major consequences: the analysis of the structure of atoms, and the determination of the atomic structure of materials. The aim of this book is to give a fair account of the events surrounding the discovery itself. It relates the successive stages of the concept of space-lattice, from Kepler to Haüy, Schoenflies, and Fedorov, and the discussions about the nature of X-rays, wave or corpuscular. Within fifteen years or so of the discovery, the main bases of X-ray analysis were established, and our understanding of the nature of chemical bonds in solids had to be revised. The number of publications devoted to X-ray crystallography increased exponentially, and only the main results achieved during these first years are outlined in this book.

Historians of science have pointed out that, for a variety of reasons, the accounts by scientists as to how they reached a certain conclusion, or made a certain discovery, are not always reliable. A typical example in our story is that of R.-J. Haüy telling how the sight of a calcite crystal falling and breaking into small rhombohedra was at the origin of his investigations. This is how a legend was born, to be repeated later in every textbook. The tale of the discovery of X-ray diffraction by its discoverers has similarly been contested. As noted by Kragh (1987), scientists have a tendency to rationalize after the events in the light of later developments. An error also made sometimes in historical writings by scientists is to ascribe more recent concepts to ancient authors. Another point underlined by Whitaker (1979) is that scientists usually 'start from modern ideas and attempt to explain how they came about rather than trying to understand the approach of former generations, they rewrite history so that it fits in step by step with the physics'. This is sometimes called 'quasi-history', and I have done my best to avoid its pitfalls and anachronisms.

When considering a discovery, there are two possible approaches. The first, favoured by historians, is to view the discovery solely in the light of the knowledge of the time. The other, which is difficult to avoid, is to situate it within the framework of the development, usually not linear, of a scientific concept. It is indeed difficult for anyone writing about a given discovery to ignore the advances that followed. Our present knowledge may even help us to make a critical analysis of the action of the discoverer and to understand, for instance, why a particular reasoning went wrong.

In writing this book, my aim has been simply to outline the successive steps that our understanding of the inner structure of crystals went through, and to tell the story of the early days of X-ray crystallography, putting the events into the context of their time.

André Authier
Peyrat-le-Château, October 2012

ACKNOWLEDGEMENTS

This book would never have seen the light of day without the unfailing and constant help and support of Helmut Klapper. Not only did he read it critically as I was writing it, but he provided me with a great number of the earlier German texts which are not accessible via the Internet. I am greatly indebted to Jenny Glusker for her detailed remarks and useful suggestions on every chapter. I thank them both heartily. Francesco Abbona, John Edwards-Moss, Dieter Hoffmann, Shaul Katzir, Peter Paufler, and Marjorie Senechal read several chapters and sent me very useful comments, I am very grateful to them. I am also indebted to Francesco Abbona, Leonid Aslanov, Jean-Claude Boulliard, Vladimir Dmitrienko, Michael Eckert, Timothy Fawcett, Howard Flack, Hartmut Fuess, Hugo Heinemann, John Helliwell, Dieter Hoffmann, Thomas Kaemmel, Joseph Lajzerowicz, Anthony North, Yuji Ohashi, Peter Paufler, Bjørn Pedersen, and Ullrich Pietsch, who provided material or illustrations for the book.

Many thanks are due to Mickael Eckert of the Deutsches Museum in Munich for very fruitful discussions, and to Moreton Moore, who introduced me to the Collections of the Royal Institution in London, as well as to Frank James, Head of the Collections, and Jane Harrison for their efficient help with the Bragg manuscripts.

Sönke Adlung, the Physics Senior Editor of Oxford University Press, encouraged me all along. His constant help was very much appreciated.

Last, but not least, I would like to express my deep gratitude to my wife, Irena, without whose love and patience this book would never have been written.

CONTENTS

SIGNIFICANCE OF THE DISCOVERY OF X-RAY DIFFRACTION

*The history of X-ray diffraction ranks as one of the epoch-making discoveries
in the history of science.*

Sir W. H. Bragg and Sir W. L. Bragg (1937)

1.1 April 1912: a major discovery

April 1912. Munich, Institute for Theoretical Physics. The photographic plate, developed after a few hours of exposure, 'betrayed the presence of a considerable number of deflected rays, together with a trace of the primary ray coming directly from the anticathode' (Laue, Nobel lecture, 12 November 1915). This was the first photograph to show the diffraction of X-rays by a crystal. In this preliminary experiment, a crystal of copper sulphate pentahydrate was put in the X-ray beam with a natural face roughly normal to it, but otherwise arbitrarily oriented. This particular crystal was chosen because of its good shape, and also because it contained copper and it was thought that the expected result 'would have something to do with fluorescence' (Friedrich *et al.* 1912). Fig. 1.1, *Left*, shows the very first X-ray diffraction photograph, with rather diffuse spots and Fig. 1.1, *Right*, one of the next pictures, with narrower slits for the beam to pass through, and exhibiting much sharper spots. The story of the discovery is told in Chapter 6. Its importance was immediately felt by the three participants: the conceiver of the experiment, Max Laue, and the two experimentalists, Walter Friedrich and Paul Knipping, as well as by the circle of A. Sommerfeld's and W. C. Röntgen's co-workers meeting regularly at the Café Lutz in Munich. The feeling of excitement spread very rapidly to their contemporaries in Germany and abroad (see Section 6.7).

The significance of the discovery was two-fold.

1. *Firstly*, it might throw some light on the nature of X-rays: are they electromagnetic waves or corpuscles? This point had been debated hotly since the discovery of X-rays by W. C. Röntgen in 1895 and is the topic of Chapter 5. As early as 1896, O. Lodge (1896c), G. G. Stokes (Stokes 1896a, b), and E. Wiechert (1896a, b) had put forward the hypothesis that Röntgen rays were electromagnetic waves. The same suggestion was made by J. J. Thomson (1898a). C. G. Barkla had discovered the polarization of X-rays in 1904 (Barkla 1905a, b). H. Haga and C. H. Wind (1903) and B. Walter and R. Pohl (1908, 1909) had observed the diffraction of X-rays by a slit (see Section 5.9). Their experiments, controversial as they were, led to estimates of the X-ray wavelengths by A. Sommerfeld (1912). Another evaluation was made by W. Wien (1907) who, generalizing Planck's theory of light quanta, reckoned that the maximum energy transferred by X-rays to secondary electrons of velocity v and mass m is $(1/2)mv^2 = hc/\lambda$ (h Planck constant, c velocity of light, λ X-ray wavelength), and deduced the X-ray wavelength from the measurement of the velocities of the electrons. Independently, J. Stark (1907), also using Planck's relation, deduced the energy of the light quanta from the energy of the primary electrons accelerated through a potential V (see Section 5.10).

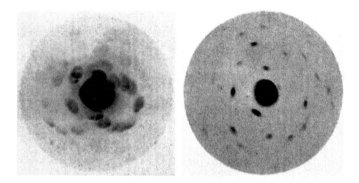

Fig. 1.1 *Left*: very first picture of a copper sulphate crystal. *Right*: new picture obtained with narrower slits. After Friedrich, Knipping, and Laue (1912).

On the other hand, William Henry Bragg in England was a staunch defender of the corpuscular theory. He held the view that the X-rays were material in nature, rather than æther pulses, and consisted of neutral pairs, such as pairs consisting of one α or positive particle and one β or negative particle (W. H. Bragg 1907; Bragg and Madsen 1908a, b, see Section 5.8). The fact that the ionization observed in an ionization chamber was clearly due to electrons produced by the X-rays and not by the X-rays themselves was considered to point to the corpuscular nature of X-rays. Stark (1909a, b) was led to the same conclusion by the photoelectric effect. W. H. Bragg's reaction to Laue's experiment was reserved (W. H. Bragg 1912a, see Section 6.10): he saw 'some remarkable effects', and 'a curious arrangement of spots some of them so far removed from the central spot that they must be ascribed to rays which make large angles with the original pencil'. He concluded that 'it is difficult to distinguish between various explanations which suggest themselves. It is clear, however, that the diagram is an illustration of the arrangement of the atoms in the crystal'. He added that, as his son (William Lawrence Bragg) pointed out to him, the directions of the secondary rays follow 'avenues' between the crystal atoms, a view also held by J. Stark (1912, see Section 6.9). This interpretation was in line with the interpretation of X-rays as corpuscles. Lawrence Bragg was soon to change his own attitude, however, after some unsuccessful experimental attempts (Ewald 1962b), and, in a paper read by J. J. Thomson to the Cambridge Philosophical Society on 11 November 1912, he explained Laue's experiment by the reflection of electromagnetic waves in a set of lattice planes (W. L. Bragg 1913a, see Section 6.11). His father wondered whether the effects observed were due to the X-ray themselves or to electromagnetic radiation associated with corpuscular rays (W. H. and W. L. Bragg 1937), but he came round reluctantly (W. H. Bragg 1912b), insisting that 'the properties of X-rays point clearly to a quasi-corpuscular theory'. 'The problem', he added, is not 'to decide between the two theories, but to find one theory which possesses the capacities of both'. It is only after the discovery of the Compton effect (Compton 1923b, see Section 9.5) and the formulation by L. de Broglie of the relations between the properties of light and those of the atom that the dual nature, corpuscular and wave, of X-rays was really understood.

Lawrence Bragg (1912), at the suggestion of his former optics teacher, C. T. R. Wilson,[1] quickly checked his explanation of the diffraction of X-rays by crystals by observing on a

---

[1]Charles Thomson Rees Wilson, born 14 February 1869 in Scotland, died 15 November 1959 in Scotland, was a British physicist and meteorologist who received the physics Nobel Prize in 1927 for his invention of the cloud chamber.

photographic plate the reflection of X-rays by a sheet of mica, while his father recorded it with an ionization chamber (W. H. Bragg 1913a). Recording the diffracted intensities with the ionization chamber proved in the future to be a much more powerful means for examining crystal structures than Laue photographs.

2. *Secondly*, the discovery confirmed the space-lattice hypothesis. At the beginning of August 1912, the English crystallographer A. E. H. Tutton[2] visited Paul von Groth, Director of the Institute of Mineralogy in Munich, and Max Laue, at Arnold Sommerfeld's Institute of Theoretical Physics. In his account of the visit, published in *Nature* under the title *The crystal space lattice revealed by Röntgen rays* (Tutton 1912b, 14 November), Tutton reported that Groth and Laue had showed him 'extraordinary' photographs obtained a few months before by W. Friedrich and P. Knipping, and had expressed the opinion that they formed 'an interference (diffraction) photograph of the Bravais space lattice'. He concluded that 'they do in reality afford a visual proof of the modern theory of crystal structures built up by the combined labours of Bravais, Schoenflies, Fedorov, and Barlow. Moreover, they emphasize in a remarkable manner the importance of the space lattice'. He added that, in his opinion, Friedrich and Knipping's photographs 'may form a crucial test of the accuracy of the two rival theories, now being discussed as to the nature of X-rays, the corpuscular and the wave theory'.

In his Becquerel Memorial Lecture to the Chemical Society on 17 October 1912, Sir Oliver Lodge[3] commented (Lodge 1912): 'This, if it be a fact, will have to be recognized as a striking and admirable case of scientific production, the various crystalline structures and accuracy of characteristic facets having been indicated by theory long before there was any hope of actually seeing them; so that once more—always assuming that the heralded discovery is substantiated—the theoretical abstraction will have become concrete and visible'.

Ewald, in the accounts he gave much later of the discovery of X-ray diffraction (Ewald 1932, 1962a and b), wrote that the concept of space lattice was then discredited, on one hand because the Cauchy relations between elastic constants, derived assuming a space lattice, were not verified by experiment, and on the other hand because there was no physical property which could be related to it. This position was strongly contested by the historian of science, P. Forman (1969), who accused Laue and Ewald of having created a myth out of the discovery of X-ray diffraction (see Section 6.12). Section 1.2 describes the views of the crystallographers of the time concerning the space-lattice concept. The problem of the validity of the Cauchy relations is discussed in Section 2.3.

He started work on the cloud chamber in 1895 in the Cavendish Laboratory in Cambridge and observed in early 1896 the 'rain-like' condensation after exposure to the newly discovered X-rays. In 1911, he was the first to observe the tracks of individual alpha- and beta-particles and electrons, confirming W. H. Bragg's predictions. He was lecturer and demonstrator at the Cavendish Laboratory at the time of W. L. Bragg's first experiments. He was awarded the 1927 Nobel Prize for Physics, 'for his method of making the paths of electrically charged particles visible by condensation of vapour'; the prize was divided between him and A. H. Compton.

[2]Alfred Edwin Howard Tutton, born 22 August 1864 in England, died 14 July 1938 in England, educated at the Royal College of Science in London, was an English crystallographer. He was a lecturer and instructor in chemistry and, at the same time, inspector of Technical Schools in London, Oxford, and Devon successively. All his life was devoted to precise measurements of crystals of several series of isomorphous salts, such as sulphates, selenates and their double salts.

[3]Sir Oliver Lodge, born 12 June 1851 in England, died 22 August 1940 in England, was an English physicist. He obtained his BSc degree and his DSc from the University of London. After being Professor of Physics and Mathematics at University College, Liverpool, he became in 1900 the first principal of the new Birmingham University, remaining there until his retirement in 1919. He elaborated on Maxwell's æther theory and is well known for his work on electrolysis, electromagnetic waves, and wireless telegraphy.

Both M. Laue's interpretation of the diagrams by means of the Laue equations (Friedrich *et al.* 1912) and W. L. Bragg's with Bragg's law (W. L. Bragg 1913*a*) established a fundamental relationship between X-ray wavelengths and the parameters of the crystal lattice. This had two major impacts for the future development of the science:

1. *Atomic structure of materials*: Knowing the X-ray wavelength, one can determine the lattice parameters of crystals; more generally, the analysis of the X-ray diagrams leads to the determination of the arrangement of atoms in materials. This 'became one of the powerful weapons of modern science, of which it has affected nearly every branch', in the words of Sir W. L. Bragg (1959), as will be outlined in Section 1.3. The first steps of crystal structure determination will be described in Chapter 8.
2. *Structure of atoms*: Conversely, knowing the lattice parameter of a crystal, one could determine X-ray wavelengths. Measurements with W. H. Bragg's ionization spectrometer showed that the spectrum contained both a continuous spectrum of 'Bremstrahlung' pulses and characteristic wavelengths, which corresponded to those discovered by Barkla (1906*b*). The relations between the wavelengths of the emission spectra determined by means of X-ray diffraction and Barkla's absorption spectra were generalized by H. G. J. Moseley (1913, 1914) who assigned atomic numbers to the elements (see Section 7.8). Moseley's pioneer investigations formed the starting point for the experimental researches on X-ray spectra, in particular by the Siegbahn group, and the theories about the structure of the atom by Kossel, Sommerfeld, Pauli, and others. The need for improvements in the theory followed the increase in the accuracy of the measurements of X-ray frequencies, and these measurements 'were and remain the most direct and accurate method to determine the energy levels of the atoms' (Cauchois 1964). The early stages of the development of X-ray spectroscopy are briefly recalled in Section 10.6.

1.2   Crystallography on the eve of the discovery of X-ray diffraction

The concept of space lattice (point-systems) introduced by Bravais (1848) was the result of a long process (see Chapters 11 and 12). It was generally accepted by crystallographers in the beginnings of the twentieth century, and it was also known by many chemists, for instance, by the American physical chemist Harry Clary Jones (1865–1916) who wrote in his *Elements of physical chemistry* (Jones 1903): 'In crystals the particles are arranged in a perfectly orderly manner, and fulfil the condition that the arrangement about one point is the same as about any other point'. Modern theories were described in several writings. Tutton's book (Tutton 1911) had presented a simplified approach, and Hilton (1903) had made a synthesis in English of Fedorov's, Schoenflies', and Barlow's formal derivations of the 230 groups. The definition of a crystal had evolved, from Haüy's simple one, that 'every crystal has a regular shape and its faces can be represented by a geometrical figure', to the sophisticated definition given by P. Groth in a lecture to the British Association for the Advancement of Science during a meeting held in Cambridge on 19 August 1904: 'A crystal—considered as indefinitely extended—consists of interpenetrating regular point systems, each of which is formed from similar atoms; each of those point systems is built up from interpenetrating space lattices, each of the latter being formed from similar atoms occupying parallel positions. All the space lattices of the combined system are geometrically identical and characterized by the same elementary parallelepiped' (Groth 1904).

Haüy and those who followed him had implicitly assumed that crystals are what we would call now 'molecular crystals', namely crystals where there are recognizable molecules in the structure, linked by weak forces. Groth (1888) made a very thorough discussion, based on considerations about dimorphism and isomorphism, on what could be the nature of the chemical molecule in the crystal. But, what with Barlow's work on close-packed structures (see Section 12.13) and the recent theories of crystal structure involving the repetition of elemental domains (*Fundamentalbereiche*, Schoenflies 1891) by the symmetry operations of the space groups (see Section 12.14), new ideas began to appear as to the possible nature of the crystal molecules. In his 1904 lecture, Groth noted that, while molecules in an amorphous substance represent individual entities, in crystals they are assemblages of atoms belonging to different point systems. For instance, in sodium chloride, one cannot tell to which particular sodium atom a chlorine atom is linked, a view which, at the time, was not accepted by all chemists (see Section 10.1.1). This idea was proposed by W. Barlow and H. Pope (see Section 12.13), but was still found to be 'more than repugnant to the common sense, not chemical cricket' by the influential chemist H. E. Armstrong (1927), fourteen years after W. L. Bragg (1913*b*) had determined the structure!

Some people, however, considered the derivation of the 230 space groups as a purely mathematical theory, not as a physical one. They criticized the fact that the theory did not make any hypothesis as to the nature of the content of the repeating units, and felt that it did not add anything to the principle of the triple periodicity of crystals (Friedel 1907). The well-known specialist of crystal physics, W. Voigt, a student of F. E. Neumann's (see footnote 18, section 2.3), wrote in 1910 that the works of Schoenflies and Fedorov are predominantly mathematical in nature and that 'it is too early to judge their physical value' (Voigt 1910). Ewald (1962*b*) was correct in saying that 'no physicist, no crystallographer was able to assign correctly to a given crystal one, or even a small selection, of the 230 possibilities of arrangement which this theory predicted, and it was left wide open for discussion what was the nature of the particles that were arranged in the three-dimensionally periodic fashions which the theory allowed'. The situation changed of course radically after the discovery of X-ray diffraction. It was soon to appear that the only way to interpret the diffraction diagram of a crystal and to describe its structure was by reference to the space group to which it belongs (Niggli 1919).

The only crystallographic information that was available to mineralogists and chemists came, as in Haüy's time, from the observation of the crystal habit and the cleavage planes, and from the accurate measurement of angles. A great amount of data was gathered, but the general consensus was that, to have any hint about the structure, this information should be combined with the best knowledge possible of all the other properties (Groth 1904; Friedel 1907). It was generally agreed that the characteristics of the elemental parallelepiped of the lattice should be determined from the most important faces of the crystal, chosen according to their occurrence and their extent. They correspond to the planes of highest reticular density (Fedorov 1902, 1912; Groth 1904; Friedel 1907), and this is nothing else but the Bravais law, which states that the cleavage planes are the planes with the highest interplanar distance (see Section 12.11.3), and which had remained unnoticed for fifty years. Friedel (1907) and Fedorov (1902, 1912) pointed out that the order of the most important faces observed does not always coincide with the theoretical order because some faces may be missing, in particular in merohedral crystals. More

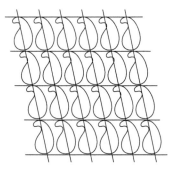

Fig. 1.2   Schematic diagram of a two-dimensional crystal structure
showing how a simple object is repeated from one unit parallelepiped to the next. After Lord Kelvin, 1894*b*.

generally, this is due to the fact that the Bravais law is based on the lattice and not on the space
groups. This was only recognized after the first studies by X-ray diffraction (Niggli 1916, 1919).

In those days there was an explosion of the number of crystallographic studies of minerals and
synthetic crystals. In a most interesting paper, already mentioned, published on the same month
as the official announcement of the discovery of X-ray diffraction, Fedorov (1912) reported that
he had compiled a table of about 10 000 substances, with, for each of them, the crystallographic
characteristics of the elemental parallelepiped, and the list of the most important faces ordered
according to decreasing reticular densities. For half of them, a stereographic projection was also
given (the projection is the same for isomorphous compounds). This table had been compiled
using the works of many crystallographers throughout the world, in particular those published in
Zeitschrift für Krystallographie und Mineralogie and in the first volumes of Groth's *Chemische
Krystallographie*. In that way, given the measurement of just a few crystal angles of an unknown
substance among these 10 000, it would be easy to identify it with a good probability. In his
review of the paper, Tutton (1912*a*) related that, like other crystallographers from Europe and
the United States, he had been invited by Fedorov to test the procedure by sending him several
substances and that all of them had been identified without hesitation by Fedorov. Tutton
enthusiastically concluded that this 'epoch-making memoir' showed that 'crystallography is
of fundamental importance to chemistry'.

There was a good deal of information about crystallography available. The history of geometrical crystallography from Guglielmini to the derivation of the space-groups was reviewed
in a report to the British Association for the Advancement of Science by Barlow and Miers
(1901). In a popular lecture to the Oxford University Junior Scientific Club, on 16 May 1894,
entitled *The molecular tactics of a crystal*, Lord Kelvin (Sir W. Thomson)[4] simulated in a
variety of ways the homogeneous assemblage of molecules which constitutes a crystal (Lord
Kelvin 1894*b*), for instance by an assembly of people seated in rows on the superposed floors
of a building or the stacking of close-packed balls. Fig. 1.2 is a schematic drawing by Lord

---

[4]Sir William Thomson, 1st Baron Kelvin, born 26 June 1824 in Belfast, Ireland, died 17 December 1907 in Scotland,
was a British physicist. He received his higher education at the Universities of Glasgow and Cambridge where he
graduated as Second Wrangler. At the age of twenty-two he was appointed to the chair of natural philosophy in the
University of Glasgow. He is best known for his formulation of the first and second Laws of Thermodynamics and
his work on electricity and magnetism. He also worked as an electric telegraph engineer and made calculations for the
submarine transatlantic telegraph cable.

Kelvin of the molecular structure of a two-dimensional crystal, namely the repetition of a given figure by the translations of a two-dimensional lattice. The illustrations used in modern-day class-rooms are no different.

1.3   Impact of the discovery on the chemical, biochemical, physical, material, and mineralogical sciences

Laue's discovery 'has extended the power of observing minute structure ten thousand times beyond that of the [optical] microscope' (W. H. and W. L. Bragg 1937). Two hundred and fifty years after R. Hooke's and A. van Leewenhoeck's first observations (see Section 11.5), X-ray diffraction produced a microscope with atomic resolution that does not show the atoms themselves, but their electron distribution. It has opened the way for entirely new developments in chemistry, solid-state physics, mineralogy, geosciences, materials science, and biocrystallography. Early applications of X-ray crystallography are described in Chapter 10.

X-ray diffraction was followed by its daughters, electron diffraction, discovered independently by C. J. Davisson and L. H. Germer (1927*a* and *b*) and by J. J. Thomson's son, G. P. Thomson, and A. Reid (1927), and neutron diffraction, discovered by H. von Halban and P. Preiswerk (1936) and first applied to crystallography by C. G. Shull and E. O. Wollan (1948). Together, they provide information on the structure of matter, crystalline and non-crystalline, at the atomic and molecular levels. As such, they are closely linked to the properties of all materials, inorganic, organic, or biological. Their impact can be felt in every aspect of the modern world: materials science with semiconductors, superconductors, super magnets and super alloys, biology with the development of new drugs, biotechnology, nanotechnology.

An indication of the importance and diversity of applications of diffraction by crystals is given by the large number of Nobel Prizes awarded for studies involving X-ray, neutron, or electron diffraction:

- *X-ray, electron or neutron diffraction*: M. von Laue, 1914 Physics (discovery of X-ray diffraction); W. H. and W. L. Bragg, 1915 Physics (crystal structures); M. Siegbahn, 1924 Physics (X-ray spectroscopy); A. H. Compton, 1927 Physics (Compton effect), C. J. Davisson and G. P. Thompson, 1937 Physics (electron diffraction); J. Karle and H. A. Hauptman, 1985 Chemistry (direct methods); C. G. Shull 1994, Physics (principles of neutron diffraction).
- *Chemistry*: P. J. W. Debye, 1936 Chemistry (molecular structure); L. C. Pauling, 1954 Chemistry (nature of the chemical bond).
- *Structures of biological interest*: J. C. Kendrew and M. Perutz, 1962 Chemistry (globular proteins); F. H. C. Crick, J. D. Watson and M. H. F. Wilkins, 1962 Physiology or Medicine (DNA); D. C. Hodgkin, 1964 Chemistry (vitamin B-12), A. Klug, 1982 Chemistry (nucleic acid-protein complexes); J. Deisenhofer, R. Huber, and H. Michel, 1988 Chemistry (three-dimensional structure of a photosynthetic reaction centre), V. Ramakrishnan; T. Steitz and A. Yonath, 2009 Chemistry (ribosome); R. Lefkowitz and H. Hughes, 2012 Chemistry (crystal structure determination of G-protein-coupled receptors).
- *Structures of inorganic materials*: W. N. Lipscomb, 1976 Chemistry (boranes), J. G. Bednorz's and K. A. Müller's, 1987 Physics (superconductivity of ceramic materials); R. Curl, H. Kroto, and R. Smalley, 1996 Chemistry (fullerenes); D. Shechtman's, 2011 Chemistry (quasicrystals).

The impact of the discovery of X-ray diffraction on the various branches of science has been developed in many publications, such as the books by W. L. Bragg (1975), P. P. Ewald (1962*a*), J. Lima de Faria (1990), and the special issues of *Naturwissenschaften* (1922, **10**) for the tenth anniversary of the discovery of X-ray diffraction, Current Science, Indian Academy of Science (1937) for the 25th Anniversary, *Beiträge zur Physik und Chemie des 20. Jahrunderts* (1959), for the eightieth birthday of Lise Meitner, Otto Hahn and Max von Laue, *Zeitschrift für Kristallographie* (1964, **120**), for the fiftieth anniversary of Laue's discovery, *Acta Crystallographica*, **A54**, 1998 (*Crystallography across the Sciences*) and **A64**, 2008 (*Crystallography across the Sciences 2*) for the fiftieth and sixtieth anniversaries of *Acta Crystallographica*. See also the review by W. L. Bragg (1968*b*).

2

THE VARIOUS APPROACHES TO THE CONCEPT

OF SPACE LATTICE

For just as all things of creation are,
In their whole nature, each to each unlike,
So must their atoms be in shape unlike.

Lucretius, quoted by Burke (1966)

The spots of a diffraction diagram occur because the diffracting crystal is arranged according to a space lattice. But how did the concept of space lattice arise before the discovery of X-ray diffraction? How did it develop, from the early perceptions of the Ancients to its rigorous definition by A. Bravais and the derivations of groups of movement by L. Sohncke, E. S. Fedorov, A. Schoenflies, and W. Barlow? The evolution of the concept was not linear and followed different routes. There are various ways to analyse them. Burke (1966) distinguished two theories of matter, which emerged in the second half of the eighteenth century:

1) The 'molecular' theory, which assumed 'the existence of primitive indestructible particles or atoms'. They combined to form molecules or corpuscles which exerted attractive and repulsive forces on one another. These molecules could be either spherical or polyhedral and their aggregation by purely mechanical means served to explain their macroscopic chemical and physical properties. This is the approach followed by the French school of R.-J. Haüy (Section 12.1), G. Delafosse (Section 12.10) and A. Bravais (Section 12.11), and by L. A. Seeber (Section 12.7) in Germany.

2) The 'polar' theory, 'which denied the existence of ultimate atoms, believing instead that matter was divisible to infinity'. The equilibrium of matter was dynamic and resulting from the reciprocal action of forces both attractive and repulsive and centred at mathematical points, or poles. This is the concept that was adopted by the German school of C. S. Weiss and F. Mohs. It will be discussed in Sections 12.3 and 12.5, respectively.

In the present chapter, we shall consider successively the geometrical aspects of the molecular theories, which characterize the two main approaches to the space-lattice concept, space-filling by polyhedra and close-packing of spheres, and the molecular theories of the elasticity of solids, the difficulties of which partially discredited that concept.

Chapters 11 and 12 at the end of the book are intended to throw a spotlight on some of the actors of the elaboration of the space-lattice hypothesis, without attempting to be exhaustive.

For further reading on the development of the space-lattice concept, the reader may consult the books by H. Metzger (1918), P. von Groth (1926), P. P. Ewald (1962*a*, Chapter II-3), J. G. Burke (1966), I. I. Shafranovskii (1978, 1980), J. J. Burckhardt (1988), J. Lima de Faria (1990), H. Kubbinga (2002) and the articles by P. von Groth (1925), M. von Laue (1952*a*),

W. F. de Jong and E. Stradner (1956), J. J. Burckhardt (1967–68), M. Senechal (1981), K. H. Wiederkehr (1981), F. Abbona (2000), M. I. Aroyo *et al.* (2004), J. N. Lalena (2006), and J.-L. Hodeau and R. Guinebretière (2007).

2.1 The space-filling approach

The problem of filling space with congruent polyhedra without leaving any gaps is one of the oldest ones and was, of course, solved in part by stone masons when they built walls. It was first discussed theoretically by Plato's most famous student, Aristotle (384–322 BC), in relation to the former's theory of matter and to the five regular convex polyhedra known as the Platonic solids (Fig. 2.1). Their discovery is attributed to the Pythagorean school (*ca* 550 BC), and their properties were demonstrated by Euclid, in book XIII of his *Elements*, around 300 BC. Much later, Kepler tried to match the orbits of the planet with them (Section 11.4.1).

Plato (*ca* 427–347 BC) associated the tetrahedron with fire, the cube with earth or solid matter, the octahedron with air, and the icosahedron with water. In his dialogue, *Timaeus*, he considered that all matter was constituted by combinations of particles of these four polyhedra. The fifth one, dodecahedron, was associated with the cosmos or æther.

Aristotle thought that the particles constituting a given substance should be packed in such a way as to fill space completely, without leaving any voids, for, according to him, voids cannot exist in nature. In the plane, equilateral triangles, squares, and hexagons could pack without leaving voids. In space, only two of the five regular polyhedra can fill space without leaving any gap: the cube and the tetrahedron. Here, Aristotle made an error, since it is not possible to pave space with tetrahedra (Senechal 1981). The properties of the three polygons were well-known to the geometricians of the Ancient times; in particular, they knew that it is the hexagon which has the largest area, as does the honeycomb cell (Sections 11.2.1 and 11.2.3). The first to discover other space-filling polyhedra was Kepler: the rhomb-dodecahedron and the hexagonal prism (Section 11.4.3). Such polyhedra are called parallelohedra. The domains introduced by Dirichlet (1850) when deriving a new proof of Seeber's theorem in relation to the properties of ternary quadratic forms belong to the parallelohedra (Section 12.7.2). They were further discussed by Voronoï (1908), and the Wigner–Seitz cells and the Brillouin zones of today are also Dirichlet/Voronoï domains. Parallelohedra were studied systematically by Fedorov (1885), who showed that they can be classified in five topological types (Section 12.14). The division of space with minimum partitional area was also discussed by Lord Kelvin, who considered two of

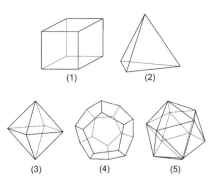

Fig. 2.1   The five Platonic solids: (1): cube; (2): tetrahedron;
(3) octahedron; (4): dodecahedron; (5): icosahedron.

Fedorov's parallelohedra, the rhomb-dodecahedron and the cubooctahedron (Lord Kelvin 1887, 1894*a* and *b*).

The term 'crystal' was introduced by the Greeks and the Romans to designate quartz (Section 11.1). Quartz and diamond were among the first minerals for which the polyhedral shape and, for quartz, the symmetry was noticed. Pliny the Elder (Section 11.2.2), Albertus Magnus (Section 11.2.4), Agricola (Section 11.2.5), Cardano (Section 11.2.6), for instance, noted that they were always found with six faces. Drawings of various shapes of quartz crystals are given in Boece de Boot's (1609) book on precious stones (Fig. 11.1). The first to have distinguished the various types of shapes of minerals was Agricola, but in a very rough way. The first to have shown that crystals could be characterized by their shape was D. Guglielmini (1688, 1705, Section 11.9): common salt by a cube, vitriol by a rhombohedral parallelepiped,[1] nitre by a hexagonal prism,[2] and alum by an octahedron.[3] The first to have based his classification of minerals on the shape was Cappeller (1723, Fig. 11.2).

The first major step towards understanding the geometrical structure of crystals was the observation of the 'constancy of interfacial angles'. It was implicit in all the observations of the symmetry of quartz crystals from antiquity to the Middle Ages, but it was in the seventeenth century that it was clearly established. It was implied by R. Hooke's reconstruction of the faces of alum crystals (Section 11.5.2, Fig. 11.17), by R. Bartholin's (Section 11.6, Fig. 11.21) and C. Huygens' (Section 11.8) measurements of the angles of calcite faces, and also by the observation of the hexagonal snowflakes by many authors (Kepler, Descartes, Gassendi, Hooke, Bartholin, Steno, Section 11.4.9), but it was N. Steno, however, who stated it in a general way for quartz (Section 11.7, Fig. 11.23). It was also Steno who gave the first correct ideas about the layer by layer growth of crystals, while in the Middle Ages (Albertus Magnus, Section 11.2.4) and until the beginning of the seventeenth century (A. Boece de Boot 1609, Section 11.4.6; J. Kepler (1611), Section 11.4; P. Gassendi 1658, Section 11.4.9), the Aristotelian view prevailed that it was due to the 'formative' power of Nature. The constancy of angles was also observed on crystals of nitre by M. V. Lomonosov (1746), but the generalization of the law to all crystals had to wait for J.-B. L. Romé de l'Isle and the measurements made with the contact goniometer developed by his student, A. Carangeot, at the end of the eighteenth century (Section 11.11).

The idea that crystals could be built up from an assemblage of identical parallelepipeds was suggested by the easiness with which calcite can be broken into small rhombohedra (Fig. 2.2). The cleavage of calcite was first observed by R. Bartholin (Section 11.6) and C. Huygens (Section 11.8); it led the latter to propose a close-packed arrangement of ellipsoids for its structure. It is interesting for the history of science that nearly simultaneously, in the second half of the eighteenth century, four mineralogists independently proposed that calcite crystals could be built up of small rhombohedra: the German C. F. G. Westfeld (1767), the Swedes J. G. Gahn and T. O. Bergman (1773), the Englishman W. Pryce (1778), and the Frenchman R.-J. Haüy (1782*b*). Westfeld and Pryce did not push their idea (Section 12.1), and Bergman's

---

[1] Vitriol, copper sulphate pentahydrate, $CuSO_4$, 5 H_2O, is, in fact, triclinic, space group $P\bar{1}$.

[2] Nitre, or saltpeter, KNO_3, is, in fact, orthorhombic pseudo-hexagonal, space-group $Pmcn$, with the aragonite structure.

[3] Alum, hydrated potassium aluminium sulfate, $KAl(SO_4)_2$, 12 H_2O, is cubic, space group $m\bar{3}m$.

Fig. 2.2   Calcite rhombohedra. Photo by the author.

theory was, in fact, a lamellar theory which only worked well for the interpretation of the dog-tooth scalenohedron (Section 11.10). Haüy knew Bergman's original paper (1773), but does not seem to have paid attention to Gahn's chance observation. In any case, he is the one who developed the idea of the triple periodicity of crystals (Section 12.1).

Haüy was an atomist, and his fundamental hypothesis was that by pushing the cleavage process to its ultimate limit one would obtain a small polyhedron that could not be broken without destroying its physical and chemical nature. He initially named these polyhedra 'constituting molecules' but later changed the name to 'integrant molecules'. His theory was based on the idea that a crystal is a tri-dimensional regular assembly of parallelepipedic building blocks. These blocks were either the integrant molecules themselves if their shape was parallelepipedic or else a combination of integrant molecules, called 'subtractive molecules'. These periodic assemblies are built up by stacks of parallel layers, themselves composed of parallel rows (Fig. 2.3). Their fundamental property is that by decreasing the number of rows on the successive topmost layers of the stack, one can generate a plane having the orientation of a crystal face (Section 12.1.2). By repeating this operation along several directions of faces or

Fig. 2.3   Application of the law of decrements to reconstruct a calcite scalenohedron. Two successive layers of the structure are shifted with respect to one another by a step of width *s* and height *h*. Redrawn after Haüy (1801).

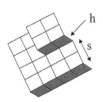

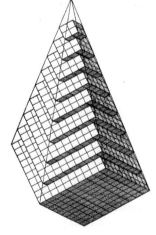

around a summit, one may thus generate all the faces of a given crystal form (Figs. 12.5 and 2.3). The width s of the shift and the height h of the step between two successive layers (Fig. 2.3) defines a certain law, called 'law of decrements'. The shift and the height of the steps are integral numbers of rows and layers, and they are usually small. The law is therefore also called the 'law of simple rational truncations'. Haüy's regular arrangement is in fact described by a lattice. If one refers it to a set of axes, the law becomes the 'law of simple rational intercepts', which is the fundamental law of geometrical crystallography. Haüy's seminal theory constitutes therefore the fundamental introduction to modern crystallography.

Haüy's influence on his contemporaries was considerable. He was, however, also much criticized. His angle measurements were not very accurate and were sometimes slightly adjusted so that the ratio between his parameters (sines and cosines) was simple. This was not accepted by the mineralogists of the time, one of whose main activities was the measurement of crystal angles. This was particularly true after the invention of the reflection goniometer by Wollaston (1809, Section 12.2), which significantly increased the accuracy of the measurements. Haüy considered, for simplicity, his integrant molecule to be the physical molecule, although he was the first to admit that this assumption was not necessary for his theory. For him, the form of the crystal was characteristic of the substance, and two different substances could not have the same form. This is why he was not ready to accept Mitscherlich's major contribution to crystallography, the theory of isomorphism, which appeared when Haüy was already in his later years (Mitscherlich 1819). All these points entailed a loss of credibility for Haüy, which he regained later in the century when the concept of crystal periodicity became firmly established.

The introduction of axes to express the properties of crystals is due to C. S. Weiss (Section 12.3). An adept of the German *romantische Naturphilosophie* of the late eighteenth century, he was not an atomist and he developed, on the contrary, a dynamistic theory of crystalline matter. According to this theory, crystals are in equilibrium under the reciprocal action of attractive and repulsive forces exerted along directions related to the main growth faces and the cleavage planes. There are in general three main directions of forces in a crystal. The direction of a face may be defined by considering its intercepts with these axes; they are the Weiss numbers or indices (Section 12.3.6). Weiss's student, F. E. Neumann, preferred to use their reciprocals; this notation was taken up by W. Whewell and the latter's successor, W. H. Miller, and is now known as the Miller notation (Section 12.6).

In the habit of a crystal, there are, in general, groups of faces which have a direction in common; they constitute zones with a zone axis. Any crystal face can be determined if one knows two zone axes parallel to it. It is the zone law (*Zonenverbandgesetz*). In modern notation, it is expressed by the so-called Weiss law, $hu + kv + lw = 0$, expressing that the direction $[uvw]$ of a zone axis is parallel to the crystal face (hkl).

Weiss's (1815) most important contribution was the classification of crystals in crystal systems, based on a consideration of their symmetry around their main axes (Section 12.3.5). He defined seven systems: five referred to three orthogonal axes, *tesseral* (cubic), *viergliedriges* (tetragonal), *zwei-und-zwei-gliedriges* (orthogonal), *zwei-und-ein-gliedriges* (monoclinic) and *ein-und-ein-gliedriges* (triclinic), and two referred to four axes, three at 120° of one another and the fourth one normal to them, *sechsgliedriges* (hexagonal) and *drei-und-drei-gliedriges* (trigonal). Erroneously, Weiss referred the monoclinic and triclinic systems to three orthogonal axes, but this error was corrected by F. Mohs (Section 12.5).

Interestingly, Weiss's and Mohs's crystal systems were corroborated by the Scottish physicist D. Brewster (1818, Section 12.4) who, at roughly the same time, was studying systematically the double refraction of a large number of crystals. His classification in biaxial, uniaxial, and non-birefringent crystals matched exactly the crystallographic one. It is rather surprising that this correlation between physical and crystallographic properties went unheeded for many years.

Weiss and Mohs had noted that, for some crystals, their symmetry was lower than that of the crystal system to which they belonged; Mohs' student C. F. Naumann coined the terms 'holohedry', 'hemihedry', and 'tetartohedry' (Section 12.5), but it is M. L. Frankenheim (Section 12.8.1) and J. F. C. Hessel (Section 12.9) who showed that there are in fact 32 classes of symmetry. Their work remained unnoticed for many years except for that of Frankenheim who was quoted by A. Bravais (1849). Bravais derived the groups of symmetry in a different way, finding 31 only. The reason was that he discarded $\bar{4}$ because it had not been observed in nature (Section 12.11.3). The 32 groups were again derived by A. Gadolin (1871), using stereographic projections (Section 12.11.3), and by E. S. Fedorov (1883), P. Curie (1884), A. Schoenflies and F. Klein (1887), B. Minnigerode (1887), and L. Sohncke (1888) using group theory (Section 12.14).

The next important step in the development of the concept of space lattice was made by L. A. Seeber (Section 12.7). He tried to reconcile Haüy's atomistic views and Weiss's dynamistic ones. Solid bodies are composed of minute particles, which is confirmed by their cleavage properties. Their cohesion results from the reciprocal action of attractive and repulsive forces exerted on one another by these particles. For them to be in equilibrium, the resultant of the forces acting on any one particle must be equal to zero and this requires the arrangement of the particles to be regular. Seeber showed that they cannot be in contact with one another, to allow for expansion, and replaced Haüy's polyhedra by small spherical particles located at their centres of gravity. They are, therefore, at the nodes of a lattice, although Seeber did not use these terms; he speaks of a 'parallelepipedic arrangement' instead.

In order to express the forces that two particles exert on one another, Seeber was led to calculate the distance between any two particles, and showed that its square is a ternary quadratic form. In a book on the properties of ternary quadratic forms (Section 12.7.2), Seeber (1831) derived empirically that their determinant has a lower limit. Gauss (1831), in his review of the book, proved Seeber's theorem and gave a geometric interpretation that has an interesting crystallographic implication. A ternary quadratic form can in general be interpreted as the square of the distance between two lattice points of a lattice, the parameters of which depend on the coefficients of the form. Its determinant is the volume of the unit cell, and its lower limit corresponds to a rhombohedron with an angle of 60°. The modern crystallographer recognizes that this is the unit cell of a face-centred cubic lattice. The face-centred cubic packing is thus shown to be the densest lattice, a result already announced by Kepler and known as the 'Kepler conjecture' (Section 11.4.8).

M. L. Frankenheim (Section 12.8.2) also introduced the notion of a lattice, without using the term. Starting from a molecular conception of matter, he thought, like Seeber, that it consists of small particles separated from one another and in equilibrium under the action of forces. A line drawn through any two particles will intersect other particles at regular intervals. It is a lattice row! From this assumption, he deduced (without any proof) that the molecules can be arranged according to fifteen families, which he called 'orders'. They corresponded

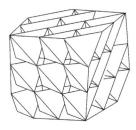

Fig. 2.4   Delafosse's proposal for the structure of boracite. Redrawn after Delafosse (1843).

in fact to Bravais' lattice modes, and Bravais (1849) showed that two of them are actually identical.

The concept of space lattice appears as such in the work of Haüy's last pupil, G. Delafosse (Section 12.10). As in the case of Seeber, whose work he was not aware of, Delafosse's conception of the nature of crystals was a synthesis of Haüy's and Weiss's. He distinguished clearly between physical molecules, which constitute the unit cells (*mailles*) of a lattice (*réseau*), and Haüy's integrant or subtractive molecules, the centres of gravity of which are located at the intersections of the families of lattice planes (the nodes of the lattice). Weiss's axes have a physical meaning and lie along rows (*rangées*) of molecules. The physical molecules always have the same orientation in a crystal (Fig. 2.4), which implies a symmetry of translation, and they impose their symmetry to the external shape of the crystal. Delafosse's attitude towards crystals was markedly different from that of his predecessors. For Haüy, the external shape and the external symmetry of the crystal came first, and the integrant molecules second. For Delafosse, it was the other way round: the internal structure was the most important factor and the external symmetry was simply a consequence of the internal symmetry. Delafosse used this concept to propose a structural interpretation of hemihedry. A crystal is hemihedral when the symmetry of the physical molecule and of the assembly of molecules is lower than that of the lattice. Boracite, magnesium borate chloride ($Mg_3B_7O_{13}Cl$) represented in Fig. 2.4, provides a good example. The structure proposed by Delafosse was suggested to him by its habit, which is cubic with small tetrahedral facets; actually, its point group is $mm2$.

It is to A. Bravais (1848, see Section 12.11), however, that one owes the mathematical theory of space lattices, based on his postulate that, 'given any point of the system, there is in the medium a discrete, unlimited number of points, in the three directions of space, around which the arrangement of atoms is identical, with the same orientation'. He derived all their geometrical properties, and determined the fourteen lattice modes. He calculated the unit area of a given lattice plane and showed that it is the inverse of the reticular area of that plane. He then ordered the lattice planes according to their decreasing reticular density for each one of the lattice modes, and made the important hypothesis that the most important planes of the crystal habit are those with maximum reticular density. This is the so-called Bravais law, in fact an empirical rule, which was taken up by the mineralogists of the early twentieth century. Another major contribution by Bravais was the introduction of the polar lattice, which is homothetic to the reciprocal lattice, and which he used for crystallographic calculations in the same way as we do now. This aspect of Bravais' work was not referred to when the reciprocal lattice became common practice in the early days of X-ray diffraction, probably because it was by reference to J. W. Gibbs (1881) that it was introduced by M. von Laue (1914).

With Bravais, the concept of space lattice was firmly established, and one may concur with Schoenflies (1891, p. 314), who summarized its development: 'the fact that Haüy's theory presupposed the reticular arrangement of the elemental bricks of the crystal was expressed for the first time in more precise form by Delafosse. He achieved that as a result of speculations which aimed at replacing Haüy's purely geometrical conceptions by real physical molecules. In Delafosse, we can see the actual founder of the lattice theory. Nonetheless it is only right and proper that the theory should bear Bravais' name, since Bravais is the one who proved for the first time that the lattice structures obey the same symmetry relations as can be found in crystals'.

The French school of the first half of the nineteenth century thus played a major role in the elaboration of the concept of space lattice, while it is to the German School of the early nineteenth century that we owe the concept of crystal systems.

Bravais' postulate included an important constraint, namely that the molecules in a crystal are parallel. This condition was unnecessary and Sohncke dropped it (Section 12.12). This led to the determination of the 230 groups of movement, or space-groups, by Fedorov, Schoenflies, and Barlow (Section 12.14).

2.2   The close-packing approach

The other approach to the concept of space lattice is that of close packing, but it leads essentially to the cubic and hexagonal packings. One may think that the three- or two-dimensional models of close packing are too crude to represent reality, but they can in fact be observed with an electron microscope (d'Anterroches 1984), as shown in Fig. 2.5, *Right*. It can be simulated with bubbles on the surface of a soap solution (Fig. 2.5, *Left*) or with ball-bearing balls between two plates of plexiglass (Fig. 2.5, *Middle*).

The first attempt at considering close packing in a crystal in order to interpret their structure was probably made by G. Cardano (Section 11.2.6). He compared the six-fold symmetry of quartz to that of the honeycomb cell and imagined quartz to be constituted of spheres surrounded by fourteen others (the number of summits of the rhomb-dodecahedron, the shape of the honeycomb cell). The first explicit description of close packing is due to T. Harriot (Section 11.3.1) who considered the stacking of canon-balls. But the most detailed analysis of

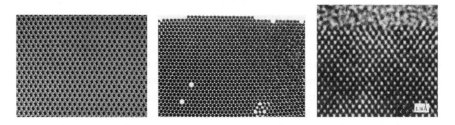

Fig. 2.5   *Left*: bubble model. After Bragg and Nye (1947). *Middle*: balls between two plexiglass plates. Photograph
by the author. *Right*: Electron micrograph of a silicon crystal capped by amorphous silicon oxide. Courtesy C.
d'Anterroches.

all the possible arrangements one can obtain by close-packing of spheres was made by J. Kepler when he tried to understand the six-cornered shape of snowflakes (Section 11.4). As stressed by M. von Laue (1952a), it is to Kepler that we owe the first graphical representations of space lattices. Very interestingly, it is from a closed-packed structure that he deduced the shape of space-filling polyhedra: by compressing the close-packed spherical seeds of pomegranate, one obtains rhomb-dodecahedra (Section 11.4.3). The same image was used more than two centuries later by W. Barlow (Section 12.13).

Under Kepler's influence, many of the philosophers and scientists of the first half of the seventeenth century, for instance R. Descartes and P. Gassendi (Section 11.4.9), and R. Bartholin (Section 11.6), tried, in a similar way, to explain the structure of snowflakes by the packing of six globules around a central one. R. Hooke (Section 11.5.2) went further and gave an interpretation of the external faces of alum crystals by changing the boundaries around heaps of small spheres. He suggested that this could be extended to other types of crystals. C. Huygens (Section 11.8) explained the double refraction of calcite by assuming a stacking of prolate ellipsoids. One problem that worried all the authors of close-packing theories was the problem of vacuum. Usually, they got round it by assuming that the pores between neighbouring particles was filled with some fluid which was the transmitting vector of light in transparent bodies: undefined fluid for Hooke and Bartholin, 'subtle matter' for Descartes, 'ethereal matter' for Huygens.

Robert Boyle[4] was one of the main adepts of the corpuscular theory. Although strongly influenced by Descartes and Gassendi, he always kept his independence of thought. Descartes did not believe in the existence of vacuum, but Boyle's experiments with the air pump when he derived the gas law led him to believe in it. He rejected Aristotle elements. For him, all bodies are made of a universal extended matter, 'often divided into imperceptible corpuscles or particles' (Boyle 1666), themselves made up of *minima naturalia*, which cannot be broken by any natural process. He did not use the words atoms or molecules, but thought that the corpuscles 'remain entire in a great variety of perceptible bodies under various forms or disguises'. He also assumed some sort of porousness to account for certain physical properties.

Close packing was again considered, nearly a century later, by the Russian scientist and poet M. V. Lomonosov[5] (Fig. 2.6) in his dissertation on saltpeter, or nitre (Lomonosov 1749). He assumed the elementary particles of matter to be spherical: 'from a round figure [corpuscle] come all the figures of salts, snow, and the like' (quoted by Shafranowskii 1975), and rediscovered the simple cubic packing and the face-centred close-packing. He could, in that way, explain why saltpeter grows into six-sided crystals (see footnote 2): 'suppose that six corpuscles are resting near one another such that the straight lines joining their centers form equilateral triangles. The resulting figure will be bounded by six lines, just like a cross section of the prisms comprising saltpeter' (quoted by Shafranowskii 1975).

---

[4]See footnote 20, Chapter 3.

[5]For details on Lomonosov's life, see, for instance, J. D. Bernal (1940). M. V. Lomonosov (1711–1765). *Nature* **146**, 16–17.

Fig. 2.6   Mikhail Vasilyevich Lomonosov. Portrait by L. S. Miropol'sky (after G. Prenner). Russian Academy of Sciences, Saint Petersburg.

Mikhail Vasilyevich Lomonosov: born 19 November 1711 in Denisovka, near Kholmogory, Arkhangelsk Governorate, Russia, the son of a peasant fisherman; died 15 April 1765 in Saint Petersburg, Russia, was a Russian physicist, chemist, mineralogist, and poet. He was educated first at the Slavic Greek Latin Academy in Moscow and then at the Imperial Academy of Sciences in Saint Petersburg. In 1737 he received a scholarship to go to Marburg University, in Germany, where he studied chemistry and mining. After spending some time in Freiburg, he came back to Saint Petersburg in 1741 to work on his dissertation. In 1745, he was appointed Professor of Chemistry at the Saint Petersburg Academy of Sciences, of which he also became a member. At the time, he was writing poetry and plays. In 1754–1755 he took part in the founding of Moscow University, and in 1760 became Director of the University and Gymnasium at the Saint Petersburg Academy of Science. His scientific work was very varied. He worked on combustion, on the kinetic theory of gases, and on the wave theory of light. He was also the first person to record the freezing of mercury.

MAIN PUBLICATIONS

1741 *Elements of Physical Chemistry.*	1752 *Real Physical Chemistry.*
1748 *Short Guide to Rhetoric.*	1752 *On the Utility of Glass.*
1749 *On the Genesis and Nature of Saltpeter.*	1757 *Russian Grammar.*

The close-packing approach was very popular all through the nineteenth century. These models were not merely geometrical, but tried to take into account the action of attractive and repulsive forces between molecules. J. J. Prechtl[6] (1808), in the wake of the dynamistic school, thought that no individual molecule pre-existed in the fluid. At the start of crystallization, globules are formed, which agglomerate. Under the force of attraction they exert on one another, they become flattened, taking up a polyhedral shape, and build up reticular crystals that look exactly like Haüy's drawings shown in Fig. 12.5. W. H. Wollaston (Section 12.2) came up with the same difficulties as Haüy with the question of what could be the primitive form of a crystal of which the smallest cleavage polyhedron is an octahedron or a tetrahedron, as in fluorspar. He solved the problem by assuming the crystal to be composed of small spherical particles that could agglomerate to form tetrahedra or octahedra (Fig. 12.14), in the manner of R. Hooke, of whom he was not aware at the start of his work. He also considered the packing of spheroids

---

[6]Johann Josef Ritter von Prechtl was an Austrian scientist of German origin, born 16 November 1778 in Bischofsheim, Germany, died 28 October 1854 in Vienna, Austria. He studied law at the University of Würzburg and was at first a private tutor in a rich family in Brünn (now Brno, Czech Republic). This is where he wrote his paper on crystallization. Later he went to Vienna where he taught physics, chemistry, and natural science and was one of the founders of the Technical University, of which he was the Director from 1814 until 1849.

and compared his results with those of Huygens. Wollaston's theory was further developed by J. F. Daniell[7] (1817) who studied the shapes taken by crystals when put in a solvent.

D. Brewster (1830) tried to correlate the optical properties of crystals with the arrangement of the molecules, by reference to Huygens' work. Molecules are arranged along main axes of attraction, and are uniaxial, biaxial, or non-birefringent depending on the number and relative importance of these axes (Section 12.4).

The well known American mineralogist, J. D. Dana[8] (1836), considered the molecules to be spherical, and to have six poles, at the extremities of three orthogonal axes, three of them 'positive' and three of them 'negative'. In the packing of spheres Dana took into account the attraction of unlike poles and the repulsion of like poles; in this way he could account in a satisfactory manner for the cubic structures, but not the hemihedral forms. Dana's theory was improved by the English mineralogist Robert T. Forster (1855) who considered eight or twelve poles.

It is W. Barlow who really developed the theory of close packing, first alone (1883, 1886), and later with the English chemist W. H. Pope. Together (Barlow and Pope 1906, 1907, 1908), they represented atoms by their spheres of influence and made predictions of real structures, such as those of the alkali halides, ZnS or calcite, which were confirmed by X-ray diffraction by W. L. and W. H. Bragg (Section 12.13).

2.3   The molecular theories of the early nineteenth century physicists

Ever since C. Huygens and I. Newton, physicists have considered matter to consist of a triply periodic regular assembly of particles. In *Opticks* (1704), Book III, query 31, Newton wrote: 'all bodies seem to be composed of hard particles, for otherwise fluids would not congeal. Since the particles of Iceland spar [calcite] act all the same way upon the rays of light for causing the unusual refraction, may it not be supposed that in the formation of this crystal, the particles not only ranged themselves in rank and file for concreting in regular figures, but also by some kind of polar virtue turned their homogeneal [*sic*] sides the same way'. The forces exerted on one another by these particles are of the same type as the gravitational forces.

R. Boscovich[9] (1758) reduced the particles to finite material points (*puncta*) interacting in pairs under the influence of attractive and repulsive forces (we would say now, central forces), thus avoiding the question of the interstices between atoms. All natural phenomena could be accounted for by the arrangement and motion of these points.

[7]John Frederic Daniell, born 12 March 1790, died 13 March 1845, was an English chemist and physicist who became in 1831 the first Professor of Chemistry at the newly founded King's College London. He is best known for the invention, in 1831, of an electric battery called the Daniell cell.

[8]James Dwight Dana, born 12 February 1817, in Utica, NY, USA, died 14 April 1895 in New Haven, CT, USA, was an American mineralogist, geologist, and zoologist. Educated at Yale College, he was at first assistant to B. Silliman in the chemical laboratory at Yale, and then, acted as mineralogist and geologist of the United States Exploring Expedition. In 1850 he succeeded Silliman as Professor of Natural History and Geology in Yale College. In 1835 he constructed a set of crystal models in glass, and, in 1837 compiled the first edition of his famous *System of Mineralogy and Crystallography*. He is also well-known for his pioneering work in volcanism.

[9]Ruder Boscovich, born 18 May 1711 in Ragusa, now Dubrovnik, Croatia, died 13 February 1783 in Milan, Italy, was a Jesuit theologian, mathematician, and physicist who made important contributions in astronomy, and developed an atomic theory in his work, *Theoria Philosophiae Naturalis*, based on the ideas of Newton and Leibnitz.

At the beginning of the nineteenth century, physicists made use of the molecular theory in optics and elasticity. A. Fresnel (1827), in his interpretation of double refraction, assumed that light was propagated by vibrating molecules and explained the optical rotation of certain crystals by an helicoidal arrangement of the molecules (Sections 3.6 and 12.10.2).

H. Navier[10] (1827) was the first to deal with the problem of the elasticity of solid bodies. In a memoir read to the French Academy of Sciences on 14 May 1821, he expressed the elastic relations in terms of the variations in the intermolecular forces which result from modifications of the molecular arrangement, and showed that the elastic behaviour of a body can be characterized by a single parameter.

A.-L. Cauchy[11] showed in 1822 that stress can be expressed in terms of six components and strain likewise in terms of six components, and generalized Hooke's law (Section 11.5) by replacing it by a set of linear relations between the two sets of components.[12] His calculation included the hypothesis of central forces between material points. He found that, in the general case, the relations between stress and strain components involve twenty-one constants, out of which fifteen are true elastic constants, and the other six express the initial stress. They reduce to one in the isotropic case, as in Navier's work. Cauchy then extended his treatment to the propagation of light in crystalline and isotropic media.

S. Poisson[13] (1829, 1831) assumed bodies to be composed of separate molecules, which are very small and separated by voids. By using slightly different hypotheses than Cauchy, he arrived at the same equations, and also at one independent elastic constant only for isotropic bodies.

Cauchy's theory was challenged by G. Green[14] (1839a and b) for optics and by G. G. Stokes[15] (1849a) for elasticity by calculations based on a continuous theory of matter. Stokes obtained the same elastic relations as Cauchy, but with twenty-one true elastic constants, which reduced to two in the isotropic case. These two coefficients are the rigidity modulus and Poisson's ratio,[16] which had a fixed value of 0.25 in the molecular theory. A big discussion ensued between those in favour of the molecular theory and those in favour of the continuum

[10]Henri Navier, born 10 February 1785, died 18 August 1836, was a French engineer, well known for his work in mechanics of fluids and the Navier–Stokes equation.

[11]Augustin-Louis Cauchy, born 21 August 1789, died 23 May 1857, was a French mathematician, an early pioneer of analysis. His work covered a wide range of mathematics and physical mathematics.

[12]Cauchy's work was communicated to the Academy of Sciences in September 1822, but was only published in 1828 (*Exercices de mathématiques*, **3**, 188–212 and 213–236).

[13]Siméon Denis Poisson, born 21 June 1781, died 25 April 1840, was a French mathematician and physicist, author of important works in electricity and elasticity.

[14]George Green, born 14 July 1793, died 31 May 1841, was an English mathematician and physicist who made important contributions to the mathematical theory of electricity and magnetism and introduced new concepts now known as Green's functions and Green's theorem.

[15]Sir George Gabriel Stokes, born 13 August 1819 in Ireland, died 1 February 1903 in England, was a British mathematician and physicist who made important contributions to fluid dynamics (the Navier–Stokes equations), optics, and mathematical physics.

[16]The rigidity modulus, or shear modulus, is defined as the ratio of shear stress to the shear strain. Poisson's ratio is the ratio, when a sample object is stretched, of the contraction or transverse strain (perpendicular to the applied load), to the extension or axial strain (in the direction of the applied load). See, for instance, Authier, A. (2013). *Physical properties of crystals, International Tables of Crystallography*, vol. **D**, John Wiley & Sons, Ltd.

theory, which lasted for more than half a century; they have been called the 'rari-' and 'multi-constant' theories, respectively.[17]

The first experimental evidence in favour of the multi-constant theory appeared in 1848, when the German-born French physicist Guillaume Wertheim (1815–1861) measured the deformation of brass and glass tubes and found values of Poisson's ratio significantly larger than 0.25 and often close to 0.3 (Wertheim 1848). Wertheim himself did not question the validity of Cauchy's theory and tried to find other explanations for the discrepancy. Measurements by other authors, however, in the course of the second half of the nineteenth century confirmed it. In particular F. E. Neumann's student, W. Voigt[18] (1889), measured the elastic constants of several crystals, and found that Poisson's ratio took values different from 0.25 for all of them. The validity of the molecular theory was not trusted any longer and A. E. H. Love (1863–1940), Professor of Mathematics at Oxford University, could write in 1906: 'the hypothesis of material points and central forces does not hold the field' (Love 1906).

It is for that reason that Ewald wrote in his accounts of the discovery of X-ray diffraction that the space-lattice theory was completely discredited by 1912 (for instance, in Ewald 1962a and b). This was, however, an overemphasis, as stressed by Forman (1969), and later admitted by Ewald (1969a) himself (Section 6.12). In fact, Voigt (1887) had pointed out that 'it is not the molecular representation itself which is contradicted by observation, it is an arbitrary assumption relative to the mode of interaction of the molecules'. From Voigt's theoretical results, Groth (1888) concluded that the reason why Cauchy's relations did not hold was that it assumed the interatomic forces to be direction-independent, and that it was wrong to restrict molecules to material points. On the other hand, Lord Kelvin, who had been one of the contributors to the multi-constant theory, and who did not have any doubt as to the existence of the space lattice, as we have seen, tried to reconcile the two theories by using a Boscovich-type model (Lord Kelvin 1893). It is clear now that Cauchy's relations hold if 1) the crystal is isotropic, and 2) if every material point is a centre of symmetry.

The difficulties raised by Cauchy's treatment were solved when M. Born[19] derived the full theory of the propagation of elastic waves in crystalline media (Born and Kàrmàn 1912; Born 1915), a work started just before the discovery of X-ray diffraction. In an interview of M. Born, made by P. P. Ewald for the American Institute of Physics in July 1960,[20] Max Born recalled that, when he and T. von Kàrmàn heard the news of Laue's discovery, they were *ashamed*

---

[17]I. Todhunter and K. Pearson, *A History of the Theory of Elasticity and the Strength of Materials*. (1886). Cambridge University Press.

[18]Woldemar Voigt, born 2 September 1850, died 13 December 1919, was a German physicist. A former student of F. E. Neumann, he taught at the Georg August University of Göttingen, where he was succeeded by P. Debye. He is well-known for his work on crystal physics, thermodynamics, and electro-optics, and he is the one who introduced the term 'tensor' in its present meaning in 1899.

[19]Max Born, born 11 December 1882, died 5 January 1970, was a German-born English physicist and mathematician who made important contributions in the fields of solid-state physics and optics, and played a major role in the development of quantum mechanics. He was awarded the Nobel Prize for Physics in 1954, for 'his fundamental research in quantum mechanics, especially for his statistical interpretation of the wave function'.

[20]<http://www.aip.org/history/ohilist/4522_1.html>.

not to have thought of it themselves because of the similarity between optical and elastic waves!

It is nevertheless true to say that prior to the discovery of X-ray diffraction, there was no physical property directly related to the space lattice. It is precisely in order to try and find such a correlation that one of topics proposed by A. Sommerfeld in 1910 to Ewald for his thesis was to investigate whether an anisotropic periodic array of dipoles (oscillating electrons) could produce observable double refraction by the mere fact of this arrangement, and, if this was the case, whether the values obtained were comparable to the measured ones (see Section 6.2).

3

THE DUAL NATURE OF LIGHT

... and teach me how
To name the bigger light, and how the less,
That burn by day and night ...

Shakespeare (*The Tempest*, Act 1, scene 2.)

3.1 The existing theories of light before Newton and Huygens

By the middle of the seventeenth century the law of refraction was well established. Angles of incidence and emergence were measured with a very ingenious experimental set-up by the Greek astronomer Ptolemy (*ca* 90–*ca* 168 AD), in Alexandria, Egypt. He established the first tables of refraction for his book *Optics*, but failed to find the law of refraction. Optics was a frequent object of study for the Arab and Persian philosophers and scholars. A geometrical construction of the law of refraction was given around 984 by the Arab mathematician and physicist Ibn Sahl (Abu Sa'd al-'Ala' ibn Sahl), associated with the Abbasid court of Baghdad, in his treatise *On the Burning Instruments* devoted to burning mirrors and lenses (Rashed 1990). Tables of angles of incidence and refraction were given by the famous Arab mathematician and physicist Alhazen,[1] considered as the 'father of modern optics' and ophthalmology. He correctly believed the speed of light to be different in different media, and so did the other famous Arab scientist, Avicenna. He also made the first experiments on the dispersion of light into its constituent colours. Alhazen was the author of the *Treasury of Optics*, which was inspired by Ptolemy's work.[2]

Alhazen's works in turn inspired many scientists of the Middle Ages, such as the English bishop, Robert Grosseteste (*ca* 1175–1253), and the English Franciscan, Roger Bacon (*ca* 1214–1294). Erasmus Ciołek Witelo, or Witelon (*ca* 1230– 1280), a Silesian-born Polish friar, philosopher and scholar, published in *ca* 1270 a treatise on optics, *Perspectiva*, largely based on Alhazen's works. The book, which also included tables of angles of incidence and refraction, was extended by J. Kepler in his *Ad Vitellionem Paralipomena quibus Astronomia pars Optica traditur* (Appendix to Witelo's optics and the Optical Part of Astronomy), published in 1604. Kepler spent some time searching for a law of refraction, but only obtained some empirical

---

[1]Alhazen is the Latin name of Ibn al-Haytham (Abū Ali al-Hasan ibn al-Hasan ibn al-Haytham), born *ca* 965 in Basra (Iraq), died 1039 in Cairo (Egypt). Educated in Basra, he started his scientific career in Baghdad, but was later called to Cairo where he taught at Al-Azhar University and where he spent most of his life.

[2]For *A History of Optics from Greek Antiquity to the Nineteenth Century*, see O. Darrigol (2012), Oxford University Press.

rules. He introduced the concept of a 'luminous surface' expanding with an infinite velocity along rays of light (Shapiro 1973).

Thomas Harriot's[3] manuscripts show tables of accurate incident and refracted angles for different substances, and he is credited with having been the first scholar to establish the sine law of refraction in modern times, around 1601 (Lohne 1959; Shirley 1951). He had an atomistic approach to refraction phenomena, about which he had a controversy with J. Kepler, briefly related in Section 11.3.2.

Twenty years after Harriot, the law of refraction was expressed in terms of the secants by the Dutch astronomer, Willebrord Snellius (1580–1626), but this work was not published.

René Descartes[4] rediscovered the law of refraction in 1637 and expressed it in the form of the sine law (Descartes 1637). He first established the law of reflection by comparing the light rays to streams of balls reflected by the ground (Fig. 3.1, *Left*), as Newton later did with tennis balls, and he is for that reason sometimes considered as the precursor of the corpuscular theory of light. He is, however, often misquoted, and Fig. 3.1, *Right* shows that he was in fact rather thinking in terms of something like waves. His reasoning was that balls are slowed down in soft media and propagate more rapidly in hard materials: a ball 'rolls less easily on a carpet than on a naked table'. The stream of balls would therefore follow path BV in water and would not penetrate in glass. On the other hand, light is 'a certain movement or action of a very subtle

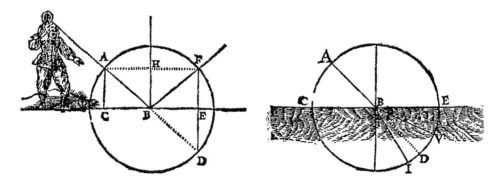

Fig. 3.1   *Left*: Law of reflection; light rays or balls following path AB are reflected at surface CBE and follow path BF. *Right*: Law of refraction. AB: incident beam. CBE surface separating the two media. BV: path followed by a stream of balls, with a larger velocity in the second media. BI: path of the light rays, with a smaller velocity in the second media. After Descartes, *Dioptrique, Discours second, de la réfraction*, in Descartes (1637).

---

[3]See Section 11.3.

[4]René Descartes, born 31 March 1596, in La Haye-en-Touraine (now La Haye-Descartes), Indre et Loire, France, died 11 February 1650 in Stockholm, Sweden, was a French philosopher, mathematician, and physicist. After schooling at the Jesuit College at La Flèche, Sarthe, he obtained a BA in law in 1616 at the University of Poitiers. He then travelled in Europe, and was engaged in various military campaigns, with Maurice of Nassau in the Dutch Republic, Duke Maximillian I of Bavaria, and the French Army at the siege of La Rochelle. In 1629 he returned to Holland where he spent twenty years. During his travels he met the leading scholars of his time, in particular Father Mersenne at Paris, with whom he corresponded and exchanged ideas all his life. His modern approach to philosophy had a profound influence on the thinkers of his time. His most famous work is the *Discourse on the Method of Reasoning Well and Seeking Truth in the Sciences*.

matter that fills the pores of other bodies' and propagates instantly. Descartes' argument was somewhat confused and contradictory but, essentially, 'the particles of air, being soft and badly joined, offer less resistance than the particles of water, and the particles of water less resistance than those of glass or crystal'. As a consequence, light rays follow path BI, with a smaller velocity, in the second media. Clearly, for Descartes, if streams of balls provide a good image of the propagation of light in a straight line, motions of the subtle matter are better suited to represent the refraction phenomena.

The French mathematician Pierre Fermat (first decade of the seventeenth century to 1665) gave a new derivation of the law of refraction in 1662, using his principle of 'shortest optical path' (or 'principle of least time'). At first, it was not found convincing by Christiaan Huygens[5] who called Fermat's principle, a 'pitiable axiom';[6] but, after a second reading, Huygens found the demonstration 'very good and clever'.[7]

The first wave theory of the propagation of light is due to the English philosopher and scholar Thomas Hobbes.[8] It was presented under the name *Tractatus Opticus* as the seventh book, *Optics*, of Father Marin Mersenne's[9] *Universae geometriae* (1644). Hobbes' theory is a continuum theory in which the phenomena of reflection and refraction are interpreted in terms of pulses. Light is propagated by 'ray fronts', called by him 'propagated line of light', perpendicular to the light rays. Hobbes disagreed with Descartes' conceptions and entered in a correspondence with him at the instigation of their common friend, Father Mersenne. Details are given in Shapiro (1973).

Hobbes' theory was widely known in the second half of the seventeenth century through the works of the French friar Emmanuel Maignan (1600–1676) and of Isaac Barrow (1630–1677), I. Newton's professor at Cambridge. They improved his proof of the sine law of refraction but abandoned the idea of a continuum theory of light. Maignan rejected both Hobbes' interpretation of light as a pulse of the æther and Descartes' image of balls, and proposed his own effluvial theory of light (Shapiro 1973). Barrow in his *Lecciones opticae* (1669) was more concerned with geometrical optics. He considered light to be a local motion, but found that its interpretation as either a corporeal emanation or an impulse through a medium presented difficulties (Shapiro 1973).

Robert Hooke's *Micrographia*,[10] Observation IX, *Of the colours observable in Muscovy glass, and other thin Bodies* (1665) contains a wave theory of light that he imagined when he was studying the interference colours of light reflected from thin films. By splitting or cleaving pieces of 'muscovy-glass' (muscovite mica), he obtained thin flakes exhibiting pleasant colours,

---

[5]See Section 3.4.

[6]8 March 1662, letter to his brother Lodewijk, *Oeuvres complètes de Christiaan Huygens, Société Hollandaise des Sciences*, The Hague: Martinus Nijhoff, Tome IV (1891) p. 71.

[7]22 June 1662, letter to his brother, N° 1025, *Oeuvres complètes de Christiaan Huygens*, Tome IV, p. 157.

[8]Thomas Hobbes, born 5 April 1588, died 4 December 1679, was a philosopher and political theorist whose 1651 treatise *Leviathan* effectively kicked off the English Enlightenment. He travelled frequently in Europe, and was a regular participant in Father M. Mersenne's debating group in Paris.

[9]Father Marin Mersenne, born 8 September 1588, died 1 September 1 1648, was a French theologian, philosopher, mathematician, and acoustician who joined the convent of Minim friars in Paris in 1611 and became its head in 1625. He held a continued correspondence with R. Descartes, T. Hobbes, C. Huygens, and most of the scholars of his time. R. Descartes, P. Fermat, B. Pascal, and others used to meet regularly around Mersenne for debates on philosophical matters.

[10]See Section 11.5.

which he observed with the naked eye and, more conspicuously, under the microscope. He noted that 'the position of these colours, in respect to one another, was the very same as in the rainbow'. He also observed such colours with peacock feathers, thin glass lamellae pressed together, a lens pressed on a glass plate (producing what is now known as Newton rings), soap bubbles, pearls, mother-of-pearl, or the oxide layer at the surface of hardened steel (which he called a 'thin skin'). For Hooke, the propagation of light requires the motion of 'parts of the luminous body', æther. It must be a very short 'vibrating motion', which propagates with a high but finite velocity. In successive transparent media, it follows the law of refraction of the 'most acute and excellent philosopher Descartes'. In free space, the vibration is 'orbicular and at right angles with the direction of propagation' but 'oblique' to it after refraction in the medium. Hooke's guess as to the transverse character of the vibration is therefore only partial and differs from Hobbes'.

A theory of colours is developed by R. Hooke in Observation IX and in Observation X, *Of metalline and other real colours*, of *Micrographia*, which is not very clear and in which there are only two basic colours. This led him to oppose Newton's theory in 1672 (see Section 3.3).

In Observation XV, *Of Kettering-stone, and of the pores of inanimate bodies*, of *Micrographia*, Hooke's description of the 'Kettering stone' or Ketton stone, a kind of limestone coming from Kettering,[11] is the occasion for Hooke to explain his conception of the mode of propagation of light. According to him, all bodies contain pores and 'there are in all transparent bodies such atomic pores'. They are filled with a 'fluid body...which is the medium, or instrument, by which the pulse of light is coney'd from the lucid body to the enlightn'd'. It is also the occasion for him to refer to Mersenne's *Hypotheses* by which the undulating pulse of light is always carried at right angles with the ray. For comments on Hooke's theory of light, see, for instance, Shapiro (1973) and Hall (1990).

R. Hooke was also one of the first to observe diffraction, which he called 'inflection'. He presented his results in two discourses to the Royal Society, one on 27 November 1672, 'containing diverse optical trials made by himself, which seemed to discover some new properties of light, and to exhibit several phenomena in his opinion not ascribable to reflexion or refraction, or any other till then known properties of light' and another one two years later on the same topic. One of the observed effects is that 'colours begin to appear when two pulses of light are blended so very well and near together, that the sense takes them for one'. Hooke did not mention Grimaldi but his results are remindful of some of Grimaldi's, whose book had been presented to the Royal Society some time before (see below, Section 3.2).

A further wave theory of light was developed at the end of the 1660s by the Jesuit Ignace Pardies (1636–1673), who was a Professor of Mathematics and Physics at the Jesuit Collège de Clermont at Paris. Its approach was both kinematic and dynamic; Pardies considered the propagation of spherical waves and expanded the idea of physical rays and wave fronts normal to them. Huygens knew and appreciated Pardies, whom he met in Paris, and read his theory in draft form. He was in correspondence with him about the double refraction of calcite, and deplored his death.[12] Pardies's theory was unfortunately not published. It is known to us

---

[11]See D. Hull (1997). Robert Hooke: A fractographic study of Kettering-stone. *Notes and Records of the Roy. Soc.* **51**, 45–55.

[12]Letter from C. Huygens to H. Oldenburg, 24 June 1673, *Oeuvres complètes de Christiaan Huygens, Société Hollandaise des Sciences*, The Hague: Martinus Nijhoff, Tome VII (1897), p. 313.

because its ideas were taken up in a book on optics by another Jesuit, P. Ango (1640–1694), in 1682 (*L'Optique divisée en trois livres*). Ango, however, only kept parts of Pardies's manuscript and mixed them with developments of his own, so that it is difficult to tell what were the contributions of each one. For a discussion of Pardies and Ango's theories, see Shapiro (1973); for a short biography of I. Pardies, see Ziggelaar (1966).

3.2 F. M. Grimaldi and the diffraction of light, 1665

The Jesuit priest, Francesco Maria Grimaldi, born 2 April 1618 in Bologna, Italy, died 28 December 1663 in Bologna, Italy, was an Italian mathematician, physicist and astronomer who taught at the Jesuit college in Bologna. He is the discoverer of the phenomenon of diffraction, which is described in his book, *Physicomathesis de lumine*, published posthumously in 1665.

In the first experiment (Fig. 3.2, *Left*) he let the light from the sun penetrate in a dark room through a small opening, *AB*. He then put a dark object, *EF*, in the cone of light coming from the hole. To his surprise, he observed that the shadow of the object was wider than the expected geometrical shadow, *IGHL*, and that light was diffused in the areas *IM* and *LN*. Furthermore, the shadow was bordered by several coloured bands, in general three. In the second experiment (Fig. 3.2, *Right*) he let a pencil of light delimited by two successive slits, *CD* and *GH*, propagate in a dark room, and he observed light diffused in the areas *GIN* and *HOK*. As in the first experiment, this could not be explained in the usual way. Grimaldi also observed coloured fringes in the shadow of thin objects such as needles, and curved fringes in the shadow of the tip of a rectangular object, called 'crested' fringes by T. Young (1804). He concluded that light bends when it passes in the vicinity of objects, and that light propagates or diffuses not only directly, by refraction or reflection but also by a fourth mode, *diffractio*.

He found a still more remarkable effect; if he let the sun come in through two small holes, and put a screen in the region where the two cones of light coming from the holes merged, there was a spot more obscure than when illuminated by either of the two cones alone. Without knowing, he had discovered the interference of light which T. Young was to study some 140 years later (see Section 3.5).

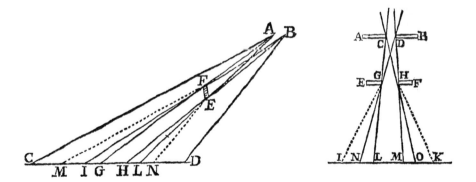

Fig. 3.2   Diffraction of light. *Left*: First experiment, a dark object, *EF* is placed in a cone of light. Diffused light is observed outside the geometrical shadow, *IGHL*. *Right*: Second experiment: two slits, *CD* and *GH* are placed in a beam of light. Diffused light is observed outside the geometrical shadow, *NLMO*. After Grimaldi (1665).

The English term, 'diffraction', was coined by the Secretary of the Royal Society, H. Olden-burg,[13] in his presentation of Grimaldi's book to the Royal Society.[14] 'Diffraction', explained Oldenburg in a rather confusing way, 'is done, according to Grimaldi, when the parts of light separated by a manifold dissection, do in *the same medium* proceed in different ways' (italics are Oldenburg's). Grimaldi's work was made known earlier through the writings of the French Jesuit, Honoré Fabri (1607–1688), *Dialogi physici* (1669).

For Grimaldi, light was wave-like and propagated by transverse vibrations; these vibrations were different for different colours, but he did not specify that the corresponding frequencies were different. Like Hobbes, Pardies, Newton, and Huygens, he compared the propagation of light to that of sound.

3.3   I. Newton and the emission theory, 1672

Sir Isaac Newton (Fig. 3.3) was an English physicist, mathematician, astronomer, natural philosopher, alchemist, and theologian, and is considered to be one of the scholars who had the greatest influence on the development of science. He was born prematurely on Christmas Day 1642 (Julian calendar), three months after the death of his father, a farmer. Newton's mother remarried when he was three, and he was brought up by his maternal grandmother. At school he showed a taste for mechanical inventions and built, for instance, a water-clock and a windmill (Brewster 1855). His mother wanted him to be a farmer like his father, but his maternal uncle, a Cambridge graduate himself, recognized his abilities and persuaded her to send him to Cambridge, where he was admitted at Trinity College in June 1661.[15]

At Cambridge, Newton at first studied chemistry. He then started learning mathematics with Euclid's *Elements*, which he threw aside 'as a trifling book' (Brewster 1855), and continuing with R. Descartes' *Géométrie*, W. Oughtred's *Clavis mathematicae*, F. van Schooten's edition of F. Viète's works, and J. Wallis's *Arithmetica Infinitorum*. He also read the astronomical works of N. Copernicus, G. Galileo and J. Kepler, the philosophical works of T. Hobbes, P. Gassendi and R. Descartes, and J. Gregory's *Optica Promota*. In 1664, the mathematician Isaac Barrow (1630–1677) moved from London to Cambridge to become Lucasian Professor of Geometry. Newton followed his lectures and contributed to their editing (*Lectiones opticae*, 1669 and *Lectiones geometriae*, 1670).

In 1665, as Newton had just obtained his BA, the University was closed on account of the Great Plague (1665–1666), the last major outbreak of the plague in England. He went back home in Lincolnshire, and only returned to Cambridge in 1667. This was one of the most fruitful periods of his scientific career. It is then that he introduced the method of 'fluxions' (derivatives) and laid the foundations for differential and integral calculus, several years before its independent discovery by Leibnitz. At the same time, he started developing his theories of optics and gravitation, and formulated the three laws of motion (published in the *Principia*, Newton 1687).

---

[13]Henry Oldenburg (1619–1677) was a German theologian who settled in England in 1653. He was the founding editor of the *Philosophical Transactions of the Royal Society*, and the Secretary of the Royal Society from its foundation in 1660 till his death, when he was succeeded by Robert Hooke.

[14]*Phil. Soc. Trans. Roy. Sec.* **6** (1971) 3063–3065.

[15]For details about Newton's life and works, see, for instance, Brewster (1855).

Fig. 3.3   Sir Isaac Newton in 1689, portrait by Godfrey Kneller. Source: Wikimedia commons.

Isaac Newton: born 4 January 1643 (25 December 1642, Julian calendar) in the manor-house of Woolsthorpe-by-Colsterworth, Lincolnshire, England, the son of a prosperous farmer also named Isaac Newton; died 31 March 1727 in Kensington, Middlesex, England. After schooling at the King's School, Grantham, he was admitted in 1661 to Trinity College, Cambridge. He was elected to a fellowship of Trinity College in 1664 and took his BA degree in 1665, and, when the University was closed because of the Great Plague, he returned home to Lincolnshire. When the University was reopened, in 1667, he came back to Trinity College as a Minor Fellow. In 1668 he was elected a Major Fellow, and in 1668 obtained his MA. In 1669 he succeeded Isaac Barrow as Lucasian Professor of Mathematics at Cambridge University, and he kept that position for thirty years. He was elected a Fellow of the Royal Society in 1672. From 1689 to 1690 and again from 1701 to 1705, Newton represented the University of Cambridge at the Parliament of England. In 1696 he moved to London to take up the post of Warden of the Royal Mint, of which he became Director in 1699. From 1703 until his death in 1727, he was President of the Royal Society. Newton was also very interested in alchemy, and during the last years of his life, in theology.

MAIN PUBLICATIONS

1671 *Method of Fluxions.*

1684 *De Motu Corporum in Gyrum.*

1687 *Philosophiæ Naturalis Principia Mathematica.*

1704 *Opticks.*

1707 *Arithmetica Universalis.*

Newton bought a glass prism at the beginning of 1666 (Newton 1672*a*) in order to study the 'celebrated phenomena of colours'. After darkening his room, he let in the sunlight through a small hole he made in the shutters, placed his prism in the beam, and observed the spectrum on the opposed wall. Looking closely at the path of the light rays after the prism, he found them to be curved. This observation led him to make a comparison with the path of tennis balls struck by a racket, and to ask whether 'the rays of light should possibly be globular bodies'. We have here the first occurrence of Newton's hypothesis of a corpuscular nature of light. Such a conception of the nature of light had in fact already been proposed by the French philosopher, Pierre Gassendi (1592–1655), whose works Newton had read. Gassendi, who was an atomist, considered light to be constituted of atoms, *atomi lucificae*, identical with heat atoms.[16] Pliny the Elder had mentioned in his *Natural History* (77 AD) that when a crystal of iris quartz receives a beam from the sun, it casts the colours of the rainbow on neighbouring walls.

In a second step, Newton experimented with two prisms in succession, and, in his *Experimentum Crucis* (crucial experiment), showed that the 'refrangibility' (index of refraction) was different for each colour: light is a 'heterogeneous mixture of differently refrangible rays'.

[16]Gassendi, P. (1658). *Opera Omnia*, Vol. 1.

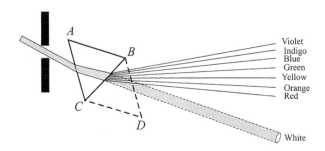

Fig. 3.4   Reconstruction of white light after passage through two opposite prisms: the rays refracted by the first prism,
ABC, are refracted back into white light by the second prism, *BCD*. Redrawn after Brewster (1865).

Colours are not 'qualifications' of light, but 'original and connate' (innate) properties. The least
refrangible colour is red, and violet is the most refrangible. Newton distinguished 'original
or primary' colours, red, orange, yellow, green, blue, indigo, and violet, and combinations of
these. The most 'surprising and wonderful' combination is white, which is the combination of
the primary colours. This statement he proved by putting behind the first prism, *ABC* (Fig. 3.4),
a second prism, *BCD*, identical to the first one, but rotated 180°. The rays refracted by the first
prism were all refracted back into a beam of perfectly white light by the opposite refraction of
this second prism. He also obtained white by recombining the rays refracted by the prism at the
focus of a lens. In the same communication, Newton also improved on Descartes' interpretation
of the rainbow.

From the differences in the index of refraction of the components of the spectrum, Newton
concluded that the perfection of telescopes is necessarily limited by aberrations, and he decided
to build a reflecting telescope with a metallic receptor, which he did in 1668. This was brought
to the attention of the Royal Society and, in 1671, its members decided to elect Newton a
Fellow. As a mark of gratitude, he donated the telescope to the Society. It is at this occasion that
he submitted his *New Theory about Light and Colours* in the form of a letter sent to the Royal
Society on 6 February 1672[17] (Newton 1672*a*).

Newton's communication was on the whole well received, but his theory of colours was
challenged from various sides. Father Pardies (1672*a*) at first did not believe in Newton's
explanation of the shape of the spectrum but was readily convinced by Newton's answer
(Pardies 1672*b*). Newton's theory was also contested by a physician from Liège, Francis Linus,
and the dispute went on for several years around 1674–1676.

More serious was the opposition that came from R. Hooke and C. Huygens, who objected
to Newton's attempt to prove, by experiment alone, that light consists of the motion of small
particles rather than waves.

Robert Hooke was a Fellow of the Royal Society and the Curator of Experiments of the Royal
Society when Newton presented his theory of light and colours to the Society. He was also, as
we have seen in the preceding section, the author of a wave theory of light and of a theory of
colours. It is not surprising that he strongly opposed Newton's views. Newton refuted Hooke's

---

[17]At the time, in England, the year started in March, so that February was in fact still in 1671.

objections in a letter to the Secretary of the Royal Society, H. Oldenburg, dated 11 July 1672 (Newton 1672*b*).

Newton at first stated more distinctly his hypothesis of the 'corporeity' of light, speaking of the 'corpuscles of light' which make 'various mechanical impressions on the organs of senses'. But Newton noted as well that this hypothesis 'has a much greater affinity with [Hooke's] own hypotheses than he [Hooke] seems to be aware of, the vibrations of the æther being as useful and necessary in *this* [hypothesis] as in *his*'. Indeed, adds Newton, 'assuming the rays of light to be small bodies emitted every way from shining substances, those, when they impinge on any refracting or reflecting superficies, must as necessarily excite vibrations in the æther, as stones do in water when they are thrown in it'. These vibrations 'cause a sensation of light by beating and dashing against the bottom of the eye'. They are of 'various depths or bignesses' and, if they are separated (by the prism), 'the largest beget a sensation of a red colour, the least or shortest, of a deep violet, and the intermediate ones of intermediate colours'.

As a further argument, the corpuscular hypothesis accounts for the propagation of light in straight lines, which Newton thinks waves do not: vibrations cannot be propagated 'without a continual and very extravagant spreading and bending every way into the quiescent medium'. In his later writings, he considered that the bodies which constitute the rays of light are subject to forces of attraction and repulsion, like material bodies.

In the second part of his reply, Newton showed that Hooke was wrong in Observation X of *Micrographia*, when he declared that there are only two basic colours.

The other opponent to Newton's theory of colours was C. Huygens. The Secretary of the Royal Society, H. Oldenburg, informed Huygens of Newton's theory by a letter dated 21 March 1672.[18] On 9 April, Huygens acknowledged, judging Newton's theory 'ingenious', but requiring to be tested experimentally. In his next letter, dated 1 July, Huygens found Newton's hypothesis relative to colours highly likely and confirmed by his *experiment crucis*, but raised some objections relative to the aberrations of lenses. Huygens seems to have had second thoughts since on 27 September he wrote a new letter to H. Oldenburg, saying that Newton's results could be explained in a different way. Maybe he felt that Newton's theory was an attack against the wave theory. Oldenburg then sent Huygens Newton's answer to Hooke's objections Newton (1672b). That letter prompted Huygens to send on 14 January 1673 yet another letter saying that, as proposed by Hooke, two colours should be enough: blue and yellow (Huygens 1673). This drew a firm reply from Newton (1673) on 3 April 1673, after which Huygens withdrew from the dispute, 'seeing that he [Newton] maintains his opinion with so much concern'. For a discussion of the comments of Huygens and Newton on each other's theory, see Shapiro (1973).

In 1675, Newton made a second important contribution to the field of optics by studying the various colours exhibited in thin plates and bubbles, which he presented on 9 December 1675 in a letter to the Royal Society.[19] The colours in bubbles and feathers had first been observed

---

[18]The exchange of letters between C. Huygens and H. Oldenburg can be found in *Oeuvres complètes de Christiaan Huygens*, Tome VII (1891).

[19]*Hypothesis explaining the properties of light by Isaac Newton*, <http://www.newtonproject.sussex.ac.uk/view/texts/normalized/NATP00002>.

by Robert Boyle[20] (1664) and by Boyle's former assistant, Robert Hooke (*Micrographia*, Observation IX) who also observed coloured rings in mica and in glass lenses pressed together. Newton went further and made a careful quantitative analysis of the successive orders of the coloured rings, now called 'Newton rings'. Hooke of course claimed priority, and Newton, who did not like quarrels, in the end postponed the publication of his optical works until after Hooke's death in 1703 (Newton, *Opticks*, 1704).

Newton's emission theory was not very convenient to explain the interference phenomena in the coloured rings. In order to do that, he had to assume that the luminous bodies could undergo periodic modifications of their state. These bodies could at certain moments be easily reflected and at others easily transmitted. He called these modifications 'fits' (he called them *vices* in Latin, turns): 'the reason why the surfaces of all thick transparent bodies reflect part of the Light incident on them, and refract the rest, is, that some rays at their incidence are in fits of easy reflexion, and others in fits of easy transmission (*Opticks* (1704), Second Book, part III, Prop. XIII).

In the Third Book of *Opticks*, Newton makes his opposition to the wave theory very clear: 'are not all hypotheses erroneous, in which Light is supposed to consist in pression (pressure) or motion, propagated through a fluid medium'? (*Opticks*, Third Book, Part I, Quest 28). He found that Huygens's explanation of double refraction by 'two vibrating mediums within that crystal' [of Iceland spar] unsatisfactory. On his part, he supposes 'that there are in all space two æthereal vibrating mediums, and that the vibrations of one of them constitute Light, and the vibrations of the other are swifter, and as often as they overtake the vibrations of the first, put them into those fits (*Opticks*, Third Book, Part I, Quest 28)'. This hypothesis of vibrations faster than light is of course impossible.

Considering the refraction of the usual (ordinary) ray, Newton used a construction similar to that of Huygens, but assumed that the sines of the incidence and emergence angles are proportional to the velocities of the luminous bodies in the two media, and therefore that these velocities are greater in denser media than in rarer media, which is the opposite of what was assumed by Huygens. Considering the refraction of the unusual (extraordinary) ray of Iceland spar, Newton used a construction that R.-J. Haüy found less accurate than Huygens'.

In his conclusion, Newton summarizes his views:

'Nothing more is requisite for producing all the variety of colours, and degrees of refrangibility, than that the rays of light be bodies of different sizes. ... Nothing more is requisite for putting the rays of light into fits of easy reflexion and easy transmission, than that they be small bodies (*Opticks*, Quest 29).

Newton knew Grimaldi's work on diffraction from the book of the Jesuit geometer, H. Fabri, which he had in his library. He also knew Hooke's 1672 observations on inflection of light, which he criticized. It is probably around 1675 that he reproduced Grimaldi's experiments (Newton 1687, *Principia* Book I, Section XIV), but most of his observations were only published in Book Three, part 1, of *Opticks*. He made quantitative studies of the phenomenon, measuring the enlarging of the shadows of various inflecting (diffracting) objects such as knife edges, straws, pins and human hair, at different distances and with different colours of light,

---

[20]Robert Boyle, born 25 January 1627 in Freshwater, Isle of Wight died 31 December 1691, in England, was an Anglo-Irish natural philosopher, chemist and physicist. He is best known for his work as a chemist and for Boyle's gas law. His air-pump was built together with R. Hooke.

but he did not discuss the fringes within the shadow. Newton also made observations with two knives whose sharp edges were brought very close to one another. When their distance was about 1/400th of an inch a dark line appeared in the middle (Observations 6 to 10). This experiment was repeated by T. Young and was at the base of the latter's interpretation of the interference phenomena (see Section 3.5). For Newton, the phenomenon of inflection was to be explained by the attractive and repulsive forces which the molecules of the object exerted on the corpuscles of light.

3.4   C. Huygens and the wave theory, 1678

The real author of the wave theory of light is of course C. Huygens. Christiaan Huygens (Fig. 3.5) was a Dutch mathematician, astronomer and physicist, probably one of the two most prominent men of science of the second half of the seventeenth century, along with Isaac Newton. His range of interests was very wide; he invented the pendulum clock and discovered Saturn's rings, played a major role in the development of modern calculus and is famous for his wave theory of light. For us, he is also well-known for having thought up a close-packed structure of ellipsoids to explain the properties of calcite (see Section 11.8). He was a remarkable mathematician and, at the same time, had an experimentalist's view on physical phenomena.

Through his father's friends, René Descartes and Father Marin Mersenne, Huygens became acquainted at a young age with the modern developments of science.[21] Descartes, who was staying in the Netherlands at the time, paid occasional visits at his home and was impressed by his skill in mathematics. Even before entering university, Huygens entered in a correspondence with Father Mersenne and, prompted by him, worked on the problem of the shape of a rope hanging from its two ends under its own weight (the catenary). He only fully solved the problem in 1690, but his 1646 study of the equilibrium polygons of force struck even Descartes. In his young years, Huygens was an admirer of Descartes's philosophy but became progressively more critical, for instance in relation to the laws of impact between elastic bodies and to the theory of light. In 1663, Huygens was elected a Fellow of the newly founded Royal Society and, in 1666, he was appointed by the minister of King Louis XIV, Colbert, a member of the newly formed French Académie Royale des Sciences. Huygens was one of its main figures and helped to organize it along the lines of the Royal Society. He was in contact with most of the scientists of his time and was perfectly at ease in the intellectual circles of Paris. There, he met P. Gassendi at the end of his life and was probably influenced by his atomistic ideas, the mathematician G. de Roberval (1602–1675), and the philosopher and mathematician B. Pascal (1623–1662) whom he admired greatly and who encouraged him to study the probability theory. In 1672 he met the German mathematician G. W. Leibnitz (1646–1716) when the latter first came to Paris. Huygens was Leibnitz's mentor in mathematics and physics and kept a regular correspondence with him all his life. Another regular correspondent of Huygens' was H. Oldenburg, who kept him abreast of the developments of science in Great Britain. It is thanks to Oldenburg that Huygens could read in 1670 the works of Newton's professor, I. Barrow, and of the mathematician J. Wallis. H. Oldenburg also served as intermediary between C. Huygens and I. Newton in 1672–1673.

---

[21]For a biography of Huygens, see, for instance, A. E. Bell (1947). *Christiaan Huygens and the Development of Science in the Seventeenth Century*. Edward Arnold & Co., London.

CHRISTIANUS HUGENIUS
natus 14 Aprilis 1629.
denatus 8 Junii 1695.

Fig. 3.5   Portrait of Christiaan Huygens. *Opera varia* (1724).
Source: Gallica.bnf.fr.

Christiaan Huygens: born 14 April 1629 in The Hague, Netherlands, the second son of Constantijn Huygens, a Dutch poet,
scholar, and diplomat; died 8 June 1695 in The Hague, Netherlands. Until the age of 16, Huygens and his brother Constantijn
were educated by private tutors. In 1645, Christiaan and his
brother entered the University of Leyden where they studied
Mathematics and Law; the mathematician F. van Schooten was
one of their professors and introduced Christiaan to Descartes's
Geometry. In 1647, he joined Collegium Auriacum in Breda
where he studied law, after which he travelled in Denmark and
Italy. From 1651 to 1660, he stayed mostly in Holland, with a
first visit to Paris in 1655. His first scientific work, on squaring
the circle, was published in 1651 and won him immediate fame
as a mathematician. He made his own telescope and, on 25 March 1655, discovered Saturn's satellite
Titan. After improving the grinding and polishing of lenses, he built a larger telescope, and, with it, was
able to detect Saturn's ring in 1656, explaining correctly its nature. At the same time, he worked on the
properties of the collisions between elastic bodies. In the years 1656–1660 he invented the pendulum
clock, derived the formula of the ideal pendulum and that for the centrifugal force, and discussed the
properties of the cycloid. From October 1660 he was most of the time in Paris, spending a few months
in London in 1661 to meet the scientists of the newly founded Royal Society, and in 1663 when he
was elected a Fellow of that Society. In 1666, he was appointed a member of the newly formed French
Académie Royale des Sciences. His interests after 1666 were mainly with mechanics and optics; he
developed his theory of light and, in 1677, found the explanation of the double refraction of Iceland
spar. Meanwhile, in 1675, he invented a regulator for clocks using a spiral spring combined with a
balance wheel, for which Hooke claimed he had a priority (see Section 11.5). Huygens' health was not
very good and he had to go back to the Hague from March 1676 to June 1678 and, finally, for good in
1681, never to return to Paris. In 1689 he made a final small trip to England to meet Newton and old
friends.

MAIN PUBLICATIONS

1651 *Theoremata de quadratura hyperbolae,*
     ellipsis & circuli.
1653 *Tractatus de refractione et telescopii.*
1654 *De circuli magnitudine inventa.*
1656 *De motu corporum ex percussione.*
1657 *De ratiociniis in aleae ludo.*
1659 *Systema saturnium.*

1666 *De aberratione radiorum a foco.*
1673 *Horlogium oscillatorum sive de motu pen-*
     dularium.
1685 *De telescopiis et microscopiis.*
1690 *Traité de la lumière.*
1690 *Discours de la cause de la pesanteur.*

3.4.1   The genesis of Huygens' wave theory of light

Huygens' interest in the study of the refraction of light arose very early. In 1653 he published a
first work on the refraction at planar and spherical surfaces, and on lenses and telescopes, *Tractatus de refractione et telescopis*, intending to write an important treatise on optics, *Dioptrica*,
which he only completed much later, in 1692. He wrote the second part in 1663, on spherical

aberrations, *De aberratione radiorum a foco*, and continued this work after he came to Paris in 1666 to become a member of the Académie Royale des Sciences.

In 1665 Huygens read R. Hooke's *Micrographia* just after its publication. Soon after, he came to know the work of Grimaldi through the writings of another Jesuit, G. Riccioli (1598–1671), under whose direction Grimaldi had worked.[22] In 1672 he met Father I. Pardies, whose wave theory he read in draft form, and he was informed of Newton's emission theory by Oldenburg.

Huygens' views on the aberrations were modified in 1672, after he had read Newton's theory of colours, which he accepted after some hesitations. It was also in 1672 that Huygens got hold of crystals of calcite, which he called 'Iceland crystal', or 'Iceland talc' (because of its cleavage properties), and which had been brought from Denmark. In a letter to his brother Lodewijk, dated 4 September 1672,[23] he announced his first observations. He was immediately very interested because it was the only crystal he knew presenting double refraction; he was later to observe also this effect in quartz. At first Huygens repeated more accurately Bartholin's measurements of the incidence and refraction angles (see Section 11.6). It is roughly at that time that his wave theory formed in his mind, as shown by his manuscript pages written around 1672/1673.[24] By putting two crystals of calcite one behind the other, he discovered the phenomenon of polarization,[25] but he did not use that term, which was introduced much later by the French physicist E. L. Malus[26]. Huygens guessed that there should be two kinds of waves, but he was initially unable to explain the double refraction. He found the explanation on Friday, 6 August 1677, in The Hague to where he had returned back a year before, due to ill health. Fig. 3.6 reproduces a page from Huygens's manuscript with the inscription *EYPEKA* (EUREKA) and the date of the discovery. Huygens's wave theory is based on the existence of a finite speed of light. He was therefore elated when he belatedly read in the July issue of the *Philosophical Transactions of the Royal Society* a translation of a paper by the Danish astronomer Ole Rømer (1644–1710) demonstrating that the speed of light was finite,[27] and he immediately wrote to Rømer, asking for more details.[28] Many philosophers and scientists, from Aristotle to Kepler, Descartes and Hooke, had believed the speed of light to be infinite, but not so Alhazen, Galileo, or Newton. Rømer's conclusion was not accepted at once, but it was very important for Huygens since it justified his hypothesis

[22]See *Oeuvres complètes de Christiaan Huygens*, Tome XV (1925), p. 382.

[23]*Oeuvres complètes de Christiaan Huygens*, Tome VII (1897), p. 218.

[24]*Oeuvres complètes de Christiaan Huygens*, Tome XIX (1937), p. 407.

[25]*Oeuvres*, p. 413.

[26]Etienne Louis Malus, born 23 July 1775, died 23 February 1812 was a French engineer and physicist, known for the discovery of polarization and his theory of double refraction. He was awarded the Rumford Medal by the Royal Society in 1810.

[27]The original paper, *Démonstration touchant le mouvement de la lumière*, had been published in the *Journal des Scavans*, 1676, p. 229. Rømer, former pupil and son-in-law of R. Bartholin, was at the time working in the Paris Royal Observatory. The discovery was made by observing Jupiter's satellite Io coming in or out of Jupiter's shadow over a period of years. According to Bobis and Lequeux (*J. Astron. Hist. Heritage*, 2008, **11**, 97–105), the first account of the discovery was actually given by the Director of the Paris Observatory, G.-D. Cassini (1625–1712), in August 1676; but in fact he was not convinced and let his collaborator Rømer present his own derivation on 21 November to the *Académie*, published on 7 December in the *Journal des Scavans*.

[28]On 16 September, *Oeuvres complètes de Christiaan Huygens*, Tome VIII (1899), p. 30.

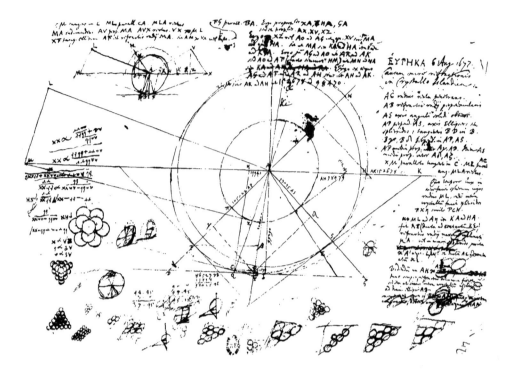

Fig. 3.6   Huygens's discovery of the explanation of the extraordinary refraction of calcite: *EYPHKA*, 6 August 1677 (top right). Note the draft sketches of the structure of calcite (bottom and left). After Huygens, *Oeuvres Complètes*, Vol. XIX (1937).

regarding the finite character of the speed of light.[29] He regarded his explanation of the double refraction as the most beautiful confirmation of his new wave theory, and he wrote so to Colbert on 14 October 1677.[30] On the formation of Huygens' wave theory, the reader may also consult Ziggelaar (1980) and Dijksterhuis (2004).

3.4.2   *Traité de la lumière (Treatise on light)*

Huygens' wave theory was developed in the *Traité de la lumière* (Treatise on light), which he wrote in 1678, read to the members of the *Académie Royale des Sciences* in the course of the summer 1679, but only published in 1690, in French[31] (see Fig. 3.7).

In the first part of Chapter **I**, Huygens explained his concept of spherical waves and how they propagate. His aim was to give, in terms of the 'principles of today's philosophy, clearer and more probable reasons of the properties of propagation, reflection and refraction of light', and,

---

[29]The first non-astronomical measurement of the speed of light was made by the French physicist H. Fizeau (1819–1896) in 1849 with a cog-wheel.

[30]*Oeuvres complètes de Christiaan Huygens*, Tome VIII (1899), p. 36.

[31]English translation by S.V. Thompson (1912), London: Macmillan and Co., Ltd.

TRAITE
DE LA LVMIERE.
Où font expliquées
Les caufes de ce qui luy arrive
Dans la REFLEXION , & dans la
REFRACTION.
Et particulierement
Dans l'etrange REFRACTION
DV CRISTAL D'ISLANDE.
Par C. H. D. Z.
Avec un Difcours de la Caufe
DE LA PESANTEVR.

A LEIDS,
Chez PIERRE vander Aa, Marchand Libraire.
MDCXC.

Fig. 3.7　　Huygens' treatise on optics. After Huygens (1690).

after that, to 'examine the properties of the refraction of a certain crystal brought from Iceland'. He immediately introduced the general framework of his wave theory by asserting that 'one cannot doubt that light consists in the movement of a certain matter'. He justified it by noting on the one hand that light is produced by 'fire and flame which no doubt contain bodies which are in rapid motion', and on the other hand that 'light has the property, when concentrated by concave mirrors, of burning and melting bodies, disuniting its particles'. He added: 'this is surely the mark of motion, at least in the True Philosophy in which one conceives the cause of all natural effects as due to mechanical reasons. This is what one has to do unless one gives up the hope of ever understanding anything in Physics ... Light consists in a movement of the matter that lies between our eyes and the luminous body ... Considering the extreme speed with which light propagates, it cannot be through the transport of matter. It is therefore propagated in another way and the propagation of sound in air can help us understand how'. Sound propagates in air at the same speed in every direction and spherical surfaces must therefore be formed that propagates with ever increasing radii until they reach our ears. In the same way, light, which propagates at a high but finite speed, without any transport of matter, must spread out in a spherical manner. Huygens called them *waves* by analogy with the rings formed at the surface of water when you throw a stone.

To justify his assumption of a finite speed of light, Huygens made an estimation of it, first to about 100 thousand times that of sound, by consideration of the eclipses of the moon, and secondly to about 600 thousand times that of sound using Rømer's derivation (see Section 3.4.1). Huygens concluded that the *successive movement of light being thus confirmed, it follows, as I have already said, that it spreads by spherical waves, like sound*. According to Huygens, the origin of the light waves lies in the rapid movement or agitation of the particles which compose the luminous bodies—movement which is communicated to the much smaller particles of the ethereal matter. This matter is of course different from the matter in which sound is propagated,

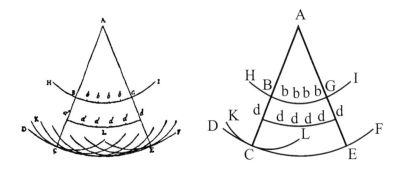

Fig. 3.8   Huygens's principle: a spherical wave can be considered as a sum of wavelets. *HBbbbbGI, dddddd, DCEF*: successive wave-fronts of the wave emitted at *A*. *KCL*: wavelet emitted at *B*. *Left*: After Huygens (1690). *Right*: redrawn for clarity.

the air which we breathe. This is proved by the fact that sound is not propagated in a vessel from which air has been removed with Boyle's vacuum pump,[32] while light is. To explain the propagation of light, Huygens assumed that the oscillations of the minute particles of ethereal matter, which fills as well the glass as the vacuum, are transmitted through collisions between each particle and its neighbours, in the manner of hard spheres. The æther acts as a spring and the oscillations of its particles propagate thus progressively from the luminous body to the eye, with a finite speed.

In the second part of Chapter **I**, Huygens stated his principle. Every particle of the luminous body emits a spherical wave. But, if it is located very far away, on the sun for instance, 'it seems very strange and, even incredible, that the undulations produced by so small movements or corpuscles might spread out to such large distances, since the strength of the waves must become weaker as they move farther away from their source'. Most importantly, each particle in the (ethereal) matter in which the wave spreads does not communicate its movement only to the next one in the straight line drawn from the luminous point, but also to its neighbours. In that way, 'a wave must be formed around each particle, of which that particle is the centre. Thus, if *DCEF* is a spherical wave emitted by *A* [see Fig. 3.8], a particle *B* within the sphere *DCEF* will emit its own spherical wavelet, *KCL*, which will touch wave *DCEF* at *C* at the same time that the main wave emanating from *A* arrives at *DCEF*, with *C* lying on the straight line *AB*. The same holds for all the particles within the sphere *DCEF*, such as *bb*, *dd*, etc., which all of them emit wavelets. Each of these wavelets may be very weak with respect to the main wave, *DCEF*, but they all contribute to it by their portion which lies farthest from *A*' (the tangential point of wave *DCEF* and each wavelet). 'Furthermore, wave *DCEF* is determined by the distance reached in a certain time by the movement emanating from point *A*; there is no movement beyond that wave-front, while there is within the space it encloses. This is what was not recognized by those who have started to consider light waves, among whom are Mr Hooke and Father Pardies'.

---

[32]This was first shown by the German physicist and politician, Otto von Guericke (1602–1686). Huygens made his own experiment in 1674; see *Oeuvres complètes de Christiaan Huygens*, Tome XIX (1937), p. 239.

Expressed in modern terms, Huygens' principle, which follows immediately from the above quotation, states: 'each element of a wave-front may be regarded as the centre of a secondary disturbance which gives rise to spherical wavelets; the position of the wave-front at any later time is the envelope of all such wavelets'. This principle was extended by A. Fresnel by considering the interference of the secondary wavelets. It is called the Huygens–Fresnel principle (see Section 3.6).

Huygens moved on to explain why light propagates in straight lines. He first noted that 'each portion of a wave should spread out in such a way that its extremities always remain within the same straight lines drawn from the luminous point. Thus, the portion *BG* of the wave having *A* as its centre will spread into the arc *CE* limited by the straight lines *ABC* and *AGE*. For, although the wavelets emitted by the particles located within the space *CAE* do also spread out of that space, at a given instant, they only concur to the main wave at their tangential contacts with the main wave, along arc *CE*'. Huygens added that, if the light emitted at *A* is limited by a slit *BG*, *HB* and *GI* being opaque bodies, only the portion *CE* will be illuminated, the parts of the wavelets extending outside this portion being too weak to produce light. By this, Huygens neglects the diffraction effects first observed by Grimaldi (see Fig. 3.2).

In Chapter II, Huygens applied his wave theory to the reflection of light by a polished surface and, in Chapter III, to the refraction of light by transparent liquids and solids. He first considered how light waves may propagate in liquids and solids and envisaged three possibilities. The first one is in the case the ethereal mater does not penetrate them. Then, *their own particles* (of the liquids and solids) 'could transmit the movement of the waves, in the same way as do those of the æther, being supposedly of a nature able to act, like them, as springs. The second possibility is that the ethereal matter occupies the pores and interstices of the transparent bodies and transmits the light waves as in air. The third possibility is that the light waves are transmitted both by the ethereal particles and by the particles which constitute the solid body—a hypothesis that Huygens will use in his explanation of the double refraction of calcite.

Huygens then proceeded to explain the phenomenon of refraction, assuming that light waves propagate in transparent bodies with a lesser speed than in air. He did so using the well-known construction shown in Huygens' original figure (Fig. 3.9). The ratio of the distances between two successive wave-fronts, *HK* in air and *AO* in the transparent medium, is equal to that of the speeds of light in the two media, respectively. The sine law of refraction is easily derived from the construction. In the last part of the chapter, Huygens showed that, in agreement with Fermat's principle, 'a ray of light, in order to go from one point to another, when these points lie in different media, is refracted in such a way at the plane surface which separates these two media that it takes the least possible time'.

Chapter IV is devoted to the refraction of light in air and Chapter V to the 'strange refraction of Iceland crystal'. After recalling Bartholin's work (see Section 11.6), Huygens first noted that the crystals of calcite form an oblique parallelepiped by cleavage (he did not use the term) and that they split as easily along its three sides. The obtuse angle of the parallelograms was measured by him to be exactly 101° 52′, while Bartholin gave a value of 101°.[33] He then carefully analysed the refraction of the ordinary and the extraordinary rays. Turning towards the explanation of the phenomenon, it appeared to him that since there are two types of refraction,

[33]Newton (1704) gave the same value as Huygens, de La Hire (1710), 101° 30′ and Haüy, 101° 52′.

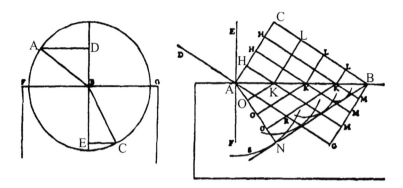

Fig. 3.9　Refraction of light waves according to Huygens' wave theory. *Left*: construction of the refracted ray; *AD* and *EC*, sines of the incidence and refracted angles, are in proportion to the refraction indices in the two media. *Right*: successive wave-fronts *AC, KL, KL, KL* in air and *OK, OK, OK, NB* in the refractive medium; *CL* and *AO* are in proportion to the speed of light in the two media. After Huygens (1690).

there must also be two types of wave. The first one arises in the ethereal matter within the crystal; it is a spherical wave that propagates at a speed slower than outside the crystal. It corresponds to the *regular* refraction. As to the other one, corresponding to the *irregular* refraction,[34] Huygens supposed that it is an *ellipsoidal* or rather, *spheroidal*, wave and that it propagates as well in the ethereal matter as in the particles which constitute the crystal (the third possibility mentioned above). He then related these spheroidal waves to the structure of the crystal: 'it seemed to me that the disposition or regular arrangement of these particles could contribute to form spheroidal waves (nothing more being required for this than that the propagation of light should be a little more rapid in one direction than in the other) and I scarcely doubted that there was in this crystal such an arrangement of equal and similar particles, because of its figure and of its angles with their determinate and invariable measure'.

Huygens was confirmed in his hypothesis by observing the 'ordinary crystal that grows in the form of a hexagon' (quartz) and which, 'because of this regularity, seems also to be composed of regularly arranged particles of a definite figure'. He found that it presents double refraction as well, although less marked than in calcite. In order to explain how ethereal matter may have enough room to transmit the optical waves in the crystal structures, Huygens made the hypothesis that the particles which constitute calcite or quartz could be very small ('of a rare texture') or themselves composed of much smaller particles between which the ethereal matter could pass freely.

From his measurements of incident and refracted angles in various situations, Huygens was able to determine the shape of the spheroidal wave, a prolate ellipsoid of revolution, and to calculate the ratio of its diameters. He showed that the spherical wave is inscribed within the ellipsoid and that the straight line that joins the points of contact of the sphere and the ellipsoid has special properties. According to his calculations, it is oriented at an angle of 45° 20′ with respect to the faces of the cleavage rhombohedron. If the incident direction is along that axis,

---

[34]It was called 'unusual' by Newton, *Opticks*, Book III, query 25, *insolita* by Bartholin (Section 11.6), and is now called extraordinary.

there is no extraordinary ray. Any plane inclined to the axis at the same angle has the same optical properties as the three faces of the cleavage rhombohedron. Huygens has thus discovered what we now know as the optical axis! He showed also that the direction of the extraordinary ray could be obtained by a construction analogous to that of Fig. 3.9, and that a similar sine law applies, but that the orientation of the refracted wave and its speed depended on the conditions of incidence. Huygens was particularly happy that he could thus understand why the propagation of the extraordinary waves was not perpendicular to the wave-front. The modern theory of double refraction is due to A. Fresnel (1827, see Section 3.6).

3.4.3 The fate of Huygens' wave theory

Huygens' theory was well received at the time, but was opposed, as we have seen, by Newton. In the eighteenth century, Newton's emission theory prevailed and Huygens' theory was all but forgotten except by a few scientists such as the Swiss mathematician and physicist L. Euler[35] (1707–1783), who compared the propagation of light and that of sound.[36] By the year 1800, Newton's theory was 'almost universally admitted' in England, and 'but little opposed in others' (Young 1800). For instance, in France, the mineralogist R.-J. Haüy and the physicists P.-S. de Laplace (1749–1827), J.-B. Biot[37] and S. D. Poisson[38] were affirmed 'Newtonists'.

Huygens' wave theory was resuscitated in the early nineteenth century by an increasing number of scientists, first of all by Young (1800, 1802, 1804, 1807) with his interference experiment (Section 3.5). W. H. Wollaston repeated and confirmed Huygens measurements of the double refraction of calcite (Wollaston 1802b, see Section 12.2). E. L. Malus also confirmed Huygens' measurements and gave a new proof of Huygens' construction for the double refraction by translating its geometry into algebra[39] (see, for instance, Shapiro 1973 and Buchwald 1980). D. Brewster confirmed Huygens' account of the double refraction of Iceland spar[40] and correlated the optical properties of crystals and their morphology (Brewster 1818, see Section 12.21). R.-J. Haüy, who was initially a Newtonian, had to admit at the end of his life, in the 1822 edition of his *Traité de Minéralogie*: 'Newton and all the physicists after Huygens rejected his law because it was derived within the framework of the wave theory of light, which was judged inadmissible. I had myself dismissed it, substituting another law as an approximation. [But] M. Malus has confirmed the truthfulness of Huygens' law by a series of very accurate experiments'.

The next step was made by A. Fresnel with his theory of diffraction (1819, see Section 3.6). F. Arago,[41] in his eulogy of A. Fresnel read to the French Academy of Sciences on 26 July

---

[35]Euler is largely responsible for the notations used today in mathematics. He adopted the symbol π in 1737, and introduced the notation $i = \sqrt{-1}$ in 1777 (Boyer 1991).

[36]See, for instance, K. M. Pedersen, Leonhard Euler's Wave Theory of Light, *Perspectives on Science*. (2002). **16**, 392–416.

[37]See footnote 52 in Section 12.10.2.

[38]See footnote 13 in Section 2.3.

[39]*Procès-Verbaux des Séances de l'Académie des Sciences*, Paris, Monday 19 December 1808; Malus E. L. (1810).

[40]*Edinburgh Philosophical Journal*, 1820, **II**, 167–171; **III**, 277–285.

[41]François Arago, born 26 February 1786, died 2 October 1853, was a French physicist and astronomer. A former student of Ecole Polytechnique, he was Professor of Analytical Geometry at that School from 1810 to 1830, and was

1830, could say: 'Huygens' law, despite its elegance and its simplicity, was unrecognized ...
Newton himself was among its opponents and, from that time, progress in optics was stopped
for more than a century'.

A further confirmation of the wave theory came from measurements of the velocity of light
in various media. In a communication to the French Academy of Sciences on 11 December
1815, F. Arago showed that the velocity of light is smaller in a denser than in a rarer medium,
which he did by showing that it is smaller in the liquid phase than in the gas phase of the
same substance.[42] This result came against the Newtonian hypothesis that refraction is due to
an attractive force exerted on the luminous bodies by the surface separating two media, and
that, consequently, the velocity of light is greater in denser media. As a final touch, the French
physicist and astronomer Léon Foucault (1819–1868), using the Fizeau–Foucault rotating-
mirrors apparatus, measured the speed of light in air and water, showing that it is smaller in
water.

3.5 T. Young and the interference experiment, 1804

The major contribution of the English physicist T. Young (Fig. 3.10) was the introduction of the
concept of light interference. This concept implied that light had an undulatory nature. A careful
comparison of Newton's and Huygens's theories led T. Young to convert 'the prepossession
which he before entertained for the undulatory system of light, into a very strong conviction of
its truth and sufficiency' (Young 1802). Both theories have in common the laws of reflection
and refraction, and both admit that light propagates in straight lines, but this is better accounted
for by the emission theory. Both also admit that the velocity of propagation is always the same
in the same medium, but this is a difficulty for the emission theory.

The two theories have major differences, which result from their basic hypotheses; for
the Newtonian theory, 'light consists in the emission of very minute particles from luminous
substances, attended by the undulations of an etherial medium' (Newton grants that there
are undulations but denies that they are light), while for the Huygenian theory it consists in
the excitation of an undulatory motion. T. Young (1802) preferred the term 'undulations' to
'vibrations' because 'an undulation is supposed to consist in a vibratory motion without any
tendency in each particle to continue its motion':

1) According to C. Huygens, the velocity of light is assumed to be the greater, the rarer the
medium, for instance greater in air than in water, while it is the reverse for I. Newton (Young
1801).

2) The explanations of the phenomenon of refraction are quite different; for Newton, the
surface between the two media possesses an attractive force capable of acting on the particles
of light, while for Huygens it is the limit of a medium through which the undulations are propagated with a diminished velocity. Partial reflection from a refracting surface arises, according
to Newton, from certain periodical retardations of the particles of light (the 'fits' referred to in
Section 3.3), while in the undulatory systems it follows as a necessary consequence.

associated with the Paris Observatory from 1805 till his death. He was elected at the French Academy of Sciences
in 1809.

[42]F. Arago et Petit (1816). Sur les puissances réfractives et dispersives de certains liquides et des vapeurs qui le
forment. *Annales de Chimie et de Physique*. **1**, 1–9.

Fig. 3.10   Thomas Young by Sir Thomas Lawrence. Source: Wikimedia commons.

Thomas Young: born 13 June 1773 to a Quaker family, in Milverton, Somerset, England; died 10 May 1829 in London, England, was an English physicist, physician, and egyptologist, well-known, among other things, for his definition of the tensile modulus in elasticity, Young's modulus, for the interference experience which interests us here, and for having partially deciphered the Egyptian hieroglyphics of the Rosetta Stone. He learned foreign languages at a young age. After beginning medical studies in London in 1792 and in Edinburgh in 1794, he moved to Göttingen in Germany where he obtained the degree of doctor of physics in 1796. In 1800, he completed his medical studies at Cambridge and settled as a practising physician in London. In 1801, he was appointed Professor of Natural Philosophy (Physics) at the Royal Institution, but after two years he resigned to devote himself to the practice of medicine. It is during these two years that he published the papers on the properties of light. In 1802, he became foreign secretary of the Royal Society of which he had been elected a Fellow in 1794. In 1811 he became physician to St George's Hospital. From 1816 to 1821 he was Secretary of the Royal Commission on Weights and Measures.

MAIN PUBLICATIONS

1801 *On the mechanism of the eye.*

1802 *On the theory of light and colours.*

1804 *Experiments and calculations relative to physical optics.*

1807 *A course of lectures on natural philosophy and the mechanical arts.*

The theory of light developed by T. Young (1802) is in full agreement with Huygens's. The undulatory nature of light is confirmed by the clear and simple explanation that Young gave of the colours of thin plates. They are due to the interference of light rays that have followed different paths. To each colour is associated a particular wave.

The notion of light interference is proved experimentally in his Bakerian lecture read on 24 November 1803 (Young 1804). Very interestingly, it is not based on the famous double-slit experiment quoted in every optics textbook.

In the first experiment he placed a narrow slip of card in the cone of light coming from a small hole. Besides the coloured fringes outside the shadow of the slip, he also observed fringes inside the shadow itself (inner fringes). These are due to the interference between rays diffracted into the shadow from each side of the slip, as was proved very simply by T. Young. He observed that these fringes disappear if a piece of card is put in front of one of the margins of the slip of card, and reappear if the card is removed.[43] The 'crested' (curved) fringes observed by Grimaldi in the shadow of the tip of a rectangular object can be explained in the same way by the interference of the rays diffracted by each side of the corner of the object (Young 1804).

---

[43]The same experiment was made, independently, by A. Fresnel (1816). Furthermore, F. Arago (1816) showed that the interposition of a thin transparent plate displaces the fringes and shifts them towards the side where the transparent plate is placed.

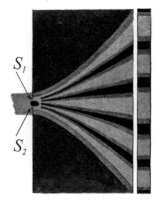

Fig. 3.11   Double-slit experiment. *Left*: path of the beams; S_1, S_2: slits.
Right: interference fringes on the screen. After Young (1807).

The second experiment was a repeat of an experiment made by I. Newton (*Opticks*, Third Book, part I, observations 8 and 9). Two razor-sharp knives were placed with their edges meeting at a very acute angle, in a beam of the sun's light, admitted through a small aperture. They constituted thus a single slit of varying width. The shadows of the two razor edges on a screen are bordered by diffraction fringes. I. Newton and T. Young observed the region around the point where the two first dark lines bordering the shadows of the respective knives meet, at various distances between the knives and the screen, and Young interpreted the result of these observations by calculating the optical paths of the interfering beams.

For interference to occur between two portions of light, they must be derived from the same origin and arrive at the same point by different paths. The simplest way to achieve this is the double-slit experiment in which one observes the interference of two beams of light having passed through two slits, S_1 and S_2 (Fig. 3.11). It was described by T. Young in his *Course of Lectures on Natural Philosophy and the Mechanical Arts* (Young 1807).

After leaving his position at the Royal Institution, T. Young devoted most of his time to his medical practice, without leaving his interest for optics. In a series of articles for the *Quarterly Review* between 1809 and 1818, he reviewed the recent progress in the field. He was very interested by Malus' 1810 work on polarization and he read Arago's and Fresnel's early papers. He was also very interested in Brewster's observations on the double refraction of biaxial crystals (Brewster 1818, see Section 12.4). Like Huygens, Young had at first considered that the vibrations of light were longitudinal, but in order to explain polarization he was now led to believe that they had a minute transverse component, and he used the expression 'imaginary transverse motion'.[44]

For further reading about Thomas Young's life and works, the reader may consult his biography by A. Wood.[45]

---

[44]In his article *Chromatics, Supplement to the British Encyclopaedia*, 1817, quoted in the 'Introduction' of *Oeuvres Complètes d'Augustin Fresnel*, (1866) by E. Verdet. Imprimerie Impériale, Paris.

[45]A. Wood (1954). *Thomas Young: Natural Philosopher 1773–1829*. Cambridge University Press.

3.6   A. Fresnel and the theory of diffraction, 1819

By combining Huygens' wave theory and Young's principle of interference, the French physicist and engineer Augustin Fresnel (Fig. 3.12) confirmed and firmly established the wave nature
of light.

Fresnel's interest in optics started in 1814 when he was living in the country. He began by
observing diffraction fringes and measuring their position. Instead of using a small hole, as the
other investigators did, he employed a convex lens of short focal length which collected the
solar rays into a focus, from which they again diverged, as if they had proceeded from a small
aperture. He studied the diffraction fringes through an eye-glass equipped with a micrometer,
which allowed him to measure their position and their breadth with a great accuracy, to one- or
two- hundredth part of a millimetre (Fresnel 1816).

Despite Young's publications and A. Fresnel's first results on the diffraction fringes (Fresnel
1816), physicists such as Academicians P.-S. de Laplace and J.-B. Biot were still convinced
of the validity of Newton's emission theory, and did not think that the diffraction experiments
of Grimaldi, Hooke, Newton, Young, and Fresnel had received a satisfactory explanation. The
French Academy of Sciences therefore decided on 17 March 1817 to open a competition for
a prize of the Academy on the following topic: 1) 'determine by accurate experiments all the
effects of the diffraction of direct or reflected light rays by opaque bodies', and 2) 'deduce
mathematically from these experiments the paths of the rays after passing in the vicinity of
the bodies'. Fresnel submitted his memoir on 29 July 1818, just before the closing date,
1 August 1818. The membership of the commission included Laplace, Biot, and Poisson, all
three supporters of the emission theory, Arago, who was a supporter of the wave theory, and

Fig. 3.12   Augustin Fresnel. Source: *Oeuvres complètes d' Augustin
Fresnel* (1866).

Augustin Jean Fresnel: born 10 May 1788 in Broglie, Eure, France,
the son of an architect; died 14 July 1827 in Ville-d'Avray, Hauts-de-
Seine, France, was a French physicist and engineer. A former pupil of
Ecole Polytechnique and Ecole des Ponts et Chaussées, from 1809 he
served as an engineer in various parts of France. In 1816 he received
an appointment in Paris, where he remained all his life. He invented
a special type of lens, now called a Fresnel lens, as a substitute for
mirrors in lighthouses. This lens has a large aperture and a short focal
length and is divided into a set of concentric annular sections known
as 'Fresnel zones' so as to reduce the mass and the volume of the lens.
In 1819 he became a commissioner of lighthouses. In 1823 he was elected a member of the French
Academy of Sciences, and in 1825 he was elected a foreign member of the Royal Society of London.
In 1827, just before his death by tuberculosis, the Royal Society awarded him the Rumford Medal.

MAIN PUBLICATIONS

1819 *Mémoire sur la diffraction de la lumière.*   1866 *Oeuvres complètes d' Augustin Fresnel.*
1821 *Mémoire sur la double réfraction.*   (posthumous)
1822 *Mémoire sur un nouveau système*
     déclairage des phares.

Gay-Lussac who was not very familiar with the problems of optics but was open-minded.
Fresnel won the prize with his memoir on the Diffraction of Light, which was awarded in
1819 (Fresnel 1821).

Fresnel's first step was to make use of the principle of interference, 'an immediate consequence of the wave-theory'. He confirmed Young's double-slit experiment with similar
experiments with two mirrors (the 'Fresnel mirrors') and two prisms (the 'Fresnel biprism').
He noted that 'two systems of vibrations of equal wavelength and of equal intensity, differing
in path by half a wave, interfere destructively at those points in the æther where they meet',
and determined the radii of the diffraction fringes observed outside and inside the shadow of an
opaque disk. For the outer fringes, he assumed there is simple interference between a direct ray
CP from a point source *C* to the point of observation *P* and a ray *CAP* having been reflected at
the edge *A* of the opaque disk (Fig. 3.13, *Middle*). The radius of the first fringe comes out to be

$$x = \sqrt{\frac{\lambda b(a + b)}{a}} \tag{3.1}$$

where λ is the wavelength, a is the distance between the source R and the plane of the opaque
disk and b the distance between the opaque disk and the screen. But observation showed that
there was a black fringe where a bright one was expected and vice-versa. This led Fresnel to
admit that there was a phase change of half a wavelength due to reflection. As a consequence,
the radius of the first dark fringe is:

$$x = \sqrt{\frac{2\lambda b(a + b)}{a}} \tag{3.2}$$

A similar calculation gave the radii of the inner fringes (Grimaldi's 'crested' fringes, also
observed by Young). A closer examination of the fringe pattern, both outside and inside
the geometrical shadow, showed, however, that it was more complex and that Fresnel's first
explanation, although approximately correct, was not satisfactory.

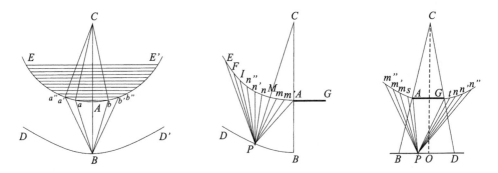

Fig. 3.13   *Left*: Fresnel's zone construction. *Middle*: Waves contributing to the wave-front after passing in the vicinity
of an opaque body *AG* (after Fresnel 1821). *Right*: Formation of the inner fringes in the shadow of an opaque body
AG. Redrawn after Fresnel 1821.

The second step, therefore, was to combine the principle of interference and Huygens's principle, which states that each element of a wave-front may be regarded as the centre of a secondary disturbance which gives rise to a spherical wavelet; the position of the wave-front at any later time is the envelope of all such wavelets (see Fig. 3.8, Section 3.4.2). Fresnel's contribution was to take into account the interference of all those wavelets.

To be able to add the vibrations of many waves, Fresnel expressed the amplitude u of the æther particles at any point in space after an interval of time t. It is proportional to that of the point-source at the instant $t - x/\lambda$, x being the distance of this point from the source of motion and λ the length of a light-wave, and is given by

$$u = a \sin[2\pi(t - x/\lambda)]. \tag{3.3}$$

Fresnel was thus able to extend Huygens' principle in the following way: 'The vibrations at each point in the wave-front may be considered as the sum of the elementary motions which at any one instant are sent to that point from all parts of this same wave in any one of its previous positions, each of these parts acting independently the one of the other'.

It follows from the principle of the superposition of small motions that the vibrations produced at any point in an elastic fluid by several disturbances are equal to the resultant of all the disturbances reaching this point at the same instant from different centres of vibration. The primary wave is thus reconstructed out of partial (secondary) disturbances. Fresnel's argument implies that all the wavelets propagate in the forward directions. G. Stokes[46] (1849b) showed that if the primary wave is resolved as proposed by Huygens, no "back wave" will be produced provided the proper law of disturbance is adopted for the secondary wave.

Consider a primary wave emitted by a point source, C, and two successive wave-fronts, EAE' and DbD' (Fig. 3.13, Left). In order to calculate the amplitude of the vibration at B on the second wave-front, Fresnel divides the first wave-front into an infinite number of small arcs, Aa. aa', aa'', Ab, bb', $b'b''$, etc., chosen such that the paths differences $Ba-BA$, $Ba'-Ba$, $Bb-BA$, $Bb'-Bb$ are equal to half a wavelength. The vibration at B is the resultant of all the secondary waves coming from all the small arcs. This division of the wave-front in small arcs, or zones, is called the Fresnel zone construction. When the rays delimiting the small arcs are very oblique, their contributions cancel out and the vibration at B only results from the arcs nearest to A. It is thus shown that the propagation of light as a straight line is a direct consequence of the wave theory, which refutes an argument of the supporters of the emission theory.

If an opaque body, AG, is put in the beam (Fig. 3.13, Middle), the vibration at any point P of the second wave-front is obtained by summing over the contributions of the small arcs in the portion AE of the primary wave intercepted at A by the opaque object. Again, the small arcs are chosen so that the path difference between the rays delimiting them are equal to half a wavelength. The same construction is used to calculate the inner fringes inside the geometrical shadow of an opaque object, AG, Grimaldi's 'crested' fringes (Fig. 3.13, Right). The same

---

[46]See footnote 15 in Section 2.3.

principle is applied for finding the fringe pattern due to a small aperture *AG*. In all cases, Fresnel obtains the value of the intensity of the fringes by integration, using what was later called 'Fresnel integrals'. All the details of the exterior and inner fringes are quantitatively accounted for by Fresnel's theory, with an excellent numerical agreement between calculation and observation

It is interesting to note that S. D. Poisson, one of the members of the Commission which judged Fresnel's memoir, found that Fresnel's integral for the calculation of the intensity could easily be computed for the centre of the geometrical shadow. The paradoxical result was that it corresponded to a bright spot. Experiment showed that this was indeed the case, which strongly impressed the Commission in favour of Fresnel.[47]

Prompted by Arago, there was an exchange of letters between T. Young and A. Fresnel after the latter had been awarded the Academy's prize. On 19 September 1819, Fresnel wrote to Young: 'it was reserved for you to enrich science with the fruitful principle of interference', to which Young replied on 16 October 1819, describing Fresnel's memoir as deserving a distinguished place among the writings which have contributed most to the progress of optics, and withdrawing his own explanation of the production of the diffraction fringes in favour of that of Fresnel.[48]

Fresnel's method of integration was revised by the German physicist G. Kirchhoff (1824–1887), who gave it an analytical basis by deriving Huygens's principle in the form of integral equations (Kirchhoff 1883). The exact mathematical solution was derived by A. Sommerfeld in 1896 (Section 5.9).

As a further proof of the validity of the wave theory, Fresnel (1822) gave a new demonstration of the law of refraction using wave-fronts and the interference principle, instead of rays.

Very early on, Fresnel became interested in the phenomena of polarization and the interference of polarized rays.[49] Together with F. Arago, Fresnel showed that light rays polarized at right angles do not interfere and that their intensities simply add up (Arago and Fresnel 1819). This is what led him to think that the molecules of the æther oscillate under the action of the light vibrations, instead of being pushed forward, as was generally believed. He proved mathematically that it was the only way to explain the experimental results (Fresnel 1827).

The principle of transverse vibrations is at the base of Fresnel's theory of double refraction for biaxial crystals, which rests on two fundamental hypotheses (Fresnel 1827):

1) a linearly polarized wave is transverse and results from the combination of a right-hand and a left-hand circularly polarized wave; in optically active media, the rotation of the plane of polarization is due to a difference in the indices of refraction of the two waves;

2) *the vibrating molecules* that propagate the optical waves *do not present the same mutual dependence in every direction*; in other words, the medium is anisotropic.

Fresnel had reached the idea of transverse vibrations independently of Young, and went much further, but he acknowledged the latter's priority (see Wood, *Thomas Young*).

[47]F. Arago (1819). *Ann. Chim. Phys.*, **11**, 5–30.

[48]Quoted by Wood, *Thomas Young*.

[49]Memoir to the Academy of Science, 7 October 1816.

The transverse character of light vibrations was confirmed by M. Faraday[50] (1846a) from quite different considerations. Faraday introduced the concept of electromagnetic field, and was probably the first to associate light with electric and magnetic forces. Radiation, he writes, is 'a high species of vibration in the lines of force which are known to connect particles and also masses of matter together'. The kind of vibration 'which can alone account for the wonderful, varied, and beautiful phenomena of polarization is lateral' (transverse). This opinion was supported by Faraday's (1846b) discovery of the 'Faraday effect' in which the plane of polarization was rotated by a magnetic field.

J. C. Maxwell[51] derived the electromagnetic theory of light in the wake of Faraday's work (Part VI of his seminal article on the theory of the electromagnetic field, Maxwell 1865). He wrote about Faraday's contribution: 'the conception of the propagation of transverse magnetic disturbances to the exclusion of normal ones is distinctly set forth by Professor Faraday in his *Thoughts on Ray Vibrations*. The electromagnetic theory of light, as proposed by him, is the same in substance as that which I have begun to develop in this paper, except that in 1846 there were no data to calculate the velocity of propagation'. The electromagnetic field of light was in fact also expressed, independently, by the German mathematician B. Riemann (1867) and by the Danish mathematician and physicist, L. Lorenz (1867).

3.7   A. Einstein and the photoelectric effect, 1905

The nature of light as a transverse vibration had just been firmly established by Maxwell's electromagnetic theory when new events took place. Heinrich Hertz[52] (1887a) in Karlsruhe had successfully achieved the emission and detection of electromagnetic waves. For the emission, he used a high-voltage induction coil to produce a spark discharge between two pieces of brass and, for the detection, a piece of copper wire bent into a circle with a small brass sphere on one end, and the other end of the wire being pointed toward the sphere. A small spark was observed in the receiver when the induction coil was activated, even if the emitter and transmitter were at a distance of several metres, but it was very weak. Hertz tried in various way to improve the visibility of the spark and came up with a very surprising result, namely that the visibility was enhanced if the receiver was exposed to light (Hertz 1887b). Further investigation showed that it was the absorption of the ultraviolet component by the copper wire which produced the effect, but Hertz could not offer any explanation for it.

[50]Michael Faraday, born 22 September 1791 in Newington Butts, England, died 25 August 1867 in Hampton Court, England, was an English physicist and chemist who was Professor of Chemistry at the Royal Institution, London. He is well known for his study of electromagnetism and electrochemistry, the Faraday effect, the Faraday cage, the discovery of benzene, etc. He was elected a Fellow of the Royal Society in 1824.

[51]James Clerk Maxwell, born 13 June 1831 in Edinburgh, Scotland, the son of an advocate, died 5 November 1879 in Cambridge, England, was a Scottish physicist and mathematician. A former student at Edinburgh and Cambridge Universities, he was successively Professor at Aberdeen University, King's College, London, and Cambridge University. He formulated the classical electromagnetic theory and contributed to the kinetic theory of gases. He was awarded the Rumford Medal in 1860.

[52]Heinrich Hertz, born 22 February 1857 in Hamburg to a well-to-do Hanseatic family, Germany, died 1 January 1894 in Bonn, Germany, was a German physicist. He obtained his PhD in Berlin and worked with H. von Helmholtz. He was appointed Professor at the University of Karlsruhe in 1885, and then at the University of Bonn in 1890. Hertz is well-known for having observed electromagnetic waves, thus confirming Maxwell's theory, and for having discovered the photoelectric effect.

Following Hertz, W. Hallwachs[53] in Leipzig devised a simpler experiment to reveal the influence of light: a clean circular zinc plate, eight centimetres in diameter, hanging from an insulating stand, was linked to a gold-leaf electroscope which was charged negatively. It lost its charge slowly, but much faster if the zinc plate was illuminated with ultraviolet light. There was no effect if the electroscope was charged positively (Hallwachs 1888, 1889). Again there was no explanation offered.

More information was obtained by J. J. Thomson[54] (1899) who showed that, when a negatively electrified metal plate in a gas at low pressure is illuminated by ultraviolet light, negatively charged particles are emitted (cathode rays).

The first quantitative studies of the photoelectric effect are due to Hertz's former assistant, P. Lenard,[55] who showed that a metal plate illuminated with ultraviolet light emits cathode rays (Lenard 1900) and who studied how the energy of the emitted photoelectrons varied with the intensity of the light produced by a bright carbon arc (Lenard 1902). The emitted photoelectrons were collected in a second metal plate, which was connected to the electrode through a sensitive ammeter. The intensity of light could be increased by bringing the carbon arc closer to the electrode. This caused the number of emitted photoelectrons to increase, as could be expected, but did not increase their average energy, which was a big surprise. Lenard also showed that the response depended on the type of light used.

In order to solve a contradiction between the wave theory of light and measurements of the electromagnetic spectrum emitted by thermal radiators, or black bodies, which puzzled physicists, Max Planck[56] elaborated in 1900 a revolutionary theory asserting that black bodies emit radiation as discrete bundles or packets of energy, approximately in the same way as matter is made up of particles. In the presentation of Planck's Nobel Prize for Physics in 1918, it was said, 'the product $h\nu$ is the smallest amount of heat which can be radiated at the vibration frequency ν'.

Einstein[57] (1905) gave a simple interpretation of the photoelectric effect, and revealed the dual nature, wave and corpuscular, of light. He assumed that the incoming radiation should be

---

[53]Wilhelm Hallwachs, born 9 July 1859 in Darmstadt, Germany, died 20 June 1922 in Dresden, Germany, was a German physicist. He obtained his PhD in Strasbourg University in 1884, and, after a short stay in Würzburg, moved in 1886 to Leipzig where he made his experiments on the photoelectric effect. In 1888, he returned to Strasbourg, and in 1890 was appointed Professor at the *Technische Hochschule* in Dresden.

[54]See Section 5.1.

[55]See Section 4.7.

[56]Max Planck, born 23 April 1858 in Kiel, Germany, the son of a Law Professor, died 4 October 1947 in Göttingen, Germany, was a German physicist, the founder of quantum theory. In 1877, he went to Berlin to study with H. von Helmholtz, G. Kirchhoff, and K. Weierstrass. He obtained his PhD in 1879, and the habilitation in 1880. He was appointed Professor of Physics at the University of Kiel in 1885 and in 1889 at the University of Berlin, as L. Boltzmann's successor. He retired in 1926 to be succeeded by E. Schrödinger, and was elected the same year a member of the Deutsche Akademie der Naturforscher Leopoldina. He had Max von Laue as one of his students.

[57]Albert Einstein, born 14 March 1879 in Ulm, Germany, the son of a salesman and engineer, died 18 April 1955 in Princeton, USA, was a German-born theoretical physicist. After schooling in Munich, he passed the Swiss *Matura* in 1896 and enrolled at the Swiss Federal Polytechnic. He graduated in 1900 and in 1902, obtained a post at the Federal Office for Intellectual Property in Bern, Switzerland. In 1905 he obtained his PhD at the University of Zürich. That year was his *annus mirabilis* when he published his four ground-breaking papers on the photoelectric effect, Brownian motion, special relativity, and the equivalence of matter and energy. He was appointed lecturer at the University of Bern in 1908 and in 1909 docent at the University of Zürich. From 1914 to 1932, he was Professor at the Humboldt University

thought of as quanta of energy $h\nu$ (*Lichtquanten*), with ν the frequency. A photoelectron is emitted because it has absorbed one quantum of energy, and its energy E will be at most equal to that quantum ($E = h\nu$). Einstein was awarded the Nobel Prize for Physics in 1921 for 'his discovery of the law of the photoelectric effect'. The Planck–Einstein theory was at the base of Bohr's theory of the structure of atoms, for which X-ray spectroscopy brought the experimental evidence (See Section 10.6). Einstein's photoelectric equation was confirmed experimentally by R. A. Millikan (1916) who deduced from it a value for the Planck constant: $h = 6.57 \times 10^{-27}$ erg. sec.

It was only after the formulation in 1924 by L. de Broglie of the relation between the properties of light and those of the atom that the dual nature, corpuscular and wave, of light was really understood. The name 'photon' was coined later, by the chemist G. N. Lewis[58] (1926).

in Berlin, after which he emigrated to the United States. He was a member of the Prussian Academy of Sciences and of the Deutsche Akademie der Naturforscher Leopoldina.

[58]Gilbert Newton Lewis, born 23 October 1875 in Weymouth, Mass., USA, died 23 March 23 1946 in Berkeley, CA, USA, was an American physical chemist. After obtaining his PhD in 1899 from Harvard University under T. Richards, he remained for one year at Harvard as instructor. He then spent a semester at Leipzig with W. Ostwald and another at Göttingen with W. Nernst, after which he returned to Harvard as instructor. He was appointed Assistant Professor at M.I.T. in 1907, and full Professor in 1911. In 1912 he became Professor of Physical Chemistry at the University of California at Berkeley. In 1902, while Lewis was trying to explain valence to his students, he depicted atoms as constructed of a concentric series of cubes with electrons at each corner (the cubic atom). He developed a theory of chemical bonding in which a chemical bond is a pair of electrons shared by two atoms (See Section 10.1.2). The term 'covalent bond' was coined by I. Langmuir (see footnote 22 in Chapter 8).

4

RÖNTGEN AND THE DISCOVERY OF X-RAYS

The discovery by Professor Röntgen of a new kind of radiation from a highly exhausted tube through which an electric discharge is passing has aroused an amount of interest unprecedented in the history of physical science.

J. J. Thomson (1896*b*)

4.1 8 November 1895: first observation

On Friday evening, 8 November 1895, Wilhelm Röntgen (Fig. 4.1) remained long hours in his laboratory and was late for dinner—so the story goes. He had been kept by a most puzzling observation he made while repeating some of Heinrich Hertz's (1892) and Philipp Lenard's (1894, 1895) recent experiments on cathode rays.

His apparatus was very simple and standard; it consisted of a Ruhmkorff spark coil with a mercury interrupter and a Hittorf discharge tube (see Fig. 4.2). That evening, in preparing for his next experiment, he had carefully covered the tube with black cardboard and drawn the curtains of the windows. He hoped to be able to detect some fluorescence coming from the tube with a fluorescent screen made of a sheet of paper painted with barium platinocyanide. That screen, which he intended to bring close to the tube later on, was lying on the table at some distance. Röntgen wanted to test the tightness of the black shield around the tube. He operated the switch of the Ruhmkorff spark coil, producing high-voltage pulses of cathode rays and looked for any stray light coming from the glass tube. He then happened to notice out of the corner of his eye a faint glimmer towards the end of his experimentation table. He switched off the coil, the glimmer disappeared. He switched the coil back on, the glimmer reappeared. He repeated the operation several times, the glimmer was still there. He looked for its source and found that it came from the fluorescent screen.

In the interview he granted in March 1896 to H. J. Dam, a London-based American reporter for the American magazine, *McClure's*,[1] Röntgen was asked: 'What did you think?'. His answer was: 'I did not think, I investigated. I assumed that the effect must have come from the tube since its character indicated that it could come from nowhere else'. Röntgen found that the intensity of the fluorescence increased significantly as he brought the screen close to the discharge tube. More baffling, the propagation of this 'radiation' was not hampered if he put a piece of cardboard between the screen and the tube, or other objects such as a pack of cards, a thick book or a wooden board two or three centimetres thick. Then he moved the screen farther and farther away, even as far as two metres, and, his eyes being well accustomed to obscurity,

---

[1]H. J. W. Dam (1896). The new marvel in photography. *McClure's Magazine*. **6**, April, 403–415.

Fig. 4.1 Wilhelm Conrad Röntgen in 1901. Source: Wikimedia commons.

Wilhelm Conrad Röntgen: born 23 March 1845 in Remscheid-Lennep, Germany, the son of a cloth merchant; died 10 February 1923 in Munich, Germany. Soon after his birth, his family moved to Holland, his mother's homeland. He attended the Technical School at Utrecht, but was not admitted to the University as a regular student. From 1865 to 1868 he studied at the Polytechnic School in Zürich, Switzerland, where he had R. Clausius as Professor of Physics. He received the diploma of mechanical engineer in 1868, and, a year later, his PhD for studies on gases. He was then appointed assistant to the experimental physicist A. Kundt, first in Würzburg (1870–1874), then in the newly founded University of Strasbourg. In 1874, after his habilitation, he became *Privatdozent* there, and, in 1875, Professor at the Academy of Agriculture at Hohenheim. In 1876, he came back to Strasbourg, as Professor of Mathematics and Physics, and continued his collaboration with A. Kundt. In 1879 he was appointed Professor of Physics at the University of Giessen, and, in 1888, at the University of Würzburg, where he was also Director of the new Institute of Physics. He became rector of that University in 1894. In 1900 he moved to Munich where he was Director of the Institute of Experimental Physics. He had many students, among whom, E. v. Angerer, E. Bassler, W. Friedrich, R. Glocker, P. P. Koch, E. Wagner, P. Knipping, J. C. M. Brentano, and P. Pringsheim. He was awarded the first Nobel Prize in Physics, in 1901, 'in recognition of the extraordinary services he has rendered by the discovery of the remarkable rays subsequently named after him'. He retired in 1920, but continued to work until his death from cancer in 1923.

MAIN PUBLICATIONS

1879 *Nachweis der electromagnetischen Drehung der Polarisationsebene des Lichtes im Schwefelkohlenstoffdampf.* With A. Kundt.

1888 *Über die durch Bewegung eines im homogen elektrischen Felde befindlichen Dielektrikums hervorgerufene elektrodynamische Kraft.*

1895 *Über eine neue Art von Strahlen.*

1896 *Über eine neue Art von Strahlen. 2. Mitteilung.*

1897 *Weitere Beobachtungen über die Eigenschaften der X-Strahlen.*

he could still see the very faint glimmer. As an added fortunate circumstance, according to H. H. Seliger (1995), Röntgen being colour-blind, his eyes had enhanced sensitivity in the dark.

After dinner, Röntgen went back down to his laboratory[2] and repeated his experiment, now putting various sheets of materials such as aluminium, copper, lead or platinum in front of the screen. Only lead and platinum absorbed the radiation completely, and lead glass was found to be more absorbing than ordinary glass. Röntgen held a small lead disk in front of the screen and was very surprised to see not only the shadow of the disk, but also the shadows of the bones of his own hand! He also found that photographic plates were sensitive to this unknown radiation.

[2]His apartment was on the top floor of the building.

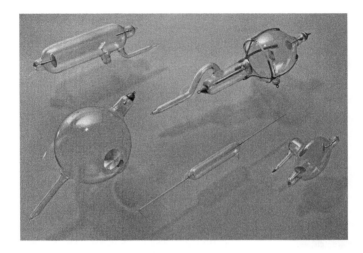

Fig. 4.2   Hittorf discharge tubes used by Röntgen in 1895/1896 to produce X-rays. Photo Deutsches Museum, Munich.

Fig. 4.3   Radiographs made by Röntgen. *Left*: radiograph of his wife Bertha's hand, 22 December 1895. Photo Deutsches Museum, Munich. *Right*: radiograph of A. von Kölliker, Professor of Anatomy at Würzburg University, 23 January 1896. Source: Wikimedia commons.

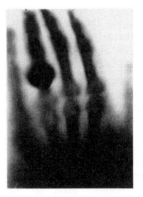
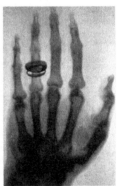

In the days that followed, Röntgen told no one of his startling observations, neither his assistants nor his wife. He was morose and abstracted, according to his wife (Underwood 1945), and often ate and even slept in his laboratory. The discovery was so astounding, so unbelievable, that he would not disclose it before he had fully convinced himself of its reality by repeated observations and had determined the properties of this new radiation.

In the same interview for *McClure's Magazine* mentioned above, he said: 'It seemed at first a new kind of light. It was clearly something new, something unrecorded'. 'Is it light'? 'No, it can neither be reflected nor refracted'. 'Is it electricity'? 'Not in any form known'. 'What is it'? 'I do not know. Having discovered the existence of a new kind of rays, I of course began to investigate what they could do'. Indeed, being a careful experimenter, he made in the following seven weeks very systematic studies of the properties of the new rays, *X-Strahlen*, as he called them. During all that period, he remained uncommunicative, but, shortly before Christmas, he invited his wife to his laboratory and showed her his work. He even took a radiograph of her hand (Fig. 4.3, *Left*). The results of his investigations are recorded in the preliminary report he handed to the president of the Würzburg Physikalisch-medicinische Gesellschaft (Würzburg Physical Medical Society) on 28 December (Röntgen 1895). On account of its outstanding importance, the President of the Society agreed that the report should be printed at once, even though it had not been presented orally at a meeting (Fig. 4.4).

4.2 Before the discovery

Röntgen was an excellent experimenter with a deep theoretical knowledge. Although he did not use mathematical formulae much himself, he was fully convinced of the importance of mathematics. His experimental skills were acquired while he was training as a mechanical engineer at the Polytechnic Institute of Zürich in the 1860s. He built his own apparatus, usually without the aid of an assistant, and he had an enormous capacity for work. He was a typical classical physicist, planning his experiments very carefully, with a critical honesty in his measurements, never bringing an experiment to a close until he was entirely satisfied with the results. One of his later students in Munich, J. C. M. Brentano, commented: 'For him, nothing could pass by without his establishing a definite reason' (in Ewald 1962a, p. 540). His first works at the Universities of Strasbourg and Giessen won him an excellent reputation. His exceptional intellect, the originality of his ideas, and his accurate methods of investigation were recognized by all his colleagues. As a person, he was modest and unassuming, with simple tastes. He liked nature and hiking trips in Switzerland.

By the time of the discovery of X-rays, he had published nearly fifty articles on many different topics. Of particular interest are his studies on the specific heat of gases, on the rotation of the plane of polarization of light (with the well-known experimental physicist A. Kundt), on the physical properties of crystals (heat conductivity, pyro- and piezoelectricity), and on dielectrics.

The study on dielectrics is the most important one, and Röntgen was as proud of it as of the discovery of X-rays (Laue 1946). In it, he showed that electromagnetic effects are induced when a dielectric is moved in an homogeneous electric field (Röntgen 1888). This constituted an experimental proof of the Faraday-Maxwell electromagnetic theory and had far-reaching consequences. It led to Lorentz's theory and to considerations as to the observability of the effects resulting from the movement of the earth. The current induced in Röntgen's experiment was called 'Röntgen current' by H. A. Lorentz.

In the years that followed, Röntgen became interested in cathode rays and in the works of H. Hertz (1892) and his assistant, P. Lenard (1894, 1895). He even wrote to Lenard and asked him some practical questions about his special cathode tubes with an aluminium window ('Lenard tubes').

Von Helmholtz had predicted from Maxwell's theory of electromagnetic radiation that high-frequency radiation would accompany electric spark discharges. It was believed at the time, in Germany, that cathode rays were some kind of æther phenomena, and it was supposed that high-frequency electromagnetic waves would be emitted in cathode-ray discharges. Röntgen was looking for this radiation, encouraged in that direction by H. Hertz's demonstration of the propagation through space of electromagnetic waves produced by electric spark discharges (Hertz 1887a). When asked in July 1896 by the English surgeon, James MacKenzie Davidson, a future President of the Röntgen Ray Society of London (1912–1913): 'What were you doing with the Hittorf tube when you discovered the X-rays?' Röntgen answered 'I was looking for the invisible rays' (quoted by Glasser 1931*a*, *b*).

It is sometimes written that the discovery of X-rays was made fortuitously. It was not. What was fortuitous about it was the way Röntgen noticed the faint glimmer on the barium platinocyanide screen; he would have found X-rays eventually in any case.

4.3 28 December 1895: Röntgen's preliminary communication

Röntgen's first communication (Fig. 4.4), written in a precise and matter-of-fact way, reveals what a thorough and meticulous investigation he made of the properties of the new rays.

1. Many other bodies besides barium platinocyanide exhibit fluorescence when submitted to the action of X-rays: calcium sulphide, uranium glass, Iceland spar, rock-salt, etc.
2. X-rays pass through all bodies, as shown by Lenard (1894, 1895) for cathode rays. Röntgen compared the attenuation of X-rays through various materials. For instance, the radiographs of

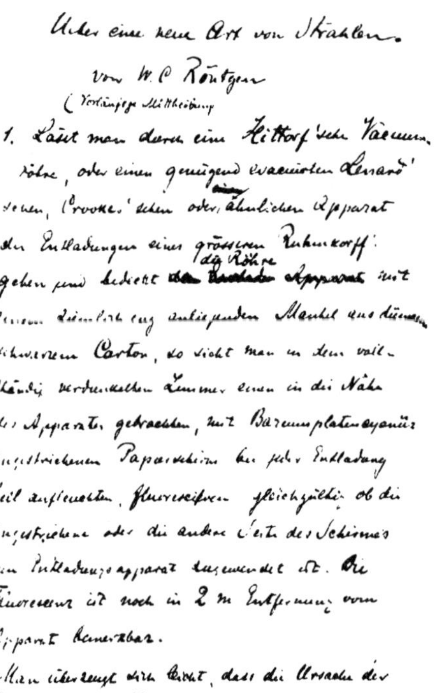
Ueber eine neue Art von Strahlen.
Von W. C. Röntgen
(Vorläufige Mittheilung)

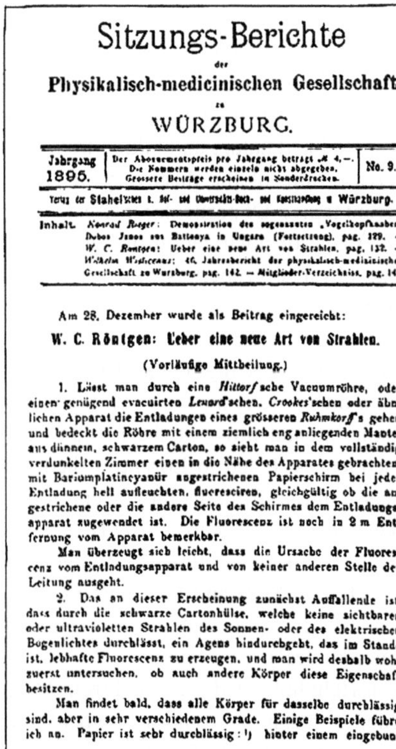
Sitzungs-Berichte

der

Physikalisch-medicinischen Gesellschaft

zu

WÜRZBURG.

Jahrgang 1895.	Der Abonnementspreis pro Jahrgang beträgt ℳ 4,—. Die Nummern werden einzeln nicht abgegeben. Grössere Beiträge erscheinen in Sonderdrucken.	No. 9.

Verlag der Stahel'schen k. Hof- und Universitäts-Buch- und Kunsthandlung in Würzburg.

Inhalt. *Konrad Rieger*: Demonstration des sogenannten „Vogelkopfknaben" Dobos Janos aus Battonya in Ungarn (Fortsetzung). pag. 129. — *W. C. Röntgen*: Ueber eine neue Art von Strahlen. pag. 132. — *Wilhelm Wislicenus*: 46. Jahresbericht der physikalisch-medicinischen Gesellschaft zu Würzburg. pag. 142. — Mitglieder-Verzeichniss. pag. 146.

Am 28. Dezember wurde als Beitrag eingereicht:

W. C. Röntgen: Ueber eine neue Art von Strahlen.

(Vorläufige Mittheilung.)

1. Lässt man durch eine *Hittorf*'sche Vacuumröhre, oder einen genügend evacuirten *Lenard*'schen, *Crookes*'schen oder ähnlichen Apparat die Entladungen eines grösseren *Ruhmkorff*'s gehen und bedeckt die Röhre mit einem ziemlich eng anliegenden Mantel aus dünnem, schwarzem Carton, so sieht man in dem vollständig verdunkelten Zimmer einen in die Nähe des Apparates gebrachten, mit Bariumplatincyanür angestrichenen Papierschirm bei jeder Entladung hell aufleuchten, fluoresciren, gleichgültig ob die angestrichene oder die andere Seite des Schirmes dem Entladungsapparat zugewendet ist. Die Fluorescenz ist noch in 2 m Entfernung vom Apparat bemerkbar.

Man überzeugt sich leicht, dass die Ursache der Fluorescenz vom Entladungsapparat und von keiner anderen Stelle der Leitung ausgeht.

2. Das an dieser Erscheinung zunächst Auffallende ist, dass durch die schwarze Cartonhülse, welche keine sichtbaren oder ultravioletten Strahlen des Sonnen- oder des elektrischen Bogenlichtes durchlässt, ein Agens hindurchgeht, das im Stande ist, lebhafte Fluorescenz zu erzeugen, und man wird deshalb wohl zuerst untersuchen, ob auch andere Körper diese Eigenschaft besitzen.

Man findet bald, dass alle Körper für dasselbe durchlässig sind, aber in sehr verschiedenem Grade. Einige Beispiele führe ich an. Papier ist sehr durchlässig:[1]) hinter einem eingebun-

[1]) Mit „Durchlässigkeit" eines Körpers bezeichne ich das Verhältniss der Helligkeit eines dicht hinter dem Körper gehaltenen Fluorescenzschirmes zu derjenigen Helligkeit des Schirmes, welcher dieser unter denselben Verhältnissen aber ohne Zwischenschaltung des Körpers zeigt.

Fig. 4.4 Röntgen's first communication to the Würzburg Physical Medical Society, 28 December 1895. *Left*: first page of the manuscript. Photo Deutsches Museum, Munich. *Right*: After Röntgen (1895).

a hand showed that bones were more absorbing than flesh. Generally speaking, the absorption of X-rays increases with the density and the thickness of the bodies. Röntgen made quantitative estimates and found roughly the same attenuation for metallic foils of platinum, lead, zinc, and aluminium, 0.018 mm, 0.050 mm, 0.100 mm, and 3.500 mm thick, respectively. He also checked the increase of absorption with thickness by means of photographs taken through tin foils of gradually increasing thicknesses.

3. X-rays are not deflected by a prism. Röntgen used water and carbon disulphide in mica prisms of 30°, and prisms of ebonite and aluminium, but found no effect. There was no refraction by lenses either, and this 'shows that the velocity of X-rays is the same in all bodies'.

4. X-rays are diffused by turbid media, like light.

5. Likewise, no conclusive reflection of X-rays by a mirror was observed.

6. After these negative observations, Röntgen thought that maybe, nevertheless, 'the geometrical arrangement of the molecules might affect the action of a body upon the X-rays for instance according to the orientation of the surface of an Iceland spar plate with respect to its [optical] axis', but the experiments with quartz and Iceland spar on this point also led to a negative result.

7. Despite all his efforts, Röntgen could not find any interference effects.[3] He attributed this negative result to the very feeble intensity of the X-rays. Laue (1946) noted that he was right in this, since, having shone X-rays on quartz and calcite crystals, he would have observed interference fringes if the intensity had been higher. But Röntgen told him that in any case he would never have imagined interference effects to be like those seen by Friedrich and Knipping!

8. X-rays are much less absorbed than cathode rays, and unlike them, are not deflected by magnets. They are a different kind of radiation.

9. The intensity of the rays decreases as the inverse square of the distance between the discharge tube and the screen.

10. X-rays cast regular shadows, as shown by many photographs of shadows of various objects, as well as by pinhole photographs. This indicates a rectilinear propagation, hence the term 'rays'.

In conclusion, Röntgen noted that 'a kind of relationship between the new rays and light appears to exist' and suggested tentatively 'Should not the new rays be ascribed to longitudinal waves in the æther'?

4.4 The news of the discovery spread round the world

There were no illustrations in the report, but Röntgen made copies of nine of the most important radiographs, such as a set of weights in a closed wooden box, a piece of metal whose lack of homogeneity was revealed by the X-rays, and a wooden door with lead paint, the most striking and extraordinary one being, of course, the radiograph of a hand showing the bones. He mailed them on New Year's Day 1896, together with preprints of his paper, to ninety leading physicists in Germany, Austria, France, and England. One of the recipients was F. Exner, the Director of the Institute of Physics at Vienna University, whom he knew from his younger days at the Polytechnic Institute in Zürich. Professor Exner showed the report and the photographs to some friends, among whom was E. Larcher. Larcher's father happened to be the editor of the journal *Die Presse* in Vienna. As a good journalist, he immediately felt the importance

---

[3]In a letter to W. C. Röntgen in July 1896, the French mathematician and physicist H. Poincaré (1854–1912) asked him how he could look for interference effects, since this would require two beams issued from the same point and having followed different paths. How could he achieve that if the rays could not be refracted nor reflected, nor bent by a magnet? Röntgen's answer is not known.

of Röntgen's discovery and wrote without waiting an article which made the front page of that journal on Sunday, 5 January 1896, under the headline *Eine sensationelle Entdeckung* (a sensational discovery). This was indeed sensational news. They were cabled immediately by foreign correspondents to their home journals, and, from then on, they spread round the world with the speed of lightning. The discovery was reported next day in the dailies, on 6 January in the *Frankfurter Zeitung* and in the London *Daily Chronicle*, on the 7th in the *Standard*, on the 13th in the Paris *Le Matin*, on the 16th in the *New York Times*, and on the 31st in the *Sydney Telegraph*. For an account of how the news reached North America, see Linton (1995).

The professional journals followed suit immediately, the *Electrical Engineer*, New York, on 8 January, under the title 'Electrical photography through solid matter', the *Electrician*, London, on the 10th, the *Lancet*, London, on the 11th, and the *British Journal of Medicine* on the 18th, with a note by the English physicist, A. Schuster,[4] one of the recipients of Röntgen's mailings. It was announced at the French Academy of Sciences on 20 January. An English translation of Röntgen's communication was published in *Nature*, London, on 23 January, along with short articles by A. A. Campbell Swinton and A. Schuster, and in *Science* (USA) on 14 February. A French translation appeared in *L'Eclairage électrique* on 8 February. The imagination of the general public was naturally inflamed and it is no surprise, in that Victorian age, that some advertisements appeared for 'X-ray proof underclothing—especially for the sensitive woman'.[5]

Röntgen's experiment was immediately reproduced in many laboratories throughout the world; for instance, by A. A. Campbell Swinton (1896, 23 January) and by J. Perrin[6] (1896, 27 January). Perrin reproduced many of Röntgen's experiments on the absorption of X-rays, and on their rectilinear propagation; he also tried to see if they could be refracted by a prism, reflected by a mirror, or diffracted by a slit, with a negative conclusion.

Medical applications in bone pathology were proposed by Lannelongue *et al.* (1896, 27 January), also reported in the *British Medical Journal* on 1 February. *The Lancet* reported the observation by A. A. C. Swinton of a piece of glass embedded in a hand (25 January), and the application of X-rays to discover a bullet lost in a wrist was described by Jones and Lodge (1896, 23 February). For more details, see, for instance, Underwood (1945) and Posner (1970). For early medical applications in the Russian Empire, see Savchuk (2007).

Non-medical applications of X-rays were also reported at a very early stage. For instance, the *Electrical Engineer*, New York (29 January) and *Scientific American* (1896, **74**, p. 103) showed an X-ray photograph by A. W. Wright of Yale University of the seam of a welding which could not be seen with the naked eye. The *Electrical Engineer* also reported practical tests by the Carnegie Steel Works in Pittsburgh (2 February). Röntgen (1897) himself had taken a photograph of his hunting-rifle in which the faults in the damask barrels could be observed. For more examples, see, for instance, Glasser (1931*a*) and Jauncey (1945).

Very soon, preliminary comments as to the possible nature of X-rays were published by some of the leading physicists, H. Poincaré (1896, 30 January), J. J. Thomson (1896*a*, 1 February), Sir O. Lodge (1896*a*, *b*, 7 February, 1 March), Lord Kelvin (Lord Kelvin 1896, 10 February). By the end of the year, more than a thousand scientific articles and books had been published, not to speak of the numerous papers in popular journals and magazines!

---

[4]See footnote 9, Section 5.1.   [5]*Electrical World*, 1896, **27**, 339.   [6]See footnote 10 in Section 5.1.

On 13 January 1896, Röntgen was invited to give a demonstration of his new rays before the Emperor, Kaiser Wilhelm II, in Postdam, and on 23 January he made a public demonstration in front of a large and enthusiastic audience at the Würzburg Physical Medical Society. He spoke very simply and modestly, without indulging in problematic speculations. During the lecture, he took a photograph of the hand of A. von Kölliker, a famous anatomist of the University (Fig. 4.3, *Right*), and showed it to the audience after development, receiving a tremendous applause. One of the surgeons present, however, 'warned against too much optimism, since the method scarcely promised to be of much, if any, value in the diagnosis of internal disturbances' (Glasser 1931*a*). At the conclusion of the meeting, A. von Kölliker proposed that the new rays, called 'X-rays' by Röntgen, should be called 'Röntgen rays', as a tribute to their discoverer.

Röntgen was showered with honours, invitations, and prizes, the most prestigious one being the very first Nobel Prize in Physics, awarded in 1901, but, being shy of nature, he declined many other invitations to speak again in public. He did not even give a lecture after receiving his Nobel Prize. The Prince Regent of Bavaria bestowed on him the Royal Bavarian Order of the Crown, which entitled the recipient to be called *von*. Röntgen accepted the decoration, but declined the nobility (Glasser 1931*a*). He did not take any patent, and gave his discovery to the world without deriving any personal profit from it.

In his inaugural speech as the newly elected Rector of the University of Würzburg, on 2 January 1894, Röntgen quoted the Jesuit scholar Athanasius Kircher (1601 or 1602–1680), who was Professor of Mathematics at the University of Würzburg: 'Nature often reveals surprising marvels in even the most ordinary events but they can only be recognized by those who with a keen sense trained to investigation ask for guidance from experiment—the teacher in all things'. How well this sentence applies to Röntgen himself!

Röntgen did not publish much after his discovery, but had many students in the Institute of Experimental Physics that he headed in Munich since 1900, among whom W. Friedrich and P. Knipping (See Section 6.1.2). Being a patriot, he was very unhappy at the end of the War, and his health declined. He died of cancer in 1923. For further reading on Röntgen's life and career, see, for instance, Wien (1923),[7] Glasser (1931*a*, *b*), Jauncey (1945), Seliger (1945), Underwood (1945), and Laue (1946). There are Röntgen Museums in his home town of Lennep-Remscheid[8] and in Würzburg.[9]

4.5 Further investigations on X-rays by W. C. Röntgen, 1896, 1897

In the two months that followed his first communication, Röntgen worked very hard to continue the study of the properties of X-rays, not letting himself be distracted by all the honours which were bestowed on him and the many unwelcome visitors. During that period, he concentrated on two points, which had been briefly mentioned in the first report, and which are described in his second communication (Röntgen 1896).

[7]Wien, W. (1923). Röntgen. *Ann. Phys.* **375**, i–iv.

[8]Deutsches Röntgen-Museum, Schwelmer Strasse 41, 42897 Remscheid, Germany, <http://www.roentgenmuseum.de/>.

[9]Röntgen-Gedächtnisstätte   Würzburg,   Röntgenring   8,   97070   Würzburg,   Germany,   <http://www.wilhelmconradroentgen.de/>.

1. The first point is the property of X-rays to discharge electrified bodies. In order to be able to observe this phenomenon in a space that is completely protected, he 'had a chamber made of zinc plates soldered together, which was airtight and large enough to contain himself and his apparatus'. He found that 'electrified bodies in air, charged positively or negatively, conductors or insulators, are discharged when X-rays fall on them'. With his customary meticulousness he detailed the conditions under which this property appears, and recognized that it is due to a change in the air, namely air is ionized by the passage of X-rays. He recognized the effect, but did not name it.
2. The second point was that X-rays could be produced in many materials other than glass, for instance if the beam of cathode rays fell on a plate of aluminium or platinum. Röntgen found that the greatest intensity was obtained with platinum. For that he used 'a discharge-apparatus in which the cathode is a concave mirror of aluminium and the anode a plate of platinum at the centre of curvature of the mirror', a usual set-up at the time.

After spending some holidays in Italy with his wife in March 1896, Röntgen continued his study of the properties of X-rays, using a discharge-tube of the type just described, with a platinum anode. He made the following observations, recorded in his third communication (Röntgen 1897):

1. Any matter, air or any other, when submitted to X-rays, itself emits X-rays; he could not ascertain whether this occurs by diffusion or by some kind of fluorescence.
2. In his first communication he had noted that a body hit by cathode rays emits X-rays in every direction. He now proceeded to study how the intensity of the X-rays varied with direction. For that, Röntgen placed the platinum anode at 45° with respect to the cathode rays, and used a spherical discharge tube. The measurements were made both photometrically and with photographic film. The result was that the distribution was isotropic in a hemisphere around the platinum plate.
3. Röntgen made a systematic and quantitative study of the 'transparency' of substances and its variation with thickness. Doing this, he made very important observations: a) if one piles up plates of equal thickness, the transparency of the upper plate decreases when the number of plates increases (obvious if one thinks of the exponential law of absorption); b) the relative transparencies of two substances depends on the thickness or nature of a body before the X-rays pass through these substances (this is because the softer radiations of the beam have been absorbed out and the X-rays have become harder[10]); c) the transparency of substances for X-rays increases if the potential applied to the discharge-tube increases, because the tube has become harder; d) a 'soft' tube is made 'harder' by improving the vacuum in the tube.
4. Repeatedly, Röntgen tried to obtain diffraction phenomena with narrow slits. On several occasions he did see some effects which were remindful of diffraction effects, but since they could not be observed again after changing the conditions of the experiment, he concluded that he could not prove with 'sufficient certainty' (*mit einer mir genügenden Sicherheit*) the existence of diffraction of X-rays.

4.6 Prior observations

While Röntgen was not the first to observe X-rays, he was, without any doubt the one who recognized that they were a new type of radiation. He studied their properties in such great detail that, for several years, nothing more was found on the properties of X-rays than the detailed

---

[10]Soft X-rays have long wavelengths; hard X-rays have short wavelengths, and are more penetrating.

descriptions in his three communications; he thus foreshadowed the future development of the studies on X-rays.

Among the earliest investigators to have, maybe, produced X-rays, were the English scientist, Francis Hauksbee (1666–1713), F. R. S., the first to have produced an electrical discharge in vacuum and William Morgan (1750–1833), in fact an actuary and amateur scientist, who observed in 1785 the development of a greenish colouration as air began to enter into a vacuum tube in which he had been experimenting with the passage of electricity (Underwood 1945). Sir William Crookes, F. R. S.,[11] the designer of the Crookes tube in the early 1870s, observed that photographic plates stored near his cathode tubes were fogged; this may have been due to X-rays (Glasser 1931a).

The American physicist and medical doctor at the University of Pennsylvania, Arthur W. Goodspeed, operated a Crookes tube on the evening of 22 February 1890, while there were several photographic plates nearby in their holders. When they were developed later by his friend, W. N. Jennings, several plates were found to be fogged, and one of them showed the shadow of two coins. After the announcement of Röntgen's discovery, A. Goodspeed realized that this one photograph must have been the first X-ray picture ever taken, and he convinced himself by repeating the experiment. But he made no claim for the discovery of X-rays, and said that his finding had just been an 'accident' (Goodspeed 1896).

The English chemist Sir Herbert Jackson, F. R. S., (1863–1936) observed X-rays in January 1894 while exposing fluorescent material to electric discharges in a vacuum tube fitted with a focusing concave cathode and an inclined platinum anode. After Röntgen's announcement in 1896, he used his equipment to study X-rays, but, like Goodspeed, he never made claims for priority.[12]

The Ukrainian physicist Yvan Pulyui, or Puluj, (1845–1918), a Professor at the Prague Polytechnic Institute, worked on cathode rays produced in discharge tubes of his own design (the 'Puluj lamp'). In the late 1880s he observed radiation produced by X-rays focused on a photographic plate, but did not recognize it to be X-rays. Just after the publication of Röntgen's work he resumed work on his tubes and developed medical applications of X-rays (Puluj 1896). For details on Puluj's work on X-rays, see Savchuk (2007).

4.7 'Lenard rays' and 'Röntgen rays'

In a normal Hittorf–Crookes discharge tube cathode rays are stopped by the glass walls, but H. Hertz[13] (1892) showed that cathode rays could pass through metallic flakes. This gave Philipp Lenard,[14] who joined Hertz at that time as his assistant, the idea of building a special

---

[11]See footnote 7, Chapter 5.

[12]*Sir Herbert Jackson. 1863–1936*. H. Moore (1938). *Obituary Notices of Fellows of the Royal Society*. **2**, 306–314.

[13]See footnote 52 in Chapter 3.

[14]Philipp Lenard, born 7 June 1862 in Poszony (Pressburg), Hungarian-Austrian Empire, now Bratislava, Slovakia, died 20 May 1947 in Messelhausen, Germany, was a Hungarian-German physicist. He studied physics and chemistry in Vienna and Budapest. In 1883, he moved to Heidelberg where he studied with R. Bunsen and H. von Helmholtz, and obtained his doctoral degree in 1886. From 1892, he worked as a Privatdozent and assistant to H. Hertz at the University of Bonn. After Hertz's death in 1894, he was successively appointed to the Universities of Breslau, Aachen, Heidelberg, and Kiel, and finally came back to Heidelberg in 1907. He at first worked on phosphorescence and luminescence, but

discharge tube with a very thin aluminium window (the 'Lenard tube'). This allowed him to study cathode rays in air, outside the tube (Lenard 1894). He detected them with a fluorescent screen made of thin tissue-paper soaked with the organic scintillator pentadecylparatolylketon. He found that photographic plates placed outside the tube became blackened and showed that the beam was deflected by a magnet, just as were the cathode rays inside the discharge tube. He also found that the rays were absorbed out by air after six to eight centimetres, and that this absorption decreased if the pressure of the air was decreased. From the fact that the rays are not stopped in vacuum, he concluded 'that the rays are processes going on in the æther' (Lenard 1896).

Lenard measured the absorption of cathode rays by various materials (Lenard 1895), and observed that this absorption depended on the degree of vacuum in the discharge tube. When the discharge tube was more exhausted, he found that both the deflectibility of the rays by a magnet and their absorption coefficient by matter were smaller than if the tube was less exhausted. According to him, with a very exhausted tube, one could thus produce rays which were not deflected in a noticeable way by a magnet, and he concluded that 'the Röntgen rays are of the same nature as the cathode rays' (Lenard 1896). He considered therefore that he was the real discoverer of X-rays, and remained bitter that this was not recognized. Lenard certainly produced X-rays during his experiments, but the pentadecylparatolylketon screen he used was not sensitive to X-rays as was the barium platinocyanide used by Röntgen, and he did not recognize that he had, in fact, produced X-rays at that time. Lenard was never reconciled with the idea that it was Röntgen who had discovered Röntgen-rays and he preferred to call them 'high-frequency rays' in order to avoid using the name Röntgen, even after Laue's discovery (Ewald 1948)[15].

In his presidential address to the Mathematical and Physical Section of the British Association for the Advancement of Science on 17 September 1896, J. J. Thomson concluded that what he called 'Lenard rays'—namely, the rays observed by Lenard in air after the aluminium window—were similar to the cathode rays in the discharge tube, and of a different nature than the Röntgen rays (Thomson 1896*b*).

The Rumford Medal was awarded by the Royal Society in 1896 to both Lenard and Röntgen, 'for their investigations of the phenomena produced outside a highly exhausted tube through which an electrical discharge is taking place'.

---

his major contribution was the study of cathode rays. He was awarded the 1905 Nobel Prize in Physics for 'his work on cathode rays'. A critic of Einstein's relativity theory, he founded, with J. Stark, the *Deutsche Physik* movement in the early 1920s, and joined the National Socialist Party. He was scientific adviser to Hitler.

[15]Draft of Ewald's talk at the first meeting of the International Union of Crystallography, Harvard, USA, 28 July–3 August 1948. Archives of the Deutsches Museum, Munich. The present author is grateful to Michael Eckert for a copy of the manuscript.

5

THE NATURE OF X-RAYS: WAVES OR CORPUSCLES?

*On Mondays, Wednesdays, and Fridays we use the wave theory; on Tuesdays, Thursdays,
and Saturdays we think in streams of flying energy quanta or corpuscles.*

Sir W. H. Bragg (1922)

5.1 The nature of cathode rays

The nature of cathode rays, or electrons, like that of light and that of X-rays, was a matter of
longstanding debate. In England, the notion of particles of electricity goes back to Faraday's
laws of electrolysis; in his 1873 *Treatise on Electricity and Magnetism*, J. C. Maxwell spoke
of 'molecules of electricity'. In Germany, where the first high-vacuum discharge tubes were
operated, the remarkable fluorescent glow they produced evoked processes in the æther.

In 1855 the German glassblower and physicist H. Geissler[1] invented the mercury vacuum
pump with which he could produce a very good vacuum. With this new vacuum pump, he
built in 1857 a glass discharge tube, which was the prototype of all the Hittorf, Crookes, etc.
discharge tubes to come (See Fig. 5.1). Geissler's friend at the University of Bonn, J. Plücker,[2]
made use of Geissler tubes to study the conditions that affect the appearance of the glow
caused by the discharge and the action of a magnet on the electric discharge in rarefied gases.
In particular, he found that the glow could be made to shift by applying an electromagnet to
the tube.

W. Hittorf,[3] one of Plücker's former students, investigated various forms of discharge tubes
and improved upon Geissler's design (the 'Hittorf tube', used by Röntgen, see Fig. 4.2). He
observed both 'negative' glow and a 'positive' light, of different colours, that propagated
in opposite directions (Hittorf 1869). He noted that the propagation is along straight lines
(*geradlinige Bahnen*, straight tracks, or *Strahlen*, rays) He studied the deviation of these rays
by a magnet and compared them to electric currents. His work, however, remained unnoticed
and was unknown to Crookes when the latter designed his own discharge tubes.

---

[1]Heinrich Geissler, born 26 May 1814 in Igelshieb, Thuringia, Germany, to a family of glassblowers, died 24 January
1879 in Bonn, Germany. He worked in various German Universities as instrument maker, in particular in Bonn, where
he met J. Plücker.

[2]Julius Plücker, born 16 June 1801 in Elberfeld near Wupperthal, Germany, died 22 May 1868 in Bonn, Germany,
was a German mathematician and physicist. He was educated at the universities of Bonn, Heidelberg and Berlin, and was
a Professor of Mathematics at the University of Bonn in 1828.

[3]Johann Wilhelm Hittorf, born March 27 1824 in Bonn, Germany, died November 28, 1914 in Münster, Germany,
was a German physicist and chemist. He studied at the University of Bonn with J. Plücker. In 1857 he was appointed
Professor of Physics and Chemistry at the University of Münster. Hittorf's early work was on the allotropes of phosphorus
and selenium. He was the first to compute the electricity-carrying capacity of charged atoms and molecules.

Fig. 5.1   Discharge tubes. *Left*: Geissler tubes. *Right*: Crookes' Maltese Cross tube. Courtesy of The Cathode Ray
Tube Site.

The radiation produced during the discharge was further studied by E. Goldstein[4] who
used the terms *Kathodenlicht*, 'cathode light' (Goldstein 1876) and *Kathodenstrahlen*, 'cathode
rays'—the first occurrence of the word (Goldstein 1880). He confirmed the rectilinear propagation of the rays and the influence of a magnet. Goldstein also discovered another effect. He
bored holes into the cathode, and was very surprised to see a golden-yellow light appear on the
wall of the glass vessel situated on the other side of the cathode. He found that it was due to
rays of a new kind propagating in the direction opposite to that of the cathode rays. He called
these rays *Canalstrahlen*, 'canal rays', referring to the holes (canals) in the cathode (Goldstein
1886).

The nature of the cathode rays was, of course, a mystery. The general belief among German
physicists was that they were a kind of 'æther wave' (Wiedemann 1880) or a 'motion of the
free æther' (Goldstein 1881). Discussions with E. Goldstein led H. Hertz[5] (1883) to undertake
experiments to find whether an electric field had any action on cathode rays, but he failed to
see any such effect. Hertz concluded that cathode rays were 'electrically indifferent' (*electrisch
indifferent*). Hertz and his predecessors had observed that the colour of the glow depended on
the operating conditions of the tube and this led him to believe that, to each colour, corresponded
a different type of cathode rays.

In 1892, Hertz (1892) and his assistant, P. Lenard[6] (1894), observed that cathode rays could
pass through thin metal foils (see Section 4.7) and this led Lenard to conclude that cathode rays
were processes of the æther (*Vorgänge im Aether*).

An entirely different view as to the nature of the cathode rays prevailed on the other side of
the English Channel. In England, W. Crookes[7] was able to produce a very good vacuum, and,

---

[4]Eugen Goldstein, born 5 September 1850 in Gleiwitz, Silesia (now Gliwice, Poland), died 25 December 1930
in Berlin, Germany, was a German physicist. He studied at Breslau (now Wrocław, Poland) and later, under H. von
Helmholtz, in Berlin where he obtained his PhD on the topic of gas discharges. Goldstein worked at the Berlin
Observatory from 1878 to 1890, and, from then on, at the Potsdam Observatory.

[5]See footnote 52 in Section 3.7.   [6]See footnote 14 in Section 4.7.

[7]Sir William Crookes, born 17 June 1832 in London, England, the son of a tailor, died 4 April 1919 in London,
England, was an English chemist and physicist. After attending a grammar school, he entered the Royal College of
Chemistry in London as an assistant. In 1855, he was appointed lecturer in chemistry at the Chester Diocesan Training

after inventing the radiometer[8] in 1873, he designed discharge tubes very similar to Hittorf's, and gave a public demonstration of them on 22 August 1879. He called 'radiant matter' the 'molecules of the gaseous residue in highly exhausted vessels that are able to dart across the tube with comparatively few collisions, and which radiate from the pole with enormous velocity' (Crookes 1879). Crookes borrowed the name 'Radiant Matter' from the title of an 1816 lecture in which M. Faraday conceived a state of matter 'as far beyond vaporisation as that is above fluidity'. For Crookes, the molecules in the discharge tube assumed properties so novel and so characteristic as to entirely justify the application of the term borrowed from Faraday. Crookes studied the properties of radiant matter and showed that it was absorbed by solid bodies and cast a shadow. He proved this by putting a Maltese cross cut out of a sheet of aluminium inside the X-ray tube, in the path of the radiant matter (Fig. 5.1, *Right*).

J. J. Thomson in Cambridge (Fig. 5.2) and A. Schuster[9] in Manchester became interested in the discharge of electricity through gases in the early 1880s. In his first communication to the Royal Society, A. Schuster assumed that the molecules of the rarefied gas in the vessel are broken up by the discharge, and reported on the influence of a magnet on the glow. He concluded that the discharge is carried by negatively charged particles (Schuster 1884). In his second communication (Schuster 1890), he obtained upper and lower limits for the mass to charge ratio, m/e, of the particles.

The news of Hertz's and Lenard's experiments showing that metal foils placed between the cathode and the walls of the discharge tube do not entirely stop the glow cast some doubts about Crookes hypothesis that the glow is due to the impact of charged molecules on the sides of the tube. J. J. Thomson undertook, for that reason, to measure the velocity of the particles and found that it was slower than what would have been expected for ethereal waves (Thomson 1894).

A significant experiment in late 1895 by the young French graduate student J. Perrin[10] confirmed the corpuscular nature of cathode rays. In order to settle the difference between the opposing views of Goldstein, Hertz, and Lenard on the one hand, and Crookes and Thomson on the other hand, he collected the cathode rays in a metal cup, which acted as a Faraday cylinder, just before they struck the glass wall, and found that they gave a negative charge, which he

College. As a chemist, he worked on compounds of selenium and discovered a new element, thallium. As a physicist, Crookes was a pioneer of vacuum tubes. In the later part of his life his research was done mainly in his home in London. Crookes was elected a Fellow of the Royal Society in 1863, and was knighted in 1897.

[8]The Crookes radiometer, or light mill, consists of an airtight glass bulb, containing a partial vacuum, and in which a set of vanes are mounted on a spindle. The vanes rotate when exposed to light, faster when the light is more intense.

[9]Sir Franz Arthur Friedrich Schuster, born 12 September 1851 in Frankfurt am Main, Germany, to a Jewish family of merchants and bankers that emigrated to Manchester, England in 1869, died 17 October 1934 in Yeldall, Berkshire, England, was a German-born English scientist. He studied mathematics and physics at Owens College, now Manchester University, and, after a year in Heidelberg, obtained his PhD in 1873. After returning to Manchester for two years, he spent five years at the Cavendish Laboratory with J. C. Maxwell and Lord Rayleigh. In 1881 he was appointed to a chair of Applied Mathematics at Owens College. Schuster is known for his work in spectroscopy, electrochemistry, optics, and X-radiography. He was elected to the Royal Society in 1879, and knighted in 1920.

[10]Jean Baptiste Perrin, born 30 September 1870 in Lille, France, the son of an artillery captain, died 17 April 1942 in New York, USA, was a French physicist. A former student of *Ecole Normale Supérieure*, he obtained his PhD in 1897. He was appointed lecturer at the University of Paris the same year and Professor in 1910. After working on cathode rays and X-rays, he was a pioneer of the study of Brownian motion and measured Avogadro's number. He was elected to the French Academy of Sciences in 1923 and was awarded the Nobel Prize for Physics in 1926 'for his work on the discontinuous structure of matter, and especially for his discovery of sedimentation equilibrium'.

Fig. 5.2   Sir Joseph John Thomson. Reproduction of a steel engraving originally published in *The Electrician* (1896).

Sir Joseph John Thomson: born 18 December 1856 near Manchester, England; died 30 August 1940 in Cambridge, England, was an English physicist. After schooling at Owens College in Manchester, he entered Trinity College, Cambridge, in 1876, as a minor scholar. He became a Fellow of Trinity College in 1880, when he was Second Wrangler, and remained a Fellow all his life. From 1884 to 1918, he was Cavendish Professor of Experimental Physics at Cambridge University, where he succeeded Lord Rayleigh. In 1884, J. J. Thomson was elected a Fellow of the Royal Society, and in 1906 was awarded the Nobel Prize for Physics 'in recognition of the great merits of his theoretical and experimental investigations on the conduction of electricity by gases'. He was knighted in 1908, and was President of the Royal Society from 1916 to 1920. Among his most notable students, one may mention C. G. Barkla, M. Born, W. H. Bragg, B. Davis, P. Langevin, J. R. Oppenheimer, E. Rutherford, C. T. R. Wilson. His son, Sir George Paget Thomson, was awarded the Nobel Prize for Physics in 1937, for the discovery of the diffraction of electrons by crystals.

MAIN PUBLICATIONS

1884 *Treatise on the Motion of Vortex Rings.*

1886 *Application of Dynamics to Physics and Chemistry.*

1892 *Notes on Recent Researches in Electricity and Magnetism.*

1895 *Elements of the Mathematical Theory of Electricity and Magnetism.*

1897 *Discharge of Electricity through Gases.*

1903 *Conduction of Electricity through Gases.*

1907 *The Structure of Light.*

1907 *The Corpuscular Theory of Matter.*

1913 *Rays of Positive Electricity.*

1923 *The Electron in Chemistry.*

measured with an electroscope (Perrin 1895). To confirm his experiment, Perrin showed that if the cathode rays are deflected by a magnet no charge is detected by the electroscope. He was also successful in carrying out the same experiment for the rays travelling in the opposite sense, and showed that they carried a positive charge. In an additional experiment, performed in 1896, but published in his thesis, Perrin showed that cathode rays were deflected by an electric field (Perrin 1897). He concluded, as did Schuster, that the discharge had broken the molecules into negative and positive ions.

A slight improvement on Perrin's experiment was made by J. J. Thomson (1897). He directed a beam of cathode rays into a Faraday cylinder *after* they had been previously deflected by a magnet. This proved beyond any doubt that they were charged corpuscles. Thomson then showed that the path of the cathode rays is deviated by the action of an electrostatic field, which Hertz had failed to observe, probably because his vacuum was not good enough. Thomson next measured the magnetic deflection of the cathode rays in different gases and their velocity, which he found to be larger than 10^9 cm/sec. As a result of his experiments, Thomson was able to calculate the charge to mass ratio e/m of the cathode rays, and found that it was 1000 times smaller than the value measured for a hydrogen ion in electrolysis. The following year, J. J. Thomson (1898b) measured the charge e of the cathode rays.

J. J. Thomson has often been credited with having been 'the'discoverer of the electron. However, as pointed out by many historians of science,[11] independently and at roughly the same time, several scientists in Germany determined the relation of the electric charge of the cathode rays to their mass, and concluded that they were corpuscles: W. Kaufmann[12] (1897) in Berlin, E. Wiechert[13] (1897) in Königsberg, and W. Wien[14] (1897) in Aachen.

The term 'electron' was coined by the Irish physicist G. J. Stoney (1826–1911) to designate the unit quantity of electricity (Stoney 1894). Another Irish physicist, Sir J. Larmor (1857–1942), took over the term 'electron' from Stoney to designate the notion of 'free electron', a charged material particle whose origin was a centre of rotational strain in the æther (Larmor 1894).

In 1895, the Dutch physicist H. A. Lorentz[15] developed an electromagnetic theory of light derived from Maxwell's, in which light vibrations are caused by the motion of ions (Lorentz 1895, 1916). It received an immediate application when Lorentz's former assistant, P. Zeeman,[16] observed the broadening and splitting of the spectral lines of a sodium flame under the influence of a magnetic field, known as the Zeeman effect (Zeeman 1897). The equations of movement of the light-emitting oscillating particles, derived by Lorentz, involve their charge, e, and their mass, m. The broadening of the lines could thus be directly related to the e/m ratio, which Zeeman calculated before Thomson did. Zeeman's experiment proved that the oscillating particles at the source of light emission were negatively charged, and were a thousand-fold lighter than hydrogen atoms. Lorentz and Zeeman were jointly awarded the 1902 Nobel Prize in Physics, 'in recognition of the extraordinary service they rendered by their researches into the influence of magnetism upon radiation phenomena'.

The term electron was first used for cathode rays by E. Wiechert (1897), and by G. J. Stoney's nephew, the theoretician G. FitzGerald[17] (1897). Soon, electron was used rather

---

[11]See, for instance, A. Pais (1986). *Inward Bound: Of Matter and Forces in the Physical World*. Oxford University Press., I. Falconer (1987). Corpuscles, Electrons and Cathode Rays: J. J. Thomson and the 'Discovery of the Electron'. *Brit. J. Hist. Sci.* **20**, 241–276, and J. Z. Buchwald and A. Warwick (2001). *Histories of the Electron: The Birth of Microphysics*. The MIT Press. Cambridge, Massachusetts.

[12]Walter Kaufmann, born 5 June 1871, died 1 January 1947, was a German physicist, known for his first experimental proof of the velocity dependence of mass.

[13]Emil Wiechert, born 26 December 1861, died 19 March 1928, was a German physicist and geophysicist.

[14]See footnote 42 in Section 5.10.

[15]Hendrik Antoon Lorentz, born 18 July 1852 in Aarnheim, the Netherlands, died 4 February 1928 in Haarlem, the Netherlands, was a Dutch physicist, well known for the theory of dispersion and the 'Lorentz transformation equation'. He studied physics and mathematics at the University of Leiden and obtained in 1875 his PhD on the theory of the reflection and refraction of light. In 1878 he was appointed Professor of Theoretical Physics at the University of Leiden. He was awarded the Nobel Prize in Physics in 1902, jointly with P. Zeeman, and the Rumford Medal in 1908.

[16]Pieter Zeeman, born 25 May 1865, Zonnemaire, the Netherlands, died 9 October 1943 in Amsterdam, the Netherlands, was a Dutch physicist who discovered the 'Zeeman effect'. He studied physics at the University of Leiden, under Kamerlingh Onnes and Hendrik Lorentz, and obtained his PhD on the Kerr effect in 1893. He was appointed lecturer at the University of Amsterdam in 1897 and Professor in 1900. In 1902 he shared the Nobel Prize with H. A. Lorentz, and in 1908 succeeded van der Waals as Director of the Physics Institute in Amsterdam.

[17]Born 1851, died 1901.

than cathode rays in many publications,[18] but Thomson himself used the term 'corpuscles', and only admitted much later that cathode rays were the same as the electrons of the theoreticians of electromagnetism.

The dual nature of the electron, corpuscular and wave, was revealed thirty years later, by diffraction experiments: C. J. Davisson and L. H. Germer observed in April 1927 at Bell Laboratories, New York, USA, the diffraction pattern of low-energy electrons by a small crystal of nickel (1927a, b); independently, J. J. Thomson's son, G. P. Thomson and his research student, A. Reid in Aberdeen, Scotland, observed in June 1927 the powder diffraction diagram of higher-energy (20 to 60 keV) electrons by thin films of platinum and celluloid (Thomson and Reid 1927). The Nobel Prize in Physics 1937 was awarded jointly to C. J. Davisson and G. P. Thomson 'for their experimental discovery of the diffraction of electrons by crystals'.

5.2   The first hypotheses concerning the nature of X-rays

X-rays, or Röntgen rays, were extremely baffling for the first observers: they were rays, because they cast shadows (such as by showing bone structure in Röntgen's photographs), but they were not ordinary light because they could not be refracted, reflected, or diffracted. They were not cathode rays because they travelled long distances, and were not deflected by a magnetic field. The general idea, nevertheless, was that they were waves of some kind. Röntgen himself, as described in Section 4.3, thought at first that they were longitudinal waves (like sound). This possibility was considered by the Scottish physicist from the University of Glasgow, J.-T. Bottomley, in the 23 January issue of *Nature*, London, echoing a lecture by Lord Kelvin in 1894 in Baltimore, USA. It was further developed by Lord Kelvin himself in a communication to the Royal Society on 10 February 1896, and was also considered by J. J. Thomson in February (Thomson 1896a).

When, on 20 January 1896, H. Poincaré presented Röntgen's startling discovery to the French Academy of Science, the French Physicist H. Becquerel[19] asked where exactly the new rays were emitted. On being told that it was the glowing wall of the discharge tube, he wondered whether the rays were not due to the vibratory movements at the origin of the glow (Becquerel 1903). He discussed this possibility with Poincaré who then suggested that all fluorescent bodies also emit X-rays (Poincaré 1896). It was while he was trying to find whether phosphorescent bodies illuminated by light could emit penetrating radiation that Becquerel discovered radioactivity (Becquerel 1896, 24 February).

In his first communications to *The Electrician*, 7 February 1896 and 1 March 1896, O. Lodge[20] (Lodge 1896a, b) suggested two hypotheses; Röntgen rays could either be transverse electromagnetic, 'light', vibrations of the æther, or longitudinal 'sound' vibrations, as mentioned by Lord Kelvin. What most disturbed physicists was, however, that X-rays were neither refracted nor reflected. This led O. Lodge, in his third communication to *The Electrician* on 17 July, to look up H. von Helmholtz's theory of anomalous dispersion (Helmholtz 1875)

---

[18]See, for instance, O. Lodge, *Modern Views on Matter* (The Romanes Lecture 1903). Oxford: Clarendon Press.

[19]Henri Becquerel, born 15 December 1852 in Paris, France, died 25 August 1908 in Le Croisic, France, was a French physicist who discovered radioactivity. He shared the 1903 Nobel Prize in Physics with P. and M. Curie, 'in recognition of the extraordinary services he has rendered by his discovery of spontaneous radioactivity'.

[20]See footnote 3 in Section 1.1.

Fig. 5.3   Representation by Sir G. G. Stokes of an X-ray pulse: it consists of two halves distinguished by a difference
in shading. After Stokes (1897).

and to suspect that X-rays were transverse ethereal waves akin to light, of very short wavelength
(Lodge 1896*c*).

Then came the 'impulse theory'. In May 1896, E. Wiechert (1896*a*, *b*) suggested that X-rays
were 'high frequency electrodynamic waves', or, more probably, electrodynamic waves with
an 'impulse character' (*stossartig*); these impulses are localized disturbances, caused by the
sudden impact of an electron, that are projected into space. The same suggestion was made a few
months later by G. G. Stokes[21] (1896*a*, *b*, 1897), in order to explain the absence of refraction
and interference: 'the simplest sort of pulse', he wrote, 'in order to distinguish it from a periodic
undulation, would be one consisting of two halves, one positive and one negative, in which the
disturbances were in opposite directions' (Fig. 5.3). A pulse, as it were, is a single period of a
wave.

J. J. Thomson (1898*a*) reached the same conclusion by considering the way X-rays are
generated: when a moving charged particle is stopped, a magnetic field is produced, 'which for
a moment compensates for that destroyed by the stopping of the particle'. The new field thus
created is not in equilibrium and 'moves off as an electric pulse'. J. J. Thomson calculated in his
paper the magnetic force and electric intensity carried by the pulse in this way, and formulated
the basic theory of X-ray generation at the anticathode of an X-ray bulb (see Section 5.5).[22]

Quite a different view was held by the German physicist Bernhard Walter (1861–1950)
in Hamburg, who thought that X-rays were cathode rays that had lost their charge (Walter
1898).

5.3   Discovery of γ-rays

The γ-rays were discovered in 1900 by P. Villard[23] while studying the refraction of radiation
emitted by a radium source (radium-rich barium chloride) as a result of its radioactive decay.
He found that the beam was divided into two parts by the action of a magnetic field. One part
was deflected, as expected for cathode rays, but the other part was not. Villard identified this
second part with very penetrating X-rays, of a very high frequency[24] (Villard 1900*a*, *b*), but that
suggestion was not accepted immediately by the other physicists. The name γ-ray was coined

---

[21] See footnote 15 in Section 2.3.   [22] In the early years, X-ray tubes were called 'X-ray bulbs'.

[23] Paul Villard, born 28 September 1860 in Saint-Germain-au-Mont-d'Or, near Lyon, France, died 28 September 1934
in Bayonne, France, was a French physicist and chemist. A former student of Ecole Normale Supérieure, he passed the
physics *agrégation* in 1884. He was for a while lecturer at the University of Montpellier, and, in 1892, became attached to
the Chemistry Laboratory of the Ecole Normale Supérieure, without pay. He worked mainly on cathode rays and X-rays.
He was elected a member of the French Academy of Sciences in 1908.

[24] On the discovery of γ-rays, see, for instance, L. Gervard (1999). Paul Villard and his Discovery of Gamma Rays.
Phys. Persp. **1**, 367–383.

by Rutherford[25] (1903), who distinguished three components in the radiation from radium: highly-absorbed α-rays, negatively-charged β-rays, similar to cathode rays, and γ-rays, which are non-deviable by a magnetic field.

The proof that γ-rays are electromagnetic radiation was given by Rutherford and Andrade (1914*a*, *b*) employing crystal diffraction (Section 7.10.3).

5.4   Secondary X-rays

When primary X-rays fall on a substance, they are on the one hand scattered, elastically (coherently) and inelastically (incoherently), and on the other hand absorbed, which may lead, depending on the energy of the X-rays and the nature of the elements in the substance, to the emission of fluorescent X-rays. The early investigators did not distinguish between scattered and fluorescent X-rays. Diffuse scattering of X-rays by air and other materials was first observed by Röntgen himself (Section 4.3), and was reported by various scientists, for instance, A. Battelli and A. Garbasso (1897) in Italy, Lord Blythswood (1897) in England, A. Imbert and H. Bertin-Sans (1897) in France, A. Voller and B. Walter (1897) in Germany, and others. G. Sagnac[26] (1897) at the Sorbonne, in Paris, was the first to identify secondary rays (*rayons secondaires*) having properties different from those of the primary rays, and to investigate their properties for various metals (Au, Ag, Zn, Cu, Pb, and Sn). He showed that the secondary rays were less penetrating than the primary ones and that their penetrating power depended on the nature of the metal. Sagnac (1901) also showed that by allowing secondary rays to fall on a metal, tertiary rays are formed, which are even more easily absorbed than the secondary rays.

Similar but more limited results were obtained by P. Langevin[27] in J. J. Thomson's laboratory in Cambridge. They were not published at the time, but later, in his thesis, where he acknowledged Sagnac's priority. Sagnac may therefore be considered as the discoverer of X-ray fluorescence.[28]

---

[25]Lord Ernest Rutherford, 1st Baron Rutherford of Nelson, born 30 August 1871 in Brightwater, New Zealand, the son of a farmer, died 19 October 1937 in Cambridge, England, was a British physicist who has become known as the father of nuclear physics. After obtaining his BSc at Canterbury College, University of New Zealand, he went to England in 1895 for postgraduate study at the Cavendish Laboratory, Cambridge, where he worked with J. J. Thomson on the propagation of electromagnetic waves. In 1898 he was appointed to the chair of physics at McGill University in Montreal, Canada, where he did the work that gained him the Nobel Prize in Chemistry in 1908. It is at that time that he found that there were two types of radiation emitted by uranium, which he called α and β. In 1903, he called γ-rays the new radiation emitted by radium, found by P. Villard in 1900. In 1907, Rutherford took the chair of physics at the University of Manchester, and, in 1919, returned to the Cavendish where he succeeded J. J. Thomson. He was elected a Fellow of the Royal Society in 1903 and was knighted in 1914. The list of his former students or one-time co-workers is impressive, and includes P. Blackett, H. Geiger, N. Bohr, C. G. Darwin, H. G. J. Moseley, C. von Hevesy, O. Hahn, P. Kapitsa, J. Cockroft, K. Chadwick, E. Marsden, E. da Andrade, E. V. Appleton, K. Fajans, and D. Hartree.

[26]Georges Sagnac, born 14 October 1869 in Périgueux, France, died 26 February 1928 in Meudon, France, was a French physicist. A former student of Ecole Normale Supérieure, he obtained his PhD in 1900 on the secondary X-rays. He was successively Associate Professor in Lille University (1900–1904) and Professor at Paris University (1904–1926).

[27]Paul Langevin, born 23 January 1872 in Paris, France, died 19 December 1946 in Paris France, was a French physicist, known for his work on magnetism and on ultrasounds. A former student of Ecole de Physique et Chimie Industrielle de Paris, and Ecole Normale Supérieure, he spent a year at the Cavendish in Cambridge (1897–1898) and obtained his PhD in Paris in 1902. In 1905 he was appointed Professor at the Ecole de Physique et Chimie Industrielle de Paris.

[28]M. Quintin (1996). Qui a découvert la fluorescence X? *Journal de Physique IV*; **6**, C4-599–C4-609.

Fig. 5.4   Charles Glover Barkla. George Grantham Bain Collection (Library of Congress).

Charles Glover Barkla: born 7 June 1877 in Widnes, Lancashire, England; died 23 October 1944 in Edinburgh, Scotland, was an English physicist. He was educated at the Liverpool Institute and in 1894 entered University College, Liverpool, where he studied physics with Oliver Lodge. He obtained his Master's degree in 1899, and, the same year, went to Trinity College, Cambridge, to work in the Cavendish Laboratory under J. J. Thomson. Barkla was an excellent musician, and in 1900 he migrated to King's College, where he joined the chapel choir. In 1902 he returned to Liverpool where he obtained a DSc degree in 1904, prepared with J. J. Thomson and O. Lodge as advisors. From 1905 to 1909, he was lecturer at the University of Liverpool, and from 1909 to 1913 he was Professor of Physics at London University. In 1912, Barkla was elected a Fellow of the Royal Society, and in 1913 was appointed Professor of Physics at the University of Edinburgh. In 1917 he was awarded the Nobel Prize in Physics 'for his discovery of the characteristic Röntgen radiation of the elements'.

MAIN PUBLICATIONS

1905 *Polarized Röntgen radiation.*
1906 *Secondary Röntgen radiation.*
1908 *Homogeneous secondary Röntgen radiations.* With C. A. Sadler.

1909 *The absorption of X-rays.* With C. A. Sadler.
1911 *The spectra of the fluorescent Röntgen radiation.*

The topic of *secondary radiation* was very much in J. J. Thomson's mind; he introduced the term in English (Thomson 1898c), and, after Langevin's return to France, he suggested further studies, in particular on the property of secondary X-rays to ionize gases, as the primary rays had already been shown to do (J. S. Townsend 1900, H. S. Allen 1902). In 1902 he encouraged another of his students, C. G. Barkla (Fig. 5.4), to work on secondary X-rays, with the success we know. Barkla, however, moved back the same year to his home town of Liverpool, and it is there that he published his first work, on secondary radiation from gases (Barkla 1903). In this study, Barkla showed that the secondary radiation from various gases had the same absorbability as the primary radiation. It is in fact *scattered* radiation, and its energy is proportional to the density of the gas from which it has come. In the second paper Barkla (1904) confirmed that secondary radiation from gases and light solids is scattered radiation, while that of heavier metals is different. In all cases, the energy of the secondary radiation is proportional to the quantity of matter passed through by a primary beam of given intensity.

5.5   J. J. Thomson and the theory of X-ray scattering, 1898, 1903

J. J. Thomson (1898c) suggested a mechanism for the generation of secondary X-ray pulses by scattering of the primary (incident) radiation: 'when the pulse of intense electric intensity which constitutes a Röntgen ray falls on charged atoms, it will suddenly change the velocities of these atoms'. This 'sudden change in the velocity of a charged atom will generate a secondary pulse of electric and magnetic intensity which constitutes a secondary Röntgen ray'. Thomson (1903) calculated the energy of the pulse scattered by an ion, and derived the expression of the amplitude scattered by a corpuscle of charge e and mass m:

$$E = -E_o \frac{e^2}{mc^2} \frac{\exp -2\pi i \nu t - kr}{r} \sin \psi \qquad (5.1)$$

where E_o and E are the amplitudes of the incident and scattered waves, respectively, c the velocity of light, ν the frequency of the pulse, $k = 1/\lambda$ its wave number, ψ the angle between the electric field vector $\mathbf{E_0}$ of the incident wave and the direction of observation of the scattered wave (Fig. 8.3), and r the distance between the scattering electron and the point of observation.

It was shown later that Thomson's result is independent of the assumption that the incident radiation consists of pulses; this is why the Thomson formula is still in use today. The electric and magnetic forces of the secondary pulse are at right angles to each other, and at right angles to the direction of propagation of the pulse. The pulses are therefore transverse.

5.6   C. G. Barkla and X-ray polarization, 1905

C. G. Barkla's theory of X-ray polarization stems directly from J. J. Thomson's theory of X-ray scattering: 'each electron in the medium through which Röntgen-ray pulses pass has its motion accelerated by the intense electric fields in these pulses, and consequently is the origin of a secondary radiation, which is most intense in the direction perpendicular to that of acceleration of the electron, and vanishes in the direction of that acceleration' (Barkla 1905a, b). The direction of electric intensity of the electromagnetic radiation is perpendicular to the direction of propagation and is in a plane passing through the direction of acceleration of the electron. The secondary radiation is therefore expected to be plane polarized, and this can be observed by a variation with direction of the intensity of a tertiary beam generated when the secondary beam hits a target.

In Barkla's first experiment with gases, this tertiary beam was too weak to allow accurate measurement, but consideration of the method of production of the primary beam led him to think that the primary beam was partially polarized in a direction perpendicular to that of the propagation of the cathode rays: 'there is probably at the antikathode a greater acceleration along the direction of propagation of the kathode rays than in the direction at right angles'. Barkla chose for that reason a beam of X-rays proceeding in a direction perpendicular to that of the cathode rays as the primary radiation. The intensity of secondary radiation was measured in two directions perpendicular to each other and perpendicular to that of the primary radiation. The ionizations due to the radiations were measured with a gold-leaf electroscope, and the deflection of the gold-leaf was observed through a microscope with a graduated eye-piece.[29] Air or sheets of paper or aluminium were used as radiators to generate the secondary radiation. The X-ray bulb was turned around the axis of the primary beam, and it was found that the intensity of the secondary beam was at a maximum when the direction of the stream of cathode rays was perpendicular to that of the secondary beam. The experiment was repeated many times and the

---

[29]Such measurements are very delicate. The present author recalls using a similar instrument to measure the intensity of the X-rays diffracted by a crystal of zinc-blende, back in 1953, in the Sorbonne, in Paris. If a smoker entered the room, the deflection of the gold-leaf was larger than the signal due to the X-rays!

evidence for partial polarization was conclusive. If a heavier metal was used as a radiator, the secondary radiation was not scattered radiation, and no polarization could be observed.

In a next series of experiments, Barkla (1906*a*) used a mass of carbon as radiator. It generated a secondary beam of sufficient intensity to produce a measurable tertiary radiation. By measuring its intensity for different orientations of the X-ray bulb, Barkla was able to show that the secondary radiation was totally polarized. Barkla's experiments clearly proved the transverse character of X-rays, but did not provide any clue as to whether they were waves or pulses.

Barkla's work on polarization was continued in other countries, by H. Haga (1907) in Groningen, the Netherlands, E. Bassler (1909), one of Röntgen's students, in Munich, Germany, J. Herweg (1909) in Würzburg, Germany, W. M. Ham (1910), in Chicago, USA, and L. Vegard (1910) in Oslo, Norway.

5.7 Characteristic X-ray lines

It is in 1905 that C. G. Barkla clearly distinguished between the properties of secondary radiation from heavy elements and those of the scattered radiation by lighter elements (Barkla 1905*c*). The secondary radiation from iron, copper, zinc, and lead does not vary in intensity with direction and is considerably less penetrating than the primary radiation producing it. It is not scattered radiation. Barkla (1906*b*) next measured the absorption of this secondary radiation in sheets of aluminium for a little more than twenty elements, and found 1) that the relation between absorption and atomic weight exhibits a periodicity which is connected to the periodicity of the chemical properties, and 2) that the atomic weight of an element could be determined by measuring the absorption of its secondary radiation and interpolating. The secondary radiation is therefore an *atomic* property.

Repeated experiments with different absorbers showed that the relation between absorption of the secondary radiation and the atomic weight of the radiator that emits it is a true one, and a very sensitive one. This enabled C. G. Barkla and a co-worker, C. A. Sadler, a demonstrator of physics at the University of Liverpool, to show that the atomic weight which had been attributed to nickel was wrong and was probably around 61.4, *larger* than that of cobalt (59), while it was usually thought to be smaller and equal to 58.7 (Barkla and Sadler 1907).

In their next paper, Barkla and Sadler (1908) distinguished the case of the lighter elements, hydrogen to sulphur, from that of the heavier elements. In the former case, the heterogeneity and the penetrating power of the secondary radiation were not very different from those of the primary radiation, while, in the latter case, the secondary radiation was very homogeneous, even if the primary radiation was heterogeneous, and the penetrating power of this secondary radiation varied from element to element. They could thus establish clearly that the secondary radiation emitted by the heavier elements was constituted of a 'homogeneous radiation characteristic' of the element.[30] They found this a general rule for all elements of atomic weight greater than that of sulphur which they had examined. This homogeneous radiation, now known as 'characteristic radiation', had the following properties:

1. The penetrating power of this radiation is independent of the intensity or the penetrating power of the primary radiation producing it; it is characteristic of the element emitting it.

---

[30]In the literature of the time, 'homogeneous' radiation is often used for 'characteristic' radiation.

2. The penetrating power of this radiation is a periodic function of the atomic weight of the radiating element.
3. This radiation is invariably more easily absorbed than the primary radiation producing it.
4. In all cases, when a primary radiation was used which was softer than the characteristic homogeneous radiation, this radiation was not emitted.

As to the origin of the homogeneous radiation (characteristic X-rays), they suggested that it is 'set up by disturbance of electrons produced directly or indirectly by the passage of the primary pulses'.

In order to quantify the absorption measurements, Barkla and Sadler (1909) introduced the linear absorption coefficient, which they called λ, through the expression of the intensity of a beam having passed through a thickness d of material, $I = I_o \exp -\lambda d$, and the mass absorption coefficient λ/ρ, where ρ is the specific mass of the element. The observation of discontinuities in the variation of absorption with atomic weight was one of the most important results of their investigation (the absorption edges).

The next major step was C. G. Barkla's identification of the harder K series and the softer L series in the emission spectra of the elements. In a preliminary communication read to the Cambridge Philosophical Society on 17 May 1909, C. G. Barkla (1909) reported that for a few elements, such as Sb and I, the secondary radiation consisted of *two* radiations— a very penetrating one and a weakly penetrating one. In the same paper, Barkla compares the homogeneous radiation to the fluorescence observed with light. Barkla (1911) measured carefully the absorption of the secondary radiation through a series of aluminium foils for nearly thirty elements. He found that he could classify these elements into two groups according to the penetrating power of the secondary radiation emitted by them. There were twenty elements, from calcium to barium, in the first group, which he called the K series, and nine elements, from silver to bismuth, in the second group, the L series. Four elements, Ag, Sb, I, and Ba, emitted both a very penetrating and a less penetrating radiation, and belonged to the two groups. For both series, the secondary radiations become stepwise harder when the atomic weight of the element that emits them increases, and the K-radiation is roughly 300 times more penetrating than the L-radiation. From the similarity of the behaviour of all elements, Barkla was led to admit the probable existence of an M series and the possibility of further series, N, etc.

Barkla's investigation marked the beginnings of X-ray spectroscopy (see Section 10.6). It showed also the great similarity of X-rays with light.

5.8   W. H. Bragg and his corpuscular theory of X-rays, 1907

Sir William Henry Bragg (Fig. 5.5), was a good student. He was head of his school on the Isle of Man, and was placed Third Wrangler in the Mathematical Tripos at the University of Cambridge. A chance remark by J. J. Thomson, who in 1884 had just been appointed Cavendish Professor and whose lectures W. H. Bragg had been attending, led to Bragg's application, with 'J. J.'s encouragement, for a Professorship in physics and mathematics at the University of Adelaide. W. H. Bragg was successful in obtaining this position and arrived in Australia in 1886. He took up his new subject of research, physics, with enthusiasm, but found practically no apparatus in the physics laboratory. W. L. Bragg (1975) recalled that his father apprenticed

Fig. 5.5   William Henry Bragg at age 22. From *The Graphic*, 12 July 1884. © Royal Institution, London, Great Britain, with permission.

Sir William Henry Bragg: born 2 July 1862 in Wigton, Cumberland, England, the son of a former merchant marine officer and then farmer; died 21 March 1942 at the Royal Institution in London, England, was an English physicist. He was educated at Market Harborough Grammar School and afterwards at King William's College on the Isle of Man. In 1881 he won a scholarship at Trinity College, Cambridge, where he was elected a minor scholar, and in 1882 he obtained a major scholarship. He graduated in 1884 as Third Wrangler, and in 1885 was awarded a 'first' in the mathematical tripos. The same year, he was appointed Professor of Mathematics and Experimental Physics at the University of Adelaide, Australia. His career in research really started in 1904, and he was elected a Fellow of the Royal Society in 1907. He remained in Australia until 1908, when he came back to England to take the chair of physics at the University of Leeds in 1909. In 1915 W. H. Bragg was appointed Professor of Physics at University College London, but he did not take up his duties there until after World War I. In 1923 he was elected Fullerian Professor of Chemistry at the Royal Institution, and succeeded Sir J. Dewar as head of the laboratory at the Royal Institution. Among his students there, one may mention W. T. Astbury, J. D. Bernal, R. E. Gibbs, K. Yardley (later K. Lonsdale), J. M. Robertson, J. P. Mathieu, A. L. Patterson, and W. G. Burgers. The 1915 Nobel Prize in Physics was awarded to W. H. Bragg and his son Lawrence 'for their services in the analysis of crystal structure by means of X-rays'. W. H. Bragg was knighted in 1920 and served as President of the Royal Society from 1935 to 1940.

MAIN PUBLICATIONS

1912 *Studies in Radioactivity.*

1915 *X-rays and crystal structures.* With Sir W. L. Bragg.

1920 *The World of Sound.*

1925 *The Crystalline State.*

1925 *Concerning the Nature of Things.*

1926 *Old Trades and New Knowledge.*

1928 *An Introduction to Crystal Analysis.*

1933 *The Universe of Light.*

himself to a firm of instrument makers, and 'learnt to use a metal lathe so that he could make his own instruments. I think this gave him the love of good design.' This ability proved to be particularly helpful later. He liked sports and enjoyed life in Australia; he married in 1889, and his first child was William Lawrence. His main academic interest shifted from mathematics to physics, and he published a few papers on electromagnetism and Hertzian waves. One day, in 1895, he was visited by a young visitor from New Zealand on his way to England, Ernest Rutherford[31], with whom he would have a life-long friendship.[32]

[31]See footnote 25 in Section 5.3.

[32]For details on Sir W. H. Bragg's life and works, the reader may consult, for example: E. N. C. Da C. Andrade (1943). William Henry Bragg, *Obituary Notices of Fellows of the Royal Society*. **4**, 277–300. W. L. Bragg and G. M. Caroe (1962). Andrade in Ewald (1962*a*), Ewald (1962*b*), G. M. Caroe (1978). *William Henry Bragg, 1862–1942. Man and Scientist*. Cambridge University Press. J. Jenkin (2008).

The news of Röntgen's discovery reached Australia on 31 January 1896, with articles in the *South Australia Register* and the *Sydney Telegraph*. W. H. Bragg immediately set about building an X-ray apparatus with his assistant A. L. Rogers. Rogers was successful and produced the first X-ray tubes in South Australia. On 17 June 1896, W. H. Bragg delivered his first lecture on X-rays.[33] In those days, W. H. Bragg had become a popular teacher, he was a good experimentalist, and he had acquired a very solid knowledge of fundamental physics, but he was not, at that time, carrying out any original investigations.

The turning point in his career came in 1904. The Australasian Association for the Advancement of Science met in New Zealand, and Bragg was asked to give the presidential address in the section dealing with astronomy, mathematics, and physics. The title of his talk was 'Some recent advances in the theory of the ionization of gases'. It included a discussion of the ranges and ionizing powers of α and β particles. It took Bragg several months to prepare for his talk, reading many articles. The research he did on that occasion triggered his wish to do experiments by himself. Some time after his return to Adelaide, some radium bromide was put at his disposal, and, with his assistant R. D. Kleeman, Bragg began experimental work on the ranges of α particles. This led to his interest in radioactivity and in the γ-rays associated with radioactivity.

In 1907, W. H. Bragg read two communications to the Royal Society of South Australia entitled 'A comparison of some forms of electrical radiation' on 7 May, and 'The nature of Röntgen rays' on 4 June[34] in which he developed for the first time his idea that X-rays and γ-rays are being corpuscular in nature and consist of neutral-pair particles. The two papers were reproduced in the October issue of *Philosophical Magazine* (W. H. Bragg 1907). According to Bragg, these neutral pairs were constituted by one α or positive particle and one β or negative particle. Being neutral, they would have 'great penetrating but weak ionizing powers', would be 'uninfluenced by magnetic or electric fields', and would 'show no refraction'. This hypothesis would also provide a satisfactory explanation for the ejection of secondary cathode rays when a substance is hit by X-rays—a phenomenon discovered by the German physicist Ernst Dorn (1848–1916) in Halle (Dorn 1900), and by P. Curie[35] and G. Sagnac (1900) in Paris. One experimental measurement went against Bragg's views, that of the velocity of X-rays, by the German physicist Erich Marx (1874–1956) in Leipzig. He showed that this velocity is equal to that of light (Marx 1906). If true, this would argue strongly against any material nature of X-rays. But Bragg dismissed this evidence, as he was not certain of what Marx had actually measured. For a modern measurement of the velocity of X-rays, see Zolotoyabko and Quintana (2002).

In the next step, W. H. Bragg compared his hypothesis with the æther pulse theory: 'It is true, of course, that the æther pulse theory has been most ably developed, and is now widely accepted. Nevertheless the evidence for it is all indirect: and indeed some of it is, I think, a little over-rated'. Bragg then proceeded to give an explanation for Barkla's polarization effects in which he assumed that the neutral α-β pairs rotated: when a cathode particle in the X-ray bulb 'strikes an atom so as to make it throw off a pair, the plane of rotation of the pair will

---

[33]These details are given in *Some Reflections on Physics at the University of Adelaide* (1986). Edited by E. H. Medlin. Adelaide: University of Adelaide, Depts. of Physics and Mathematical Physics, Mawson Institute for Antarctic Research.

[34]*Trans. Roy. Soc. S. Aust.* (1907). **31**, 79–93 and 94–98.   [35]See footnote 78 in Section 12.14.

be the same as that of the atom from which it has come, and will contain the direction of the translatory motion of the pair. The pair will therefore be able to show polarization effects'.

In a letter to A. Sommerfeld, dated 7 February 1910, prompted by the debate between J. Stark and A. Sommerfeld (See Section 5.11), W. H. Bragg admitted that the neutral pair hypothesis 'fails only in being unable to give an obvious explanation of polarization, but it is only just to say that the existence of polarization is by no means fatal to it'. 'On the other hand', Bragg added, 'the pulse theory fails in a far more important and fundamental particular, *viz.* that to which you allude in your last paper' (Sommerfeld 1910)—the emission of secondary electrons.[36] In his reply, Sommerfeld admitted a 'weakness in his position relative to the generation of secondary radiation'. 'Nevertheless' he added 'the agreement between the impulse theory and Bassler's (1909) results [on polarization] are so good that they should be further pursued'. Furthermore, 'the diffraction experiments by Walter and Pohl (I [Sommerfeld] have the originals with me) are in good harmony with the pulse theory'. Bragg was not convinced and, in his reply dated 17 May, he asked Sommerfeld 'how are you going to account for the production of a β-ray by a γ-ray?' The tone of the letter was, however, very congenial and Bragg added: 'I am very far from being averse to a reconcilement of a corpuscular and a wave theory'. In another letter, dated 25 June 1911, Sommerfeld could hope 'that the time is not too far away, of reconcilement of a corpuscular and wave theory'.[37]

Bragg's remark on polarization drew an immediate reply from Barkla in the 31 October 1907 issue of *Nature* and in *Philosophical Magazine* (Barkla 1908). A lively debate ensued, through letters to *Nature*, like a table-tennis game across the oceans, neither of two men convincing the other. Bragg's theory, Barkla argued, would lead to an isotropic distribution of scattered radiation, while Thomson's theory led to an anisotropic distribution, in rough agreement with his observations. Bragg retorted that this was not the case. Furthermore, with his colleague J. P. V. Madsen, they showed that the distribution of cathode rays generated during γ-ray irradiation was asymmetric, which would not be the case if γ-rays were æther pulses (Bragg and Madsen 1908*a, b*). And so the discussion went on, with both parties doing new experiments to prove the other wrong.[38] At this point, it may be appropriate to mention the well-known, but usually misquoted, quotation of Sir W. H. Bragg in the Robert Boyle Lecture at Oxford University for 1921:[39] 'There must be some fact of which we are entirely ignorant and whose discovery may revolutionize our views of the relations between waves and æther and matter. For the present we have to work on both theories. On Mondays, Wednesdays, and Fridays we use the wave theory; on Tuesdays, Thursdays, and Saturdays we think in streams of flying energy quanta or corpuscles. That is after all a very proper attitude to take. We cannot state the whole truth since we have only partial statements, each covering a portion of the field'. This text was written, however, after the discovery of X-ray diffraction! For W. H. Bragg's reaction to the discovery, see Section 6.10.

[36]Deutsches Museum Archive, Munich. The present author is grateful to Dr M. Eckert, of the Deutsches Museum, for copies of Bragg's letters to Sommerfeld.

[37]Sommerfeld's letters to Bragg are in the archives of the Royal Institution, London (WHB 6B/2 and WHB 6B/3), where the author could consult them.

[38]The details of the controversy between W. H. Bragg and C. G. Barkla are discussed in R. H. Stuewer, 'William H. Bragg's corpuscular theory of X-rays and γ-rays'. (1971). *Brit. J. Hist. Sci.* **5**, 258–281.

[39]In the February issue of *Scientific Monthly*, New York **14** (1922) 153–160.

In 1908, W. H. Bragg accepted the offer of a professorship at the University of Leeds, and he sailed back to England in early 1909. He was still fully convinced of the validity of the corpuscular theory; see, for instance, W. H. Bragg (1910).

5.9 Diffraction by a slit: estimation of X-ray wavelengths

After several unsuccessful attempts by Röntgen (1897) himself and by a number of other investigators, for instance L. Fomm (1896) and G. Sagnac (1896), the first, rather uncertain, results of the diffraction of X-rays by a slit were obtained in the physics laboratory of the University of Gronigen, the Netherlands, by the Director of the laboratory, Hermann Haga, and his co-worker Cornelis H. Wind (1899), who observed the broadening of the image of a slit due to diffraction. Their slit was wedge-shaped, with a width varying from 2 to 14 μm, and an exposure time ranging from 30 to 200 hours. They assumed X-rays to be electromagnetic waves, and they made a rough estimate of the X-ray wavelength from the broadening they observed, using ordinary diffraction theory. The result, however, varied depending on the width of the diffracting slit and on the conditions of the experiment. They arrived at values of the wavelength ranging from 0.12 to 2.7 Å.

The German physicist, Bernhard Walter (1902), in Hamburg, Germany, sharply criticized Haga and Wind's experiments, and did not believe they showed any evidence of diffraction. He asserted that their wavelength estimations were invalid. As a result, Haga and Wind (1903) repeated their earlier experiments with an improved set-up. This time, they obtained values of 1.6, 0.5 and 1.2 Å for slits 7, 4 and 6 μm wide, respectively.

For his habilitation, Sommerfeld[40] (1896) had developed a mathematical theory of the diffraction of light. The starting point was the Huygens–Fresnel–Kirchhoff principle, but he derived an exact solution of the problem from Maxwell's equations, making use of Riemann's concept of double space, and without the simplifications made by Kirchhoff. At the suggestion of his friend, E. Wiechert, he undertook the extension of his diffraction theory to the case of X-ray pulses. Sommerfeld (Fig. 5.6) was fully convinced of the truth of the Wiechert–Stokes pulse theory: 'Röntgen-rays consist of pulsed disturbances of the æther propagating in time and space according to Maxwell's equations' (Sommerfeld 1901). He applied his diffraction theory to the diffraction of an X-ray single impulse, first by a half-plane (Sommerfeld 1899), then by a slit (Sommerfeld 1900, 1901). If one considers a pulse, one does not expect diffraction fringes, as in the case of light, but a shadow. From his analysis of Haga and Wind's results, he obtained for the width of the pulse a value of the order of 1.35 Å. The question was still open, though, and Sommerfeld could deplore, in 1905, in a letter to Wien, that 'it is a shame that, ten years after Röntgen's discovery, one still doesn't know what Röntgen rays really are'[41].

B. Walter and his co-worker, R. Pohl (1908, 1909), repeated Haga and Wind's experiment in Hamburg with an improved set-up. They came to the conclusion that there was not any observable diffraction and that if the wave nature of X-rays was proved by other means, their wavelength should be smaller than 0.12 Å. Nevertheless, A. Sommerfeld, who was a firm defender of the impulse theory, described in Section 5.2, and who was engaged in a controversy

---

[40]See Section 6.1.3.   [41]Quoted in Eckert 2012.

Fig. 5.6   Arnold Sommerfeld in 1897. Source: Wikimedia commons.

Arnold Johannes Wilhelm Sommerfeld: born 5 December 1868 in Königsberg, Province of Prussia, now Kaliningrad, Russia, the son of a practising physician; died 26 April 1951 in Munich, Germany, in a traffic accident, was a German theoretical physicist. After schooling at the Altstädtisches Gymnasium in Königsberg, Sommerfeld entered the University of Königsberg in 1886, where D. Hilbert (1862–1943) and E. Wiechert (1861–1928) were two of his teachers. As a student, he was a member of the student fraternity Deutsche Burschenschaft, which resulted in a fencing scar on his face. He obtained his PhD in 1891 on mathematical physics, and, after a year of military service, moved to Göttingen in October 1893. There he spent at first one year as assistant at the Mineralogical Institute, which he did not enjoy. He was at heart a mathematician and he preferred to become assistant to F. Klein under whose direction he prepared his habilitation thesis on diffraction theory, presented in 1895. From 1895 to 1897, he taught partial differential equations as *Privatdozent*. In 1897, he was appointed Professor of Mathematics at the Mining Academy at Clausthal, where he succeeded W. Wien. In 1900, Sommerfeld again succeeded Wien, this time as Professor of Applied Mechanics at the *Technische Hochschule* Aachen (now RWTH), where he applied his mathematical knowledge to engineering problems. In 1906 he became Director of the newly established Institute of Theoretical Physics in Munich. In 1918, Sommerfeld succeeded Einstein as chair of the Deutsche Physikalische Gesellschaft, and one of his first steps was to encourage the establishment of *Zeitschrift für Physik*. Sommerfeld was a renowned teacher, and, among his numerous students, one may mention H. Bethe, L. Brillouin, P. Debye, P. P. Ewald, P. Epstein, H. Fröhlich, E. R. Fues, W. Heisenberg, W. Heitler, H. Hönl, L. Hopf, A. Landé, W. Lenz, W. Pauli, L. Pauling, and R. Peierls.

MAIN PUBLICATIONS

1891 *Die willkürlichen Functionen in der mathematischen Physik.*

1896 *Mathematische Theorie der Diffraction.*

1897–1910 *Über die Theorie der Kreisels.* With F. Klein.

1919 *Atombau und Spektrallinien.*

1929 *Wellenmechanischer Ergänzungsband für Atombau.*

1943 *Mechanik, Vorlesungen über theoretische Physik Band 1.*

1945 *Mechanik der deformierbaren Medien, Vorlesungen über theoretische Physik Band 2.*

1947 *Partielle Differentialgleichungen der Physik - Vorlesungen über theoretische Physik Band 6.*

1948 *Elektrodynamik - Vorlesungen über theoretische Physik Band 3.*

1950 *Optik, Vorlesungen über theoretische Physik Band 4.*

1952 *Thermodynamik und Statistik, Vorlesungen über theoretische Physik Band 5.*

with J. Stark (see below, Section 5.11), asked Röntgen's chief assistant, P. Koch, to analyse again their photographs very carefully with an accurate optical microphotometer. This was done in Röntgen's Institute (Koch 1912), and, from the broadening of the image of the slit, Sommerfeld (1912) recalculated the width of the X-ray impulses to be 0.4 Å. The paper was submitted 1 March 1912!

5.10 Derivation of X-ray wavelengths from the consideration of energy elements

5.10.1   W. Wien, 1907

The mechanism by which secondary electrons are produced when γ- or X-rays are stopped by atoms was a topic of hot debate in the years that followed their discovery by Curie and Sagnac (1900) and by Dorn (1900). The Scottish physicist, P. D. Innes (1907) measured photographically the velocities of the secondary electrons, and found that these velocities were independent of the intensity of the primary rays. His conclusion was that the origin of the secondary electrons was atomic disintegration.

W. Wien[42] was convinced by the impulse hypothesis of the nature of X-rays, and had tried to relate their energy to that of electrons (Wien 1904). In a communication to the *Königlische Gesellschaft der Wissenschaften* at Göttingen on 23 November 1907, Wien proposed that the energy of the secondary electrons came entirely from that of the X-rays. The expression of the energy of the X-rays was obtained by generalizing Planck's radiation theory: the elemental quantum energy of X-rays is hc/λ, where h is Planck's constant, c is the velocity of light, and λ the wavelength of the X-rays, which Wien assimilated to the width of the pulses (Wien 1907). Wien used an electrical method to determine the velocity v of these cathode rays. Their energy is $1/2mv^2$, where m is their mass. By writing Planck's relation, one deduces:

$$\frac{hc}{\lambda} = \frac{1}{2}mv^2, \tag{5.2}$$

where hc/λ is the energy associated to an energy quantum of X-rays. Wien was in that way able to estimate the X-ray wavelength to be about 0.675 Å, not too far off Haga and Wind's estimated values.

Bragg and Madsen (1908*b*) considered three hypotheses: 1) 'the energy and the material of the β radiation is furnished by the atom alone, the γ-ray is a pulse which merely pulls the trigger', 2) 'the energy of the β radiation is entirely derived from that of the æther pulses', 3) 'both the energy and the material of the secondary electron are derived from the primary ray', γ-rays being neutral pairs. This last hypothesis was of course the one they preferred. They wrongly associated Wien's experiment and his estimation of the X-ray wavelength with the first of their three hypotheses and concluded: 'it seems to us to be clear that the application of Planck's theory is not justified'. One of Wien's students, J. Laub (1908), who repeated Wien's experiment, pointed out that Bragg and Madsen had misunderstood Wien's argument, which was that the energy of the secondary cathode rays came from the energy quanta contained in the X-rays. Laub added that the experiment 'supported the light quantum theory'. The pulse theory was progressively discredited in favour of the quantum hypothesis (see, for instance B. Davis 1917, A. H. Compton 1926).

---

[42]Wilhelm Wien, born 13 January 1864 in Fischhausen, East Prussia, the son of a landowner, died 30 August 1928 in Munich, Germany, was a German physicist, well-known for his Law of the Distribution of Black-body Radiation. After schooling in Heidelberg, he attended the Universities of Berlin, where he was a student of H. von Helmholtz, and Göttingen. He obtained his PhD in 1886, and remained in Helmholtz's laboratory until 1896, when he succeeded P. Lenard as Professor in Aachen. In 1890 he succeeded W. C. Röntgen as Professor in Würzburg. He was awarded the Nobel Prize in Physics 1911 'for his discoveries regarding the laws governing the radiation of heat'.

5.10.2   J. Stark, 1907

Like Wien, Stark confused impulse width and wavelength of X-rays (an impulse is limited in time, while 'wavelength' implies a continuous wave). Independently of Wien, J. Stark[43] had submitted on 26 October 1907 an article to *Physikalische Zeitschrift*, published on 1 December (Stark 1907), where he applied Planck's elementary law on the emission of Röntgen rays by stopped cathode ray particles. The energy of the cathode rays accelerated through a potential V is eV. If the cathode rays are stopped within half a wavelength of the emitted X-rays and if their energy is entirely converted into energy elements, the expected minimum wavelength of the X-rays is given by:

$$\lambda = \frac{2hc}{eV} \tag{5.3}$$

For an accelerating voltage of 60 000 Volts, Stark obtained a value of 0.6 Å for the X-rays. The calculation of the X-ray wavelength was republished in *Nature*, the following February (Stark 1908). Stark (1907) then considered the process of the generation of the secondary electrons by the braking of X-rays. Their maximum energy is given by a relation identical to that in equation 5.2.

5.11   J. Stark's atomic constitution of the X-rays, 1909

In his next articles, Stark (1909*a, b*) discussed the two modes of generation of X-rays, by fluorescence, as described by Barkla, and by collision of primary electrons, in terms of the two hypotheses for the nature of X-rays: æther waves or light quanta. He took a decisive step and considered that X-rays differ from light in the spectrum range, $\lambda = 10^{-7} - 10^{-9}$ cm, by being material bodies (*materielle Körper*). To justify his position, he measured the asymmetric distribution of the X-rays emitted by the braking of electrons from a carbon anticathode, and he claimed that it could not be explained by the impulse theory.

Sommerfeld (1909) replied immediately in defence of the impulse theory. He also distinguished X-rays emitted by fluorescence (*Fluoreszenzstrahlung*) and by the braking of electrons (*Bremsstrahlung*). The former is direction-independent, and the latter is polarized and depends on direction when the braking takes place along a straight line. Sommerfeld then calculated the angular distribution of the *Bremsstrahlung* and showed that it is emitted mostly in the forward direction, and that it could be explained by the impulse theory.

Stark (1910) disagreed, saying that his observations on the angular distribution of X-rays were in contradiction with Sommerfeld's calculations that invoked the impulse theory. They

---

[43]Johannes Stark, born 15 April 1874 in Schickenhof, Germany, the son of a landowner, died 21 June 1957 in Traunstein, Germany, was a German physicist. After schooling at the Bayreuth Gymnasium, he studied physics, mathematics, chemistry, and crystallography at the University of Munich. He obtained his PhD in 1897, and, after three years as assistant in Munich, he was appointed *Privatdozent* at Göttingen in 1900. There, he discovered the Doppler effect of the canal rays in 1905. In 1906, he became Professor at the *Technische Hochschule* Hanover, and in 1908 Professor at the *Technische Hochschule* (now RWTH) Aachen. During his time in Aachen, he discovered the splitting of spectral lines under the influence of an electric field, now called the Stark effect. In 1917 he was appointed Professor at the University of Greifswald. Stark was awarded the 1919 Nobel Prize in Physics, for his two discoveries. In 1934 he was elected a member of the Deutsche Akademie der Naturforscher Leopoldina. Aggressively anti-Semite, he was with P. Lenard one of the founders of the *Deutsche Physik* movement in the early 1920s.

could, however, be explained in terms of light quanta; furthermore, the light quanta hypothesis accounted for the emission of secondary electrons, which the impulse theory could not do.

Sommerfeld (1910) pointed out some inaccuracies in Stark's interpretation of his own measurements and in his presentation of historical facts. He maintained without hesitation his view regarding the impulse theory. The disagreement between them continued and, at the time of the discovery of X-ray diffraction by Friedrich and Knipping in 1912, Stark was still convinced of the validity of his corpuscular theory. He tried at first to explain the diffraction by crystal lattices in terms of the propagation of corpuscles along crystal avenues, rather than by the diffraction of waves (Stark 1912, see Section 6.9).

The corpuscular nature of X-rays was shown by the discovery of the Compton effect (Compton 1923b, see Section 9.5), and their dual nature was accounted for by L. de Broglie's formulation in 1924 of the relation between particles and waves. The history and the 'empirical roots' of wave–particle dualism are discussed in detail in a book by B. R. Wheaton (1983), entitled *The Tiger and the Shark*, this title coming from a quotation by J. J. Thomson in 1925.

6

1912: THE DISCOVERY OF X-RAY DIFFRACTION

AND THE BIRTH OF X-RAY ANALYSIS

I congratulate you for your wonderful success. This experiment belongs to
the most beautiful moments in physics.

Albert Einstein (postcard to M. Laue, 10 June 1912. Archives of the Deutsches Museum, Munich)

6.1 Munich in 1912

The discovery of X-ray diffraction took place in Munich, the capital of Bavaria, and the cultural and artistic centre of Southern Germany. It was an attractive city, nested at the foot of the Bavarian Alps, with its historic buildings, its museums, and its parks, such as the Hofgarten (Fig. 6.1) and the Englische Garten. Its professors and its numerous students made the Ludwig-Maximilians University a celebrated centre of learning. Many famous physicists lived in Munich (see, for instance, Teichmann *et al.* 2002). In 1912, three Institutes dominated the scientific life in physics and crystallography in Munich, the Institutes headed respectively by P. von Groth, W. C. Röntgen, and A. Sommerfeld.

6.1.1 Groth's Institute for Mineralogy and Mineral Sites

The Director of Munich's Institute for Mineralogy and Mineral Sites in 1912 was Paul von Groth (Fig. 6.2). Before occupying this position, P. von Groth had been Director of the Mineralogy Institute of the University of Strasbourg, which he had built from scratch. He had also established a notable teaching mineral collection, a model for future collections. When J. C. Poggendorff retired as Editor of the *Annalen der Physik und Chemie*, Groth feared that his successor would not be so open and would not continue to include articles on mineralogy and crystallography. He therefore decided in 1877 to found the *Zeitschrift für Krystallographie und Mineralogie*, which remained the prominent journal on these topics until the second World War.[1]

Groth was called to Munich in 1883 to succeed the mineralogist Wolfgang Xavier Franz Baron von Kobell (1803–1882) as Director of the Munich Institute of Mineralogy. There, he promptly reorganized the teaching of mineralogy and installed in a new location the extensive royal Bavarian mineral collections, of which he was made curator. A renowned teacher, he had an encyclopedic knowledge of minerals. He was the leading figure of mineralogy, at home and internationally. Students and visitors flocked from round the world to learn from the

---

[1]For details on the history of *Zeitschrift für Krystallographie*, see Steinmetz and Weber (1939) and Authier (2009).

Fig. 6.1   The Hofgarten in 1912. Private collection of the author.

master, and he stimulated and supervised countless numbers of investigations dealing with all aspects of crystallography and mineralogy. He was 'a small and very lively man, wearing strong glasses, with a round face surrounded by a clipped shaggy grey beard' (Ewald 1948).[2] He was 'communicative and ever ready to discuss any scientific item'. P. P. Ewald, who consulted Groth as to which crystal would best be suited to test the theoretical results of his thesis, described, in a very vivid and picturesque way, a visit to Groth's office: 'At the wall opposite to where the visitor entered he would finally detect the old-fashioned desk with a small worthy old gentleman facing the wall and turning his back to the visitor while eagerly entering the end of a sentence in a manuscript or a correction in a galley proof. This was the *Geheimrat*,[3] P. von Groth, then in his early seventies' (Ewald 1962a).

P. von Groth's contributions to mineralogy and crystallography were immense and touched every field of these sciences. His first great achievement was the classification of minerals according to chemical relationships. P. von Groth is also well-known for his work on isomorphism and its opposite, morphotropy, which describes the changes in crystal structure due to the replacement of an atom of the crystal by an atom of an element from a neighbouring group in the Periodic Table, while isomorphism concerns crystals whose structure and morphology are not changed by such a replacement (see Section 12.1.5). The relations between crystal structure and chemical composition were a constant concern for him (see, for instance, Groth 1888, 1904). He had a physical understanding of the notion of space lattices, while Schoenflies had a mathematical one. His most remarkable work was, without doubt, the *Chemische Kristallographie* whose five volumes were a compilation of all the properties of 9000 to 10 000

[2]Draft of Ewald's talk at the first Meeting of the International Union of Crystallography, Harvard, USA, 28 July–3 August 1948. Archives of the Deutsches Museum, Munich, NL 089/027. The present author is grateful to Michael Eckert for a copy of the manuscript.

[3]*Geheimrat*, or privy councillor, was the title of advising officials at the Imperial or Royal courts of the Holy Roman Empire. In the case of Groth, Sommerfeld, or Wien, it was an honorary title.

Fig. 6.2   Paul von Groth. After Steinmetz and Weber (1939), reproduced with permission from Oldenburg Wissenschaftsverlag.

Paul Heinrich Ritter von Groth: born 23 June 1843 in Magdeburg, Germany, the son of a portrait painter; died 2 December 1927 in Munich, Germany, was a German mineralogist and crystallographer. He received his education at the Bergakademie in Freiberg and at the College of Engineering in Dresden. In 1865, he moved to the University of Berlin, where he obtained his PhD in 1868, and habilitation in 1870, under the direction of physicist A. Magnus, on the 'relations between crystal habit and chemical constitution'. From 1870 to 1872, von Groth was Dozent at the Bergakademie in Berlin. In 1871 he was appointed Professor of Mineralogy at the newly founded University of Strasbourg, where he created an Institute of Mineralogy. In 1877 he founded the *Zeitschrift für Kristallographie* of which he remained Editor until 1921, and, in 1883, he became Director of the Munich Institute of Mineralogy and curator of the State Collection. From the material he collected in 55 Volumes of the *Zeitschrift für Kristallographie*, and in many other journals, he produced the *Chemische Krystallographie*, which appeared in five volumes between 1906 and 1919, a complete dictionary of the physical and chemical properties of crystalline substances known at that time.

MAIN PUBLICATIONS

1874 *Tabellarische Übersicht der einfachen Mineralien.*

1887 *Grundriss der Edelstein-Kunde.*

1888 *Ueber die Molekularbeschaffenheit der Krystalle.*

1905 *Physikalische Krystallographie und Einleitung in die krystallographische Kenntniss der wichtigsten Substanzen.*

1906–1919 *Chemische Krystallographie.*

1921 *Elemente der physikalischen und chemischen Krystallographie.*

1926 *Die Entwicklungsgeschichte der mineralogischen Wissenschaften.*

crystals and minerals described in 55 volumes of *Zeitschrift für Kristallographie*, and in many other journals, from Germany and from abroad. Groth was very meticulous and, whenever he did not feel sure of the correctness of the data reported in the articles, he asked his assistants, B. Gossner and H. Steimetz, to remeasure them!

6.1.2   Röntgen's Institute of Experimental Physics

Another of the world-famous institutions in Munich was the Institute of Experimental Physics. At the request of the Bavarian government, the Philosophical Faculty of the University of Munich invited W. C. Röntgen to succeed the physicist Eugen von Lommel (1837–1899) as Professor of Physics. Röntgen (Fig. 4.1, Section 4.1) accepted and became in 1900 the Director of the Institute of Experimental Physics. He took this step only after very careful considerations because he had grown very fond of his work and life in Würzburg, but once he took over the Institute he organized the work with his usual thoroughness.

The Institute of Experimental Physics was the largest of these three Munich Institutes, attended by many students, among them medical students. This required a large number of assistants and lecturers, among whom were P. Koch, chief assistant, E. Wagner and E. von Angerer. Röntgen had usually some ten to twelve PhD students who were looked after by assistants and by himself. As thesis adviser, he was very exacting; three to four years work were usually necessary for a thesis (Ewald 1962a). As a man, he was 'very seclusive and forbidding, but a most unselfish character' (Ewald 1948, see footnote 2), 'retiring, self-effacing, almost shy, creating the impression of being unapproachable. Actually he was kind, and keen on encouraging the work of his students' (J. C. M. Brentano, in Ewald 1962a).

Röntgen's administrative load resulted of course in a slowing down of his publications. In Munich, he resumed his studies on the physical properties of crystals, 'which he considered to be better defined objects than other materials' (J. C. M. Brentano, in Ewald 1962a)— in particular, electrical conductivity, pyro- and piezo-electricity, and the influence of X-ray irradiation on these properties. This work was done in cooperation with his Russian pupil, A. F. Ioffe,[4] who came from St Petersburg in 1902, obtained his PhD in 1905, and returned to Russia in 1906. Due to Röntgen's fastidiousness, the work was only published a few years later (Röntgen and Ioffe 1913). Röntgen naturally also maintained his interest in the nature of X-rays, as shown by the following titles of some of the doctoral works performed under his direction:

E. v. Angerer, 1905: Bolometric (absolute) energy measurement of X-rays.
E. Bassler, 1907: Polarization of X-rays.
W. Friedrich, 1911: Emission by a platinum target.
R. Glocker, 1914: X-ray interferences and crystal structure.

Among the other PhD students, one may mention P. Koch, 1901, E. Wagner, 1903, P. Pringsheim, 1906, P. Knipping, 1913, and J. C. M. Brentano, 1914.

6.1.3   Sommerfeld's Institute of Theoretical Physics

The chair of Theoretical Physics at the University of Munich was established in 1890 for the Austrian physicist Ludwig Boltzmann (1844–1906), founder of statistical mechanics and statistical thermodynamics. But Boltzmann left for Vienna only after four years and the chair had remained vacant since then. It was due to W. C. Röntgen's efforts that a new interest was created in 1905 for re-establishing the chair. Röntgen even went to Leyden to try to persuade H. A. Lorentz to accept the position. Lorentz, however, declined, and the chair was taken by Sommerfeld in 1906, on the recommendations of Boltzmann, Lorentz, and Wien.

Sommerfeld (Figs. 5.6 and 6.3, *Left*) had started his career as a pure mathematician, and he considered F. Klein,[5] whose assistant he had been in Göttingen, as his 'true' teacher. But he soon became interested in applications of mathematics to physics, as shown by his diffraction

---

[4]Abram Fedorovich Ioffe, born in 1880, died in 1960, was a prominent Russian/Soviet physicist who studied electromagnetism, crystal physics, thermoelectricity, and photoelectricity. In 1918 he became head of the Leningrad Physico-Technical Institute (LPTI), now the Ioffe Institute.

[5]Christian Felix Klein, born 25 April 1849, died 22 June 1925, was a German mathematician. He obtained his doctorate in 1868 at the University of Bonn, Germany, under the supervision of J. Plücker. He was professor successively at the University of Erlangen in 1872, at Munich's *Technische Hochschule* in 1875, at the university of Leipzig in 1880, and, finally at the university of Göttingen in 1886. He is best known for his work in group theory, function theory, and non-Euclidean geometry.

Fig. 6.3   *Left*: A. Sommerfeld around 1910. Photo Deutsches Museum, Munich. *Right*: Laue in 1904 in the garden of
Strasbourg University. Photo Deutsches Museum, Munich.

theory (See Section 5.9), and his electron theory. It is therefore not surprising that, when he
was appointed Professor of Theoretical Physics at Munich University, he insisted on having an
Institute with experimental facilities. His chair was associated with the position as one of the
curators of the Bavarian Academy of Sciences, with the duty of taking care of a collection of
physical instruments (Ewald 1968*b*). A workshop, an assistant, and a technician were attached
to this position. Sommerfeld's Institute, which was opened in 1909, was the smallest of the
three Munich Institutes and was close to Röntgen's, in the Amalienstrasse. It consisted of a
small lecture room, a museum room with the old Sohncke models (Section 12.12), which came
probably from the collection of the Bavarian Academy of Sciences,[6] a workshop, a dark room
and four rooms for experiments or storage. The first experiments performed there were on
turbulence, by one of Sommerfeld's PhD students, L. Hopf.[7]

Peter Debye[8] followed Sommerfeld from Aachen to Munich, and was his first assistant. He
was, 'no less than Sommerfeld himself, a centre for the senior students and graduates' (Ewald
1962*a*). He left, however, in 1911, to succeed Einstein at the University of Zurich, Switzerland.

[6]Ewald interview by G. Uhlenbeck with T. S. Kuhn at the Rockefeller Institute, New York, N.Y., 29 March
and 8 May, 1962, Niels Bohr Library & Archives, American Institute of Physics, College Park, MD USA,
<http://www.aip.org/history/ohilist/4523.html>.

[7]Details about Sommerfeld's Institute can be found in Ewald (1962*a*, 1968*b*) and in M. Eckert (1999).

[8]See Section 7.9.

That same year, 1911, Sommerfeld appointed two assistants, Wilhelm Lenz (1888–1957), for theoretical matters, and Walter Friedrich to carry out experiments in order to test Sommerfeld's views on the properties of X-rays, views expressed in an important study of *Bremsstrahlung* and the properties of γ-rays.[9]

Sommerfeld was quick to adopt the new developments in theoretical physics, and he was one of the first to defend Einstein's relativity theory. He realized immediately the importance of the quantum theory, and, as early as 1911, he foresaw that it would be a key to an understanding of the structure of the atom.[10] Bohr's breakthrough papers on the constitution of atoms appeared in 1913, and they were soon known in Munich. Bohr sent a reprint to Sommerfeld (Heilbron 1967). Ewald, who attended the fall meeting of the British Association for the Advancement of Science in Birmingham (Section 7.10.1) where Bohr's theory was discussed (Ewald 1913c), gave an account of it in the Munich colloquium on 19 November 1913. Sommerfeld was highly interested in it and he soon started to work on the generalization of Bohr's atom theory. This led to his quantum theory of spectral lines, in particular of X-rays (Sommerfeld 1919). The atomic quantum theory was then further developed by Sommerfeld's students W. Heisenberg and W. Pauli. Sommerfeld was Linus Pauling's guide to quantum physics when he visited Munich in 1927 (Section 10.1.2).

Sommerfeld was an exceptional lecturer, recognized as such by all: 'an incomparable teacher from whose lectures a powerful inspiration emanated' (M. von Laue in Ewald 1962a), a 'brilliant lecturer as well as an inspiring leader of research' (J. C. M. Brentano in Ewald 1962a), his talents as a lecturer and a debater left a fascinating impression (Born 1928), 'he had a unique gift for detecting good students, of inspiring them, and of making them feel they could do useful work (Laue 1952b). He was 'the Teacher', but he also liked to be called Herr Geheimrat. Sommerfeld's superior knowledge, the clear-cut luminosity of his explanations, his personal charm and the great personal interest he took in his co-workers attracted students from all over the world. He founded one of the foremost Schools of Theoretical Physics, which became the nerve centre for new ideas about the structure of atoms. M. Born (1928), writing about 'Sommerfeld founder of a school', estimated that at least ten professorships on theoretical physics in German-speaking universities were held by former students of Sommerfeld, not counting the numerous foreigners who had spent time in Munich. For details about Sommerfeld's school, see, for instance, Eckert (1999).

An important institution in Sommerfeld's Institute was the weekly physics colloquium, which provided a place for discussion of theoretical and experimental subjects. It was started thanks to Ewald, who felt the need for the younger students to hear discussions about modern developments in physics (Ewald 1968b). He prompted his friend D. Hondros, who had been with Sommerfeld half a year longer than he (Ewald) had, to ask P. Debye to organize seminars for the younger students; Debye agreed, and Sommerfeld contributed a box of cigars. The first visiting speaker was Max Laue, who gave a talk on the 'behaviour of light waves at a focal point'. Laue gave another talk after his appointment in Munich, this time on his work on 'entropy of radiation'. A larger room became available for the colloquium after

---

[9]Sommerfeld (1911). Über die Struktur der γ-Strahlen. *Sitzungsber. König. Bayer. Akad. Wiss.* 1–60.

[10]Heisenberg (1951). Arnold Sommerfeld. *Naturwiss.* **38**, 337–338.

Fig. 6.4   Café Lutz today. Photo by the author.

Sommerfeld's move to the new building, in the autumn of 1910. The colloquium was then also attended by members of Röntgen's group, Wagner, Koch, v. Angerer, Friedrich, Knipping, Glocker, Brentano, but never by Röntgen himself. Sometimes K. Fajans or P. von Groth also attended.

There was another celebrated institution for the young physicists in Munich, one that was more informal: the gatherings in the 'Café Lutz' in the *Hofgarten* (Fig. 6.4). According to the physicist Paul S. Epstein (1883–1966), who joined Sommerfeld's group in 1909, the tradition had been founded by Röntgen's assistants, P. Koch and A. F. Ioffe.[11] P. Debye always went to lunch at 12 o'clock with P. Koch and then went with him to one of the cafés[12]. P. P. Ewald (1962*a*) recalled: 'This was the general rallying point of physicists after lunch for a cup of coffee and the tempting cakes. Once these were consumed, the conversation which might until then have dealt with some problem in general terms, could at once be followed up with diagrams and calculations performed with pencil on the white smooth marble tops of the cafe's tables—much to the dislike of the waitresses who had to scrub the tables clean afterwards'. Another famous

[11]P. S. Epstein interviewed by A. Epstein (22 November 1965 to 8 February 1966), California Institute of Technology archives, Pasadena, California, USA.

[12]Interview of P. Debye by T. S. Kuhn and G. Uhlenbeck at Rockefeller Institute, New York City, New York, 3 May 1962. Niels Bohr Library & Archives, American Institute of Physics, College Park, MD USA, <http://www.aip.org/history/ohilist/4568_1.html>.

X-ray physicist, M. Siegbahn, who spent some time in Munich in 1909, recollected that 'many of the calculations and formulæ which have later become cornerstones of the edifice of science, have first been jotted down on the marble tops of café tables'.[13]

6.2   Ewald's thesis, 1912

Paul Ewald (Figs. 6.5 and 9.9, *Left*) was one of Sommerfeld's students. He came from an educated family. His father had been an historian, his mother was a portrait painter, his paternal grandfather studied philosophy and history, but took up painting, a great-uncle had been a mathematician, another great-uncle a geologist, one uncle was Professor of Medicine at the University of Berlin, and another Professor of Physiology at the University of Strasbourg. Paul's father died three months before he was born, and he was educated by his mother, 'a remarkable woman of many gifts' (Ewald 1968*a*). She took him on many travels abroad when he was young, to Paris where she took painting lessons, to Switzerland, and, more frequently, to Cambridge, England. During these stays, he picked up French and English easily and rapidly. His first acquaintance with science came when he was 11 years old, during a visit to Cambridge. One day, a Fellow of Caius College and Professor of Inorganic Chemistry took a Bunsen burner and blew a large glass sphere. After cooling it, he poured two liquids in it and produced a perfect silver mirror inside the sphere. Back home in Berlin, Ewald tried to make a Bunsen burner by unscrewing the (Auer) gas light and attaching a long rubber tube to it. He lit the gas coming out of the tube, producing a loud bang and a long flame which started burning the draperies! Ewald recounted that it is to the credit of his mother that she nevertheless encouraged his experiments (Ewald 1968*a*). Ewald was well read. Among other things, he had read Helmholtz's biography and had been fascinated by his theory of the dispersion of light and of molecular resonances.[14]

In 1905–06, Ewald spent a very happy winter as an undergraduate at Gonville and Caius College in Cambridge. Apparently, he did not learn much during that stay, but he later recalled that its 'formative value far exceeded any academic gain' (Ewald 1968*a*). Serious things started in the spring of 1906 when he moved to Göttingen. He did not like chemistry, but eagerly attended D. Hilbert's course on calculus. He became fascinated as he watched 'the working of a mathematical mind' (Ewald 1968*a*) and decided to make mathematics his goal. He then went to Munich to follow A. Pringsheim's lectures which were reputed to present a different approach to mathematics from that offered in Göttingen. He was impressed by Pringsheim's lectures, delivered with humour and wit, but, for him, there was something lacking, and this he found in Sommerfeld's lectures. Somewhat by chance he had been forced by a friend, the Greek physicist D. Hondros (1882–1962), to attend Sommerfeld's hydro-dynamics course. He was immediately 'spell-bound' and attracted to the 'interplay between

---

[13]Quoted in H. Atterling (1991). Karl Manne Georg Siegbahn. 3 December 1886–24 September 1978. *Biogr. Mems. Fellows R. Soc.* **37**, 428–444.

[14]For biographical details of Ewald's life, see Ewald (1968*a*, 1968*b*); H. Bethe and G. Hildebrandt (1988). *Biographical Memoirs of Fellows of the Royal Society.* **34**, 134–176; G. Hildebrandt in Cruickshank *et al.* (1992); for Ewald's role in founding *Acta Crystallographica* and the International Union of Crystallography, see Ewald (1977), Kamminga (1989), Cruickshank (1999), and Authier (2009).

Fig. 6.5   Peter Paul Ewald in 1932. Source: Archives of the Max-Planck-Gesellschaft, Berlin, with permission.

Peter Paul Ewald: born 23 January 1888 in Berlin, Germany, the son of a historian, Privatdozent at the University of Berlin; died 22 August 1985 in Ithaca, New York, USA. He received his early education at the Realgymnasium and the Wilhelms-Gymnasium in Berlin and, from 1900, at the Victoria Gymnasium in Potsdam. After his Abitur, obtained in 1905, he spent the winter of 1905–1906 in Caius College, Cambridge, England. In spring 1906 he moved to Göttingen, and, from there, to Munich in the autumn of 1907. Ewald obtained his PhD in February 1912 and returned to Göttingen to be physics assistant to D. Hilbert. In the spring of 1913, he went back to Munich where he shared an assistantship with W. Lenz. After the breakout of the war, he enlisted in the army as an X-ray technician. During quiet moments on the Russian front he developed the dynamical theory of X-ray diffraction, which was the topic of his *Habilitationsschrift*. After his habilitation in 1917, he was appointed Privatdozent at the Ludwig-Maximilians University in Munich. In 1921 he became Assistant Professor at the Stuttgart Technical University, where he succeeded E. Schrödinger, and, in 1922, was promoted to full Professor. In 1924, Ewald became co-editor of *Zeitschrift für Kristallographie*. An Institute for Theoretical Physics was created in 1930 at the Stuttgart Technical University, after the model of Sommerfeld's Institute, with Ewald as Director. E. Fues, H. Hönl, H. Bethe were among his assistants, and M. Renninger was one of his students. Ewald was elected Rector at Stuttgart in 1932, but due to increasing difficulties with faculty members who were also members of the National Socialist party, he resigned in the spring of 1933. In the summer of 1936 he taught for two months at Ann Arbor, Michigan, USA. In Stuttgart the problems with the Nazis become more and more acute, and, finally, he was asked to stop teaching in December 1936. Thanks to Sir Lawrence Bragg, he obtained a research grant in Cambridge, where he went in the autumn of 1937. In August 1939, Ewald was appointed Lecturer at Queen's University in Belfast, and, in 1945, Professor of Mathematical Physics. The same year, he became a British citizen. In a lecture given at Oxford in 1944, Ewald stressed the pressing need for an international journal of crystallography. Subsequently, he played a leading role in founding *Acta Crystallographica*, of which he was the first Editor, and the International Union of Crystallography. In 1949, he moved to the United States, where he was appointed Professor and Director of the Physics Department of the Polytechnic Institute of Brooklyn, until his retirement in 1959. He was elected a Fellow of the Royal Society in 1958, a member of the Deutsche Akademie der Naturforscher Leopoldina in 1966. He was President of the Provisional International Crystallographic Committee (1946–1948) and President of the International Union of Crystallography from 1960 to 1963.

MAIN PUBLICATIONS

1912 *Dispersion and Doppelbrechung von Elektronengittern (Krystallen).*

1916 *Zur Begründung der Kristalloptik. I. Theorie der Dispersion. II. Theorie der Reflexion und Brechung.*

1917 *Zur Begründung der Kristalloptik. III. Die Kristalloptik der Röntgenstrahlen.*

1923 *Krystalle and Röntgenstrahlen.*

1927 *Der Aufbau der festen Materie und seine Erforschung durch Röntgenstrahlen.*

1930 *Physics of solids and fluids.* With T. Pöschl and L. Prandtl.

1931 *Strukturbericht 1913–1928.* With C. Hermann.

1955 *Properties of solids and solid solutions.*

1962 *Fifty years of X-ray diffraction.*

Fig. 6.6   Sommerfeld's group. Top row: second from left, W. Friedrich; far right, P. Epstein; second from right, P. P. Ewald; second row, third from left: A. Sommerfeld. After *Paul S. Epstein, interview by Alice Epstein*. Courtesy of the Archives, California Institute of Technology.

the mathematical formalism and the physical arguments' which was so vividly described by Sommerfeld. From then on, his 'heart was set on mathematical physics': 'Sommerfeld's courses and seminars were my main preoccupation; henceforth I considered myself a theoretical physicist'. It is from Sommerfeld's lectures that Ewald learned about vectors and vector analysis.[15]

Ewald joined Sommerfeld's group (Figs. 6.6 and 6.9 *Left*) in the fall of 1908 (Ewald 1968*b*). By the autumn of 1910, after one or two minor publications, he felt sufficiently sure of himself to ask Sommerfeld for a thesis topic. Sommerfeld took from a drawer in his cherry-wood desk a sheet of paper on which were listed ten to twelve topics suitable for a doctoral thesis, ranging from hydrodynamics to propagation of waves in wireless telegraphy. The last one down the list was 'To find the optical properties of an anisotropic arrangement of isotropic dipoles'. Sommerfeld explained that the index of refraction of a medium was accounted for by assuming that the molecules in the material react as resonators under the influence of the incident wave. These resonators were assumed to be isotropic. Could one explain the double refraction of crystals by the regular arrangements of these isotropic resonators? Sommerfeld warned Ewald that he had no definite idea how he would tackle the problem, and that he could not 'foretell to what mathematical difficulties it might lead'. But Ewald's heart had already been set on that topic, and, at the second interview, Sommerfeld gave him his blessing, along with a

[15]For accounts of Ewald's encounter with Sommerfeld and his thesis work, see Ewald 1948 (see footnote 2), 1961, 1962, 1979*a*, Ewald's 1962 interview (see footnote 6), Ewald in Authier (2003).

reprint of Planck's paper on the *Theory of Dispersion*,[16] and the recommendation that he read H. A. Lorentz's corresponding paper. Of course, Planck and Lorentz had considered the dipoles to be randomly distributed in amorphous media.

The notion of 'dispersion'—namely, the fact that the index of refraction varies with wavelength—was first established by Newton (1672, Section 3.3). Fresnel had interpreted the index of refraction as the ratio of the wave-velocity in free space to that in the medium. Double refraction had been related by Huygens (1678, Section 3.4) to the structure of calcite and by Fresnel more generally to the anisotropy of the medium (1819, Section 3.6), but no one had, so far, explained either quantitatively or qualitatively, crystal optics by the periodic arrangement of isotropic dipoles. Such was the ambitious aim of Ewald's doctoral work.

Ewald took Planck's paper with him and went hiking up the Rhine Valley for his summer vacation (Ewald 1979a). He found Planck's treatment 'incomprehensible' and could not 'disentangle the incident wave, the total field and the field of excitation' in it. He therefore decided to divide the problem into two parts—one about the propagation of waves in an infinite triply periodic assembly of dipoles, and the other about the reflection or refraction of an incident wave by a semi-infinite medium.

In the first part (*theory of dispersion*) he considered the field generated by the dipoles when excited by a plane wave of frequency ν and unknown velocity (Ewald 1912, 1916a). Each dipole is set in oscillation by that incoming field and emits a spherical wave (Hertz 1887a). This wave, which Ewald called a 'wavelet', propagates with the velocity of light, $c = \nu/k_o$ ($k_o = 1/\lambda$, wave number in vacuum), and contributes to the excitation of the other dipoles. The total wave propagating inside the crystal is the resultant of all these wavelets; it is what Ewald calls 'the optical field'. This is the same wave that excited the individual dipoles, and the problem is a self-consistent one. The balance between this optical field and the oscillations of the dipoles was called a *dynamical balance* by Ewald (1912, 1979a). As a result of this interaction with the dipoles, the phase velocity, v, of the resultant field differs from c. There appears, therefore, an index of refraction, $n = c/v = K/k_o$ ($K = n/\lambda$, wave number in the medium). The problem is therefore to find all possible values of n; the relation between the wavelength and the frequency is expressed in the *dispersion equation*.

The contribution of each dipole is obtained from Maxwell's equations. The optical field acting on a particular dipole is the sum of the contributions of all the other dipoles: a dipole does not excite itself. This creates a difficulty. It turns out that the expression of the total field involves the transformation of the sum of these individual contributions into a sum of plane waves, of wave vectors **K**. This requires interchanging a summation and an integration, which is very easy nowadays using Fourier transforms, but was very complicated at the time. Ewald (1979a) recalled that he received help from Debye during a skiing Easter vacation in Mittenwald in the Bavarian Alps. Physicists from Munich and Würzburg used to meet every year in this charming little village at the top of the Isar valley, where Wien had a cottage. Debye told Ewald about a method used by Riemann in a similar case, and this did the trick. A summary of Ewald's derivation using Fourier transforms can be found, for instance, in Authier (2003). Ewald made the calculation assuming an orthorhombic lattice, and found the following expression for the total potential, Π, (the optical field):

---

[16]M. Planck (1902). *Sitzungsberichte Kgl. Preuss. Akad. Wiss.* 470–494.

$$\Pi = -\frac{\pi}{2abc} \sum_{l,m,m}^{-\infty..+\infty} \frac{\exp -i[(l\pi + a\alpha)x/a + (m\pi + b\beta)y/b + (n\pi + c\gamma)z/c]}{k_o^2 - (l\pi + a\alpha)^2/a^2 - (l\pi + b\beta)^2/b^2 - (n\pi + c\gamma)^2/c^2} \qquad (6.1)$$

where a, b, c are the parameters of an orthorhombic lattice, l, m, n are three integers, $(l\pi + a\alpha)/a$, $(m\pi + b\beta)/b$, $(n\pi + c\gamma)/c$ are the components of a wave vector, $\mathbf{K_h}$, and x, y, z are the coordinates of a position vector, $\mathbf{r}$. In modern notation, equation 6.1 can be written:

$$\Pi = -\frac{\pi}{2abc} \sum \frac{\exp -i\pi \mathbf{K_h} \cdot \mathbf{r}}{k_o^2 - K_h^2} \qquad (6.2)$$

The factor $1/(k_o^2 - K_h^2)$ was called *resonance factor* by Ewald, where $k_o = 1/\lambda$ is the wave number in vacuum.

In the second part of his thesis (*theory of refraction and reflection*), Ewald (1916b) introduced the boundaries of the crystal and an incident wave. By a formal truncation to a half-space of the lattice sum of the radiating dipoles, he proved that the progressive wave which excites the dipoles in the crystal may be expressed as the sum of two terms—one propagating with velocity c and which cancels out exactly the incident wave, and another which satisfies the wave equation for propagation with velocity c/n. This is the so-called Ewald–Oseen extinction theorem (Bullough and Hynne, in Cruickshank *et al.* 1992) which was also proved by Oseen in 1915 for isotropic media.

Once the theoretical result had been obtained, Ewald decided to test it in a practical situation. He therefore asked Groth for a suitable orthorhombic crystal with a simple primitive lattice, P. Groth suggested anhydrite ($CaSO_4$), which he was almost certain had a simple orthorhombic

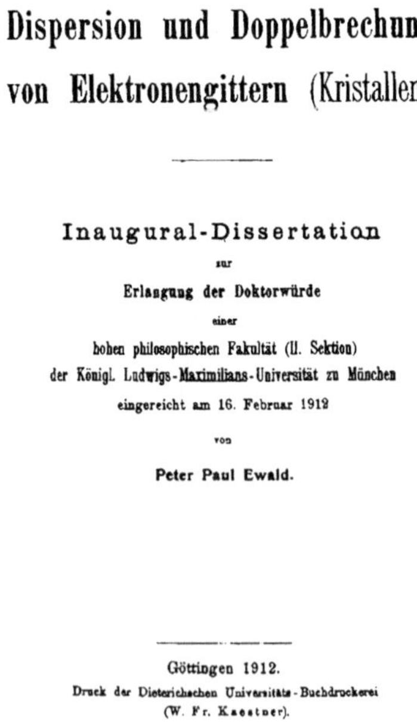

Dispersion und Doppelbrechung
von Elektronengittern (Kristallen).

———————

Inaugural-Dissertation

zur

Erlangung der Doktorwürde

einer

hohen philosophischen Fakultät (II. Sektion)
der Königl. Ludwigs-Maximilians-Universität zu München

eingereicht am 16. Februar 1912

von

Peter Paul Ewald.

———————

Göttingen 1912.
Druck der Dieterichschen Universitäts-Buchdruckerei
(W. Fr. Kaestner).

Fig. 6.7   Cover page of Ewald's thesis.

lattice: 'the three refractive indices and the dispersion have been accurately measured. The three cleavages at right angles to one another renders the simple, uncentred orthorhombic lattice almost a certainty' (quoted by Ewald 1948, see footnote 2). Ewald obtained a qualitative agreement, but a poor quantitative agreement. The calculated double refraction was too large in two directions and too small in the third one. This may have been due in part to his assumption that anhydrite had a primitive P lattice, while, in fact, its space group is $Amma$, and in part to the assumption that the dipoles were isotropic, which they are certainly not, as suggested by M. Brillouin (1913). Nevertheless, this was the first time that a physical property of a crystal had been directly related to the periodic arrangement of its structure. Ewald's thesis was submitted on 16 February 1912 and defended on 5 March 1912 in front of Röntgen, Groth, Sommerfeld, and the mathematician Pringsheim (Fig. 6.7).

The fact that the index of refraction could be calculated independently of any external excitation was something entirely new with respect to the existing dispersion theories, and Ewald had some doubts. He decided he needed the advice of a theoretician, and some time in late January 1912 he consulted Max Laue,[17] who was known to be familiar with theoretical optics.

6.3 M. Laue: Privatdozent in A. Sommerfeld's Institute

According to his autobiography Laue's interest in optics dated back to his schooldays (Figs. 6.8 and 9.9, *Right*). His first contact with science came when he was a pupil at the Wilhelms-Gymnasium in Berlin, and he happened to hear at school about the electrolytic deposition of copper from a copper sulphate solution: 'the impression which this first contact with physics made with [him] was overwhelming' (Laue 1952*b*). For several days he was lost in thought, and his mother, understanding what was wrong with him, saw to it that he visited Urania, a society that aimed to popularize science, where a large collection of physics apparatus was on show.

In fact, it was during his studies in Göttingen, under the influence of W. Voigt, that Laue found his vocation: theoretical physics. This did not please his father, who thought that an academic position would not provide any opportunity to rise in society (Zeitz 2006)! The person who maybe had the strongest influence on him was M. Planck. When Laue arrived in Berlin, he 'hurried to Planck's lectures on theoretical optics' (Laue 1952*b*). He also attended O. Lummer's[18] course on interference phenomena, especially in ruled gratings, and in plane-parallel plates. Laue wrote that it was in these lectures that he 'acquired the feeling for optics which became so very useful to [him] later on'. It is also in reference to these lectures that Planck gave him 'Theory of interference phenomena in plane-parallel plates' as a topic for his thesis, which he defended in July 1903.

After his thesis, Laue went back to Göttingen where he took courses on the 'Theory of electrons' and 'Geometrical optics'. He also took the examination to teach in secondary schools. He had good marks in chemistry, but did poorly in mineralogy. Laue (1952*b*) recalled

---

[17]M. Laue's father was raised to nobility in June 1913, and, from that date, M. Laue signed M. von Laue.

[18]Otto Lummer, born 17 July 1860, died 5 July 1925, was a German physicist who was the first to observe interference effects in glass plane-parallel plates.

Fig. 6.8   Max von Laue. ©The Nobel Foundation.

Max von Laue: born 9 October 1879 in Pfaffendorf, now part of Koblenz, Germany, the son of a high-ranking civil servant of the Prussian military administration who was raised to hereditary nobility in 1913; died in a car accident 24 April 1960 in Berlin, Germany. He received his early education in the Friedrich-Wilhelms-Gymnasium in Posen (now Poznań, Poland), the Wilhelms-Gymnasium in Berlin and in the Protestantische Gymnasium in Strasbourg. After passing his Abitur in Strasbourg at Easter 1898, he served his compulsory year of military service, and was able to attend courses at the University of Strasbourg at the same time. He then resumed his studies of mathematics, physics, and chemistry in Strasbourg. After one year he moved to Göttingen, where he stayed for two years. He then went to Munich for one semester and finally settled in 1902 at the Friedrich-Wilhems University in Berlin, where he began his doctoral work under Max Planck. He obtained his PhD in July 1903 and went back to Göttingen for two years of post-doctoral work. At the end, he took the examination required for high-school teaching, Lehramtsexamen. In 1905, Laue was offered a position of assistant to M. Planck, and in 1906 obtained his habilitation, on the 'entropy of interfering beams'. That same year, he was appointed Privatdozent. In 1909 he went to Munich, in Sommerfeld's Institute, as Privatdozent. In the summer of 1912, he was appointed to the chair of Theoretical Physics of the University of Zürich, then in 1914 Professor at the University of Frankfurt am Main, and, finally, in 1919, Professor of Theoretical Physics at the University of Berlin, where he remained until his retirement in 1943. In 1922, Laue was appointed Deputy Director of the Kaiser-Wilhelm Institut für Physik (KWIP), and, in 1933, its Director. In April 1945, French troops entered the city of Hechingen where the KWIP had been moved in 1943 for security reasons. Two days later, the Americans arrived and M. von Laue was arrested. He was interned for six months at Farm Hall estate, near Cambridge, England, along with ten other German physicists, including W. Gerlach, O. Hahn, W. Heisenberg, and C. F. von Weizsäcker (July 1945–January 1946). From 1946 to 1951, he was Professor of Physics at the University of Göttingen and Acting Director of the Max Planck Institute for Physics. In 1951 he became Director of the Fritz-Haber-Institut for Physical Chemistry of the Max-Planck Gesellschaft in Berlin-Dahlem, where he remained until 1958. M. von Laue was awarded the Nobel Prize in Physics 1914 for his 'discovery of the diffraction of X-rays by crystals'. In 1920 he was elected a member of the Prussian Academy of Sciences and, in 1926, a member of the Deutsche Akademie der Naturforscher Leopoldina. When the International Union of Crystallography was founded in 1948, he was elected Honorary President.

MAIN PUBLICATIONS

1911 *Das Relativitätsprinzip.*

1913 *Wellenoptik,* in *Enzyklopädie der Mathematischen Wissenschaften.*

1921 *Die Relativitätstheorie.*

1923 *Die Interferenz der Röntgenstrahlen.* With W. Friedrich, P. Knipping and F. Tank.

1941 *Röntgenstrahlinterferenzen.*

1944 *Materiewellen und ihre Interferenzen.*

1946 *Geschichte der Physik.*

1947 *Theorie der Supraleitung.* With C. Hermann.

1952 *Mein physikalischer Werdegang.*

1960 *Röntgenstrahlinterferenzen.* (third edition)

the amusement of the examiner, a geologist, when Laue's obvious ignorance became evident. After two years, M. Planck, who called Laue his 'cherished student' (*Lieblingsschüler*, quoted by Zeitz 2006) obtained for him an assistantship in Berlin. In 1906, Laue's interest shifted to the newly published theory of relativity and, in 1907, he met Einstein during a visit to the patent office in Bern, Switzerland. In 1910, Laue wrote the first book on relativity.

In 1907, Sommerfeld offered Laue a position as a Privatdozent in Munich. Laue preferred to finish the work he had undertaken, but accepted the offer in 1909. In 1910, Sommerfeld, who was Editor of Volume 5, Physics, of the *Encyclopedia of Mathematical Sciences*, asked Laue to write the chapter on 'Wave optics', which was to include a section on diffraction by gratings. Lord Rayleigh (1874) had shown that the characteristic of a grating is the periodic repetition of its elements and not the nature of those elements. Laue elaborated on Rayleigh's work, and arrived at an equation for the position of the diffraction maxima, which he extended to the case of cross-gratings; in the latter case two equations had to be formulated. In his 1962 interview (see footnote 12), Debye recalled that Laue also had in mind to 'put gratings in space'. Laue's contribution to the *Encyclopedia* was finally published in 1915,[19] and included a section on X-ray diffraction.

For details on M. von Laue's work and life, see, among others, Borrmann (1959a), Ewald (1960, 1979b), Hildebrandt (1987),[20] Hoffmann (2010),[21] Laue (1952b), and Zeitz (2006).

6.4 Ewald's question to Laue, January 1912

As noted by Michael Eckert (2012), there is no direct archival record of the events that led to Friedrich and Kniping's experiments and to the discovery of X-ray diffraction. One can therefore only rely on the *a posteriori* accounts by the different protagonists: Ewald 1932, 1948 (see footnote 2), 1960, 1962a, 1979a and b, Ewald in Authier 2003, Ewald 1962 interview (see footnote 6); Laue 1915, 1937, 1948, 1952a and b; Friedrich 1922, 1949. These accounts are reviewed in Hildebrandt (1993). Several historians of science have, understandably, questioned the objectivity of the accounts of the protagonists of the discovery; for instance, Forman (1969) and Eckert (2012a and b). Their arguments are discussed in Section 6.12. One should therefore try to reconstruct the events according to the logic of the scientific knowledge of the time. This is not easy, as the present-day commentator must try to forget the scientific developments that took place after the discovery.

Ewald's and Laue's accounts of their discussion are essentially in agreement. For Laue it took place in February, for Ewald in late January. Given that Ewald's defence took place in mid-February, late January seems more likely. Laue only speaks of their discussion at his home, but, according to Ewald's relation, he first asked for an appointment with Laue, 'who suggested they meet the next day in the Institute and discuss it later at his home, in the Bismarckstrasse, before and after supper. They met as arranged and took a detour through the Englischen Garten' (Laue 1962a). On the way, Ewald began telling Laue of the general problem he had been working on.

---

[19]M. von Laue (1915). Wellenoptik. *Enzyklopädie der Math. Wiss.* Volume 5–3, 359–487.

[20]G. Hildebrandt (1987). *Max von Laue, der 'Ritter ohne Furcht und Tadel'*. In: *Berlinische Lebensbilder*, Vol. 1, *Naturwissenschaftler*. Eds W. Treue and G. Hildebrandt. Colloqu. Verl., Berlin.

[21]D. Hoffmann (2010). Nicht nur ein Kopf, sondern auch ein Kerl. *Physik Journal*, **9**, 39–44.

Fig. 6.9 *Left*: Physicists in Munich. Far left: M. Laue, second from left: P. Epstein, far right: P. P. Ewald, second from right: P. Koch. End of the table, with a black coat, W. Friedrich. Photo Deutsches Museum, Munich. *Right*: Plaque commemorating Laue's discovery on old Sommerfeld's Physics Institute, Ludwigs-Maximilian University, Munich. Photo by the author.

To his astonishment, Laue did not know about his work. Ewald explained that he assumed the crystal to be a regular array of resonators.

'How do you know that?' asked Laue. Ewald answered that it was the general belief of crystallographers and that it was the subject of an elaborate mathematical theory.

Laue then asked 'what are the particles which are regularly arranged? Are they atoms?' (Ewald 1948, see footnote 2). 'It is not known and it might also be groups of atoms'.

The next question was, 'what are the distances between the particles?' 'They could probably be calculated from the density of the material, but they would be one thousandth or less of the wavelength of light for which he [Ewald] had attempted to calculate the dispersion'.

On the rest of the walk, Ewald explained the main steps of his development, which he detailed at Laue's home, showing him the main equations. Laue listened absentmindedly, but when Ewald came to equation (7) of his thesis (equation 6.1, Section 6.2), Laue pricked up his ears and asked repeatedly: 'what would happen if the wavelength was very much shorter than the wavelength of light'? Ewald replied that this expression represented the conversion of the field from spherical waves into plane waves, and that it would still be valid for very short wavelengths, but that he had to finish his thesis and did not have the time to work that situation out. Ewald never got the answer to his own question about the validity of his treatment and he went home somewhat disappointed by Laue's lack of interest for his troubles!

Whether Ewald's tale was embellished is quite possible. But what remains is what Laue got from the discussion: 1) crystals consist of a regular arrangement of identical particles; 2) the distances between these particles are very much smaller than the wavelengths of visible light; 3) Ewald's coherent summation of the wavelets emitted by the resonators when submitted to an exciting field is still valid for wavelengths much smaller than those of visible light (implied: such as X-ray wavelengths). This is when Laue's intuition came to him, which he attributed to his 'optical feeling' (Laue 1937).

Another version of the encounter has been given by Sommerfeld, in which he was present, along with Ewald and Laue.[22] In a letter to Sommerfeld, Ewald diplomatically wrote that his memory was too bad to ascertain the exact circumstances of the meeting.[23] According to J. W. M. DuMond, P. Epstein, another member of Sommerfeld's group, used to tell his students at CalTech that it was while overhearing a conversation between Ewald and Sommerfeld about Ewald's thesis that Laue had his intuition[24]—but this is second-hand information. In their later accounts, neither Ewald nor Laue ever alluded to Sommerfeld having been present.

6.5   Laue's intuition, January 1912

We now turn to Laue's accounts. Ewald had used the German word *Raumgitter* to describe the periodic distribution of the dipoles in a crystal, and this reminded Laue of diffraction gratings (*Beugungsgitter*). Interference occurs if the period of the grating and the wavelength are close: 'the ratio of wavelengths and lattice constants was extremely favourable if X-rays were to be transmitted through a crystal' (Laue, Nobel lecture), but 'one should irradiate crystals with shorter waves, i.e. X-rays. If the atoms really formed a lattice this should produce interference phenomena similar to the light interferences in optical gratings' (Laue 1952*b*). In his Nobel lecture, Laue added, 'I immediately told Ewald that I anticipated the occurrence of interference phenomena with X-rays', but that was never mentioned by Ewald.

Laue's intuition was a stroke of genius! Without it the work of the Braggs and X-ray analysis, Moseley's pioneering work on X-ray spectroscopy and the experimental study of atomic structure would have been delayed.

Röntgen could have had the idea; in fact, he did shoot X-rays at quartz and calcite crystals. Had he had more powerful X-ray sources, he might have been the one to observe X-ray diffraction.

As pointed out by Ewald (1948, see footnote 2) in his speech at the First General Assembly of the International Union of Crystallography, 'Sommerfeld had all the trumps in his hand, he was probably the most ardent defender of the wave nature of X-rays on the Continent, and it was he who had suggested the topic of Ewald's thesis! Ewald himself could have had the idea, he had already derived the appropriate equations, but he was too engrossed in the finalization of his thesis'. It took him but a few minutes to apply his equations to the X-ray case when he heard the news of the discovery.

[22]A. Sommerfeld (1924). *Dtsch Lit. Ztg.*, **45**, 458–459. Quoted in Eckert 2012.

[23]Letter from Ewald to Sommerfeld, 28 April 1924. Deutsches Museum, Munich, NL 89/007.

[24]J. W. M. DuMond (1974). Paul Sophus Epstein. *Biographical Memoirs*. Nat. Acad. Sci. Washington, **55**, 130–54.

M. Born[25] and T. von Kármán also could have had the idea. Born later recalled that when he had read about Laue's discovery he had not been surprised at all and had been ashamed not to have thought of it himself: when one has studied the propagation of oscillations in a crystal and one's topic is optics, it is a pity not to have come across this idea.[26]

J. Stark and G. Wendt (1912a and b) could have had it when they bombarded various crystal slabs with canal rays (positive ions).

But none of them had Laue's idea.

On the other hand, Laue's success was also a stroke of luck. He had no definite idea as to what the interference pattern would look like, and he conceived his experiment with a wrong premise. He did not think interference phenomena could arise from the white spectrum emitted by the X-ray bulb: 'since we thought it would have something to do with fluorescence radiation, we chose a crystal with a heavy metal in order to produce an intense homogeneous secondary radiation [characteristic radiation]' (Friedrich *et al.* 1912, p. 314). Laue thought that, by analogy with the case of a grating illuminated with white light, blackening of the photographic plate would result: 'in a pulse, there is a continuum of wavelengths within a certain range', he wrote to W. H. Bragg on 15 October 1912 (Section 6.10). 'As far as I can see, the diffraction points due to this continuous spectrum will lie close to each other in such a way that the photographic plate will be evenly blackened'. Significantly, what Laue had not foreseen, and did not understand when he interpreted the diagrams obtained by Friedrich and Knipping, *was that the crystal selected the appropriate wavelengths from the white spectrum.* In the aforementioned letter to Bragg, and which was a reply to a previous letter from Bragg, he wrote: 'You write in your letter that the fact that the interference diagram does not depend on the nature of the anticathode material can be deduced from my equations in which the coordinates of a spot are given in terms of three indices. I can not agree with that statement because these indices depend on the wavelength'. Bragg was correct: the position of the diffraction spots depends only on the orientation of the crystal, and they only occur if the appropriate wavelengths are present in the white spectrum emitted by the X-ray bulb.

Laue wrongly attributed the origin of the diffraction spots to characteristic radiation. In the theoretical part of the published articles by Friedrich *et al.* (1912, p. 311), he wonders whether this characteristic radiation was generated inside the crystal by fluorescence or whether it preexisted in the primary beam besides the impulses constituting the *Bremsstrahlung* described by Sommerfeld. He persisted for a long time in this misconception, as is clearly shown not only in his letter to W. H. Bragg, but also by his inaugural lecture at the University of Zürich on 14 December 1912:[27] the reflected beam 'is so narrow that it is to be explained by monochromatic radiation. But, to this time, it is still an open question as to where this monochromatic radiation comes from'. He changed his mind soon after, however. In the addenda to the reproduction in *Ann. Phys.* of his 1912 paper (Laue 1913a), submitted in March 1913, he wrote that 'the

---

[25]See footnote 20, Section 2.3.

[26]M. Born (1959). Erinnerungen an Max von Laues Entdeckung der Beugung von Röntgenstrahlen durch Kristalle. *Zeit. Krist.* **113**, 1–3. Interview of M. Born by P. P. Ewald in June 1960 at his home in Germany. Niels Bohr Library & Archives, American Institute of Physics, College Park, MD, USA, <http://www.aip.org/history/ohilist/4522_1.html>.

[27]M. Laue (1913). Die Wellentheorie der Röntgenstrahlen. *Zeit. Himmel und Erde.* **25**, 433–438. Reproduced in M. von Laue, *Gesammelte Schriften un Vorträge.* (1961). Vieweg & Sohn, Braunschweig, pp. 218–224.

lattice elements' (*Gitterelemente*) have chosen a finite number of wavelengths out of the infinite number of wavelengths in the incident beam', where by 'lattice elements' he meant the group of different atoms in the unit cell. As he made no mention of reflecting planes, it is not known whether this change was under the influence of W. L. Bragg's communication to the Cambridge Philosophical Society (published 13 February 1913). Laue's turnabout was underlined by Friedrich (1913*b*) who spoke of the 'earlier form' of Laue's theory in his talk at the September 1913 meeting of German scientists in Vienna (Section 7.10.2).

W. L. Bragg, in his Cambridge talk on 11 November 1912, considered Laue's estimation of the wavelengths giving rise to the diffraction spots 'unsatisfactory', and then gave the correct interpretation. Laue's misconception was judged ruthlessly, with the insolence of youth, by H. G. J. Moseley. He wrote, in a letter to his mother after a talk he had given in the presence of W. H. Bragg in November 1912 (Section 7.8.2): 'I was talking chiefly about the new German experiments of passing rays through crystals. The men who did the work entirely failed to understand what it meant, and gave an explanation which was obviously wrong. After much hard work Darwin and I found the real meaning of the experiments' (quoted in Heilbron 1974). Laue himself later admitted that his first interpretation had not been correct (Sections 7.10.2 and 7.10.3). Ewald (1962*a*) noted that 'the persistence of [Laue's] misapprehension led to speculations along a wrong line'. Similar remarks can be found, for instance, in Bijvoet *et al.* (1952), Ewald (1979*b*), and in Phillips (1979), but in general a tactful silence was cast on Laue's misjudgement.

What was then exactly on Laue's mind when he had his flash of intuition?

1) Laue had shown that, for cross-gratings, the directions of the maxima satisfy two equations, and he guessed that in the case of a three-dimensional grating one simply had to add a third equation. Debye's recollections were that he had already thought of that before (Section 6.3).

2) Barkla's paper in the November 1911 issue of *Philosophical Magazine* on the 'The spectra of the fluorescent Röntgen radiations' (Barkla 1911), and which is the paper referred to by Friedrich *et al.* (1912), clearly implies that the secondary radiation is wave-like in nature.

3) Laue's reasoning must have followed directly Ewald's derivation: the radiation emitted by the regularly arranged resonators is expected to produce an interference pattern if the wavelength is close to the distance between resonators. If the exciting radiation is Röntgen radiation, the radiation emitted by the resonators will produce a diffraction diagram. This idea is in itself perfectly sound; we *now* know that it corresponds to the interference of characteristic radiation emitted by lattice sources, and that the effect is in reality very difficult to detect. It is at the origin of the Kossel lines, and was first observed for X-rays by Borrmann (1935) and explained theoretically by Laue (1935), see Section 9.7.1. It certainly could not, however, have been seen with Friedrich's and Knipping's set-up. When he had his intuition, however, Laue did not elaborate on his idea, he simply believed that interference of some sort should occur.

What Friedrich and Knipping observed was, of course, *scattered* radiation chosen by the crystal lattice from the white spectrum of the anticathode, as shown by W. L. Bragg (1913*a*).

6.6   W. Friedrich and P. Knipping's experiment: April–May 1912

6.6.1   Preparing for the experiment

W. Friedrich (Fig. 6.10)[28] first heard about Laue's intuition in a discussion that took place during a meeting of the Institute's colloquium (Friedrich 1922, 1949), and he expressed at once his willingness to carry out a relevant test (Laue Nobel lecture). He was a very good experimentalist, and had just obtained his PhD on the azimuthal intensity distribution of X-rays emitted by a platinum anticathode (Friedrich 1912).

Laue's idea was the subject of lively debates among the younger physicists at Café Lutz. They all believed that X-rays were electromagnetic radiation,[29] and the conviction that the experiment would be a success finally prevailed. The matter was also discussed by Laue and the senior physicists, Sommerfeld, Wien, and others ('the acknowledged masters of our science'; Laue, Nobel lecture), during their usual skiing vacations in Mittenwald. This was at Easter, which fell on 7 April that year.

Fig. 6.10   Walter Friedrich, from an album presented to Sommerfeld on the occasion of his 50th birthday in December 1918. Archives of the Deutsches Museum, Munich, NL89,056.

Walter Friedrich: born 23 December 1883 in Magdeburg, Germany, the son of an engineer; died 16 October 1968 in Berlin, Germany, was a German biophysicist. Already during his schooldays at the Stephaneum Gymnasium in Aschersleben (Harz), he had the occasion to use an X-ray apparatus which had been given to him by his father, and with which he took X-ray pictures for the local clinic. After his Abitur in 1905, he studied Physics in Geneva and Munich. He obtained his PhD in Röntgen's Institute in 1911, and from 1911 to 1914, he was Sommerfeld's assistant. In 1914 he joined the University Clinic in Freiburg, where he devoted himself to the medical applications of X-rays and radium. He became Privatdozent in 1917 and professor in 1921. In 1922 he was appointed Professor of Medical Physics in Berlin and Director of the Institute for Radiation Research. There, he laid the bases of radium therapy and radium dosimetry. In 1929 he was elected Dean of the Faculty of Medicine. From 1949 to 1951 he was Rector of the Humboldt University in Berlin, and, from 1951 to 1956, President of the Academy of Sciences of the German Democratic Republic.

MAIN PUBLICATIONS

1918 *Physikalische und biologische Grundlagen der Strahlentherapie.*

1923 *Die Interferenz der Röntgenstrahlen.* With M. von Laue, P. Knipping and F. Tank.

1931 *Die methodischen Grundlagen beim Arbeiten mit spektral zerlegtem Licht.*

---

[28]The author is grateful to Michael Eckert, Deutsches Museum, Munich, for a copy of Friedrich's portrait.

[29]Interview of P. Debye, see footnote 12.

The various accounts of the preparation of the experiment differ in the details. What seems clear is that Sommerfeld was not in favour of it. Neither he nor Wien thought that the X-rays emitted by the resonators would be coherent. One reason was thermal motion in the crystal, which might impair the regularity of the grating (see Section 6.12). Thermal agitation was a popular topic at the time. Debye presented his calculation of specific heat on 9 March 1912,[30] and Born and Kármán's 1912 paper on lattice vibrations appeared on 30 March. Röntgen was also skeptical of the proposed experiment. He had himself unsuccessfully let X-rays fall on various crystals.

Furthermore, Sommerfeld was reluctant to let Friedrich do an experiment that he did not think would be successful, particularly since he really wanted Friedrich to make some tests on the directional emission of X-rays. The situation was certainly aggravated by the strained relations between Sommerfeld and Laue. They are revealed in a letter from Laue to Sommerfeld on 3 August 1920,[31] quoted in Forman (1969) and Eckert (2012a and b). In that letter, Laue recalled that 'he has often thought back to his Munich days with a certain bitterness', and that 'he went through many unpleasant moments'. One incident he resented particularly was when Sommerfeld did not invite him to celebrate the discovery of X-ray diffraction with Friedrich, Knipping and the other young members of the group. He admitted that 'he was not correct with [Sommerfeld], in particular shortly after [he] moved to Munich'. 'You well knew, however', Laue told Sommerfeld, 'in which state of mind I was then'. Laue concluded the letter by suggesting to let bygones be bygones. It is true to say that, according to his biographer, K. Zeitz (2006), Laue was a not very affable person who defended his viewpoint in a stubborn and obstinate way in discussions. The frictions between Sommerfeld and Laue were in general diplomatically left out in later accounts, but are briefly mentioned in Ewald's 1962 interview (see footnote 6). Anyhow, 'a certain amount of diplomacy was necessary' before the experiments were finally allowed to start, 'according to my plan' (Laue Nobel lecture).

Friedrich became hesitant, but Laue insisted. One of Röntgen's PhD students, P. Knipping (Fig. 6.11), dubbed the 'watchmaker' because of his experimental skill, who had just started, or was about to start, a study of the transmission of X-rays through metals (Knipping 1913), offered to help. From then on, things moved smoothly (Letter from Laue to Ewald, 1924, quoted in Forman 1969).

6.6.2   The experimental set-up

Friedrich and Knipping put together the necessary X-ray equipment in the cellar of Sommerfeld's Institute while Sommerfeld was away, and the first tests began 'a little behind the backs of the particular bosses'. As we have seen, they expected the diffracted rays to be secondary radiation, and Friedrich, from his own experience, was prepared for long exposure times. Luckily, they had at their disposal a big induction coil and a powerful X-ray bulb (Friedrich 1922). It is not clear when the first tests began. In some accounts, March is mentioned (Laue 1948), but it is more likely that it was in April. The first attempt was negative, because the photographic plate was placed in an unfavourable (*ungünstig*) location, parallel to the primary beam (Friedrich 1922), 'so that it could catch rays which were deflected at ninety degrees'

[30] See footnote 28, Chapter 7.   [31] Deutsches Museum, Munich, HS 1977–28A, 197.

Fig. 6.11   Paul Knipping. After Ewald (1979*b*).

Paul Knipping: born 20 May 1883 in Neuwied am Rhein, Germany, the son of a medical service councillor, died 26 October 1935 in a traffic accident, was a German physicist. He studied physics in Heidelberg and Munich. As a PhD student in Röntgen's Institute, he took part with W. Friedrich in the experiment leading to the discovery of X-ray diffraction. He obtained his PhD in 1913 on 'Influence of its past history on various properties of lead'. After working briefly with the Siemens Laboratories in Berlin, he transferred to the Kaiser-Wilhelm-Institut für Physikalische und Elektrochemie in Berlin-Dahlem, shortly after the outbreak of the war in 1914, where he worked with Fritz Haber. He obtained his habilitation in 1924, and was appointed the same year Privatdozent at the Technical University in Darmstadt. He became professor in 1928, and in 1929 he founded an Institute for X-ray physics and X-ray techniques.

MAIN PUBLICATIONS
1923 *Die Interferenz der Röntgenstrahlen.* With
    M. von Laue, W. Friedrich and F. Tank.

(Laue 1948), again because they were looking for secondary radiation. They then put one photographic plate behind the crystal, perpendicular to the primary beam, and several others in various locations (Fig. 6.12, *Left*).

Laue (1952*b*) recalls that success came with the second test, but, according to other accounts, it was not so rapid. Friedrich (1922) described his unforgettable experience when, one evening, alone in the darkroom, he developed the photographic plate, and he saw the traces of the deflected beams appear! Early next morning, the first thing he did was to show the plate to P. Knipping. Friedrich (1922) recalls: 'we then hurried to tell Laue and my Chief [Sommerfeld]'. Laue (1948) has a more picturesque story: that day, he 'joined the usual group for coffee in the garden of Café Lutz'. P. Koch, P. Epstein, E. Wagner, and Lenz were there, but W. Friedrich and P. Knipping were missing. 'But an unusual atmosphere prevailed. Instead of conversing as usual, each one silently read a newspaper ... One of the company merely made an obscure remark, incomprehensible to me. Shortly thereafter another man made a similarly incomprehensible remark, and so it went, right around the table. I grew more and more mystified. But finally it dawned on me what had happened: the interference experiment had had a positive result'! Laue then rushed to the Institute where he was shown the first interference diagram (Fig. 6.12, *Right*, or a similar one).

Sommerfeld was enthusiastic, and, on 4 May 1912, took advantage of the monthly meeting of the mathematical–physical class of the Bavarian Academy of Sciences to deposit a sealed note on Friedrich, Knipping, and Laue's behalf, describing their set-up and the principle of their experiment, thus securing priority for them (Fig. 6.13). The first successful experiment is dated 21 April on the document. From then on, Sommerfeld gave Friedrich and Knipping all the help they needed, and Friedrich was excused from work on *Bremsstrahlung*. Sommerfeld later estimated that this experimental success was more important than his and his students

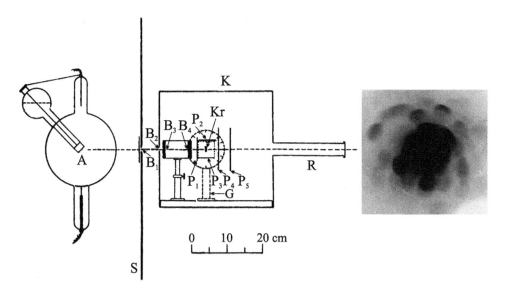

Fig. 6.12   *Left*: Set-up for Friedrich and Knipping's experiment. *A*: anticathode; *S*: lead screen; B_1 to B_4: slits; K_r: crystal; *G*: goniometer; P_1 to P_5: photographic plates (P_4 and P_5 showing diffraction spots); *K*: lead box. *R*: cathetometer. Distance anticathode-crystal: 350 mm. Distance crystal-P_4: 35 mm. Distance crystal-P_5: 70 mm. After Friedrich *et al.* (1912). *Right*: One of the first pictures taken with a copper sulphate crystal. Photo Deutsches Museum, Munich.

achievements in theoretical physics: 'The most important scientific event in the history of the Institute was Laues discovery in 1912, in which Dr Friedrich participated as the Institute's assistant and Dr Knipping [as a doctoral student of Röntgen]', Sommerfeld wrote in 1926 (quoted in Eckert 1999).

The first tests were made with a makeshift apparatus. As soon as positive results were obtained, a more elaborate set-up was built after the same design. It is described in Friedrich *et al.* (1912) and in Friedrich (1913*a*, 1922), and is represented in Fig. 6.12, *Left*. A later version, with a different X-ray tube, is shown in Figs. 6.14 and 6.15.[32]

The primary beam from the X-ray bulb *B* was first limited by hole *H* in a lead screen to eliminate secondary radiation (Fig. 6.15, *Left*), and then narrowly collimated to about 1 mm in diameter by a series of apertures, B_1 to B_4 (Fig. 6.12, *Left*) pierced in thick lead screens; this may have been the key to success. The fine beam impinged on the crystal, K_r, sealed by wax on a crystal holder held on a goniometer, *G* (Figs. 6.12, *Left*, and 6.15, *Right*). The anticathode, the apertures and the crystal were aligned with the telescope of a cathetometer, *R*, at a distance of 3 metres from the tube. Five photographic plates were put around the crystal, P_1, P_4 and P_5, perpendicular to the primary beam, and P_2 and P_3 parallel to it (Fig. 6.12, *Left*). A heavy

[32]The apparatus exhibited at the Deutsches Museum, Munich, Germany, was in fact reconstructed by W. Friedrich in 1921, at the request of the Director and founder of the Deutsches Museum, O. von Miller, from separate items donated by A. Sommerfeld; the X-ray tube is not the original one, which was broken (M. Eckert 2012, Die 'Laue-Apparatur' im Deutschen Museum. *Kultur und Technik*. Heft 3, 40–41).

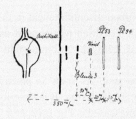

INSTITUT
FÜR THEORET. PHYSIK
MÜNCHEN, UNIVERSITÄT,
LUDWIGSTRASSE 17.

Fig. 6.13 Sealed note deposited by A. Sommerfeld with the Bavarian Academy of Sciences. Photo Deutsches Museum, Munich.

Munich, 4 May 1912. Since 21 April 1912, the undersigned have been involved with investigations on the interference of X-rays propagating through crystals. Interference phenomena take place as a result of the reticular structure of crystals because the lattice constant of the crystal is roughly ten times larger than the conjectured order of magnitude of the X-ray wavelength. As evidence, pictures Nos 53 and 54 are attached [now lost]. Material through which X-rays are transmitted: copper sulphate. Exposure time: 30 minutes. Current in the X-ray bulb: 2 milliampere. Distance between photographic plate and crystal: 30 mm for No 53; 60 mm for No 54. Distance of slit 3 (diameter 1.5 mm) from crystal: 50 mm. Distance between the source of the primary beam in the anticathode and the crystal: 350 mm.

Fig. 6.14   The 'Laue-apparatus' at the Deutsches Museum, Munich, Germany. Photo by the author.

lead box, K, hanging from a tall wooden tripod, visible on Fig. 6.14, could be lowered by means of a counterweight, C, so as to completely shut off the region round the crystal and the photographic plates (Figs. 6.12, *Left*, and 6.15, *Left*). Details of the adjustment are given in Ewald (1962a). Plates P_2 and P_3 showed even light blackening. Plates P_4 and P_5 showed diffraction spots.

Fig. 6.15   *Left*: Improved set-up for Friedrich and Knipping's experiment. *B*: water-cooled X-ray bulb; *H*: hole letting the primary beam pass through the lead screen; *K*: lead box; *C*: counterweight. *Right*: close-up showing the goniometer and the crystal holder. Deutsches Museum, Munich. Photos by the author.

6.6.3   *The first experiment with copper sulphate, 21 April 1912*

The first crystal tested was triclinic copper sulphate pentahydrate, $CuSO_4$, $5H_2O$ (vitriol), chosen because copper would provide, according to Barkla, intense secondary radiation; it was also easy to obtain, and presented a very nice shape. It was put on the crystal holder, oriented at random, but with (110) pinacoid faces roughly perpendicular to the primary beam. Figs. 1.1, *Left*, and 6.12, *Right*, show typical diagrams. One can see large spots, which were not arranged regularly, but which revealed the interference phenomena. Figure 1.1, *Right* shows a picture of copper sulphate obtained with narrower slits.

Friedrich and Knipping then made a number of tests: 1) In order to make sure that the spots were due to the crystal structure, the crystal was replaced by powdered copper sulphate in a small paper box. The spots disappeared. 2) The crystal was shifted parallel to itself, the diagram remained unchanged. 3) The orientation of the crystal with respect to the primary beam was offset by a few degrees, the position of the spots changed. 4) A photographic plate, P_5, was placed at double the distance of plate P_4 from the crystal (Fig. 6.12, *Left*), and the size of the image was doubled, proving that the spots were due to rays fanning out from the crystal. The exposure times ranged from one to several hours.

6.6.4   *Experiments with zinc-blende, sodium chloride, and diamond*

After the positive experiment with copper sulphate, Friedrich, Knipping, and Laue had the new set-up built and made preliminary tests of the secondary radiation emitted by zinc-blende, ZnS. This crystal was probably chosen because zinc, like copper, was expected to emit intense characteristic radiation. They ordered a plane-parallel slab from the firm Steeg & Reuter in Bad Homburg; it was 10×10 mm and 0.5 mm thick, cut in a good-quality crystal parallel to a cubic face, (001). They oriented this plate exactly perpendicular to the primary beam, and obtained

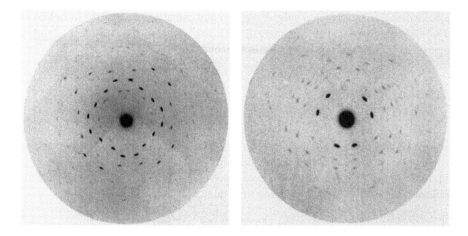

Fig. 6.16 *Left*: Laue photograph; zinc-blende (ZnS); primary beam normal to (001). After Friedrich *et al.* (1912).
Right: Laue photograph; zinc-blende (ZnS); primary beam normal to (111). After Friedrich *et al.* (1912).

a picture symmetric with respect to the four-fold axis and two symmetry planes parallel to the axis (Fig. 6.16, *Left*). The symmetry of the picture was that of the holohedry of the cubic system, $m\bar{3}m$, while in fact zinc-blende's point group is $\bar{4}3m$. The authors concluded that 1) the picture represented a beautiful proof of the existence of crystal lattices, and 2) that it was due to no other property than the lattice itself. In the version of the paper reproduced the following year in *Ann. Phys.* (Friedrich *et al.* 1913), Friedrich added a footnote saying that this latter result was contradicted by experiment on other crystals such as pyrite, FeS_2, and hauerite, MnS_2, for which the Laue patterns did exhibit hemihedral symmetry.

Next, Friedrich and Knipping took a picture with the primary beam normal to a face of the octahedron, (111), and obtained a picture with a three-fold axis (Fig. 6.16, *Right*). They also took a picture with the primary beam normal to a face of the dodecahedron, (110), and obtained a picture with a two-fold axis. In a series of further experiments, Friedrich and Knipping checked that the diagram rotated when the crystal was rotated round the primary beam, and that the diagram lost its symmetry if the crystal axis was offset by a few degrees from the direction of the primary beam.

Crystals of sodium chloride and diamond were also studied, and they provided similar diagrams. The case of diamond was particularly puzzling, since Barkla had written that characteristic radiation had not yet been observed from carbon. No spots had been expected, but they were present, not only on the photographic plate perpendicular to the beam, but also on the photographic plates P_2 and P_3 parallel to the beam. No explanation was offered.

The results were published in two papers—one submitted on 8 June 1912, with an experimental part by Friedrich and Knipping and a theoretical part by Laue (Friedrich *et al.* 1912),[33]

[33] An English translation has been provided by J. Stezowski (1981). *Interference Phenomena for X-rays*, by W. Friedrich, P. Knipping, and M. Laue. Benchmark Papers in Physical Chemistry and Chemical Physics. **4**. Structural Crystallography in Chemistry and Biology. Edited by J. P. Glusker, pp. 23–39. Hutchinson Ross Publishing Company. Stroudsburg, Pennsylvania / Woods Hole, Massachusetts.

and a second one, submitted on 6 July 1912, with a quantitative proof of the theory by Laue (1912). A bronze plaque commemorating Laue's discovery can be seen on the outer wall of Sommerfeld's old basement laboratory in the courtyard of the Physics Institute, Ludwigs-Maximilian University, Munich (Fig. 6.9, *Right*).

6.6.5 *Theoretical part*

In the theoretical part of the article by Friedrich *et al.* (1912), Laue calculated the intensity of the wave emitted by a three-dimensional arrangement of atoms, assuming that the amplitude incident on each atom was the same, thus neglecting the dynamical effects taken into account by Ewald. In the two-dimensional case, this intensity is the product of two interference functions. In the three-dimensional case, it is the product of three such functions:

$$I \propto \frac{\sin^2 \pi M A}{\sin^2 \pi A} \frac{\sin^2 \pi N B}{\sin^2 \pi B} \frac{\sin^2 \pi P C}{\sin^2 \pi C} \tag{6.3}$$

where the crystal is assumed to be a parallelepiped of sides $M\mathbf{a}$, $N\mathbf{b}$, $P\mathbf{c}$, the unit cell of the lattice being defined by the three vectors, $\mathbf{a}$, $\mathbf{b}$, $\mathbf{c}$, and

$$A = \mathbf{a} \cdot \frac{\mathbf{s} - \mathbf{s_0}}{\lambda}; \quad B = \mathbf{b} \cdot \frac{\mathbf{s} - \mathbf{s_0}}{\lambda}; \quad C = \mathbf{c} \cdot \frac{\mathbf{s} - \mathbf{s_0}}{\lambda} \tag{6.4}$$

$\mathbf{s}$ is a unit vector in the diffracted direction and $\mathbf{s_0}$ a unit vector in the direction of the primary beam, For simplification, the notations used by Laue in his talk at the October 1913 Solvay Congress (Laue 1913*b*) and in Laue (1914) have been used. Equation 6.3 is the basis of the geometrical, or kinematical, theory of diffraction.

The intensity is maximum for directions of the incident and diffracted beams such that

$$A = h_1; \quad B = h_2; \quad C = h_3. \tag{6.5}$$

where h_1, h_2, h_3 are integers.

Laue then considered the case of a cubic crystal, such as zinc-blende. The three vectors $\mathbf{a}$, $\mathbf{b}$, $\mathbf{c}$ are mutually orthogonal and have the same length, a. He took vector $\mathbf{c}$ along direction [001] of the crystal and parallel to the primary beam, that is to $\mathbf{s_0}$. The expressions 6.4 were simplified:

$$A = a\alpha/\lambda; \quad B = a\beta/\lambda; \quad C = a(1 - \gamma)/\lambda. \tag{6.6}$$

where α, β, and γ are the direction cosines of the diffracted beam with respect to the three axes. Combining 6.5 and 6.6, Laue obtained:

$$\alpha = h_1\frac{\lambda}{a}; \quad \beta = h_2\frac{\lambda}{a}; \quad 1 - \gamma = h_3\frac{\lambda}{a}. \tag{6.7}$$

These are the three conditions satisfied by the direction of the diffracted beam, now known as the three Laue relations. These directions lie on cones with [100], [010] and [001] as axes, respectively. The intersections of these cones with the photographic plates are hyperbolæ for the first two and circles for the third one.

In the second paper, Laue (1912) proceeded to index the diffraction spots on the diagram. From relations 6.7 and since $\alpha^2 + \beta^2 + \gamma^2 = 1$, it is easy to show that

$$\frac{\lambda}{a} = \frac{2h_3}{h_1^2 + h_2^2 + h_3^2} \tag{6.8}$$

and that

$$\alpha = \frac{2h_1 h_3}{h_1^2 + h_2^2 + h_3^2}; \quad \beta = \frac{2h_2 h_3}{h_1^2 + h_2^2 + h_3^2}; \quad \gamma = \frac{h_1^2 + h_2^2 - h_3^2}{h_1^2 + h_2^2 + h_3^2}; \tag{6.9}$$

These relations correspond to equations '13)' and '13 a)' of Laue (1912), respectively. The direction cosines α, β, γ are deduced from the coordinates of the spots on the photographs. The indices h_1, h_2, h_3 and the wavelength are then easily determined from the value of the parameter a of the crystal, which is obtained from the density of the crystal and the molecular weight of its constituents. Laue explained in that way that the spots observed on Fig. 6.16 were due to five different wavelengths. He made, however, an error because he assumed the zinc and sulphur atoms to be at the corners of a simple cubic lattice, and the wavelengths he found were incorrect.

Had Laue interpreted the indices h_1, h_2, h_3 as the Miller indices of a family of lattice planes, he could have found Bragg's law. The angle between the diffracted, **s**, and the incident, **s$_0$**, directions is twice the Bragg angle, 2θ. It is straightforward to show that

$$1 - \gamma = 2\sin^2\theta = \frac{2h_3^2}{h_1^2 + h_2^2 + h_3^2} \tag{6.10}$$

and equation 6.8 becomes

$$\lambda = 2d\sin\theta \tag{6.11}$$

where $d = a/\sqrt{h_1^2 + h_2^2 + h_3^2}$. It is only in April 1913 that Laue recognized the equivalence of equation 6.8 and Bragg's law (Laue 1913b), but without any reference to Bragg's seminal paper (see Section 7.3).

When Laue's 1912 article was republished on March 1913 (Laue 1913a), he appended three notes; in one of these he developed the theory of diffraction by a crystal containing several atoms per unit cell. This was the first introduction of what was to be the structure factor. His attention had been drawn by Tutton's (1912b) paper describing the structure of zinc-blende as suggested by Barlow and Pope (1907). Taking into account the fact that its lattice is face-centred cubic, he modified the value of the lattice parameter of zinc-blende, which resulted in a modification of the values of the wavelengths he had previously calculated.

6.7 The propagation of the news of the discovery and the first reactions

6.7.1 Timeline of the propagation of the news of the discovery and the first reactions during the year 1912

- 4 May 1912. Deposition by A. Sommerfeld of a sealed note at the Bavarian Academy of Sciences (Fig. 6.13).
- May 1912. W. C. Röntgen's and P. von Groth's first impressions, Section 6.7.2.

- 8 June 1912 – Submission of the first paper by W. Friedrich, P. Knipping, and M. Laue: experimental set-up, first diagrams with copper sulphate, diagrams with zinc-blende showing symmetry, expression of the diffracted intensity (geometrical theory), the three Laue equations.
- 10 June 1912 – A. Einstein congratulates M. Laue.
- 14 June 1912 – Talk by M. Laue to the Physikalische Gesellschaft in Berlin. Comment by M. Planck–Section 6.7.3.
- Mid June 1912 – Talk by A. Sommerfeld in Göttingen; Ewald introduces the Ewald sphere and the reciprocal lattice (published in 1913)–Section 6.8.
- 21 June 1912 – Einstein writes to Laue: 'The photograph is so sharp that one would have scarcely have suspected it, due to thermal agitation' (quoted by Hoddeson *et al.* 1992).
- 26 June 1912 – Letter of L. Vegard to W. H. Bragg telling him about Laue's discovery–Section 6.10.
- 6 July 1912 – Submission of Laue's theoretical paper: indexation of the diagrams and estimation of the wavelengths.
- Beginning of August 1912 – The English crystallographer, A. E. H. Tutton, visits Munich–Section 1.1.
- Late August 1912 – Publication of the papers by Friedrich, Knipping and Laue and by Laue in the *Sitzungsberichte der Kgl. Bayer. Akad. der Wiss.*
- 26 August 1912 – J. Stark writes up his interpretation of the Friedrich, Knipping and Laue diagrams as due to corpuscular rays (submitted 1 September 1912 to *Phys. Zeit.*)–Section 6.9.
- October–November 1912 – Exchange of letters between W. H. Bragg and M. Laue–Section 6.10.
- 24 October 1912 – Letter of W. H. Bragg to *Nature*: 'the directions of the secondary pencils are avenues between the crystal atoms'.
- 11 November 1912 – W. L Bragg reads his paper to the Cambridge Philosophical Society, deriving Bragg's law–Section 6.11.
- 24 November 1912 – Letter of G. Wulff to *Z. Kristallogr.* showing that Laue's indices h_1, h_2, h_3 define a crystallographic orientation (Wulff 1913*a*).
- 28 November 1912 – Letter of W. H. Bragg to *Nature* suggesting dual nature, wave and corpuscular, of X-rays–Section 7.1.
- 12 December 1912 – Letter of W. L. Bragg to *Nature* on the 'The specular reflection of X-rays', Section 6.11.
- 19 December 1912 – Letter of C. G. Barkla and G. H. Martyn to *Nature* on the Bragg reflection from sodium chloride crystals–Section 7.1.

6.7.2 *Röntgen and Groth's reactions, May 1912*

W. C. Röntgen came over as soon as he heard about the success of the experiment. 'I can still see Röntgen as he carefully studied these pictures in my laboratory' wrote Sommerfeld later. Friedrich showed him the pictures and the apparatus, and after Röntgen had studied it critically, he declared that he could find no experimental error. Röntgen then sincerely congratulated Friedrich. He was deeply impressed, and agreed that the spots were due to the crystal structure, but expressed his doubts: 'those are not interference phenomena. They look different to me' (Laue 1948). Laue sent Planck a postcard with a picture of the master (Röntgen) looking at the apparatus, still somewhat skeptical and unbelieving (Planck 1937).

Röntgen was nevertheless very helpful. The X-ray tubes of that time were operated with a Ruhmkorff induction coil and with interrupted direct current. For that, a Wehnelt interrupter was used that interrupted the current about a thousand times per second and transmitted vibrations of this frequency into the air and into the university's electric system. This led to many technical problems and induced a lot of nuisances. In particular, the arc-lights illuminating the campus at night or used in the slide projectors emitted a sound corresponding to these vibrations, lasting for five seconds, and then were silent for ten seconds. There was a general outcry and the university authorities demanded an immediate remedy or the suspension of the experiments. The solution came from Röntgen who allowed an electrical line to be drawn across the university courtyard from his Institute which had the appropriate filter coils to suppress the effect (Laue 1948). It was only after the works of W. H. and W. L. Bragg and after E. Wagner and R. Glocker's double crystal experiment in his own Institute (Wagner and Glocker 1913, Section 7.10.2) that Röntgen was finally convinced!

P. von Groth had been regularly consulted by Friedrich, Knipping, and Laue as to which crystal was best suited for their experiment. He was among the first to be informed and he was elated. He was the one who followed with the keenest interest the progress of the experiment, and it is with a 'beaming bliss' that he discovered the first successful pictures of X-ray interference. Laue (1943) remembered thirty years later that Groth had wanted to be assured that the diagrams were a definite proof of the reality of the atoms. He could only reply that it was from his conviction that atoms really existed that he had had the pluck to make the experiment! In an Editor's note appended to an article by Friedrich on the Munich experiment in the January issue of *Zeitschrift für Kristallographie* (Friedrich 1913*a*), Groth expressed the view that the experiment was in agreement with Barlow and Pope's (1907) theory of crystal structures.

6.7.3   M. Laue's talk to the Physikalische Gesellschaft in Berlin, 14 June 1912

M. Laue, who was a member of the German Physical Society, described the discovery of X-ray diffraction during a meeting of the Society in Berlin on 14 June 1912. Max Planck, who considered the discovery as one of the most striking examples of the fruitfulness of the combination of theory and experiment, remembered 25 years later the unforgettable impression left by the talk (Planck 1937): 'we were all in a great state of excitement. When, after the theoretical introduction, Laue showed the first picture obtained with copper sulphate [(Fig. 6.12, *Right*)], with a few odd spots, the audience cast a look full of attention and expectation at the image on the screen, but was not yet fully convinced. However, when the first real Laue diagram [(6.16, *Left*)] appeared, with its ordered diffraction points, an only feebly repressed *ah* went through the assembly. Every one of us felt that a remarkable feat had been achieved, and that a hole had been pierced in an hitherto impenetrable wall'.

Thirty-seven years after the event, another member of the audience, the physicist Ernest Lamla (1888–1986), who was one of the pioneers of the theory of *n*-beam diffraction (Lamla 1939), recalled the memorable 14 June 1912 session of the Physical Society and the intentness that communicated itself to the listeners, and culminated in a storm of applause at the end of the talk (Lamla, E. (1949). Max von Laue. *Naturwiss.* 36, 353–354).

Not all reactions were as positive. In a letter to E. Rutherford dated 29 October 1912, C. G. Barkla wrote 'I have had a copy of Laue's paper for some time and certainly am sceptical of any interference interpretation of the results. A number of features do not point that way ... This in no way affects my absolute confidence of the truth of the wave theory of X-rays'.[34]

[34]Quoted by Forman (1969).

6.8   Ewald introduces the reciprocal lattice and the Ewald construction, mid-June 1912

Around Whitsuntide, mid-June 1912, Sommerfeld gave a talk in Göttingen, where Ewald had been appointed physics assistant to D. Hilbert. Ewald had not yet been informed of the successful result of the experiment, but, as soon as he heard the news, he made the connection with the results of his thesis. On coming home, Ewald looked at equation (7) of his thesis (equation 6.1, Section 6.2), and remembered that he had pointed it out to Laue on that famous day when Laue had had his intuition. He found a geometrical interpretation by introducing a lattice having $\mathbf{a}/a^2$, $\mathbf{b}/b^2$, $\mathbf{c}/c^2$ as elementary vectors, which he called reciprocal lattice (*reziprokes Gitter*). The wave vectors, $\mathbf{K_h}$, of the expansion 6.2 (Section 6.2) of the optical field in plane waves are related to one another by:

$$\mathbf{K_h} = \mathbf{K_o} - \mathbf{h} \tag{6.12}$$

where $\mathbf{K_o}$ is a particular wave vector and $\mathbf{h}$ a reciprocal lattice vector. The only terms in 6.2 which are not negligible are those for which $K_h \approx k_o = 1/\lambda$ and the resonance factors $1/(k_o^2 - K_h^2)$ are very large. Ewald distinguished two cases, 1) the optical case where k_o is much smaller than the unit vectors of the reciprocal lattice and only one term in the sum (6.2) has a non-negligible amplitude; only one wave may propagate in the crystal, and there is no interference; 2) the X-ray case where several terms may have a non-negligible amplitude, and there are several waves propagating in the crystal whose wave vectors are nearly equal to k_o, so that X-ray interference occurs. Ewald represented this situation geometrically by means of the construction that bears his name: there are as many waves with a non-negligible amplitude as there are reciprocal lattice nodes close to the diffraction sphere, or 'Ewald sphere' (Fig. 6.17). In his paper, which was only published after W. L. Bragg's articles, Ewald showed that his

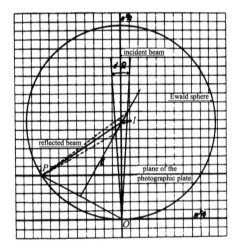

Fig. 6.17   Ewald sphere. O: origin of the reciprocal lattice. P: node of the reciprocal lattice. Diffraction takes place if the direction of the incident beam, IO, is such that the sphere of centre I and radius $1/\lambda$ passes through a reciprocal lattice node, P. There is then a reflected beam along IP. Adapted from Ewald (1913).

construction was equivalent to Bragg's law (Ewald 1913). One of the first applications of the Ewald construction was made by E. Keller (1914) in his analysis of Laue diagrams of diamond.

Ewald's definition of the reciprocal lattice was only valid for an orthorhombic lattice. It is Laue (1914) who generalized it to any type of symmetry, making use of the definitions by J. W. Gibbs (1881). At the time, Laue seems to have been unaware of the 'polar lattice' introduced by A. Bravais (1850, 1851) more than half a century before to facilitate crystallographic calculations (see Section 12.11.1), and which is homothetic to the reciprocal lattice. Laue (1960) acknowledged it later, however. The history of the introduction of the reciprocal lattice in crystallography has been told by Ewald (1936).

The concept of reciprocal lattice was not immediately familiar to crystallographers. It is Ewald (1921) who showed its use in the description of structures.

6.9   J. Stark's 'corpuscular' interpretation of the Laue diagrams

Johannes Stark was the first to react after the publication of the papers by Friedrich *et al.* (1912) and Laue (1912). Within a few days, on 26 August 1912, he submitted to *Physikalische Zeitung* a paper entitled 'Remarks on the scattering and the absorption of β-rays and Röntgen rays' (Stark 1912).

For nearly ten years, Stark had been studying the positive rays emitted by discharge tubes, and propagating in the direction opposite to that of the cathode rays, called canal rays by their discoverer, E. Goldstein (Section 5.1). These positive rays are positive ions such as hydrogen ions (the term 'proton' was coined around 1920). On 13 and 20 April 1912, at about the same time that Friedrich, Knipping and Laue submitted their own paper, Stark and a co-worker, G. Wendt, submitted a paper on the penetration and propagation of hydrogen canal rays through thin plates of crystals such as rock-salt, calcite, fluorite etc., and of glass, which is amorphous, for comparison (Stark and Wendt 1912a, b). For high voltages of acceleration, they observed etch pits or hillocks on the surface of the crystals. They further found that the penetration of the fast ions is easier parallel to the direction of important families of lattice planes such as the cleavage planes.

In his paper prompted by the publication of Laue's discovery, Stark (1912) at first recalled that the investigations of Stark and Wendt had shown that the penetration of the canal rays in crystals was easier along certain remarkable planes than along others. The next step after these studies would be, he suggested, to investigate the penetration of β-rays and Röntgen rays through thin plates, but, being 'at present hindered' from making experiments, he could only make some speculations. Stark had previously interpreted properties of X-rays in terms of light quanta (Section 5.11). The energy of X-rays could be thought of as concentrated into small volumes (*Lichtzelle*). Stark now suggested that β-rays and these packets of X-ray energy could be diffracted and absorbed selectively along directions 'free of atoms' of the crystal ('wells' or *Kristallschächte*), which W. H. Bragg (1912a) called 'crystal avenues' (Fig. 6.18). This effect is called nowadays 'channelling'. In the same paper, Stark claimed that he had already thought of that before, but there is, of course, no proof of that; he was wont to relate other people's discoveries to his own ideas. He then proceeded to explain all the features of the Laue diagrams by means of his hypothesis, but he did not attempt any quantitative analysis of the position of

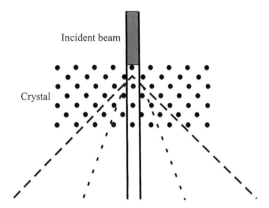

Fig. 6.18   Stark's view of the diffraction of X-rays by a crystal. The broken lines show the propagation of the deflected beams. Redrawn after Stark (1912).

the spots (Stark 1912). As an argument against Stark's corpuscular interpretation, Laue noted in a letter to W. H. Bragg, dated 13 November 1912,[35] that the atom-free directions would not be ordered in a hemihedral crystal like zinc-blende as in a holohedral crystal exhibiting the full symmetry. He also pointed out that the pictures obtained after rotation of the crystal by a few degrees were not compatible with Stark's hypothesis (Laue 1913b).

6.10   The news reaches W. H. Bragg: his first reactions

In March 1909, W. H. Bragg (Fig. 5.5, Section 5.8) and his family arrived in Plymouth, England, after the long trip from Australia. They then settled in Leeds, where Bragg had been appointed a professor at the University. He very rapidly set up a laboratory, and soon visiting researchers began to arrive, among them the Norwegian physicist Lars Vegard (Fig. 6.19, for a biography of L. Vegard, see Schwalbe 2014), who came from Thomson's laboratory in Cambridge. The topic suggested to Vegard by Bragg concerned the polarization of X-rays and the excitation of secondary cathode rays (electrons) by X-rays. It has been seen in Section 5.8 that Bragg had difficulties interpreting the polarization of X-rays in terms of his neutral-pair theory. Vegard (1910) confirmed the existence of polarization in the primary radiation, and showed that polarized X-rays have a great power of exciting high velocity cathode rays.

After his stay in Leeds, L. Vegard joined W. Wien's group in Würzburg for a year. Sometime in June, M. Laue gave a talk and described Friedrich and Knipping's experiments and his theoretical interpretation of them. On 26 June, Vegard wrote Bragg a detailed letter with an account of Laue's talk. The letter can be found in the Bragg archives in the Royal Institution, London (WHB 7A/3):

[35]Bragg archives, Royal Institution, London. WHB 4A/8.

Würzburg, 26/06/1912, Dear Professor Bragg,

During my stay in Germany this last year I have occasionally had the opportunity of discussing the Röntgen problem. The current idea here is that they are ether pulses, and by several occasions I have attempted to put forward the difficulties involved in the wave theory, and which at present no one has been able to overcome by means of mechanically intelligible conceptions.

Recently, however, certain new curious properties of X-rays have been discovered by Dr Laue in Munich. As I thought the matter would interest you, I asked Dr Laue, who gave an account of his discoveries here at Würzburg, to give me a copy of one of his photographs to send to you.

Without entering into any special conception as to the nature of this phenomenon, it may be described in the following way: A narrow pencil of primary X-rays is made to pass through a crystal (b) and then to fall upon a photographic plate (P). Without crystal he gets a single black spot, when the crystal is introduced, however, a most curious scattering of the primary beam takes place. He gets a number of very sharp regularly arranged ray-bundles surrounding the primary beam. Absorption by aluminium has shown that the rays producing the surrounding spots have a penetrating power much greater than that of the main bulk of the primary X-radiation and very much greater than that of the radiation characteristic of the anticathode. The effect is obtained by crystals of *Zinkblende* and copper sulphate. The crystals are cut as plane-parallel plates with the planes perpendicular to the principal symmetry axis of the crystals, and the rays must pass as exactly as possible along this axis. The plates maybe many mm thick.

Regarding the explanation, Laue thinks that the effect is due to diffraction of the Röntgen rays by the regular structure of the crystals which should form a sort of grating with a grating constant of the order of 10^{-8} cm corresponding to the supposed wavelength of Röntgen rays. He is, however, at present unable to explain the phenomenon in its details, and there are several difficulties from the diffraction point of view. Let me call attention to the following:

1) According to Laue the diffraction in a grating with regularities in three dimensions is most complicated and there is in such a grating a very little chance that a maximum may occur.

2) The deviated spots seem to be much more distinct than should be expected when the points should be due to diffraction. It is also very difficult to understand how the scattered points can be smaller than the middle point due to the primary rays.

3) It is not easily understood how by diffraction a heterogeneous beam can give such sharp maxima— and sharp maxima only. If the scattered rays are at all due to diffraction it must be from some homogeneous group of rays which are mixed up with the primary ones.

On the other hand as the scattered ray bundles according to Laue are made up of very penetrating rays it is not easy to see how the corpuscular theory of Röntgen rays can explain the scattering into such sharp bundles of parallel rays. If the rays were very soft, it might be possible that these scattered bundles might be due to structures in the crystal, but such an explanation seems hardly possible for so very penetrating rays.

As you will see the matter is not yet clear and it is necessary to wait for further investigations. (The first publication by Laue will appear in *Berichte der Münchener Akademie*). But whatever the explanation may be it seems to be an effect of a most fundamental nature.

Since we met at Portsmouth I have been for some weeks in Paris and from the beginning of October I have been working with Professor W. Wien at Würzburg. I have made some work on light emission of positive rays (*Kanalstrahlen*). The results will appear in a paper I have sent to *Ann. der Physik*.

I am staying here for five weeks more, then I am going back to Norway and hope to enjoy a little holiday in the country before I am taking up my work for the next term.

Give my kind regards to Mrs Bragg, and if Mr Campbell is still at Leeds will you kindly remember me to him. Yours sincerely, L. Vegard

Fig. 6.19   Lars Vegard. Courtesy Bjørn Pedersen.

Lars Vegard: born 3 February 1880 in Vegårshei, Norway, the son of a farmer; died 21 December 1963, in Oslo, Norway, was a Norwegian physicist. After schooling at Risør, he entered the Royal Frederick University of Christiania in 1899, and obtained his bachelor's degree in 1905. L. Vegard then became assistant to the physicist Kristian Birkeland, a specialist of the *aurora borealis*. In 1907 he obtained a travel fellowship, which enabled him to study in England from 1908 to 1910, first with J. J. Thomson in Cambridge, then with W. H. Bragg in Leeds, and, in 1911–1912, with W. Wien in Würzburg, Germany. Back in Oslo, he obtained his PhD in 1913. He was then docent at the University of Oslo from 1913 to 1918 and professor from 1918 to 1952. He was also the dean of the Faculty of Mathematics and Natural Sciences from 1937 to 1941. His scientific works covered a broad area. He built the first Bragg X-ray crystal spectrometer in Norway, established the empirical rule stating the linear relation between the lattice parameters and the composition of a continuous substitutional solid, known as Vegard's law, discussed the structure of the chemical elements in terms of the electron configurations of atoms, and was an authority on auroral spectroscopy. He was elected a member of the Norwegian Academy of Science and Letters in 1914.

MAIN PUBLICATIONS

1912 *Über die Lichterzeugung in Glimmlicht und Kanalstrahlen.*

1918 *On the X-ray spectra and the constitution of the atom.*

1921 *Die Konstitution der Mischkristalle und die Raumfüllung der Atome.*

This highly interesting letter shows that the formation of the diffraction spots was still unclear in Laue's mind. It was discussed by W. H. Bragg and his son Willie during their summer holidays near Scarborough on the Yorkshire coast (Fig. 6.20). W. H. Bragg was unprepared to give up his corpuscular theory, and, on their return to Leeds, he had a large wooden sphere made in the workshop, 'on the surface of which he marked the position of the spots, because it made their relationship clearer' (W. L. Bragg, in Bragg and Caroe 1962). Willie set up an experiment in his father's laboratory to test whether the spots in Laue's picture were made by corpuscles shooting down avenues between the atoms in the crystal. He built a lead-lined box; at one end he pierced a small hole and put a crystal slice in front of it; at the other end he put a photographic plate wrapped in black paper. The box pivoted on gimbals and could be rotated about a vertical and a horizontal axis, the hole remaining in the same position. An X-ray tube was placed in front of the hole and Willie tilted the box in all possible directions within a certain solid angle during the exposure, in the hope that, if the X-rays shot down crystal avenues, they would show up on the photographic plate. But he got no positive result and the investigation was never published (Ewald 1962*b*, W. L. Bragg 1975). At the end of the holidays, Willie went back to Cambridge, and, for a time, father and son worked independently. W. L. Bragg's contribution will be described in Section 6.11.

Fig. 6.20   The house where the Braggs spent the summer holidays in 1912, near Scarborough UK. Photo courtesy
A. C. T. North, with the assent of the present owner. Inset: Plaque unveiled 5 July 2013, commemorating
the discussions between W. H. and W. L. Bragg about Laue's discovery of X-ray diffraction. Photo courtesy
C. Hammond.

W. H. Bragg's first reaction can be deduced from his exchange of correspondence with Laue
in October/November 1912 and from his first letter to *Nature* (W. H. Bragg 1912*a*). Bragg's
letters to Laue are unfortunately lost, but Laue's replies can be found in the Bragg archives at
the Royal Institution, in London (WHB 4A/8, 4A/9, 4A/10). From the context of Laue's reply
on 15 October, it seems clear that W. H. Bragg, in his first letter, expressed the view that the
observed interference diagram was due to *Bremsstrahlung* and that the position of the spots
would be the same if one changed the material of the anticathode. This, he implied, would be
because the appropriate wavelengths would be present in the white spectrum. In his reply, Laue
strongly expressed his disagreement, as was told in Section 6.5. The interference spots could
not be due to *Bremsstrahlung* and were due to some characteristic radiation whose origin was
not known.

The contents of W. H. Bragg's second letter apparently followed closely those of his letter
to *Nature* (W. H. Bragg 1912*a*), a preprint of which was included. His son Willie had pointed
out to him that the diffracted beams probably followed 'avenues' between the crystal atoms. As
a consequence, Bragg the father thought, the deflected beams should follow directions along
which atoms are separated by distances being an integral number of the interatomic distance.
The crystal is cubic. Let a be the lattice parameter, and O and A two atoms separated by
an integral number of parameters, then $OA = na$, where n is an integer. If k_1, k_2, k_3 are the
numerical coordinates of A with respect to the three axes and if O is the origin of the lattice,
then $n^2 = k_1^2 + k_2^2 + k_3^2$. The directions of the deflected beams, concluded W. H. Bragg, are those
for which the sum of the squares of the coordinates is also a square. In the letter to Laue, he

determined the values of his parameters k_1, k_2, k_3 from the positions of several of the diffraction spots on Laue's diagram such that $k_1^2 + k_2^2 + k_3^2$ was a perfect square or a perfect square short one. W. H. Bragg also suggested rotating crystal and photographic plate simultaneously (as his son had done).

In his reply, dated 10 November, Laue related these three integers, k_1, k_2, k_3, to the numerators of the direction cosines of the reflected beams (equation 6.9, Section 6.6.5), the sum of the squares of which is indeed a square (it is straightforward to show that $k_1^2 + k_2^2 + k_3^2 = (h_1^2 + h_2^2 + h_3^2)^2$, where h_1, h_2, h_3 are the indices introduced by Laue). He drew Bragg's attention to Stark's recently published paper, and expressed his disagreement with the corpuscular hypothesis. Laue also suggested that Bragg publish an additional remark to his letter to *Nature* summarizing their common discussion, but Bragg did not follow up that suggestion. In a third letter, dated 13 November, Laue gave the correspondence between Bragg's parameters k_1, k_2, k_3 and his own indices in a small number of specific cases.

In a letter to *Nature* dated 14 November 1912, the English crystallographer A. E. H. Tutton (1912*b*) gave a detailed description of Laue's experiment and of its crystallographic implications. It emphasizes, he noted, 'the importance of the space lattice', and it may be a test as to the real nature of X-rays, waves or corpuscles. He also suggested that Professor Bragg revise the account of his views after considering the crystallographic information contained in the Laue diagrams.

W. H. Bragg (1912*b*) replied that his son's paper read before the Cambridge Philosophical Society on 11 November gave a theory which makes it possible to calculate the positions of the diffraction spots: 'It is based on the idea that any plane in the crystal which is *rich* in atoms can be considered as a reflecting plane'. In the same paper, W. H. Bragg admits that, 'if the experiment helps to prove X-rays and light to be of the same nature, then such a theory as that of the *neutral pair* is quite inadequate'. On the other hand, he added, 'the properties of X-rays point clearly to a quasi-corpuscular theory'. The problem, therefore, 'is not to decide between the two theories, but to find one theory which possesses the capacities of both'.

6.11 W. L. Bragg and Bragg's law

William Lawrence (Willie) Bragg's first encounter with X-rays took place when he was six. While he was riding his tricycle, he was overturned by his younger brother, Bob, and he broke his left elbow. His father had recently set up a primitive X-ray apparatus (Section 5.8), one of the first in Australia, and he used it to visualize the damage on his son's arm. W. L. Bragg (Figs. 6.21 and 10.10) remembered later that he had been 'scared stiff by the fizzing sparks and the smell of ozone'.[36] He was rather advanced at school, but was bad at games because he lacked the 'necessary drive and self-assurance'. His chemistry master used to let Willie help him to set up the class experiments. It was this, he recalled in his personal reminiscences (W. L. Bragg in Ewald 1962*a*), which gave him his first interest in the methods of science. For details about W. L. Bragg's life and works, the reader may consult, for instance, Bragg in Ewald (1962*a*), Phillips (1979), Perutz (1971, 1990), Thomas and Phillips (1990), Hunter (2004), Jenkin (2001, 2008), Thomas (2012) and Glazer and Thomson (2015).

---

[36]Quoted in Hunter (2004).

Fig. 6.21   Sir William Lawrence Bragg, about 1913. ©The Nobel Foundation.

Sir William Lawrence Bragg: born 31 March 1890 in Adelaide, South Australia; died 1 July 1971 in Ipswich, Suffolk, England. He started school at the age of 5, at Queen's Preparatory School in Northern Adelaide. From 1901 to 1914 he attended St Peter's College, Adelaide, after which he went to the University of Adelaide. He obtained his degree in mathematics with first-class honours in 1908. He came to England with his father in 1909 and entered Trinity College, Cambridge, taking first-class honours in the Natural Science Tripos in 1912. In 1914 he was elected to a Fellowship at Trinity College to prepare a thesis with J. J. Thomson and his father, W. H. Bragg, as advisors, and was appointed lecturer. During the First World War, W. L. Bragg served as Technical Advisor on Sound Ranging to the Map Section, General Headquarters, in France. He was appointed Professor of Physics at Manchester University in 1919, and held this post until 1937. He then became Director of the National Physical Laboratory in 1937–1938 and, in 1938, was appointed Cavendish Professor of Experimental Physics at Cambridge, where he remained until his retirement in 1953. In 1954, he became Resident Professor at the Royal Institution. He shared the 1915 Nobel Prize in Physics with his father, awarded 'for their services in the analysis of crystal structure by means of X-rays', and he is still the youngest person ever to receive a Nobel Prize. He was elected a Fellow of the Royal Society in 1921, and knighted in 1941. Together with P. P. Ewald, he was instrumental in the establishment of the International Union of Crystallography, of which he was the first President (1948–1951).

MAIN PUBLICATIONS

1915 *X-rays and crystal structure*. With W. H. Bragg.
1930 *The structure of silicates*.
1933 *The crystalline state, a general survey*.
1936 *Electricity*.

1943 *History of X-ray analysis*.
1965 *Crystal structure of minerals*. With G. F. Claringbull.
1967 *The start of X-ray analysis*.
1975 *The development of X-ray analysis*.

The start of W. H. Bragg's career in research coincided with Willie's undergraduate years. It became usual for him to discuss his ideas with his son and to tell him about his results. Willie remembered living in an 'inspiring scientific atmosphere'. The most important topic that occupied W. H. Bragg's mind around 1907 was that of the nature of X-rays and γ-rays. He had become skeptical of Stokes's and Thomson's views of X-rays as pulses of electromagnetic radiation. By analogy with the properties of α-rays and β-rays, he felt the properties of γ-rays would be better explained if they were neutral pairs of material particles. As described in Section 5.8, this led to a controversy with C. G. Barkla that lasted for several years.

After W. H. Bragg's appointment in Leeds and the move of the family to England, Willie entered Trinity College, Cambridge, for a course in mathematics, but his father urged him, in his second year, to change to physics (W. L. Bragg in Ewald 1962a). The Scot C. T. R. Wilson,[37] who ran the practical class and lectured on optics, 'made the strongest impression on him and left him with a love of physical optics that never left him' (Phillips 1979). One of the features of Wilson's teaching was his use of amplitude phase diagrams to illustrate diffraction and interference effects (Hunter 2004). In particular, he showed once that 'when light falls on a diffraction grating, we can either regard it as composed of light of all colours, which are sorted out by the grating into a spectrum, or we can regard it as a series of irregular pulses from which the grating manufactures the train of waves which form the spectrum' (Bragg 1975). After Willie's graduation in 1911, he started research with J. J. Thomson on the mobility of ions in gases.

Life at college was very rewarding, and Willie made many friends. Cecil Hopkinson, who came from a family of engineers, was one of his closest friends. He introduced Willie to sailing, skiing, and climbing. J. J. Thomson's son, G. P. Thomson, was also at Trinity, and there was chaffing among the students: Braggs were good at experiments, Thomsons were not (Phillips 1979). The physicist B. S. Gossling was another of Willie's close friends at Trinity. He once read to a society of young scientists a paper about the theory of crystal structures based on the ideas of Pope and Barlow about valency volumes (they are discussed in Section 12.13). This was W. L. Bragg's first introduction to the geometry of three-dimensional structures and to the work of W. H. Pope, who was Professor of Chemistry at Cambridge (Phillips 1979).

On his return to Cambridge after the 1912 summer holidays, W. L. Bragg pored over a reprint of Laue's paper, 'the drift of my thoughts being very much influenced by J. J. Thomson's lectures on the theory of X-rays as pulses of very short electromagnetic radiation, C. T. R. Wilson's course on optics, and Gossling's paper on crystal structure (Bragg in Bragg and Caroe 1962).

It was walking along the 'Backs',[38] by St John's College, that he had his 'brain wave' (Bragg 1967, 1975): X-ray diffraction could be considered in the same way as the diffraction of light by a diffraction grating, with the sheets of atoms playing the role of the lines of a grating.

W. L. Bragg started by repeating Laue's calculation of the indices h_1, h_2, h_3 of the spots on the diagram, due to five different wavelengths, λ, but he found immediately that there were other sets of three integers leading to approximately the same values of λ. Laue's interpretation was therefore 'unsatisfactory', and Bragg proposed an entirely different one. He assumed the incident beam to be composed of a number of independent pulses (a continuous spectrum over a wide range of wavelengths), as did A. Schuster[39] in his treatment of a line grating. If a pulse falls on a number of particles distributed over a plane, these particles act as centres of disturbance, and the secondary waves from them built up a wave front, as if part of the pulse has been reflected from the plane, as in Huygens' construction.

The atoms composing the crystal can be arranged in many different possible systems of lattice planes. Bragg considered one particular set of parallel planes. 'A minute part of the energy of a pulse is reflected by each plane in succession, and the corresponding interference maxima are produced by a train of reflected pulses. The pulses in the train follow each other at intervals of $d \cos \theta$ where d is the interplanar spacing and θ the angle of incidence' (the angle between

[37]See footnote 1, Section 1.1.

[38]The 'Backs' is an area, along the River Cam, at the back of the Cambridge Colleges.

[39]A. Schuster (1904). *An Introduction to the Theory of Optics*. Edward Arnold, London.

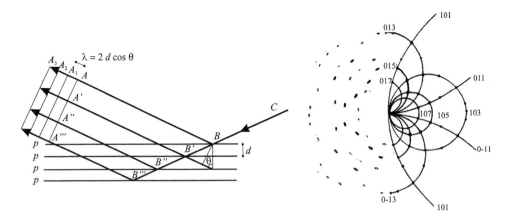

Fig. 6.22   *Left*: Derivation of Bragg's law. *CB*: incident wave. *BA*, *B′A′*, *B″A″*, *B‴A‴*: waves reflected by the successive planes, *p*. *A*, A_1, A_2, A_3: position of the successive pulses at $\lambda = 2d\cos\theta$ intervals. Redrawn after Bragg (1975). *Right*: Interpretation of Laue diagrams by the intersection of cones having a row as axis and the incident beam as a generator, with the photographic plate. Redrawn after Bragg (1913*a*).

the incident beam and the normal to the reflecting planes). 'The crystal actually *manufactures* light of definite wavelengths', λ, and its harmonics, λ/n, satisfying the relation $n\lambda = 2d\cos\theta$ where n is an integer. This equation was already in Schuster's book, as applied to line gratings, but Bragg applied it to three-dimensional gratings, and it is now known as 'Bragg's law'. It is illustrated in Fig. 6.22, *Left*, adapted from a later publication (Bragg 1975). It was not illustrated in Bragg's original publication.

W. L. Bragg then proceeded to explain the formation of the diffraction spots by reflection from lattice planes. His argument was as follows. Consider now a particular row; the angles of the incident and reflected beams with that row are equal, for any lattice plane-parallel to the row. The beams reflected by all the lattice planes parallel to the row lie therefore on a cone having the row as axis and the incident beam as one of the generators. This cone cuts the photographic plate in an ellipse. If one takes a different row, the locus of the reflected spot as the reflecting plane is turned around that second row is again an ellipse. The intersection of the two ellipses gives the position of a spot reflected by a plane-parallel to the two rows. Fig. 6.22, *Right* shows the ellipses thus obtained to match the spots on Laue's picture 6.16, *Left*.

There is not, however, full agreement between the expected and observed distributions of spots on the diagram. Laue had assumed the zinc and sulphur atoms to be located at the nodes of simple cubic lattice, but Barlow and Pope had suggested that zinc-blende had a face-centre cubic lattice. Bragg assumed that the zinc and sulphur atoms are 'identical as regard their power of emitting secondary waves'.

Suppose waves of wavelength λ fall on a crystal and the phase difference between vibrations from successive atoms along each of the three axes (O and A along Ox, O and B along Oy, O and C along Oz, Fig. 6.23, *Left*) to be an integer times 2π, respectively: $2\pi h_1$, $2\pi h_2$, $2\pi h_3$ (namely that the three Laue conditions are satisfied along the three axes). The incident wave is then reflected by a family of lattice planes of indices h_1, h_2, h_3. For the vibrations from the atoms at the centre of the cubic faces, such as D and E to be also in phase, one must have:

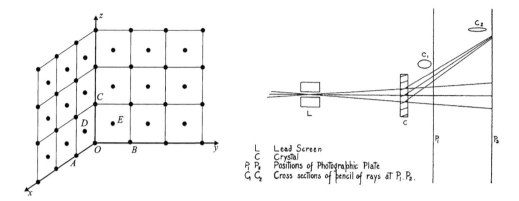

L   Lead Screen
C   Crystal
P₁ P₂   Positions of Photographic Plate
G₁ G₂   Cross sections of pencil of rays at P₁.P₂.

Fig. 6.23   *Left*: Face-centred cubic lattice. Modified after Bragg (1913*a*). *Right*: Variation of the shape of the diffraction spots with distance of the photographic plate from the crystal. There is focusing in the vertical plane, but no focusing in the horizontal plane, and the spots are elliptical. After Bragg (1913*a*).

$$\frac{h_1}{2} - \frac{h_3}{2} = an\ integer; \qquad \frac{h_2}{2} - \frac{h_3}{2} = an\ integer. \tag{6.13}$$

This condition is satisfied when the three indices, h_1, h_2, h_3, are all odd or all even. With these relations between the indices, W. L. Bragg obtained a perfect fit between the predicted and observed diffraction spots, and concluded that the lattice of zinc-blende was in fact face-centred cubic, with the atoms of zinc and sulphur at the centres of the alternating white and black squares of a chessboard. The lattice mode could be deduced from Laue diagrams. This was the start of X-ray analysis.

W. L. Bragg consulted W. H. Pope about the structure of zinc-blende. The latter explained that the atoms of zinc and sulphur are both located at the nodes of a face-centred cubic lattice and that the atoms of one type, say zinc, are at the centres of tetrahedra whose summits are occupied by atoms of the other type, sulphur. But that did not change Bragg's analysis of which reflections to expect.

There was another interesting feature of the Laue diagrams that W. L. Bragg used to prove that the spots were due to the diffraction of the incident beam by a set of lattice planes. He observed that the shape of the spots depended on the distance of the photographic plate from the crystal. They are at first round and become elliptical when the photographic plate is moved further away (Fig. 6.23, *Right*). The incident beam collimated by the lead screen, L, is divergent and is focused in the vertical plane by the reflection on the lattice planes. The vertical size of the spots decreases when the distance between the crystal and the photographic plate P_2 increases. There is no focusing in the horizontal plane, and the size of the spots remain constant in that plane. This explains why the shape of the spots becomes elliptical when the photographic plate is moved away.

The correct interpretation of Laue diagrams was quite a remarkable accomplishment for a 22-years-old man. Bragg himself attributed it to a 'concatenation of circumstances', but, as noted by M. F. Perutz (1990), Bragg's former assistant and Nobel Prize winner, Bragg's 'brilliant paper soon convinces you that its success owed more to [his] astute powers of penetrating through the apparent complexities of physical phenomena to their underlying simplicity.'

Bragg's paper was entitled 'The diffraction of short electromagnetic waves by a crystal' because he 'was still unwilling to relinquish [his] father's view that the X-rays were particles'. He thought 'they might possibly be particles accompanied by waves' (W. L. Bragg 1943). The paper was communicated to the Cambridge Philosophical Society by J. J. Thomson on 11 November 1912 and published on 13 February 1913 (W. L. Bragg 1913a). A short account appeared on 5 December 1912 in *Nature*.[40]

Since the diffraction spots on Laue diagrams could be interpreted as due to the partial reflection of the incident beam by sets of parallel planes in the crystal, it was to be expected that this would be the case for the simplest of such planes, namely cleavage planes. C. T. R. Wilson suggested to W. L. Bragg that crystals with very distinct cleavage planes, such as mica, might show a strong specular reflection. W. L. Bragg tested this idea by letting a narrow beam of X-rays, collimated by a series of stops, fall at an angle of incidence of 80° on a slip of mica about 1 mm thick, mounted on a sheet of aluminium. After a few minutes of exposure only, a well-marked reflected spot appeared after development on a photographic plate placed behind the crystal of mica. Variation of the angle of incidence and of the distance between the crystal of mica and the photographic plate left no doubt that the laws of reflection were obeyed. W. L. Bragg was very excited with the results. To his father he wrote: 'I have just got a lovely series of reflections of the rays in mica plates with only a few minutes' exposure! Huge joy. Your affectionate son, W. L. Bragg' (quoted in Perutz 1990), and his father wrote to E. Rutherford, 'My boy has been getting beautiful X-ray reflections from mica sheets, just as simple as the reflection of light from a mirror'.[41] W. L. Bragg (1967) remembered taking the photograph showing a reflection at a series of angles still wet to J. J. Thomson: 'he glared at it, thrust his spectacles up on his forehead and scratched his head in a characteristic gesture, and grinned with pleasure'. The reflection from mica was indeed a beautiful confirmation of Bragg's theory and the prototype of a new method for the study of X-ray diffraction. W. L. Bragg also showed that by bending the mica sheet into a semicircle the reflected beam could be focused. The paper was published in *Nature* on 12 December 1912 (W. L. Bragg 1912).

Laue's contribution to the field of X-ray diffraction was fundamental, but Bragg's was essential for the development of X-ray analysis. It is interesting to compare their approaches. Laue's was that of a theoretician, Bragg's that of a physicist. After his discovery, Laue was more interested in the theory of diffraction, as shown by his dynamical theory, than in the structure of materials. The analysis of structures, on the contrary, was the goal of Bragg's life.

6.12 The viewpoint of a science historian: the Forman–Ewald controversy

The accounts of the discovery of X-ray diffraction given by Laue, Friedrich, and Ewald have been vigorously contested by the historian of science Paul Forman, in an article entitled, 'The discovery of the diffraction of X-rays by crystals: a critique of the myths' (Forman 1969). In rebuttal, Ewald (1969a) retorted in the same vein, with an article entitled 'The myth of myths'.

---

[40]*Nature* (1912). **90**, 402.

[41]Royal Institution Archives. RI MS WLB94/A/1. Quoted by Hunter (2004).

Forman's article was commented on by a number of authors, among others, L. D. Gasman (1975), M. A. B. Whitaker (1979), K. H. Wiederkehr (1981), H. Kragh (1987), L. Hoddeson *et al.* (1992), and M. Eckert (2012).

The 'myth' condemned by Forman is that Laue and Ewald have grossly exaggerated the significance of the discovery of X-ray diffraction. According to him, they 'have created and elaborated an account of the conceptual situation in physics *circa* 1911 which is, in certain respects, utterly mythical', and a 'gross misrepresentation'. Forman's view is essentially based on Laue's Nobel lecture and on Ewald's book written for the fiftieth anniversary of the discovery (Ewald 1962*a*): 'This myth', Forman states, '—for that is what it is—attained its fullest elabora-tion, replete with quite fictitious details, in Ewald's retrospective account'. Forman considered also that this had been done deliberately by Laue because of 'his desire to explain how he, a man of no great originality (for so he regarded himself), came to conceive this experiment; his desire to acknowledge a debt to the Munich intellectual milieu, as distinct from Sommerfeld personally'. Another aspect of Forman's paper is that the accounts of the discovery, or 'Myth of the Origins', serve as a ritual for the 'clan of crystallographers' (*i.e.* the International Union of Crystallography) to reinforce their separate identity.

Very interesting questions are raised here: can a scientist give a fair answer to the question of 'where', 'when', and 'why' is a discovery made? Can he give a fair account of his own researches?

Forman is correct in several places, but he also makes a number of judgements of intent and some inaccurate statements. It is interesting to discuss successively the various points of his analysis.

1) *'Why Munich? The Space Lattice Hypothesis, Elaboration of a Myth'* (Forman 1969, p. 41)

Forman's first point was about the myth that Munich was the unique place where the discovery could have been made. 'The question', he states, '—*why Munich ?*—was first raised publicly, and given its now traditional answer, in Laue's Nobel lecture'. Forman quotes the following sentence from Laue' lecture:

'But no more far-reaching physical conclusion had evolved from this line of thought [the Sohncke, Fedorov, Schoenflies mathematical theory of space lattices] and thus, in the form of a questionable hypothesis, it remained a somewhat unknown quantity to physicists. But in Munich, where models of the Sohncke space-lattices were to be found in more than one university institute, it was P. Groth who expressed his defence of it, both orally and in writing, and I [von Laue], also, thus learned from it'.[42]

Forman judged that this sole sentence implied that Laue considered Munich as unique, but this view seems quite exaggerated. Furthermore, in no place did Ewald (1962*a*) stress the uniqueness of Munich. The myth, if myth there is, was rather created by Laue's contempo-raries. For instance, the Swiss crystallographer P. Niggli,[43] who was the first to apply space

---

[42]The wording follows that given in the Nobel web page. Forman used a slightly different translation of the Laue original.

[43]Paul Niggli, born 26 June 1888 in Zofingen, Switzerland, died 13 January 1953 in Zürich, Switzerland, was a Swiss crystallographer. After studying at the Polytechnic Institute of Zürich, he obtained his PhD at Zürich University. He was appointed professor at the University of Leipzig in 1915, at the University of Tübingen in 1918, and, finally, Professor of Mineralogy and Petrography at the ETH Zürich, where he remained all his life.

group theory to the X-ray determination of crystal structures, wrote: 'It is not by chance that Laue's discovery took place in Munich, where one of the few German-speaking researchers in crystallography was teaching. It is not by chance that the first crystal structure determinations were made in England, where Barlow and Pope had developed the concept of close-packing' (Niggli 1922). In a similar way, the German physicist, W. Gerlach, co-discoverer of the Stern–Gerlach effect, wrote, in his Munich Souvenirs: 'The Munich air was so to speak predestinated for the Laue discovery' (Gerlach 1963).

The second point made by Forman while stressing that sentence is that it grossly misrepresented the conceptual situation. To strengthen his point, Forman refers to Ewald (1932, 1962a) who wrote that the invalidity of the Cauchy relations 'discredited' the space lattice hypothesis. Ewald was wrong there (see Section 2.3), as he readily admitted in his reply to Forman (Ewald 1969a). Laue's reference to a 'questionable hypothesis' was certainly exaggerated, but he referred to the Schoenflies–Fedorov analysis, which he was right in saying that it had not been related directly to a physical property. Ewald's thesis was the first attempt to do so.

In the last part of this section, Forman questions whether Laue's knowledge of the periodic arrangement of particles in a crystal was as limited as he pretended. This is difficult to ascertain. One only knows that Laue did poorly in his examination of mineralogy.

2) *'The Space Lattice No Hypothesis'* (Forman 1969, p. 44)

Forman stated here that 'the existence of the space lattice was neither unknown to the physicist, nor indeed regarded by him as a hypothesis; it was an assumption which, despite the lack of any direct evidence, was made universally and implicitly, and in 1911 was regarded as far more secure than, say, the laws of mechanics'.

One cannot follow Forman here. Physicists had based theories on the assumption of a regular distribution of atoms or molecules in crystals (see Section 2.3), but without any quantitative proof. After Haüy, it was considered that the properties of a crystal such as the cleavage planes and the natural faces could only be explained by assuming such a regular arrangement (see Chapter 12). As noted by the Russian crystallographer G. Wulff (1913a), who showed the crystallographic significance of the Laue indices, the lattice structure of crystals was a 'most probable hypothesis'. One could only conjecture the nature of the particles building up the crystals, and no one knew how to find the positions of the atoms or molecules in the crystal. That was still a complete mystery. Ewald (1962b) was correct in saying that 'no physicist, no crystallographer was able to assign correctly to a given crystal one, or even a small selection, of the 230 possibilities of arrangement which Schoenflies' and Fedorov's theory predicted'. At that time, even finding the proper Bravais lattice of a crystal was difficult, as shown by the error made by Groth in assigning a P mode to anhydrite while it is in fact A.

Laue's experiment was a *tangible* proof of the periodic arrangement of crystals in that it gave an *image* of the distribution of the atoms in the crystals. It is true that this image is in Fourier space; at that time people had of course no way of knowing that, but it explains why Laue's diagrams were puzzling to many and were difficult to interpret. Fedorov, nevertheless, grasped at once the meaning of the Laue diagrams. He wrote, on 2 October 1912, to the Russian scientist N. A. Morozov (1854–1946): 'Diffraction spectra of atoms were photographed by means of X-rays. Indirectly people were able to see the immediate effect brought about by atoms, that is, in principle, they saw the atoms with their own eyes. For us crystallographers this discovery is of prime importance because now, for the first time, we can have a clear picture of that on which we have but theoretically placed the structure of crystals and on which the analysis of crystals is based' (quoted in Ewald 1962a, p. 346). It was, however, only when W. L. Bragg showed that the Bravais lattice of zinc-blende could be determined from these diagrams that their significance began to appear.

3) '*The 'Wave' Theory No Necessary Condition*' (Forman 1969, p. 52)

Forman thought that it was already clear that X-rays were waves. He pointed out that if Stark and Wendt had used X-rays in their 1912 experiments (Section 6.9), they could have discovered X-ray diffraction. But these authors had been interested in the mechanical damage produced by positive ions and in whether the penetration of these ions was easier along certain crystallographic planes. They had not thought of scattering. W. L. Bragg failed to interpret the Laue diagrams by means of channelling, and he had to use waves.

4) '*An Unpromising Proposal*' (Forman 1969, p. 55)

In this section, Forman tried to prove that Laue's motivations in planning the experiment were not justified, given the knowledge of the time, and speculated on the reasons why Sommerfeld, Wien and others did not think Laue's experiment would work. According to Laue and Ewald, the main reason was the fear that thermal motion would destroy the regularity of the arrangement of atoms and then the crystal would not behave as a regular grating. But, from his own analysis of the articles of 1910–11 on thermal vibrations, Forman judged that this was not the case. The myth, he said, derived from Friedrich *et al.*'s (1912) paper where Laue had expressed the view that thermal motion would result in displacements of the molecules by a sizeable fraction of the lattice constants, and that this would affect the value of the diffracted intensity (equation 6.3, Section 6.6.5). The myth, Forman added, was further established by Ewald (1962a) who made a calculation of the expected amplitude of the lattice variations. In 1962, however Ewald had made an error by one order of magnitude, which Forman denounced and which Ewald (1969a) acknowledged afterwards. It is clear now that Laue had made an overestimation in 1912, and he was genuinely happy, after the experimental success, that 'the influence of thermal motion was small, against all expectations' (Laue 1913b). Nevertheless, Einstein wrote on 21 June 1912: 'The photograph is so sharp that one would have scarcely have suspected it, due to thermal agitation' (quoted by Hoddeson *et al.* 1992), which proves that there was indeed a reason to worry about the effect of the lattice vibrations on the diffraction pattern, and that the problem due to thermal vibrations maybe was not a myth. A proof that Laue did not want to create a myth of this problem is that he correctly stated in his talk at the October 1913 Solvay Congress (Laue 1913d) that, according to Debye's (1913a to d) calculations, the influence of thermal motion is small for low-order reflections and increases rapidly with increasing order.

Forman next asked whether Laue and his contemporaries, who adhered to the 'wave' theory of X-rays, had reason to feel confident about the success of the experiment: a) *Bremsstrahlung*, according to Sommerfeld, is constituted by square pulses, which are not expected to produce diffraction fringes, and, b), the characteristic radiation emitted by the atoms of the sample would not be coherent. Forman concluded: 'What sort of interference effect could one then possibly hope for? Far from being a mere extension of an optical experiment from a two-dimensional to a three-dimensional transmission grating, this was an experiment without analogy or precedent. No wonder Sommerfeld refused machine time'. W. L. Bragg (1913a, 1967) had a ready answer to the first point: 'the reflection of a pulse from a series of planes one behind the other turned it into waves of definite length, like white light reflected from a opal or mother-of-pearl'. The second point was discussed in Section 6.5.

5) 'Retrospect: Mythicization' (Forman 1969, p. 67)

In the last section of his paper, Forman elaborated on the social function of myths and recapitulated which were, according to him, the various components of the myth: 'if the Munich milieu, permeated by the wave theory as well as the space lattice theory, was uniquely favourable to conceiving the diffraction experiment, why did Sommerfeld, and others, refuse Laue support and encouragement? In order then to eliminate this inconsistency the myth must be further elaborated; a stratum of metamyths of justification is thus laid over the original account':

a) The thermal motion myth.
b) The Cauchy relation myth.
c) 'Laue's myth that among crystallographers the space lattice theory was *hardly mentioned anymore*'.
d) 'The interpretation of the discovery which it shores up is also a myth. The impulse theory advocated in Munich precluded the existence of a distinguishable interference pattern, and Laue's interpretation of the phenomenon was received with scepticism by adherents of that theory'. W. L. Bragg (1913a, 1967) showed the first point not to be true, and Laue's experiment, although indeed puzzling at the time, did prove that X-rays had a wave character.

To sum up, Forman rightly pointed out a number of inconsistencies and exaggerations in the accounts of the discovery of X-ray diffraction by its authors, showing that accounts of a discovery by its authors are not always completely reliable, so that the history of the discovery has to be modified accordingly. Forman, however, overemphasized these shortcomings so as to present a 'myth' adorned with metamyths.

What is, in fact, to be remembered, one hundred years after the discovery of X-ray diffraction? 1) That it was a tangible proof a) of the regular inner structure of crystals and b) of the wave, rather than corpuscular, character of X-rays; this is not a myth; 2) that it was a question by Ewald which triggered Laue's intuition. This has not been disputed by historians (Eckert 2012).

Finally, concerning the 'clan of crystallographers', it is certainly true that there is a tightly bound community of crystallographers. What keeps them together is not occasion to celebrate a 'myth of origins', and neither is it a common technique. The inner structure of a material, ordered or not, its imperfections, at the nanoscopic, microscopic, or macroscopic scales, are directly related with its physical, chemical, mineralogical, or biological properties. The common goal of unravelling these structures or their defects by a large variety of techniques, and understanding their relations to the properties of materials, is what keeps the crystallographers together.

7

1913: THE FIRST STEPS

The X-ray spectrometer opened a new world. It proved to be
a far more powerful method of analyzing crystal structure
than the Laue photographs I had used.

Sir W. L. Bragg (1943)

7.1 First experiments in the reflection geometry

The year 1912 saw the discovery of X-ray diffraction, the derivation of Bragg's law, and the first determination of a Bravais lattice. From then on, studies on X-ray diffraction developed very rapidly in many countries, England, Germany, France, Japan, the Netherlands, and more than one hundred papers were published on that topic in 1913. The momentum was stopped for a while, however, due to the war, with less than one hundred publications in 1914, but two neutral nations joined in: the United States and Sweden.

W. L. Bragg's interpretation of the Laue diagrams was that any plane in the crystal which is 'rich' in atoms can be considered as a reflecting plane (quoted in W. H. Bragg 1913*b*). As a proof, and following a suggestion of the chemist W. H. Pope, he looked for the reflection of X-rays from cleavage planes of mica (W. L. Bragg 1912, 12 December 1912, see Section 6.11). That was the first experiment in the 'reflection geometry', now often called the *Bragg geometry*, while the transmission geometry is likewise called the *Laue geometry*. C. G. Barkla and his co-worker G. H. Martyn at London University had the same idea. They observed the reflection of X-rays from cleavage planes of rock-salt and showed that it obeyed the usual laws of reflection (Barkla and Martyn 1912, 19 December 1912). They had been about to repeat the experiment with cleavage planes of mica when W. L. Bragg's letter to *Nature* on *The specular reflection of X-rays* by mica appeared (W. L. Bragg 1912, 12 December)! W. L. Bragg's observation inspired H. G. J. Moseley and C. G. Darwin at Victoria University in Manchester, England (1913*a*, 30 January 1913, see Section 7.8), von L. Mandelstam and H. Rohmann (1913, 17 February 1913) in Strasbourg, then Germany, E. Hupka and W. Steinhaus (1913, 23 February 1913) in Charlottenburg, Germany, H. B. Keene (1913*a*, 3 April 1913) in Birmingham, England, to also study the reflection of X-rays from mica. E. A. Owen and G. G. Blake at the National Physical Laboratory, Teddington, England (1913, 10 April), and J. Herweg in Greifswald, Germany (1913*a*, 15 May) observed the reflection from a cleavage surface of gypsum.

Willie's father, W. H. Bragg, also repeated the experiment, but used an ionization chamber to follow the reflected beam from a sheet of mica as it was rotated in the incident beam (1913*a*, 23 January). So did Moseley and Darwin (1913*a*, 30 January), who deduced from that result that the reflected beam is, indeed, a beam of X-rays. Furthermore they agreed with W. H. Bragg that, in describing reflected X-rays, 'their energy is concentrated as if they were corpuscular', and that X-rays showed 'the contrary properties of extension over a wave-front and concentration in a point'.

Mandelstam and Rohmann (1913) observed that, if they scratched the surface of a sheet of mica, dark lines would appear on the reflected spot, that corresponded to the scratches that had been made. This was the first 'Berg–Barrett' reflection topograph,[1] but this fact is usually not known! They also studied the reflection of X-rays from the cleavage planes of a gypsum crystal oriented in the transmission geometry. They concluded that the spots on the Friedrich and Knipping diagrams must be due to reflections from optically invisible cleavage planes.

Reflection spots studied by Hupka and Steinhaus (1913) exhibited a fringe-like structure. Such a fine structure of diffraction spots was observed by many authors, both in the reflection geometry, for instance, by Barkla and Martyn (1913) and in Laue diagrams by Hupka (1913*b*), de Broglie (1913*b*), Wulff and Uspenski (1913*a*), and Keene (1913*b* and *c*). The crystals used by the first investigators were, in general, of poor quality, and these fringes were usually due to crystalline imperfections, such as cracks at internal cleavage planes. But, initially, it was thought that this structure of the spots might be due to further diffraction fringes. Streaks and star-shaped figures were also observed around the central spot of Laue diagrams, in particular those from metals, and were usually due to the microcrystalline structure of the materials.

7.2   Equivalence of Laue's relations with Bragg's law

The equivalence between Laue's relations 6.7 (Section 6.6.5) and Bragg's law is implicit in Bragg's seminal paper (W. L. Bragg 1913*a*). It was worked out by several authors during 1913:

1. The Russian crystallographer G. Wulff[2] (1913*b*, 3 February) rediscovered Bragg's law independently in February 1913. He showed, in the particular case of a cubic crystal, that equation 6.8 can be written, using his notations:

$$\frac{\lambda}{2} = \frac{\Delta \cos \Theta}{m} \tag{7.1}$$

where Δ in the distance between two neighbouring planes, Θ the angle between the incident direction and the normal to the reflecting planes, and m an integer. Equation 7.1 is identical to Bragg's law (Fig. 6.22), although Wulff had clearly not heard of Bragg's work. In the same article, Wulff showed that the Laue spots were distributed along ellipses corresponding to families of lattice planes with the same axis, as had already been observed by W. L. Bragg (1913*a*). The same result was described by M. de Broglie (1913*a*) and T. Terada (1913*a* and *b*).

2. Laue (1913*b*, 1 April) derived Bragg's law in the following way. He considered a given reflecting plane and took the basic lattice vectors **a** and **b** parallel to that plane. He then considered the conditions 6.5 with $A = h_1 = 0$ and $B = h_2 = 0$. The third equation 6.4 is then

---

[1]Berg, W. F. (1931). Über ein röntgenographische Methode zur Untersuchung von Gitterstörungen an Kristallen. *Naturwiss.* **19**, 391–396. Barrett, C. S. (1945). A new microscopy and its potentialities. *Trans. Metal. Soc. AIME.* **161**, 15–64.

[2]George (Yuri Victorovich) Wulff, born 22 June 1863 in Nischyn, Russia, now Ukraine, died 25 December 1925 in Moscow, USSR, was a Russian crystallographer, the founder of the Moscow school of crystallography. He is well-known for his rule for the equilibrium shape of a crystal and for the Wulff net he designed for the study of stereographic projections.

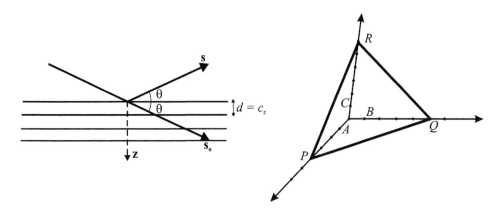

Fig. 7.1   *Left*: Laue's derivation of Bragg's law; c_z: projection of lattice vector **c** on the normal to the reflecting planes.
Right: Friedel's derivation of Bragg's law. *P, Q, R*: intersections of a lattice plane with three conjugate rows, *OA*,
OB, OC, defining the lattice.

$$h_3 = C = \mathbf{c} \cdot \frac{\mathbf{s} - \mathbf{s_0}}{\lambda} \qquad (7.2)$$

If c_z is the projection of the lattice vector **c** on the normal to the reflecting planes parallel to **a**
and **b** (Fig. 7.1, *Left*), this equation reduces to

$$h_3 = \frac{2\sin\theta}{\lambda} c_z \qquad (7.3)$$

where θ is the angle between **s** or $\mathbf{s_0}$ and the lattice planes. This is Bragg's law.
3. Ewald (1913*a*, 8 May 1913) showed the equivalence of Laue's and Bragg's approaches by noting
that equations 6.4 and 6.5 can be generalized as:

$$\mathbf{r} \cdot \frac{\mathbf{s} - \mathbf{s_0}}{\lambda} = n, \qquad (7.4)$$

which is satisfied if $(\mathbf{s} - \mathbf{s_0})/\lambda$ is a vector **h** of the reciprocal lattice defined in Section 6.8. Bragg's
law then holds for the family of lattice planes associated with reciprocal lattice vector **h**.
4. The French crystallographer G. Friedel[3] (1913*a*, 2 June) showed the equivalence between the
three Laue relations 6.7 and Bragg's law in a simple and direct way. Consider a family of lattice
planes and let *OA, OB, OC* be the three conjugate rows defining the lattice (Fig. 7.1, *Right*).
Further let *P, Q, R* be the nodes at the intersection of one plane of the family with the three axes.
If the three Laue relations 6.7 are satisfied for the three rows, *OP, OQ, OR*, the waves diffracted
by *O* and *P*, *O* and *Q*, *O* and *R*, respectively, are in phase. The waves diffracted by all the
planes of the family of lattice planes are therefore also in phase. The path difference between the

---

[3]Georges Friedel, born 19 July 1865 in Mulhouse, France, the son of the mineralogist and chemist, Charles Friedel
(1832–1899), died 11 December 1933 in Strasbourg, France. After his graduation from the Paris School of Mines, where
he was a pupil of the mineralogist F.-E. Mallard (1833–1894), he was Director of the Saint-Etienne School of Mines
and in 1919 was appointed professor at the University of Strasbourg. G. Friedel is well-known in crystallography for his
theory of twinning, his generalization of the Bravais law, his studies of liquid crystals, and his 'law'about the equality of
hkl and $\overline{hkl}$ reflections, the so-called Friedel's 'law' (Section 7.12).

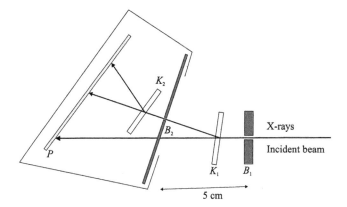

X-rays

Incident beam

5 cm

Fig. 7.2   Double crystal arrangement. B_1, B_2, slits; K_1, K_2: rock-salt crystals P: photographic plate. Redrawn after Wagner and Glocker (1913).

waves diffracted by two successive planes of the family is 'twice the projection of the interplanar spacing of the planes of the family on the incident beam', and is equal to an integral number of wavelengths. This is Bragg's law. Friedel (1913b) also showed, independently of W. L. Bragg, how to interpret Laue diagrams by means of the stereographic projection and Wulff's nets.

5. The fact that a Laue spot corresponds to the reflection of the incident beam on a particular family of lattice planes was demonstrated experimentally by Wulff and Uspenski (1913b, 24 June 1913) and by Wagner and Glocker (1913, 24 September). Wulff and Uspenski isolated the beam corresponding to a particular spot by means of a lead screen pierced by a hole and letting the beam impinge on a second crystal identical to the first one and with the same orientation. A single reflected beam was then observed with the same angle of reflection, showing at the same time that a single wavelength had thus been selected.

Like his chief, W. C. Röntgen, E. Wagner[4] was not convinced that the diffracted ray giving rise to a Laue spot contained a single wavelength. He therefore asked his student, R. Glocker,[5] to select a particular ray by a slit and have it diffracted by a second crystal, identical to the first one. The set-up is shown in Fig. 7.2. The incident beam, collimated by a first slit, B_1, falls on a first rock-salt crystal oriented with a cubic face making a certain angle, α, with the incident beam. A reflected beam is selected by the second slit, B_2, and falls on a second rock-salt crystal, K_2, identical to the first one. The diffracted spots observed on the photographic plate, P, are found to correspond to wavelength $\lambda = 2d \sin \alpha$ and its harmonics, $\lambda/2$ and $\lambda/4$. The theory is

[4]Ernst Wagner, born 14 August 1876, the son of an ophthalmologist, died 1 November 1928, was a German physicist. After attending Röntgen's lectures in Würzburg, he studied in Berlin and Munich, where he obtained his PhD in 1903 under Röntgen, and his habilitation in 1909. He was Röntgen's assistant for many years until he was appointed Assistant Professor of Physics at Munich University in 1915, and Professor in 1919. In 1922 he became Professor of Physics at Würzburg University.

[5]Richard Glocker, born 21 September 1890, died 31 January 1978, was a German physicist. He was Röntgen's last doctorate student. During World War I he was the head of the X-ray section of the Württemberg field hospital. After the war he founded an X-ray Institute at the Technische Hochschule Stuttgart for the development and applications of X-ray techniques. It was located in front of Ewald's Institute. Among other topics, Glocker developed the X-ray fluorescence analysis. Later, Glocker worked on the structure of metals, and in 1934 his Institute incorporated the Max Planck Institute for Metal Physics, under his direction.

thus confirmed. This work was part of R. Glocker's thesis (Glocker 1914), and was presented by E. Wagner at the Conference of German scientists in Vienna (Section 7.10.2).

These two experiments were the first double-crystal experiments.

7.3   M. von Laue's conversion

Scientific recognition came quickly for Laue. In October 1913 he was appointed 'extraordinary' Professor at the University of Zürich, where he succeeded Debye who had himself been appointed at the University of Utrecht, and, on 14 December 1912, gave his inaugural lecture. Laue was very busy with his teaching duties and he had little time for science. He was appointed at the University of Frankfurt am Main in 1914. He was not very happy in Zürich, and in a letter to Lise Meitner, written in 1940 (quoted in Zeitz 2005), he called his time there, 'those two bad (*böse*) years in Zürich'. He could nevertheless do some research, and had frequent encounters with Einstein who was Professor of Theoretical Physics at the ETH Zürich. Laue set up an X-ray equipment identical to that used by Friedrich and Knipping, and had two co-workers, F. Tank and J. S. van der Lingen. With F. Tank, he studied the variation of the shape of the diffraction spots with the distance between the crystal and the photographic plate (Laue and Tank 1913). They also studied quite a number of different crystals, including fluorite (CaF_2), nickel sulphate hexahydrate, $NiSO_4$, $6H_2O$, cassiterite (TiO_2), cuprite (Cu_2O), diamond, quartz, and beryl. They observed, in particular, that the diagrams of two cubic crystals oriented with a four-fold axis parallel to the incident beam are similar, which showed that the diagram is independent of the wavelength. With J. S. van der Lingen, Laue studied the influence of temperature on the Laue diagrams of rock-salt, silicon, and diamond (Laue and van der Lingen 1914*a*, *b* and *c*, respectively, Section 7.10.2).

When Laue arrived in Zürich, he still held his misconception about the formation of the diffraction spots, as shown by his letters to W. H. Bragg and his inaugural dissertation in December 1912 (Section 6.5). The first signs of a possible change in his attitude appear in one of the addenda he wrote at the occasion of the reprinting in *Annalen der Physik* of his 1912 article on the interpretation of the diffraction spots (15 March 1913): 'The amazing and still unexplained fact that the highly inhomogeneous impulses of the incident beam give rise to waves of well-defined wavelengths can be interpreted by saying that it is the crystal lattice that makes a choice out of all the available wavelengths. The photographic plate would be thoroughly blackened, had not *lattice elements (Gitterelemente)* selected a finite number of wavelengths' (Laue 1913*a*).

In an important paper written immediately afterwards (Laue 1913*b*, 1 April 1913), Laue referred to the various authors (W. H. and W. L. Bragg, Mandelstam and Rohmann) who had suggested that the diffraction spots could be interpreted as reflections from specific planes; W. H. and W. L. Bragg considered atom-rich planes while Mandelstam and Rohmann had considered cleavage surfaces present inside the crystal. Laue was not happy with the suggestion made by Mandelstam and Rohmann, but accepted Bragg's theory to be in agreement with his own. He found, however, that the actual diffraction process was yet unclear. What were the 'lattice elements' responsible for it: atoms or groups of atoms? What was the role of the complicated point systems that exist in a crystal? If, as suggested by W. L. Bragg, it is the planes of highest density of atoms which are significant, it is strange that the pictures from zinc-blende should not exhibit hemihedry.

In the next article, (Laue 1913b, 21 May 1913) considered trigonal and hexagonal symmetries to which he applied the theory he had developed for a cubic crystal. He could thus index Friedrich and Knipping's photograph of a zinc-blende crystal oriented with a three-fold axis parallel to the incident beam (Fig. 6.16, *Right*), and that of a diamond crystal oriented in the same way. For that, he considered the face-centred cubic lattice to be a hemihedry of an hexagonal lattice; this is correct, of course, but Ewald (1913b) found it was 'unnecessary'; in other words, that there was no reason to refer to the hexagonal superlattice. In the course of this investigation, Laue studied the diffraction diagrams of several trigonal and hexagonal crystals, such as quartz and beryl. In particular, he observed that the diagrams of quartz indicated a three-fold axis and not a six-fold one.

7.4   M. de Broglie and the French Schools

There were two main laboratories of crystallography in France in 1913: that of Georges Friedel at the *Ecole des Mines* in Saint-Etienne, and that of Frédéric Wallerant (1858–1936) at the Sorbonne in Paris. The second of these was the former laboratory of Haüy and Delafosse, where Pierre and Jacques Curie had discovered piezoelectricity (Curie and Curie 1880). Both Friedel and Wallerant were at the time mostly interested in the study of liquid crystals, which had been recently discovered by the German physicist O. Lehmann (1855–1922), a former student of P. von Groth. Neither of the two laboratories had equipment for X-ray work. Wallerant's former student, Charles Mauguin,[6] actively developed X-ray diffraction studies in Paris after World War I. Friedel's contributions in that year, 1913, were theoretical, and the papers mentioned above on the relation between Laue's and Bragg's diffraction conditions (Friedel 1913a) and on the interpretation of Laue diagrams with the stereographic projection (Friedel 1913b), and his article on the symmetries observable with X-rays and the so-called Friedel's 'law' (Friedel 1913c), will be described later in Section 7.12.

In 1913, the only place where X-ray diffraction could be studied was Paul Langevin's laboratory at the Collège de France in Paris, where Maurice de Broglie[7] was studying ionization in gases. Soon, de Broglie established a private laboratory in his own mansion, 29 rue

[6]Charles Victor Mauguin, born 19 July 1878 in Provins, France, died 25 April 1958 in Villejuif, France, was a French mineralogist and crystallographer. As a student, he attended Pierre Curie's lectures on symmetry in physical phenomena He obtained his PhD in 1910 on organic chemistry. He then became assistant to F. Vallerant at the Sorbonne in Paris and started the study of liquid crystals. From 1913 to 1919 he was Assistant Professor at the University of Nancy, but his research was interrupted by World War I. In 1919 he was appointed Assistant Professor at the Sorbonne, and, in 1933, he succeeded F. Wallerant as Professor of Mineralogy, a post which he held until his retirement in 1948. He determined the structure of cinnabar, calomel, and was the first to study systematically the structure of micas with X-rays. C. Mauguin had many students and he founded the most important school of crystallography in France. He is well-known for being the co-author of the Hermann–Mauguin notation of space-groups. He was elected a member of the French Academy of Sciences in 1936.

[7]Maurice de Broglie, born 27 April 1875 in Paris, France, the son of Victor de Broglie, 5th duc de Broglie, died 14 July 1960 in Neuilly, France, was a French physicist. After graduating from naval officer's school, he spent nine years in the French Navy, which he left in 1904 to pursue a scientific career. He studied under Paul Langevin at the Collège de France in Paris, and obtained his doctorate in 1908. His work on X-ray diffraction and spectroscopy earned him election to the French Academy of Sciences in 1924. In 1934 he was elected to the Académie Française and in 1942, he succeeded Paul Langevin as professor at the Collège de France. He had a close cooperation with his younger brother, Louis de Broglie, who won the 1929 Nobel Prize in Physics.

Chateaubriand, Paris 8. The laboratory was transferred to the state after the Second World War; the present author remembers that in 1956 he measured the absorption of neutrons from a Van de Graaff generator in what had been the servants' common room, where the bells that were linked to the dining room and to the Duke's and Duchess' bedrooms could still be seen!

In the first of a dozen articles published in 1913, M. de Broglie (1913*a*, 31 March) reported on the Laue diagrams of a number of cubic crystals including zinc-blende, rock-salt, fluorite, and magnetite. His observations were: 1) all the diagrams looked alike, only the intensities of the spots varied from one crystal to the other, which meant that they depended on the nature of the crystal, 2) the Laue diagram of rock-salt was unchanged if the crystal was plunged in liquid nitrogen, the spots not being any sharper, which meant that thermal motion did not have a significant effect on diffraction results, 3) the Laue diagram of magnetite remained unchanged if a magnetic field was applied to the crystal during exposure, which meant that its symmetry was not affected by the action of the field, 4) the spots were arranged on the diagrams along ellipses, as predicted by Wulff (1913*b*). In the following articles, de Broglie discussed the various systems of fringes and striations observed in the diffraction images of many different crystals. This work was done in part in collaboration with F. A. Lindemann (1886–1957), the future Lord Cherwell and scientific adviser to Winston Churchill (de Broglie and Lindemann 1913, 1914). M. de Broglie's most important contribution, the rotating crystal method for the study of X-ray spectra, will be described in Section 7.11.

7.5   T. Terada, S. Nishikawa, and the Japanese School

The first studies of X-ray diffraction in Japan were described by Torahiko Terada[8] in a letter to *Nature*, submitted 18 March and published 10 April 1913 (Terada 1913*a*). A detailed report was presented at the Tokyo Physico-mathematical Society on 3 May 1913 (Terada 1913*b*). Terada was at that time Assistant Professor at the Department of Physics of the University of Tokyo, headed by Hantaro Nagaoka (1865–1950), author of the so-called Saturnian model of the atom (Nagaoka 1904).

Terada's work was done after the news of Friedrich and Knipping's experiment and Laue's interpretation of them reached Japan, but before W. L. Bragg's paper with Bragg's law, which arrived in Tokyo just in time to be mentioned in the printed version of Terada's talk. In the course of his investigations, Terada became aware of Wulff's (1913*b*) paper, which confirmed his first conclusions. He had at his disposal a powerful Müller X-ray tube. The incident beam was collimated by a diaphragm with an aperture 5 to 10 mm in diameter. The intensity was large enough for Terada to be able to observe the diffracted beams directly on a fluorescent screen. He was the first to do this; it was also done later by de Broglie and Lindemann (1914). The spots were arranged on ellipses, and, by rotating the crystal, Terada could observe on the screen the changing patterns of the ellipses. This phenomenon led him to interpret the spots on the ellipses as due to the reflection of X-rays from net-planes having a zone axis in common. With

---

[8]Torahiko Terada, born 28 November 1878, died 31 December 1935, was a Japanese physicist and writer. He was Professor at the Imperial University in Tokyo. He headed a laboratory at the Japanese Research Institute, RIKEN, founded in 1913, from 1924 until his death, and he was also professor at the Earthquake Research Institute. He had a wide range of research interests, from astrophysics to geophysics, and was also an accomplished painter and writer, writing many essays.

Japanese politeness and understatement, Terada thought that Laue's idea that the spots were due to fluorescent radiation emitted by the atoms of the crystal was 'not absolutely necessary', and he correctly indexed the diagrams obtained with crystals of rock-salt and fluorite. Terada also studied a big variety of other crystals, including quartz, mica, borax, alum, cane-sugar, gypsum, epidote, and tourmaline.

Terada pursued his X-ray work for about one year only. One evening, he showed the movement of the diffraction spots on the fluorescent screen to one of the post-graduate students of the Department of Physics, Shoji Nishikawa.[9] Nishikawa, who had been engaged in the study of radioactivity, was fascinated and soon turned to the field of X-ray diffraction, becoming the first major Japanese X-ray crystallographer and founder of an important school of X-ray crystallography. That story was told by one of Nishikawa's former students, I. Nitta[10] (in Ewald 1962a). Nishikawa's first work, with his fellow student, S. Ono, dealt with the diffraction of fibrous, lamellar, and granular substances using Laue diagrams (Nishikawa and Ono 1913). He then noted that the elementary crystals of a fibrous substance such as asbestos or hemp arranged parallel to the axis of the fibres gave a diagram similar to that one could obtain with de Broglie's rotating crystal method (Nishikawa 1914). Nishikawa was the first to attempt to use the theory of space groups in studies of crystal structure (Nishhikawa 1915). During his stay in the United States (1916–19), at Cornell University, Ithaca, NY, he taught Ralph W. G. Wyckoff, then a graduate student, how to take Laue diagrams and how to interpret them. This was Wyckoff's introduction to X-ray crystallography, and he was for ever grateful to Nishikawa (Wyckoff in Ewald 1962a). Nishikawa's contributions were many, and bore in particular on the structure of minerals and the $\alpha - \beta$ transition of quartz. He also gave one of the first experimental proofs that Friedel's 'law' does not hold for absorbing crystals (Nishikawa and Matukawa 1928, see Section 7.12).

7.6 W. H. Bragg and the X-ray ionization spectrometer

William Pope was 'intensely interested' in W. L. Bragg's results with zinc-blende, as he saw in them a justification of his and Barlow's valency–volume theory (Barlow and Pope assumed proportionality between the volumes of the spheres of influence of atoms and their valencies, see Section 12.13). He encouraged W. L. Bragg to take Laue pictures of sodium chloride, NaCl, and potassium chloride, KCl, which he believed to be close-packed structures with the atoms at the nodes of two face-centred cubic lattices. For that purpose, Pope ordered crystals of both

[9]Shoji Nishikawa, born 1884, died 1952, was a Japanese crystallographer. He studied at the Faculty of Science of the University of Tokyo, graduating in 1910. He then did research at the Department of Physics of the University of Tokyo and at RIKEN. In 1916–17, he went abroad to the United States where he stayed at Cornell University, Ithaca, USA. After the end of the First World War, Nishikawa stayed for half a year with W. H. Bragg in London. In 1920 he returned to Tokyo where he became Director of the Nishikawa Laboratory at RIKEN. In 1924 he became professor at the University of Tokyo, where he remained until his retirement in 1945. Among his many students, one may mention I. Nitta and S. Kikuchi (1902–1974).

[10]Isamu Nitta, born 1899, died 16 January 1984, was a Japanese crystallographer. A student of chemistry in the University of Tokyo from 1920 to 1923, he was research associate in Nishikawa's laboratory until 1933, when he was appointed Professor of Physical Chemistry at the newly established Osaka University. He studied the structure and physical properties of molecular crystals such as pentaerythritol. He was Vice-President of the International Union of Crystallography from 1963 to 1969.

Fig. 7.3   W. H. Bragg's spectrometer. Museum of the Royal Institution, London. Photo by the author.

NaCl and KCl from the German firm Steeg & Reuter—the same company that had provided the zinc-blende crystals for Laue (Bragg in Ewald 1962a, p. 533).

W. H. Bragg, on the other hand, was interested in the properties of X-rays—their absorption and the ionization they produce on impact. Father and son joined forces during the winter of 1912–13 when Willie returned home to Leeds for the Christmas holidays. In order to characterize the reflected beam in detail, W. H. Bragg designed the X-ray spectrometer. In his Rutherford lecture, W. L. Bragg (1961) recalled: 'My father was supreme at handling X-ray tubes and ionization chambers'. 'He tailored every piece of the apparatus, which was constructed in the Leeds workshop by its clever head mechanic, Jenkinson' (Bragg 1975).

The apparatus resembled an optical spectrometer and is represented in Fig. 7.3. The collimator was replaced by a lead block pierced by a hole, O (Fig. 7.4), placed in front of the X-ray tube, and the crystal was mounted on a table that could turn around a vertical axis. The crystal

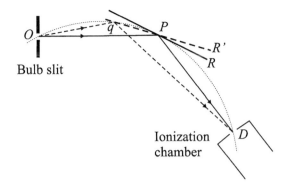

Fig. 7.4   Principle of the Bragg ionization spectrometer. PR and PR': two successive positions of a lattice plane. The incident beams OP and Oq make the same angle with the lattice plane. Redrawn after Bragg and Bragg (1913a).

holder could rock around a horizontal axis so as to permit adjustment of the orientation of the reflecting planes. The reflected beam of X-rays was received in the ionization chamber mounted so as to be able to rotate around the same axis as the crystal, and connected to a Wilson gold-leaf electroscope. The bulb slit O and the slit D of the ionization chamber were placed at equal distances of the crystal, P (Fig. 7.4). Let PR and PR' be two successive positions of a lattice plane when the crystal is rotated, and let q be the intersection of a circle passing through O, P and D with PR'. The incident beams OP and Oq make the same angle with the lattice plane, and the reflected beam is always focalized at the entrance D of the ionization chamber. This is essentially the geometry of early industrially-produced X-ray diffractometers. W. L. Bragg (1961) remembered: 'You must find it hard to realize nowadays what brutes X-ray tubes were. The measurement of ionization with a Wilson gold-leaf electroscope was quite an art too, and my father had thoroughly mastered all the techniques in his researches'.[11]

Several series of measurements of the reflected intensities were recorded with rock-salt and pyrite. This is when W. H. Bragg made the next great discovery: 'In addition to the white radiation, he found that each metal used in the X-ray tube as a source of radiation gave a characteristic X-ray spectrum of definite wavelengths, just as elements give spectra in the optical region' (Bragg 1943). The variation of the reflected intensity with angle of incidence is shown in Fig. 7.5 for an anticathode of platinum and reflections from the (100) and the (111) faces of rock-salt (Bragg and Bragg 1913a, read 17 April). Three peaks can be observed for each order of reflection, labelled A, B, and C, respectively. They were later identified with the three characteristic lines of the platinum target of the X-ray tube, $L\alpha$, $L\beta$, $L\gamma$. Each line in turn was isolated by the use of narrow slits and their absorption by aluminium sheets measured. The angles of reflection of the three lines were also measured with various crystals: potassium ferrocyanide, cleavage faces of calcite, and zinc-blende.

W. H. Bragg and W. L. Bragg (1913a) noted that the sharpness of the diffraction peaks indicated that the waves 'must occur in trains of great length', and not in short pulses. They judged that the results did not affect the use of the corpuscular theory, since the theory simply

---

[11]See footnote 29 in Chapter 5 about the sensitivity of the gold-leaf electroscope.

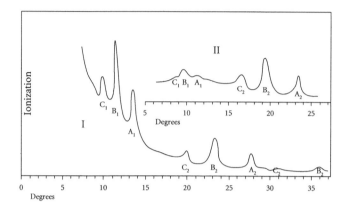

Fig. 7.5   First X-ray spectrum: reflection of platinum radiation by a sodium chloride crystal. Vertical axis: intensity
of reflected beam. Horizontal axis: angle of incidence. The three peaks, *A*, *B*, *C*, correspond to the three platinum
lines, *Lα*, *Lβ*, *Lγ*, respectively. I. 200 and 400 reflections. II. 111 and 222 reflections. After Bragg and Bragg
(1913*a*).

represented the transfers of energy from electron to X-rays and *vice versa*, but that the problem
of how the two hypotheses, corpuscular and wave, can be linked remained open.

From Bragg's law, $2d \sin\theta = n\lambda$, and an assumed structure of rock-salt with a value of $a =$
4.45×10^{-8} Å for its unit cell parameter, the authors initially estimated the wavelength of the *B*
line to be $\lambda = 0.89 \times 10^{-8}$ Å. However, after W. L. Bragg (1913*b*, read 26 June) had correctly
determined the structure of NaCl, a corrected value of the parameter, $a = 5.62$ Å, was adopted
(to be compared to the present accepted value, $a = 5.64$ Å). It was also observed that the *B* line
was in fact a doublet. Using the 200 reflection, W. H. Bragg (1913*b*, read 26 June) obtained a
value of $\lambda = 1.10 \times 10^{-8}$ Å for the average wavelength of the doublet.

W. H. Bragg also had X-ray tubes made with anticathodes of osmium, iridium, palladium,
rhodium (W. H. Bragg 1913*c*), copper, and nickel. He determined the values of the wavelengths
of their characteristic lines, based on the known structure of rock-salt, measured their absorption
coefficients, and identified the discontinuities later attributed to absorption edges (W. H. Bragg
1913*c*, *d*, 1914*a*). Whiddington (1911) had established the energies of the cathode rays required
to excite Barkla's *K* and *L* radiations in atoms of different atomic weight. W. H. Bragg was
able to link wavelength with energy, and show that the relation was in agreement with Planck's
quantum law (Bragg and Bragg 1915). He also found that the characteristic wavelengths from
successive elements in the Periodic Table were roughly proportional to the square of their
atomic weights. W. L. Bragg (1952) wrote: 'it was my father who founded X-ray spectroscopy
and first established the existence of relations between the spectra of successive elements in the
Periodic Table'.

A detailed description of the X-ray spectrometer was given in W. H. Bragg (1914*b*) and in
Bragg and Bragg (1915). W. H. Bragg wrote enthusiastically in the University's Annual Report
for the year 1912–13 that 'a new crystallography is rapidly being created' (quoted by Jenkin
2008), 'new crystallography', a term also used by W. H. Bragg in a lecture given on 5 June 1914

at the Royal Institution,[12] and taken up later by Ewald (1962*a*, *b*). W. L. Bragg (1943) noted: 'the X-ray spectrometer opened a new world. It proved to be a far more powerful method of analyzing crystal structure than the Laue photographs I had used. On the other hand a suitable crystal face could be used to determine the wavelength of the characteristic X-rays coming from different sources'.

7.7   W. L. Bragg and the first structure determinations

W. L. Bragg did not have the same facilities in the Cavendish as his father had in Leeds, and remembered (Bragg 1975): 'When I achieved the first X-ray reflections, I worked the Ruhmkorff coil too hard in my excitement and burnt out the platinum contact. Lincoln, the mechanic, was very annoyed as a contact cost ten shillings, and refused to provide me with another one for a month'. Fred Lincoln was the head laboratory assistant at the Cavendish, whose 'his fierce eye and even fiercer moustache . . . induced a very proper respect in the young research worker applying to him for apparatus or stores'[13]. During the winter of 1912–13 the younger Bragg learnt crystallography and the use of Miller indices, and visited William Barlow, at W. Pope's suggestion (Hunter 2004).

Following his preliminary study of zinc-blende, W. L. Bragg (1913*b*) made, during the first term of 1913, the first comprehensive analysis of crystal structures. These structures were simple by modern standards but were not so easy to obtain at the time. One must remember that the structure factor had not yet been introduced! The only information Bragg had at his disposal was the position of the diffraction spots and their relative intensities. The analysis was simplified for the first crystals he considered because they were all cubic and binary compounds: alkali halides, sylvine (KCl), rock-salt (NaCl), KI, KBr, and zinc-blende (ZnS). In Cambridge, he took Laue diagrams; in Leeds, with his father, he used the spectrometer.

As mentioned by G. Wulff, M. de Broglie, and W. L. Bragg himself, in a Laue diagram, the beams reflected from planes belonging to a given zone lie on a cone having as axis the zone axis. The intersection of this cone with the photographic plate is an ellipse, which W. L. Bragg found inconvenient to reproduce. He preferred to use a stereographic projection on the photographic plate of the intersections of the reflected beams with a sphere centred at the crystal. The ellipses are then replaced by circles, as shown in Fig. 7.6, *Left*, representing the Laue diagram of a potassium chloride crystal, KCl, oriented with a cubic face normal to the incident beam. Bragg indexed the spots, which lie, as expected, on circles corresponding to various zones. He found that they could be associated with the lattice planes of a primitive cubic lattice of edge a, as shown in Fig. 7.6, *Right*. This was confirmed by a Laue diagram taken with a three-fold axis parallel to the incident beam. The structure could thus be interpreted with KCl molecules at the nodes of a simple cubic lattice. Laue pictures taken with other alkali halides were, however, different, with different intensity ratios between spots and, more importantly, with some spots missing. These patterns could only be understood by assuming that the distance between neighbouring (111) planes was not $a\sqrt{3}/3$, as expected in the case of the KCl diagrams, but $2a\sqrt{3}/3$ (see Fig. 7.6, *Right*). This was confirmed directly by comparing the position of

[12]W. H. Bragg (1914). X-rays and crystalline structure. *Nature* **93**, 494–498.

[13]W. L. Bragg (1954). Mr Fred Lincoln. *Nature* **174**, 953.

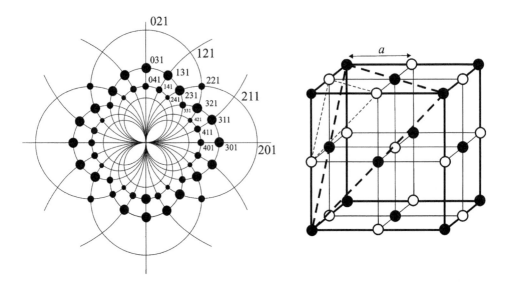

Fig. 7.6 *Left*: Stereographic projection of the Laue diagram of a potassium chloride crystal. Redrawn after W. L. Bragg (1913*b*). *Right*: Structure of alkali halides.

the diffraction lines from octahedral faces of KCl and NaCl, respectively, obtained with the spectrometer. The only possible explanation was that the structure was face-centred cubic, with one type of atom at the nodes of a face-centred cubic lattice and the other type of atom at mid-distance between two atoms of the first type along a cube edge (Fig. 7.6, *Right*). The atoms of the second type also lie at the nodes of a f.c.c. lattice.

The difference between the case of sylvine (potassium chloride, KCl) and that of rock-salt (potassium chloride, NaCl) and the other alkali halides was at first sight baffling. The solution did not come immediately. As we know now, the K^+ and Cl^- ions both have eighteen electrons, and their atomic form factors are very close. At the time, it was thought that the scattering power of an atom was roughly proportional to its atomic weight. The atomic weights of potassium and chlorine are 39 and 35, respectively, and do not differ very much. Their scattering powers were therefore expected to be very close, and the potassium and chlorine atoms to be difficult to distinguish in the diffraction process. It is for that reason that the crystal lattice seems to be simple cubic with parameter a (Fig. 7.6, *Right*). On the other hand, the atomic weight of sodium is 23, and is quite different from that of chlorine. The diffraction pattern of sodium chloride, rock-salt, is that of a face-centred cubic lattice, with lattice parameter $2a$, and looks different from that of potassium chloride. In a conference delivered to the Manchester Literary and Philosophical Society on 18 March 1914[14], W. H. Bragg still presented sylvine as having a simple cubic lattice, although his son had shown it to be face-centred cubic nearly a year before (W. L. Bragg 1913*b*).

[14]W. H. Bragg (1914). Crystalline structures as revealed by X-rays. *Nature* **93**, 124–126, 2 April 1914.

In the same paper, W. L. Bragg determined the crystal structure of zinc-blende and made preliminary considerations about the structures of fluorite, CaF_2, and calcite. The method was the same, by analyzing which diffraction spots were present and by comparing their relative intensities.

Together, father and son solved the structure of diamond (Bragg and Bragg 1913*b* and *c*, 30 July). They recorded the diffraction spectra from the (100), (110) and (111) faces of a diamond crystal with a rhodium anticathode. They showed that diamond was face-centred cubic. The lattice parameter (unit-cell edge) was measured thanks to Bragg's law. From the density of the crystals, W. H. and W. L. Bragg showed that there were eight atoms of carbon per unit cube, and, therefore, that there were two types of carbon atoms. They observed the absence of the 200, 600 and 222 reflections, while 400, 800, 111, 333, 444 and 555 were present. From this information, father and son deduced that carbon atoms of one type were located at the centres of tetrahedra whose summits were occupied by atoms of the second type (Fig. 7.7), in agreement with the model predicted by Barlow (Section 12.13). The argument runs as follows. The structure of diamond can be described as the stacking of double sheets of carbon atoms parallel to the octahedral planes, (111); the distance between the two planes, A and B, of the double sheet is one fourth of the period, $d_{111}/4$ (Fig. 7.7, *Left*). Bragg's law is, for the second order reflection, 222:

$$2d_{111} \sin\theta = 2\lambda, \tag{7.5}$$

which can also be written:

$$2\frac{d_{111}}{4} \sin\theta = \frac{\lambda}{2}, \tag{7.6}$$

which shows that the contributions of the two planes, A and B, to 222 are in opposite phase and cancel out. This was the clue to the structure determination.

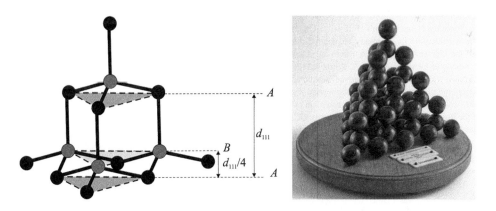

Fig. 7.7   *Left*: Structure of diamond. The two types of diamond atoms are represented with different colours. In the structure of zinc-blende, ZnS, one type of atom, for instance sulphur, occupies the positions of the black circles, and the other type, zinc, occupies the positions of the grey circles. *Right*: Bragg's model of diamond. Museum of the Royal Institution, London. Photo by the author.

The structure was confirmed by analysis of Laue diagrams taken with a three-fold axis parallel to the incident beam. The structure of zinc-blende is easily deduced from that of diamond by replacing one type of carbon atom by sulphur atoms (for instance the black circles in Fig. 7.7, *Left*) and the other type by zinc atoms (the grey circles in Fig. 7.7, *Left*). It can be described as a stacking of double layers of sulphur and zinc atoms parallel to the octahedral planes {111} (Fig. 7.17, *Right*, p. 166).

The determination of the diamond structure roused considerable interest because of the elegant way it was derived and because it corroborated both Kekulé's[15] theory of the tetravalency of carbon and van't Hoff's[16] and Lebel's[17] hypothesis of the tetrahedral character of carbon's four bonds in methane (Section 12.13). The structure of alkali halides was not so easily accepted by chemists, as it went against their molecular theory (Section 10.1.2). For instance, the Dutch crystallographer J. M. Bijvoet[18] recalled the painful memory of a visit Sir William Bragg paid to the Amsterdam laboratory: 'So persistent were they [the chemists] in pouring out the hackneyed arguments against the results of the X-ray analysis that Sir William in despair, raised his arms to heaven' (J. M. Bijvoet in Ewald 1962*a*). Bragg's model of rock-salt was still not accepted by the chemists J. Alexander, in the USA in 1924 (see Section 10.1.5) and H. E. Armstrong (see Section 1.2) in the UK!

W. L. Bragg later remembered the time when he and his father joined forces (W. L. Bragg 1943, 1961, 1970), and worked in close cooperation: 'I had the heaven-sent opportunity at this stage of joining in the work with the ionization spectrometer. My father and I worked furiously all through the summer of 1913, using the spectrometer ... It was a glorious time, when we worked far into the night with new worlds unfolding before us in the silent laboratory. It was like discovering an alluvial gold field with nuggets lying around waiting to be picked up'

[15]Friedrich Kekulé von Stradonitz was a German organic chemist, born 7 September 1829 in Darmstadt, Germany, died 13 July 1896 in Bonn, Germany. He received his higher education at the University of Giessen and was professor at the Universities of Ghent and Bonn. He is recognized as one of the main founders of the theory of chemical bonds. He discovered the tetravalency of carbon and proposed the structure of benzene with alternating single and double bonds.

[16]Jacobus Henricus van't Hoff was a Dutch chemist, born 30 August 1852 in Rotterdam, the Netherlands, died 1 March 1911 in Berlin, Germany. He studied chemistry at the Delft University of Technology, the University of Leiden, and the University of Utrecht, where he obtained his doctorate in 1874. He also studied with the organic chemists F. Kekulé in Germany and C. A. Wurtz (1817–1884) in France (Wurtz is well-known for his studies of structures of chemical compounds and of chemical reactions; the mineral wurtzite, zinc sulphide with a hexagonal close packing, a polymorph of cubic zinc-blende, was named after him). Van't Hoff was appointed Professor of Chemistry, Mineralogy, and Geology at the University of Amsterdam in 1878, and in 1896 moved to Berlin as Honorary Professor and member of the Royal Prussian Academy of Sciences. He is best known for his work on stereochemistry and osmotic pressure. He was the first Nobel Prize Laureate in Chemistry, in 1901.

[17]Joseph Achille Le Bel was a French chemist, born 21 January 1847 in Pechelbronn, France, died 6 August 1930 in Paris, France. A former student of Ecole Polytechnique, he studied with C. A. Wurtz. He is best known for his work on the optical activity of organic compounds and in stereochemistry.

[18]Johannes Martin Bijvoet, born 23 January 1892 in Amsterdam, the Netherlands, died 4 March 1980 in Winterswijk, the Netherlands, was a Dutch chemist and crystallographer. He studied Chemistry at the University of Amsterdam, and after World War I, he set up a X-ray laboratory in Amsterdam, having learnt X-ray crystallography during brief visits to N. H. Kolkmejer in Utrecht and to W. L. Bragg in Manchester. In 1928, Bijvoet was appointed lecturer in Amsterdam, and in 1939 Professor of Chemistry at the University of Utrecht. He is well-known for devising a method of establishing the absolute configuration of molecules. After World War II, he played an important role in the newly founded International Union of Crystallography, of which he was President from 1951 to 1954. For details about Bijvoet's life, see MacGillavry, C. H. and Peederman, A. F. (1980). Johannes Martin Bijvoet. *Acta Crystallogr.* **A36**, 837–838.

(Bragg 1943). W. H. Bragg's notebook for the year 1913[19] reveals the tremendous amount of activity of father and son. Besides the crystals whose structures were published (W. L. Bragg 1913*b*, 1914*a*; Bragg and Bragg 1913*a* and *b*), diffraction spectra were recorded from many crystals, particularly quartz, for which there are many entries in the notebook, but also ammonium chloride (NH_4Cl), senarmontite (Sb_2O_3), spinel ($MgAl_2O_4$), magnetite (Fe_3O_4), sulphur, and cuprite (Cu_2O). The structure determinations of fluorite (CaF_2), pyrite (FeS_2), sodium nitrate ($NaNO_3$), calcite ($CaCO_3$), dolomite ($[Ca,Mg]CO_3$), rhodochrosite ($MnCO_3$), siderite ($FeCO_3$), were submitted in November 1913 and published in early 1914 (W. L. Bragg 1914*a*), see Section 8.3.1; preliminary considerations relative to the structure of sulphur and quartz were submitted in December 1913 and published in January 1914 (W. H. Bragg 1914*d*). The minerals were lent from the mineral collection at Cambridge by A. Hutchinson, then Demonstrator and Lecturer of Mineralogy. W. L. Bragg (1949) remembered: 'I shall never forget Hutchinson's kindness in organizing a black market in minerals to help a callow young student. I got all my first specimens and all my first advice from him'. The Notebook shows that the minerals were duly returned!

W. L. Bragg was encouraged by the results he had obtained with alkali halides and the carbonates and he wrote optimistically (W. L. Bragg 1914*a*): 'we can obtain enough equations to solve the structure of any crystal, however complicated, although the solution is not always easy to find'. He summarized later the way to determine structures (Bragg 1943): 'One could examine the various faces of a crystal in succession, and by noting the angles at which and the intensity with which they reflected the X-rays, one could deduce the way in which the atoms were arranged in sheets parallel to these faces. The intersections of these sheets pinned down the positions of atoms in space'. During the year 1914, back in Cambridge, W. L. Bragg now had at his disposal a spectrometer after his father's design, and he tried to attack more complex structures, of lower symmetry than cubic, such as quartz (trigonal), wurtzite (ZnS, hexagonal) and aragonite ($CaCO_3$, orthorhombic), but he was also very busy with the book he and his father were writing (Bragg and Bragg 1915), and, apart from the structure of copper, submitted in July 1914 (W. L. Bragg 1914*b*), no more structures were completed before the outbreak of the First World War.

In 1913, W. L. Bragg had been approached about a position at the University of British Columbia, in Vancouver, Canada, but he had wisely refused. In 1914 he was elected to a Fellowship at Trinity College, Cambridge, and was appointed lecturer in Natural Science. His first research assistant was E. V. Appleton (1892–1965), the future 1947 Nobel Prize in Physics winner for his research on the ionosphere which led to the development of radar. World War I began on 28 July 1914, and, on August 1914, W. L. Bragg was sent to a Territorial battery, the Leicestershire Royal Horse Battery. In August 1915 he was appointed Technical Advisor on Sound Ranging and, in September, was sent to France, where he remained until the end of the war.

The working relationship between father and son was unique and remarkably fruitful, but, naturally, it was the father, W. H. Bragg, the recognized scientist, who was invited to present their results at conferences and meetings. W. L. Bragg wrote many years later: 'Inevitably the results with the spectrometer, especially the solution of the diamond structure, were far more

---

[19]<http://www.leeds.ac.uk/library/spcoll/bragg-notebook/pdf.htm>.

striking and far easier to follow than my elaborate analysis of Laue photographs, and it was my father who announced the new results at the British Association, the Solvay Conference, lectures up and down the country and in America, while I remained at home' (unpublished memoirs, quoted by Perutz 1990). The father always gave due credit to his son and was very proud of his 'boy'. In the preface to their common book (Bragg and Bragg 1915), he wrote: 'I am anxious to make one point clear, *viz.*, that my son is responsible for the *reflection* idea which has made it possible to advance, as well as for much the greater portion of the work of unravelling crystal structure to which the advance has led'. Nevertheless, W. L. Bragg could not help feeling some frustration, and, as wrote one of his biographers, Sir David Phillips (1979), who knew him well, and worked closely with him, W. L. Bragg's father, 'however hard he tried, did not quite avoid leaving the impression that his was the guiding part: what was no more than fair looked like parental generosity'. Phillips added: 'there is no doubt that a cloud remained that overshadowed W. L. B.'s future relationship with W. H. B., and was remembered 60 years later with pain mixed with gratitude for his father's part in making possible the rapid development of the work'.

For details about the early developments of X-ray analysis, see W. L. Bragg (1943, 1961, 1967, 1975), W. L. Bragg in Ewald (1962*a*), Phillips (1979), Hunter (2004), Jenkin (2008).

7.8   H. G. J. Moseley and the high-frequency spectra of the elements

7.8.1   *Moseley's personality*

Henry Moseley (Fig. 7.8) was an English physicist who, in his short life, made major contributions to our knowledge of the structure of the atom. Firstly, he showed that atomic numbers are not simply ordinal numbers but that they represent a fundamental property of the atoms, the number of the electrons and the charge of the nucleus. Secondly, he established the relationship between these atomic numbers and the frequencies of the X-ray spectra, a significant contribution known as 'Moseley's law'.

His remarkable personality has been described by his contemporaries: extraordinary power of work (Rutherford 1915*b*; Fajans 1916; Darwin in Ewald 1962*a*), and outstanding experimental skillfulness and ingenuity (Fajans 1916). Rutherford (1915*b*) summed this up by describing Moseley as a 'born investigator, with a great originality, endowed with unusual intellectual powers and equipped with a good mathematical training'. At the same time, his 'cheerfulness and willingness to help in all possible ways endeared him to all his colleagues (Rutherford 1915*a*), a quality also underlined by the Polish chemist K. Fajans (1887–1975), with whom Moseley collaborated in Rutherford's laboratory. According to Darwin (in Ewald 1962*a*), Moseley had two principles: 'the first was that when one starts to set up an experiment one must not stop for anything until it is set up. The second was that when one starts the experiment itself one must not stop till it is finished'. Darwin (in Ewald 1962*a*) also remembered that there was no regular meal time for him, and that work often went on for most of the night. He noted 'indeed one of Moseley's expertises was the knowledge of where one could get a meal in Manchester at 3 o'clock in the morning'. Darwin found nevertheless that 'it was most agreeable to work with him'.

Since he had a father and two grandfathers who were scientists, all of them Fellows of the Royal Society, Moseley was brought up in a scientific atmosphere. A year before his graduation

Fig. 7.8 Henry G. J. Moseley around 1910. Image © Museum of the History of Science, Oxford, UK, with permission.

Henry Gwyn Jeffreys Moseley: born 23 November 1887 in Weymouth, Dorset, England, the son of H. N. Moseley, Professor of Anatomy and Physiology at Oxford University, F.R.S.; died 10 August 1915 in Gallipoli, Ottoman Empire, was an English physicist. After schooling at Eton, he entered Trinity College, Oxford, at the age of 18, and obtained his BSc in Mathematical Moderations and in Natural Science in 1910. Immediately afterwards, he was appointed Lecturer and Demonstrator at Manchester University, where he worked with E. Rutherford. After two years, however, he resigned his lectureship to devote himself entirely to research and was awarded a John Harling Fellowship. At the beginning of 1914 he returned to Oxford to live with his mother, and to continue his experimental work in the laboratory of Professor Townsend. In the summer of 1914, he travelled with his mother to Australia, via Canada, to attend a meeting of the British Association for the Advancement of Science. When war was declared, he returned to England where he was granted a Commission in the Royal Engineers. He was killed soon after his arrival in the Dardanelles.

MAIN PUBLICATIONS

1912 *The Number of β-Particles Emitted in the Transformation of Radium.*

1913 *The reflection of X-rays.* With C. G. Darwin.

1913 *The high frequency spectra of the elements.*

1914 *The high frequency spectra of the elements. II.*

from Oxford, he had decided to undertake original work in physics and he visited Manchester to discuss the matter with E. Rutherford (Rutherford 1915*a*). Up to the end of 1912, like the rest of Rutherford's laboratory, Moseley's work had been devoted to the study of radioactivity. His first work consisted in determining the number of β-particles emitted in the transformation of radium (Moseley 1912). It is then that N. Bohr made his first visit to Manchester. He greatly influenced Moseley and his colleagues by 'his deep appreciation of the limitations of the classical theory' (Darwin in Ewald 1962*a*). Physicists at that time found it difficult to reconcile the conflicting views of classical mechanics and the quantum theory. Bohr's quantum model for the structure of the atom (Bohr 1913) had not been accepted by all at that time (see, for instance, Kragh 2011).

7.8.2 Early work with C. G. Darwin

When the news of the Laue–Friedrich–Knipping experiment and of W. L. Bragg's interpretation came round, H. Moseley proposed to 'take up the study of X-ray diffraction as being the most exciting new field in physics' (Darwin in Ewald 1962*a*), and C. G. Darwin agreed to join him, although they 'did not know in the least what would come out of it'. When approached about this new line of study, Rutherford was at first 'distinctly discouraging' (Darwin in Ewald

1962*a*). Moseley and Darwin found out Laue's errors in the interpretation of the diagrams (see Chapter 6), and presented their own analysis of Laue's experiment at the Manchester physics colloquium on 1 November 1912, in the presence of W. H. Bragg (the elder) (see Section 6.5). Bragg told them that his son had reached a similar conclusion. Moseley and Darwin did not have, however, a full explanation because they did not then know the exact structure of zinc-blende, not even its lattice-mode, which they learned after reading Tutton's (1912*b*) letter to *Nature*. While they left the theoretical analysis to W. L. Bragg, they decided to go ahead experimentally, using W. L. Bragg's reflection geometry arrangement (W. L. Bragg 1912). They also decided to employ the ionization chamber technique used by W. H. Bragg (1913*a*, 17 January) to observe the reflection of X-rays from a mica sheet. They preferred it to the photographic one, because they thought it would be more appropriate to obtain quantitative results. Their initial observations were published in a note to *Nature* dated 21 January 1913 (Moseley and Darwin 1913*a*).

They then proceeded to study systematically the reflection of X-rays by cleavage faces of rock-salt, potassium ferrocyanide ($K_3[Fe(CN)_6]$, $3H_2O$), and selenite (gypsum). At first, they used very fine slits, and only observed the white spectrum of the platinum anticathode. At about the same time, February–March 1913, the Braggs observed selective reflection at specific angles, which they attributed to the reflection of characteristic X-ray wavelengths. They informed Moseley and Darwin, and published their results in April 1913 (Bragg and Bragg 1913*a*). Moseley presented the early observations of the Manchester group at one of the Friday physics colloquia, when W. H. Bragg was visiting the laboratory. W. H. Bragg told Moseley and Darwin about the characteristic lines (Darwin in Ewald 1962*a*), and, according to another member of Rutherford's laboratory, the Hungarian chemist G. von Hevesy,[20] Moseley visited nearby Leeds to see W. H. Bragg's apparatus. Following Bragg's advice, Moseley and Darwin modified their set-up and then were able to observe the characteristic radiation lines of platinum (Darwin in Ewald 1962*a*). They made careful measurements of the positions and intensities of the selective reflections of these lines and found, in fact, two more lines than the Braggs had. They had some difficulties because W. L. Bragg had not yet solved the structure of rock-salt and they were uncertain about its lattice parameter (W. L. Bragg's paper on the structure of rock-salt appeared in June 1913). Their work was published in July 1913 (Moseley and Darwin 1913*b*).

In June 1913, N. Bohr[21] visited Manchester and had a very interesting discussion with H. Moseley, C. Darwin, and G. von Hevesy about the proper sequence for the arrangement of the elements in the Periodic Table, in particular for nickel and copper, and which sequence the X-ray spectra followed, atomic weight or rank in the Periodic Table (Hevesy 1923). In 1869 Mendeleev had listed the elements according to their atomic weight, with a few inversions, such as nickel and copper, in order to classify them more correctly into chemical families, each element being attributed an ordinal number (*Ordnungszahl*). The relation between atomic

---

[20]Interview of G. von Hevesy, American Institute of Physics, 25 May 1962 and 4 February 1963, <http://www.aip.org/history/ohilist/4670_1.html>.

[21]Niels Bohr, born 7 October 1885 in Copenhagen, Denmark, the son of Christian Bohr, Professor of Physiology at Copenhagen University, died 18 November 1962 in Copenhagen, Denmark, took his MSc in physics in 1909 and his PhD in 1911. He visited J. J. Thomson in 1911 and E. Rutherford in 1912. In 1913–1914 he was lecturer at Copenhagen University and, in 1914–1916, at the Victoria University in Manchester. In 1916, he was appointed Professor of Theoretical Physics at Copenhagen University, and from 1920 until his death in 1962, was at the head of the Institute for Theoretical Physics, established for him at that university. In 1922 he was awarded the Nobel Prize in Physics 'for his services in the investigation of the structure of atoms and of the radiation emanating from them'.

weight and the number of electrons in the atom was unclear and varied according to the model of atomic structure.

A few months after Moseley's arrival in the laboratory, Ernest Rutherford had experimentally shown the existence of the nucleus, a notion already put forward mathematically by Nagaoka (1904), but without any compelling evidence in its favour. In order to explain the high scattering angles of α-particles observed by his students, H. Geiger and E. Marsden (1909), Rutherford was led to imagine that the positive charge of the atom was concentrated at its centre, contrary to J. J. Thomson's concept that it was distributed throughout its volume. Rutherford's calculations showed that the positive charge is of the order of half the atomic weight, for instance of the order of 100 for gold, a result also reached by C. G. Barkla (1911) on the basis of J. J. Thomson's (1903) theory of scattering (in fact the value is 79).

On his return to Denmark in the summer of 1912, N. Bohr applied the quantum theory of radiation to his nucleus theory. As a result, he developed a model of the structure of the atom which substantiated Rutherford's concept. His work was published in three instalments in April, September, and December 1913 (Bohr 1913).

In January 1913 the amateur physicist Antonius van den Broek (1870–1926) proposed that the charge N in the nucleus of an atom should be exactly equal to its position, Z, on the Periodic Table: 'if all elements be arranged in order of increasing atomic weights, the number of each element in that series must be equal to its intra-atomic charge' (van den Broek 1913). The idea was discussed by N. Bohr (1913) in the second of his three papers, with a direct reference to van den Broek's work. It was, however, already present in his mind at the time of the aforementioned discussion with Moseley, Darwin, and von Hevesy.

According to Rutherford's nuclear theory, the properties of an atom depend essentially on the magnitude of the charge of the nucleus, that is on its atomic number.[22] He realized that it was therefore 'of fundamental importance to settle whether the properties of the atom are defined by the nuclear charge rather than by its atomic weight' (Rutherford 1915b).

It may never be known whether Moseley's decision to study the relationship between X-ray spectra and atomic number experimentally was the outcome of the discussion with Bohr, von Hevesy, and Darwin, as recalled by von Hevesy (1923) (*Wir werden sehen welche Grösse für die Röntgenspektra massgebend ist*—We shall see which quantity is decisive for the X-ray spectra), or whether he had already thought of it, as claimed by Bohr (1961). In any case, Darwin and Moseley now followed different paths. Darwin developed the theory of diffraction (see Section 8.4) and Moseley attacked the problem directly, by comparing the X-ray spectra of a series of elements.

7.8.3 Moseley's law

After spending some time using ionization methods (Darwin in Ewald 1962a), Moseley decided to employ a photographic technique to obtain the spectra of each element in one shot. For that, he modified the apparatus used by G. W. C. Kaye (1909) to study characteristic X-ray radiation from a variety of targets. The equipment is shown in Fig. 7.9. In the first investigation (Moseley

---

[22]The first occurrences of the designation, 'atomic number' in the literature are to be found in Moseley (1913) and Rutherford (1913). From Moseley (1914), it seems that the name was coined by Moseley, presumably after talking with Rutherford. The author is grateful to H. Kragh for discussions on this point.

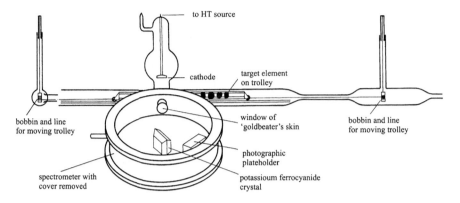

Fig. 7.9   Schematic diagram of Moseley's apparatus for recording X-ray spectra. X-rays proceed from the cathode to
the crystal and further to the photographic plate. Image © Museum of the History of Science, Oxford, UK, with
permission.

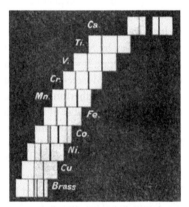

Fig. 7.10   X-ray spectra arranged on horizontal lines for each element, and placed in register. Those parts of the photographs which
represent the same angle of reflection are in the same vertical line.
After Moseley (1913).

1913), samples of twelve elements, from calcium to zinc, were arranged on an aluminium trolley
placed directly in the discharge tube, so that they could serve as an anticathode one after the
other. This trolley could be drawn to and fro by means of a silk fishing line wound on two
brass bobbins as shown in Fig. 7.9. As an analyser crystal he used a fine crystal of potassium
ferrocyanide with a good and large cleavage face. Its lattice parameter was determined by
comparison with that of rock-salt. The slit that collimated the X-ray beam from the tube, of
width 0.2 mm, and the photographic plate were placed on a cylindric spectrometer centred at
the analyzing crystal, ensuring focalization (Fig. 7.9).

The spectra of all the elements consisted of two strong lines, K-lines, which further investigation showed to be doublets, and were observed for two orders. Figure 7.10 shows the third order
spectra of ten elements chosen as forming a continuous series with only one gap, scandium
(Moseley 1913). These spectra are placed approximately in register so that the parts of each
photograph corresponding to the same angle of reflection are in the same vertical line. In a
further investigation (Moseley 1914), more than thirty heavier elements were studied, and their

L-lines photographed. Lumps of the pure elements (Mg, Al, Si, Mo, Ru, Pd, Ag, Sb, Ta), foils (Rh, W, Au), chemical deposits (Os) or alloys (ZrNi, WFe, NbTa, SnMn) were used. KCl and the oxides of the rare elements were rubbed on the surface of nickel plates.

Moseley found that the square root of the frequencies of the corresponding lines in each spectra increased by a constant amount when passing from one element to the next using the chemical order of the elements in the Periodic Table, and that, 'except in the case of nickel and cobalt, this is the order of atomic weights'. In Fig. 7.11, the square-roots of the frequencies of the X-ray spectra are arranged on horizontal lines for each element, characterized by its atomic number, spaced at equal distances. The spectra consist of two lines for the K-series of the lighter elements, $K\alpha$ and $K\beta$, and of five lines for the L-series of the heavier elements, $L\alpha$, $L\beta$, $L\gamma$, $L\delta$, and $L\epsilon$.

It can be seen that the spectra are aligned along straight lines in the figure, and Moseley showed that the frequencies, v, of the spectra are given by:

$$v = v_o A(N - b)^2 \qquad (7.7)$$

where N is an integer, v_o is the fundamental Rydberg frequency, and A and b are constants characteristic of each line:

$$K\alpha \ line: \quad A = [\frac{1}{1^2} - \frac{1}{2^2}] = 0.75; \quad b = 1; \qquad (7.8)$$

$$L\alpha \ line: \quad A = [\frac{1}{1^2} - \frac{1}{3^2}] = 0.14; \quad b \approx 7.4; \qquad (7.9)$$

Relation 7.7 is known as 'Moseley's law'. The X-ray spectrum of an element is entirely determined by the integer, N, which is equal to the charge of the nucleus, and to the rank of the element in the Periodic Table. It is the 'atomic number'. The atomic numbers were tabulated by Moseley assuming that the value of N for aluminium is 13.

Moseley interpreted the factor $1/1^2 - 1/2^2$ in equation (7.8) in relation to Bohr's model by saying that the transition was between two states in which the angular momentum of each electron was $2 \times h/2\pi$ and $h/2\pi$, respectively, and suggested that the number of electrons in the ring was equal to four, but Kossel[23] (1914) and Bohr (1915) pointed out that that was not correct.

The square root of $N = Z$ increases by a constant value when passing from one element to the next in the order of the Periodic Table, and known elements correspond to all the numbers on the vertical axis of Fig. 7.10, except three, numbers 43, 61, and 75. Moseley therefore suggested that there were three elements not yet discovered. This was in agreement with the predictions of the chemists. There was, however, some confusion concerning the rare earths; Moseley had

---

[23]Walther L. J. Kossel, born 4 January 1888 in Berlin, Germany, the son of Albrecht Kossel, Nobel Prize in Physiology in 1910, died 22 May 1956 in Tübingen, Germany. He prepared his thesis in Heidelberg under P. Lenard, whose assistant he became in 1910. In 1913 he moved to Munich, where he was Zenneck's assistant at the *Technische Hochschule* until 1920, but closely associated with A. Sommerfeld's Institute. He was then appointed professor at the University of Kiel and in 1932 at the University of Danzig, now Gdansk, Poland. He is known for his studies of X-ray spectra, his theory of the chemical bond, his model of crystal growth, together with I. Stranski, and for his discovery of the crystal sources of X-rays, or Kossel effect.

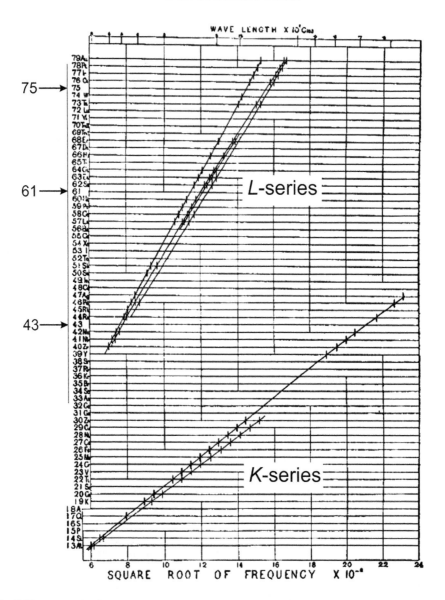

Fig. 7.11 Square-roots of the frequencies of X-ray spectra, represented horizontally for each element characterized
by its atomic number, shown vertically. Elements of atomic numbers 43, 61, and 75 correspond to elements
unknown at the time (see text). After Moseley (1914).

wrongly attributed number 72 to lutetium (71). Number 72 in fact also corresponds to a missing
element. This was made clear during a short visit to Oxford in May 1914 by the French authority
on rare earths, the chemist Georges Urbain (1872–1938), one of the discoverers of lutetium, who
brought with him several samples to be tested by Moseley (Rutherford 1915*b*; Heilbron 1966,
1974). These studies showed that there are, in fact, four missing elements between aluminium

and gold, later shown to be 43 (technetium), 61 (promethium), 72 (hafnium), and 75 (rhenium). The story of their discovery is told briefly in Section 7.8.4.

Moseley's work on the rare earths was his last. It was no doubt included in the report he presented to the British Association for the Advancement of Science on 25 August 1914, during its meeting in Australia (*High-frequency spectra*). Unfortunately, no written record has been kept.

The fundamental importance of Moseley's work was promptly recognized by the scientific world in general (Rutherford 1915*b*), except for some chemists.[24] A brilliant scientific career could be expected for him. This was cut short by his untimely death in the early stages of World War I. After the outbreak of war, he came back to England where he was granted a Commission in the Royal Engineers. He later became signalling officer to the 38th Brigade of the First Army and left for the Dardanelles on 13 June 1915. He took part in severe fighting and was killed by a sniper on 10 August 1915 (Rutherford 1915*a*). Moseley's death aroused the deepest regrets throughout the scientific community across the borders (Fajans 1916). He was nominated for the 1915 Nobel Prize in Physics (Heilbron 1974), and, had he survived, he would most probably have obtained it. Three weeks after Moseley's death, Lawrence Bragg's younger brother Bob was mortally wounded, nearly at the same place, in Gallipoli.

For further reading on H. G. J. Moseley's life and works, see, for instance: Rutherford (1915*a* and *b*, 1925), Fajans (1916), Darwin (in Ewald 1962*a*), Heilbron (1966, 1974).

7.8.4   Discovery of the missing elements

The four missing elements between Al (13) and Au (79), just mentioned, were eventually found.

Element 72, Hf, was discovered in Copenhagen by the Hungarian chemist George von Hevesy (1885–1966), a former member of Rutherford's laboratory, and the Dutch physicist Dirk Coster (1889–1950) in 1923, by X-ray spectroscopic analysis of zirconium ore. They called it hafnium, after the Latin name of Copenhagen, *Hafnia*. The discovery of hafnium was made known just before N. Bohr received his Nobel Prize, and was announced by him in his lecture. G. von Hevesy was later awarded the 1943 Nobel Prize in Chemistry 'for his work on the use of isotopes as tracers in the study of chemical processes'.

Element 75, Re, was discovered in 1925 in Germany by the German chemists Walter Noddack (1893–1960), his future wife Ida Tacke (1896–1978) and Otto Berg (1873–1939), by X-ray spectroscopic analysis of the mineral colombite. They also found it in gadolinite and molybdenite. They named it rhenium, after the Latin name, *Rhenius*, of the Rhine. They also thought that they had discovered element 43, Tc, technetium, but their experiment could not be reproduced by others and the element was only discovered later, in Italy, by Carlo Perrier (1886–1948) and Emilio Segré (1905–1989) in 1936.

Element 61, promethium, was produced and characterized in 1945 by Jacob A. Marinsky, Lawrence E. Glendenin, and Charles D. Coryell at Oak Ridge National Laboratory, who separated and analyzed the fission products of uranium fuel irradiated in a graphite reactor.

---

[24]The chemist H. E. Armstrong declared during the discussion on the Structure of Atoms in Australia that 'to conclude, on such evidence [the X-ray spectra], that all but very few of the elements are discovered is scarcely justifiable'.

There was in fact another element missing, beyond gold, element 85, At, astatine, which was discovered by Dale R. Corson, Kenneth Ross MacKenzie, and Emilio Segré in 1940 in California, USA.

7.9 P. Debye and the temperature factor

Peter Debye (Fig. 7.12) was regarded 'one of the best mathematicians of the day' (letter from E. Rutherford to G. N. Lewis, December 1915, quoted by his biographer, M. Davies[25]), but his approach to problems was not fundamentally a mathematical one, he was a 'model-builder'.[26] He had a 'predilection for the experimental control of theory' (Davies, see footnote 25) and this may have been due to his association with Sommerfeld of whom he had been the student and the first assistant, and whom he followed to Munich. In his speech when Debye was awarded the Planck Medal of the German Physical Society on 13 October 1950, Sommerfeld confirmed this judgment, speaking of P. Debye as a highly gifted mathematician who was drawn by experiment and whose motto, in science and in life, was 'it is all so terrible simple' (*Das ist alles so furchtbar einfach*)[27].

The five years Debye spent in Munich played a decisive role in his life-long interest in diffraction and in the interaction of radiation with matter.

In 1911, he succeeded A. Einstein as Professor of Theoretical Physics at Zürich University. Einstein had just published three papers on the specific heat of solids. In the third one, Einstein had shown that the oscillations of the molecules are far from monochromatic (Einstein 1911). Debye took up this problem and presented his results to the Swiss Physical Society on 9 March 1912[28] and in a paper to the *Annalen der Physik* in July 1912 (Debye 1912). Born and Kármán's paper on lattice vibrations appeared at approximately the same time (1912, 30 March). Debye's work resulted from his efforts to present the theory of radiation and specific heat in his lectures on thermodynamics at the University of Zürich during the winter of 1911–12. Debye showed that the mechanical system constituted by a lattice of vibrating atoms could be described as an 'agglomeration of a number of independent vibrating systems, each with its proper frequency' (later called 'phonons'); the question then was how to find the spectrum of frequencies. By a remarkable feat of mathematics, Debye evaluated the total spectral distribution of the elastic modes, and showed that the total number of permitted proper vibrations of a system of N atoms, is $3N$. He derived the expression of the specific heat (the 'Debye equation'), which is given in terms of a characteristic temperature, the 'Debye temperature', $T_D \approx h\nu_D/k$, where ν_D is the maximum frequency of the lattice, h and k the Planck and Boltzmann constants, respectively. The higher the Debye temperature, the harder a material is, as shown by Table 7.1 that gives the Debye temperature of a few substances.

---

[25]M. Davies (1970). Peter Joseph Wilhelm Debye, 1884–1966. *Biogr. Mem. of Fellows of the Royal Society.* **16**, 175–232.

[26]F. A. Long (1967). Peter Debye: An appreciation. *Science (USA).* **166**, 979–980.

[27]Überreichung der Planck-Medalle für Peter Debye durch Arnold Sommerfeld. (1950). *Phys. Bl.* **6**, 509–512.

[28]P. Debye (1912). Les particularités de la chaleur spécifique à basse température. *Arch. Sci. Phys. Nat. Genève.* **33**, 256–259.

Fig. 7.12   Peter Debye in 1912. Source: Wikimedia commons.

Petrus (Peter) Josephus Wilhelmus Debye: born 24 March 1884 in Maastricht, The Netherlands, died 2 November 1966 in Ithaca, New York, USA, was a Dutch physicist and physical chemist. After schooling in Maastricht he attended the Aachen University of Technology, graduating in 1905. There, he started research with A. Sommerfeld whom he followed to Munich in 1906. He obtained his PhD in 1908 on radiation pressure, and in 1910 became lecturer at the University of Munich. In 1911 he was appointed Professor at Zürich University, where he succeeded A. Einstein. He was then successively Professor at the Universities of Utrecht (1912) and Göttingen (1914), at the ETH Zurich (1920), at the University of Leipzig in 1927, and, finally, in Berlin, where, succeeding Einstein, he became in 1934 Director of the Kaiser Wilhelm Institute for Physics (now named the Max-Planck-Institut). From 1936 to 1939, he was also Professor of Theoretical Physics at the Friedrich-Wilhelms-Universität in Berlin. Debye left Germany in early 1940 for the United States, where he became professor at Cornell University, Ithaca, New York, and Chair of the Chemistry Department until his retirement in 1952. During the war he worked on polymers and synthetic rubber. He became an American citizen in 1946. Debye's research career was impressive and he made significant contributions in many topics, including specific heat and the Debye temperature, the influence of thermal motion on diffracted X-ray intensities, dipole moments, X-ray studies of electron distributions, and powder diffraction. He was awarded the 1936 Nobel Prize in Chemistry 'for his contributions to our knowledge of molecular structure through his investigations on dipole moments and on the diffraction of X-rays and electrons in gases'. Debye was Editor of *Physikalische Zeitschrift* from 1915 to 1940; he was awarded the Rumford Medal by the Royal Society in 1930, and he was elected a Fellow of the Royal Society in 1933. L. Onsager, who was awarded the 1968 Nobel Prize in Chemistry, P. Scherrer, and G. K. Fraenkel were among his best-known former students.

MAIN PUBLICATIONS

1913 *Interferenz von Röntgenstrahlen und Wärmebewegung.*

1916 *Interferenzen an regellos orientierten Teilchen im Röntgenlicht.* With P. Scherrer.

1918 *Atombau.* With P. Scherrer.

1929 *Polar molecules.*

1933 *Die Struktur der Materie.*

1935 *Kernphysik.*

Debye left Zürich in 1912 for Utrecht, in The Netherlands, where he was appointed Professor of Theoretical Physics. It is there that he calculated the intensity of X-rays diffracted by a crystal, extending Laue's derivation to take into account the vibrations of the diffracting atoms. His recent treatment of lattice specific heats provided him with a clear insight into the properties of lattice vibrations. The results were published in four publications in the summer of 1913, hardly more than one year after the discovery of X-ray diffraction (Debye 1913 *a* to *d*). His biographer, M. Davies (see footnote 25) considered this to be one of the great achievements in crystal physics. The first paper, dated 29 July 1913, begins by asking whether X-ray diffraction might give some information about the eventual existence of a zero-point energy at absolute zero, but the question remained open at the end of Debye's work, and had to wait for the paper

Table 7.1   Debye Temperature of a few substances, in Kelvin

Substance	Debye temperature
Lead	96 K
Ice	192 K
Silver	215 K
Copper	315 K
Iron-α	464 K
Silicon	640 K
Diamond	2200 K

by James *et al.* (1928), which confirmed Planck's prediction that the crystal possessed a zero-point energy that was half a quantum per degree of freedom (see Section 10.4.2).

The conclusions of Debye's calculations were:

1. Thermal agitation has an appreciable influence on the X-ray diffracted intensities.
2. The sharpness of interference maxima are not affected by thermal agitation.
3. Expression 6.3 of the diffracted intensity obtained by Laue (Section 6.6.5) should be multiplied by a factor $\exp -M$ (Debye temperature factor).
4. M is proportional to $\sin^2 \theta / \lambda^2$, where θ is the Bragg angle and λ the wavelength: the diffracted intensity decreases with increasing Bragg angle and increasing order of the reflection.
5. M depends on the ratio T_D / T between the Debye temperature and the absolute temperature. The diffracted intensity decreases with increasing temperature.
6. The interference maxima are accompanied by scattered radiation in the direction where the diffracted intensity is minimum (this is now called 'Thermal Diffuse Scattering (TDS)').

As numerical examples, Debye plotted the variations of $\exp -M$ with Bragg angle for diamond and sylvine, for which he took Debye temperatures of 1830 K and 219 K, respectively (Table 7.1 shows that the value for diamond has been reevaluated). He pointed out the very special case of diamond for which the influence of temperature agitation is very small, as a result of the abnormally high value of its Debye temperature.

The influence of temperature on the diffracted intensities was first studied experimentally by W. H. Bragg (1914c) with rock-salt. One of M. Siegbahn's students, H. Faxèn[29] (1918) showed that there were some inconsistencies in Debye's calculations. Further discrepancies between Debye's theory and experimental results were noted by G. E. M. Jauncey (1922) with calcite and rock-salt, and by E. H. Collins (1924) with aluminium, in the United States. These were confirmed by very careful measurements by R. W. James[30] (1925) on rock-salt in the temperature range between 19°C and 650°C, and by R. W. James and Miss E. Firth (1927) at the temperature of liquid air (−195°C), also on rock-salt.

---

[29]Hilding Faxèn, born 1892, died 1970, was a Swedish physicist active in the field of mechanics.

[30]See Fig. 8.13.

Another of M. Siegbahn's students, I. Waller[31] (Waller 1923, 1927), using a different method of connecting the amplitudes of the atomic vibrations with the normal vibrations of the crystal, showed in his thesis that Debye's temperature factor $\exp -M$ for the intensities should be replaced by $\exp -2M$, now called 'Debye–Waller factor', which is the temperature factor in use today.

7.10 Review of one year of X-ray diffraction: the Birmingham, Solvay, and Vienna meetings, autumn 1913

The first results obtained by diffraction of X-rays by crystals were reported at three major conferences in the fall of 1913. These were the meetings of the English scientists in Birmingham, England (September 1913), of the German scientists in Vienna, Austria (also September 1913), and at the International Solvay Conference in Brussels, Belgium (October 1913). Few new major results were announced at these conferences, but the discussions are most interesting. They reveal the concerns of the physicists of the time and show that diffraction phenomena were still far from being completely understood.

7.10.1 Eighty-third meeting of the British Association for the Advancement of Science, 10–17 September 1913, Birmingham

English scientists met in Birmingham 10–17 September 1913 for their annual meeting. X-ray diffraction was not a major topic at the conference. In his presidential address, Sir Oliver Lodge only briefly mentioned the work of Professor W. H. Bragg and his son, 'who have confirmed in a striking way' the 'theoretical anticipations' of Barlow, Pope, and Tutton in England, and of Groth and Fedorov abroad. With great forethought, he added: 'these brilliant researches, which seem likely to constitute a branch of physics in themselves, and which are continued by Moseley, Darwin, Keene and others, may be called an apotheosis of the atomic theory of matter'. He spent, however, much more time on Planck's theory, Bohr's atom model, and relativity theory!

W. H. Bragg (1913*d*) described how he obtained spectra with the spectrometer, and the determination of crystal structures with both the spectrometer and Laue diagrams. Diamond was his main example. He demonstrated on a model of diamond made of balls and rods and explained how the succession of planes of carbon atoms account for the absence of the 222 reflection (see Fig. 7.7, *Left*). P. P. Ewald was present, and he later recalled that Bragg's talk, 'with the wonderful result and the beautiful directness and simplicity of the argument, made a deep impression on all' Ewald (1962*b*).

On his return to Germany, P. P. Ewald wrote a detailed report of the most important talks in the Physics Section, in particular on Barkla's studies of characteristic radiations, Bohr's atom model, and Bragg's structure determinations (Ewald 1913*c*), and presented it orally at the Munich Physics colloquium on 19 November 1913.

---

[31]Ivar Waller, born 11 June 1898, died 12 April 1991, was a Swedish physicist. A former student of M. Siegbahn, he obtained his PhD in 1925. He was Professor of Theoretical Physics at the University of Uppsala, and was elected a member of the Royal Swedish Academy of Sciences in 1945.

7.10.2   Eighty-fifth Meeting of the German scientists, 21–28 September 1913, Vienna

The annual meeting of the German scientists took place in Vienna, Austria, 21–28 September 1913. Of interest to us is the meeting of the Chemistry and Mineralogy Section on 24 September, where Laue, Friedrich and Wagner, among others, gave a talk.

1. *M. von Laue* started his talk by deriving the three Laue interference conditions and showing their equivalence with Bragg's law (although that was not how he named it). He noted that the Braggs, Moseley, and Darwin had been able to isolate monochromatic radiation with the X-ray spectrometer. He also recalled some basic properties of Laue diagrams, such as the fact that different cubic crystals oriented in the same way give similar patterns. This was obviously still quite a novelty in September 1913. Laue then broached three interesting new topics.

   (a) **Diamond** was a particularly interesting case because diffraction spots could also be observed on a photographic plate located in front of the crystal (Plate P_1 of Fig. 6.12, Section 6.6.2), in what is nowadays called the back-reflection mode. Laue found the explanation in the papers that had just been published by Debye (1913*a* and *b*, July 1913). These papers were preliminary reports of Debye's calculation of the influence of thermal motion on the intensity of the diffracted spots. He had shown that the intensity should be decreasing with increasing diffraction angle. This explained why there were no diffraction spots at large angles. Diamond was, however, a special case because of its rigidity and its exceptionally high 'characteristic' (Debye) temperature (see Section 7.9). It was therefore to be expected that the intensity diffracted by diamond crystals should be much less affected by thermal motion than other crystals, and this explained why diffraction spots were observed at high diffraction angles in diamond. That result prompted Laue and his co-worker J. S. van der Lingen to study the influence of temperature on the Laue diagrams of rock-salt, silicon and diamond (Laue and van der Lingen 1914*a*, *b* and *c*, respectively). They found, as did M. de Broglie (1913*a*), that the sharpness of the spots did not change with temperature, and were surprised that silicon was much more affected by thermal motion than diamond (which is, in fact, normal because the Debye temperature of silicon is much smaller than that of diamond).

   (b) **Hemihedral crystals**. Laue then considered the more general case of complicated point systems, and in particular of hemihedral ones. A lattice element (*Gitterelement*) no longer represented a single atom but the group of atoms contained in an elemental parallelepiped. The position of the interference maxima was again given by the three Laue conditions, but since there might also be interference effects between the atoms of a given lattice element, the intensity of the spots could be influenced in many ways. In some cases, such as zinc-blende, cuprite (Cu_2O), or nickel sulphate ($NiSO_4, 6H_2O$), there was no effect due to hemihedry, but in other cases, such as pyrite (FeS_2) or hauerite (MnS_2), an effect was detected, as shown by Friedrich (to be published in Ewald and Friedrich 1914). The interference picture (Laue diagram) taken with the incident beam parallel to the cubic axis did not exhibit a four-fold axis, but a two-fold axis (this is normal since the group is $m\bar{3}$). Beryl exhibited six-fold symmetry, but quartz did not, it only exhibited three-fold symmetry. Laue claimed also that the difference between right- and left-hand quartz could be observed on diagrams taken with the incident beam parallel to a two-fold axis, but G. Friedel showed that this could not be so (Section 7.12).

   (c) **The phase problem**. In Laue's concluding remarks, the phase problem was for the first time raised. 'Can the structure of a crystal be deduced from the interference diagrams? If this were the case, one of crystallography's most important problems would be solved. But this is not the case because the phases cannot be measured'! Of course, Laue added, 'conclusive keys' to the structure can be obtained by comparing diagrams taken with different orientations, as has been done by the Braggs.

2. In the second talk, W. Friedrich described the equipment used for the initial experiment by Knipping and himself. He then brought up the question of the nature of the incident beam, and alluded to Laue's initial misconception: 'Laue had believed that his theory, in its initial form, was not compatible with the assumption of a continuous spectrum, and that the presence of an extended white spectrum would have led to a blackening of the photographic plate'. To each diffraction spot corresponds a value of λ/a (a, lattice parameter), Friedrich explained. When the crystal is slightly rotated, the position of the spot is slightly modified, and so is the value of λ/a, and therefore of λ. This confirms that the incident beam is polychromatic, as was beautifully shown by W. H. and W. L. Bragg, Moseley, and Darwin. This matter was obviously very important for Friedrich because he took quite a number of Laue diagrams of zinc-blende with different anticathodes to ensure that there was no change in the position and in the relative intensities of the diffraction spots and that the spots were in no way due to characteristic radiation.

   In the discussion following Friedrich's talk, Laue tried to justify himself, and protested: 'my honorable co-worker, Herr Friedrich, has somewhat decried me (*hat mich hier etwas schlecht gemacht*) by saying that I thought that the primary beam was monochromatic. I did not think anything like that, and I had myself severe reservations about the only possible explanations that I saw'. Laue then explained that, if his early explanation had been correct, the spots would have disappeared when rotating the crystal, and would have been replaced by others, which was not what was observed.

   Among the other points raised by Friedrich, one dealt with the presence of streaks radiating from the central spot after long exposures, similar to those observed by Wulff and Uspenski (1913a), but the most interesting one concerned the limits of the Bremsstrahlung spectrum.

   Friedrich pointed out that there were no diffraction spots near the central spot in Laue diagrams, 'as if there were a short wavelength limit in the spectrum emitted by the X-ray tube, or as if they could not be detected with our present means'. W. L. Bragg (1913a) had already pondered whether the shorter wavelengths were missing from the spectrum. Friedrich also noted that the minimum diffraction angle for which diffraction spots could be observed depended on the lattice constant of the crystal: it was smaller for potassium bromide, KBr, with a large lattice constant, than for potassium chloride, KCl, with a smaller one, a point that Sommerfeld had drawn his attention to. The limits of the spectrum of the incident beam were also discussed by Laue and Sommerfeld at the Solvay congress (Section 7.10.3) and by Ewald (1914a) who noted that, if a lower limit did not exist, the density of spots would increase with decreasing angle of diffraction, giving the impression of a blackening. Although it was known, in particular since the works of Wien (1907) and Stark (1907), that the maximum energy $h\nu$ that could be transferred from the electrons to the X-rays was eV, it was only after Duane and Hunt's (1915) work that the wavelength distribution of the beam emitted by the anticathode was really understood (see the full description in Section 7.13).

3. The third talk of interest for us at the Vienna conference was that by E. Wagner. Its contents (that Laue spots had selected a single wavelength of X-rays) are described in Section 7.2.

4. An interesting point was raised in the general discussion by the well-known German mineralogist and crystallographer, Friedrich Rinne.[32] He noted that the question of whether there were seven or thirty-two types of building units for crystals was answered by Laue's colleagues who showed that pyrite was indeed hemihedral, although why zinc-blende did not exhibit hemihedry was still to be explained. Laue answered that the influence of hemihedry was not felt in the interference diagrams, except in the case of pyrite and hauerite. This discussion shows how unclear the inner structure of crystals still was!

---

[32]See footnote 32 in Chapter 8.

7.10.3 Congress of the Solvay Institute on 'Structure of Matter' 27–31 October 1913, Brussels

The first meeting of the Physics Council (*Conseil de Physique*) organized by E. Solvay, following an initiative by W. Nernst,[33] was held 30 October–3 November 1911 in Brussels, Belgium, under the chairmanship of H. A. Lorentz. Its topic was 'Theory of radiation and quanta'. The second meeting was held 27–31 October 1913 in Brussels, Belgium, also under the chairmanship of H. A. Lorentz (Fig. 7.13). The topic was 'Structure of Matter', and there were talks by J. J. Thomson, Mme M. Curie, M. von Laue, W. H. Bragg, A. Sommerfeld, W. Barlow and W. H. Pope, M. Brillouin, W. Voigt, E. Grüneisen, and R.-W. Wood. The discussions were

Fig. 7.13   Second Solvay Conference, 27–31 October 1913, Brussels, Institut International de Physique Solvay. Photograph Benjamin Couprie, Institut International de Physique Solvay. Courtesy AIP Emilio Segre Visual Archives.

---

[33]Walther Nernst, born 25 June 1864 in Briesen, West Prussia/Pomerania, now Wąbrzeźno, Poland, died 18 November 1941 in Zibelle, Lausitz, Lower Silesia, now Niwica, Poland, was a German physical chemist. He studied physics and mathematics at the universities of Zürich, Berlin, Graz, and Würzburg, where he graduated in 1887. In 1894 he was appointed Professor of Physical Chemistry in Göttingen, where he founded the Institute for Physical Chemistry and Electrochemistry, and in 1905 he was appointed Professor of Chemistry, later of Physics, in the University of Berlin, becoming Director of the newly-founded Physikalisch-Chemisches Institut in 1924. He is well known as the author of the 'Third Law of Thermodynamics' for which he was awarded the 1920 Nobel Prize in Chemistry. He was elected a member of the Deutsche Akademie der NaturforscherLeopoldina in 1911.

actively animated, among others, by M. de Broglie, A. Einstein, F. A. Lindemann, P. Langevin, H. A. Lorentz, W. Nernst, E. Rutherford, and W. Wien. Lawrence Bragg was not invited but a postcard was sent, signed by Sommerfeld, Mme Curie, Laue, Einstein, Lorentz, Rutherford, and others congratulating him for 'advancing the course of natural science' (Jenkin 2008).

The confrontation between theories of crystal structure elaborated before the discovery of X-ray diffraction (W. Barlow and W. H. Pope, M. Brillouin) and the first structures actually determined by means of X-rays (W. H. Bragg) is particularly fascinating. The relevant talks and discussions are summarized here:

1. **M. von Laue**. *Interference phenomena of Röntgen rays due to the three-dimensional lattice of crystals* (Laue 1913d). In the first part of his talk, M. von Laue presented the theory of X-ray diffraction by crystals, including the three Laue conditions, Bragg's law, the Ewald construction, and the expression of the diffracted intensity. As in the discussion following Friedrich's talk in Vienna, he explained why his initial hypothesis that the Laue diagrams were due to the resonance of the atoms for a few wavelengths did not hold. He described the beam emitted by the anticathode as composed of (1) line spectra, which give rise to 'selective reflections', observed up to the sixth order by Moseley, and (2) white radiation, giving rise to 'general reflections'. Laue noted that there were no reflections associated with small wavelengths, but that it was not clear whether there was a sharp lower limit for the wavelengths (upper limit for energy or frequency). He also noted that the ratios λ/a of the diffraction spots on zinc-blende diagrams had a rational relationship, but Sommerfeld declared in the discussion that this was pure coincidence. In the second part of his talk, Laue repeated the same considerations relative to hemihedral crystals and to the influence of thermal agitation, as he had at the Vienna meeting.

Three interesting points were raised during the discussion:

(a) *H. A. Lorentz* first noted that there was no need to distinguish between selective reflection by line spectra and general reflection by white radiation. He then explained that, if the maximum intensity of a diffraction spot should indeed be proportional to N^2, where N is the number of molecules in the crystal, the width of the intensity distribution should be proportional to $1/N$, by analogy with the diffraction by optical gratings, and that the 'total' intensity (which was later called the 'integrated intensity') should therefore be proportional to $N^2 \times 1/N = N$. He also outlined a method for calculating the intensity of radiation diffracted by a plane using Fresnel zones. This is the same method that was used by Darwin (1914a), without having knowledge of Lorentz's remark (Section 8.4).

(b) *W. Nernst* asked whether something could be deduced about zero-point energy from X-ray measurements. A. Sommerfeld, A. Einstein, F. A. Lindemann, de Broglie, and W. Wien took part in the discussion, without drawing any conclusion, but this point remained always a source of interest for the X-ray crystallographers and was finally solved by James et al. (1928) (see Section 10.4.2).

(c) *A. Sommerfeld* considered molecules to be dipoles, in other words, point scatterers (as he had done when proposing to Ewald the topic of his thesis), and therefore that the decrease of intensity with increasing diffraction angle could only be due to thermal agitation. The fact that the volume of the electron density around the nuclei should be taken into account had to wait until W. H. Bragg's (1915c) and A. H. Compton's (1915) papers were published!

2. **W. H. Bragg**. *The reflection of X-rays and the X-ray spectrometer* (W. H. Bragg 1913e). W. H. Bragg a) reported the observation of the spectra from Pd, Rh, Cu, and Ni anticathodes and the measurement of their respective wavelengths using the reflection from the cleavage surface of a rock-salt crystal, and b) described the structures of alkali halides, diamond, and fluorite, as well as the first results concerning the structures of the carbonates, consisting of alternating planes

of calcium atoms and CO_3 groups (Fig. 12.48), and similarly, of sodium nitrate, consisting of alternating planes of sodium atoms and NO_3 groups. He also gave preliminary considerations about the structure of sulphur and quartz.

In the discussion, E. Rutherford noted that the platinum lines observed by Bragg and Moseley were identical to the L lines observed by Barkla, M. Brillouin asked for a confirmation that in calcite the CO_3 groups were gathered in small volumes separated from the calcium atoms, and de Broglie compared Bragg's law with the principle of Lippmann colour photography.

3. **A. Sommerfeld**. *On the four-fold and three-fold pictures of zinc-blende and the Röntgen ray spectrum.* (Sommerfeld 1913). Sommerfeld gave here a summary of a yet unpublished work by Ewald (Ewald 1914a, 26 January 1914). The following points were discussed:

   (a) *Limits of the spectrum*: According to Sommerfeld, there is neither a sharp lower limit nor a sharp upper limit to the wavelength spectrum of the incident beam. The question of the limits of the spectrum had also been raised, as has been seen above, by Friedrich and by Laue. It was also raised by Friedel (1913b). In fact, Duane and Hunt (1915) will show that there is a sharp lower limit (Section 7.13).

   (b) *Maximum of the intensity distribution of the spectrum*: a careful analysis of the Laue diagrams of zinc-blende enabled Ewald to determine the variation of the intensity of the incident beam as a function of wavelength around the maximum of the intensity distribution. This had also been done, independently, by Friedel (1913b).

   (c) *Thermal agitation*: According to Debye (1913a and b), the intensity of the diffraction spots is multiplied by a factor $\exp -M$ to take thermal agitation into account. Sommerfeld noted that M is proportional to $S = h^2 + k^2 + l^2$, where h, k, l are the indices of the reflection. The intensity of the spots therefore decreases rapidly with increasing values of S, and therefore of the diffraction angle.

   (d) *Structure factor*: In zinc-blende, the amplitude of the diffracted beam should take into account the difference between the scattering powers of zinc and sulphur and the phase difference between the waves diffracted by the two atoms. The total diffracted amplitude Ψ should therefore be proportional to a quantity:

$$F_{hkl} = 1 + A \exp -2\pi i[hx + ky + lz] \qquad (7.10)$$

where A is the ratio of the scattering powers of zinc and sulphur, x, y, z, the coordinates of one atom, for instance sulphur, with respect to the other, and h, k, l the indices of the reflection. This was the first expression of the structure factor; the term *Strukturfaktor* was coined by Ewald (1914b). If one applies this expression to diamond, for which $A = 1$, one has $F_{hkl} = 1 + \exp -2\pi i[h + k + l]/4$, and one can immediately see why reflections 200 and 222 are absent.

4. **W. Barlow and W. H. Pope**. *Relation between atomic structure and chemical composition.* Barlow and Pope developed here their valence–volume theory (Barlow and Pope 1906, 1907, 1908, see Section 12.13).

5. **M. Brillouin**.[34] *Crystal structure and molecular anisotropy. Dimorphism of calcium carbonate.* M. Brillouin started by opposing two types of crystal structure theories. The first one is purely

---

[34]Marcel Léon Brillouin, born 19 December 1854 in Melle, Deux-Sèvre, France, died 16 June 1948 in Faris, France, was a French mathematician and physicist. A former student of Ecole Normale Supérieure, he obtained his PhD in 1882. He was Assistant Professor of Physics successively at the universities of Nancy, Dijon, and Toulouse, and Professor at Ecole Normale Supérieure from 1888 to 1900, when he was appointed Professor at the Collège de France, which he remained until his retirement in 1931, when he was succeeded by his son, Léon Brillouin (1889–1969), the discoverer of Brillouin scattering and father of Brillouin zones. Marcel Brillouin was elected a member of the French Academy

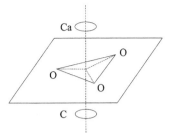

Fig. 7.14 Molecule of $CaCO_3$ in calcite, redrawn after Brillouin (1913). Note that, for Brillouin, the carbon atoms did not lie in the plane of the oxygen atoms.

geometrical and involves only the notions of volume and shape of the molecules. In that theory, an atom does not belong to a particular molecule. In the second type of theory, the molecules keep their individuality and the atoms of the molecules are supposed to be concentrated in a very small volume. The structure results from the balance of long-distance interatomic forces. This requires the knowledge of the mutual energy of each pair of molecules, which depends on the distance between the molecules, and, if they are anisotropic, on parameters defining the orientation of each molecule. Brillouin preferred this representation because its complexity makes it possible to interpret physical properties, which the first type of theories can not do. He then illustrated his viewpoint with considerations about polymorphic substances. This can be done in relation to vectorial properties, such as electrical or optical properties. The example taken by Brillouin is that of the two forms of calcium carbonate, $CaCO_3$, calcite and aragonite, and their optical properties. Calcite is trigonal and uniaxial. Aragonite is orthorhombic, pseudo-hexagonal, but strongly biaxial. Brillouin concluded that the birefringence of aragonite can not proceed from the structure, and must be due to the anisotropy of the molecule $CaCO_3$ itself. In this, he followed the general opinion of the mineralogists of the time. In calcite, on the contrary, the molecule must be isotropic, and Brillouin suggested that the molecule should be as represented in Fig. 7.14: the carbon and calcium atoms lie on a trigonal axis, with the three oxygen atoms at the summit of an equilateral triangle located in a plane perpendicular to the trigonal axis, half-way between the carbon and calcium atoms, with an equal number of molecules with the calcium on top and molecules with the carbon on top. No wonder that Brillouin asked Bragg whether the CO_3 groups were concentrated in small volumes (see above)! Barlow (1886) had at first proposed a wrong structure for calcite, but Barlow and Pope (1908) predicted the correct one (see Fig. 12.48 and Section 12.13) The structure of aragonite was determined later by W. L. Bragg (1924).

6. **General discussion.** There were two interesting contributions during the general discussion:

(a) *E. Rutherford* related the experiments he was carrying with E. N. da Andrade to show that γ-rays were electromagnetic waves. They were using γ-radiation from radium B and radium C which was allowed to fall on a rock-salt crystal. They had observed the corresponding spectra and measured their constituting wavelengths. They found that the soft γ-rays emitted by radium B were identical with the spectrum of the *L*-series emitted by element 82 (lead). The work was published the following year (Rutherford and Andrade 1914*a* and *b*). The arrangement they used was a divergent-beam technique described in Rutherford and Andrade (1914*b*). As shown in Fig. 7.15, the beams RA and RA' emitted by the γ source, R, are reflected by the rock-salt crystal and converge at O. Corresponding dark lines ('reflection

of Science in 1921. His research activities covered the fields of the kinetic theory of gases, viscosity, thermodynamics, electricity, and the physics of melting conditions.

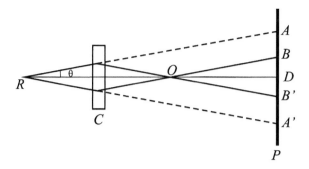

Fig. 7.15   Divergent-beam (wide-angle) technique. *R*: γ source; *C*: rock-salt crystal; *P*: photographic plate. Redrawn
after Rutherford and Andrade (1914*b*).

lines') are observed at *B* and *B'* on photographic plate *P*, while white lines showing minimum intensity ('absorption lines') appear at *A* and *A'*. The same arrangement was used by
Borrmann (1941) under the name of a wide-angle (*weitwinkel*) set-up when he discovered
anomalous transmission. It is discussed in detail in Lonsdale (1947).

(b) *W. Nernst* explained that the specific heat of diamond is that of a monoatomic substance, and
that there can not be clusters in the structure. The dissociation of KCl in separate potassium
and chlorine atoms in the structure of the solid seemed plausible to him. He added that the
structures of ice, sulphur, graphite and organic crystals were waited for impatiently. According
to him, ice could not be dissociated into hydrogen and oxygen ions. Since W. H. Bragg
had mentioned preliminary studies of sulphur, Nernst asked him whether he had found S_8
clusters. Bragg answered that it was too early to answer that question, but that there were
eight interpenetrating lattices of sulphur atoms in the structure.

7.11   M. de Broglie and the rotating crystal method

For many years one of the most useful techniques both for X-ray spectroscopy and for X-ray
structure analysis was the rotating crystal method, introduced by M. de Broglie in November
1913 (Broglie 1913*c, d*). De Broglie mounted the crystal on the drum of a recording barometer,
which gave it a rotation of 2° per hour, and the successive positions of the reflected beam as the
crystal turned were recorded on a photographic plate. 'A true spectrum was thus obtained, just
as in optics', in 15 minutes in the best case. A platinum anticathode and a rock-salt analyser
were used. De Broglie observed a number of lines which had not been observed by Bragg
or Moseley, but that was probably due to the fact that the clockwork driving the drum was not
running smoothly (Wyart in Ewald 1962*a*). In the next paper (Broglie 1913*e*), two photographic
plates separated by an absorbing screen were used. In that way it was possible to observe
separately coinciding first and higher order lines and to study the absorption of the observed
radiations. At the start of the recording, the crystal face was parallel to the incident X-ray
beam, and the drum was rotated from zero to a certain angle α, then a second spectrum was
recorded on the same film by rotating the drum from $180° - \alpha$ to 180°. The same line thus
appears twice and the corresponding angle of diffraction is obtained by halving the difference
between the angles of incidence of the two lines, thereby avoiding the uncertainty of the zero

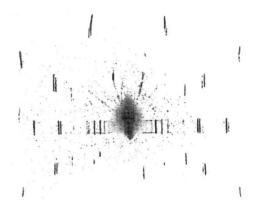

Fig. 7.16   Rotating crystal photograph, showing the spectrum of platinum, reflected by the cleavage face of a rock-salt crystal. After de Broglie (1914*a*).

position. The first spectra recorded were those of platinum and tungsten and the first rotating crystal photograph was published in January 1914 (Broglie 1914*a*, Fig. 7.16). These papers were the first of a long series of articles by de Broglie on X-ray spectroscopy.

A considerable improvement introduced by M. de Broglie was the observation of the emission spectra from substances that could not serve as a target in the X-ray tube. For that, he used the secondary radiation of these substances, excited by primary radiation (Broglie 1914*c*, *d*, *e*). He confirmed, in this way, the spectra of a large number of elements already studied by Moseley. Besides the emission lines, de Broglie also observed two sharp discontinuities in the blackening of his spectral plates. He showed that they were due to the K-absorption edges of the silver and the bromine present in the photographic emulsion, an explanation suggested to him by W. H. Bragg and M. Siegbahn (Broglie 1914*d*). This was the first observation of absorption spectra. The first detailed studies of absorption spectra were by E. Wagner (1914, 1915*b*). Developments in the rotating crystal technique are described in Section 8.6.

7.12   G. Friedel and Friedel's 'law', or rule

M. von Laue reported on several occasions that it was surprising that zinc-blende diagrams did not exhibit hemihedry, while pyrite did. G. Friedel pointed out that the interference diagram is unchanged if one reverses the sense of the incident beam. As a matter of fact, the interference diagram is identical for each of the possible directions of incidence, 1, 2, 3, 4 in Fig. 7.17, *Left*. It is therefore impossible to observe the absence of a centre of symmetry in the crystal: the phenomenon of X-ray diffraction 'adds' a centre of symmetry to the symmetry of the crystal. This is 'Friedel's law or rule'. As a result, it is impossible to distinguish by diffraction between a centrosymmetric group and its non-centrosymmetric subgroups. Out of the thirty-two point groups, one can then only distinguish between the 11 centrosymmetric point groups: $\bar{1}$, $2/m$, mmm, $\bar{3}$, $\bar{3}m$, $4/m$, $4/mmm$, $6/m$, $6/mmm$, $m\bar{3}$, $m\bar{3}m$. They have been called the 11 Laue classes. Laue had claimed that he had observed the difference between right- and left-hand quartz on diagrams of quartz taken with the incident beam parallel to a two-fold axis, but

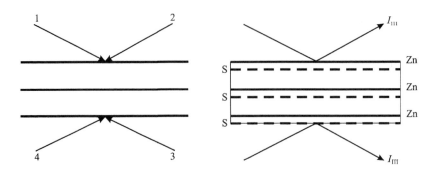

Fig. 7.17   *Left*: Friedel's law. The interference diagrams are identical for any of possible directions of incidence, 1,
2, 3, 4. *Right*: structure of zinc-blende. The reflected intensities I_{111} and $I_{\bar{1}\bar{1}\bar{1}}$ from the zinc and sulphur faces,
respectively, are different due anomalous dispersion, allowing the absolute configuration to be determined.

Friedel, who had the opportunity to examine copies of the diagrams shown to him by de Broglie,
said that this was not the case.

Ewald and Friedrich (1914) took Laue diagrams of pyrite, hauerite (MnS_2) group $m\bar{3}$, and
sodium chlorate ($NaClO_3$), group 23. They applied Friedel's 'law' to explain why all three
exhibit the hemihedry, while zinc-blende, group $\bar{4}3m$, does not. Independently, W. L. Bragg
(1914c) gave a structural interpretation of the fact that pyrite exhibits the hemihedry and zinc-
blende does not, confirmed by spectrometer recordings of the reflection from the (210) and
(120) planes of pyrite and from the (111) and $(\bar{1},\bar{1},\bar{1})$ planes of zinc-blende (Fig. 7.17, *Right*).

Friedel's paper was critically discussed by Laue (1916), but Laue reached the same conclusion, namely that one can not decide the presence or absence of a centre of symmetry by
diffraction alone.

Friedel's 'law' can also be derived from the consideration of the diffracted intensity. It
was noted in Section 7.10.3 that Ewald (1914a) showed that the diffracted amplitude, Ψ, is
proportional to a quantity which he called structure factor, and which depends on the relative
scattering powers of the various atoms in the unit cell and on their phase differences. Expression
7.10 of the structure factor can be rewritten, in a more appropriate form:

$$F_{hkl} = f_{Zn} + f_S \exp -2\pi i[hx + ky + lz] \qquad (7.11)$$

where f_S and f_{Zn} represent the scattering powers, now called atomic form factors, of sulphur
and zinc, respectively. The diffracted intensity $|\Psi|^2$ is proportional to $|F_{hkl}|^2$, and 7.11 shows
that:

$$|F_{hkl}|^2 = F_{hkl} F_{hkl}^* = F_{hkl} F_{\bar{h}\bar{k}\bar{l}} = |F_{\bar{h}\bar{k}\bar{l}}|^2. \qquad (7.12)$$

The intensities of the h, k, l and $\bar{h}\bar{k}\bar{l}$ reflections are therefore equal and the absence of a
centre of symmetry can not be observed. Relation 7.12, however, holds only if f_S and f_{Zn}
are real, which is in fact not true when one takes absorption into account. Due to anomalous
dispersion (or anomalous scattering, see Section 8.4.1), the form factors have an imaginary
part which is proportional to the absorption coefficient of the relevant element. In that case,
$F_{hkl} F_{hkl}^* \neq F_{hkl} F_{\bar{h}\bar{k}\bar{l}}$, and the intensities I_{hkl} and $I_{\bar{h}\bar{k}\bar{l}}$ reflected respectively from the sulphur

and zinc faces of a zinc-blende crystal cut parallel to (111) will be different (Fig. 7.17, *Right*). The absence of the centre of symmetry can then be detected. This was first pointed out by Ewald and Hermann (1927), and first observed experimentally by S. Nishikawa and K. Matukawa (1928) using tungsten L radiation, which is close to the K absorption edge of zinc, and is strongly absorbed. The Dutch physicists D. Coster, K. S. Knol and J. A. Prins (1930), in Groningen, using tungsten and gold L radiations, went further. They measured the phase of the structure factor, and determined the zinc and sulphur faces of a crystal of zinc-blende. The application of such measurements to obtain information about the phases of structure factors, so that one can proceed to a direct Fourier synthesis of crystal structures, and then determine absolute configurations was discussed by J. M. Bijvoet (1949, 1954). Using this technique with zirconium radiation, J. M. Bijvoet *et al.* (1951) achieved the first experimental determination of the absolute configuration of sodium rubidium tartrate.

7.13   The first X-ray spectrometer in the United States: the Duane–Hunt law, 1914

The first X-ray spectrometer in the United States was built by a young graduate student at Harvard University, David Webster.[35] Following his thesis, Webster became interested in X-rays and built an X-ray spectrometer by adapting an optical spectrometer.[36] He began observing the spectra emitted by the tube, plotting diffracted intensity against diffraction angle, and to his surprise, he found that the spectrum had a high frequency limit. At that point, an Assistant Professor of Physics in the Department, William Duane,[37] who had studied the absorption of X-rays and the relation between their wavelengths and the voltage required to produce them, borrowed the spectrometer from Webster.

Duane asked one of his students, Franklin L. Hunt, to explore the continuous X-ray spectra of tungsten with the spectrometer, making careful records of the voltages used to accelerate the electrons. They used the 100 reflection of calcite as an analyser. For a given voltage, they measured the reflected intensity with an ionization chamber and a gold-leaf electroscope as a function of the diffraction angle, namely, through Bragg's law, $\lambda = 2d\sin\theta$, as a function of the wavelength. A typical curve is shown in Fig. 7.18. It shows that there is no wavelength below a

[35]David L. Webster, born 6 November 1888, died 17 December 1976, was an American physicist. After graduating at Harvard University, he obtained his doctorate in 1913 under the direction of T. Lyman. He was then appointed an instructor, and, in 1917, Assistant Professor at the University of Michigan. After World War I, in which he served as an aviator, he was appointed first at M.I.T. and then at Stanford in 1920 as a full Professor. His main research topics was X-ray physics. He took part in the development of the klystron, and became Director of the Klystron Department at Stanford. During World War II, he was assistant chief of the Army Rocket Research Branch in Maryland. Back at Stanford after the war, he devoted himself to teaching until his retirement in 1954.

[36]Details are given in Kirkpatrick, P. (1980). Confirming the Planck–Einstein equation $h\nu = (1/2)mv^2$. *Am. J. Phys.* **48**, 803–806.

[37]William Duane, born 17 February 1872, died 7 March 1935, was an American X-ray physicist. After graduating from Harvard University in 1893, he became assistant professor for two years, then visited Germany for another two years where he obtained his PhD in 1897. In 1898, he was appointed Professor of Physics at the University of Colorado. In 1905, he visited the Curies' laboratory in Paris and, eventually, remained there for six years, working on radioactivity. In 1913, he was appointed Assistant Professor of Physics at Harvard University and Research Fellow in Physics at Huntington Hospital in Boston. In 1917 he was promoted to Professor of Biophysics, the first in America. He was the founder of one of the first American schools of X-ray crystallography. He was elected to the National Academy of Sciences in 1920.

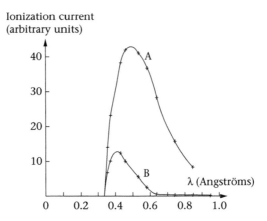

Fig. 7.18   A: Ionization current as a function of wavelength for a given voltage applied to the X-ray tube. B: the same
curve after absorption of the reflected beam through 3 mm of aluminium. Tungsten anticathode, cleavage face of
calcite as analyser. Redrawn after Duane and Hunt (1915). ©American Physical Society, 1915.

sharp cut-off limit. The value of this limit, λ_o is related to the voltage V applied to the tube by
Planck–Einstein's relation:

$$h\nu_o = hc/\lambda_o = eV \qquad (7.13)$$

where c is the velocity of light and e the charge of the electron. Duane and Hunt repeated the
measurements for six values of the voltage V, ranging from 25,000 to 39,150 volts, and deduced
from them values of Planck's constant h ranging from 6.34×10^{-34} J.s to 6.44×10^{-34} J.s,
with an average value of 6.39×10^{-34} J.s, close to the value of 6.41×10^{-34} J.s given by
Planck. The results were presented at the April 1915 meeting of the American Physical Society
in Washington (Duane and Hunt 1915). The existence of a lower limit for the wave-lengths
emitted by an X-ray tube has become known as the 'Duane–Hunt law'. This lower limit was
also observed, independently, by Rutherford et al. (1915, July) by measuring the absorption of
the white radiation from a Coolidge X-ray tube through aluminium foils, but the accuracy was
less. The determination of the value of Planck's constant h by X-ray measurements was further
refined by a number of scientists (see Section 10.4.3).

7.14   M. von Laue (1914) and the Braggs (1915) are awarded the Nobel Prize

M. von Laue was awarded the 1914 Nobel Prize in Physics, and W. H. Bragg and his son,
W. L. Bragg, shared the 1915 Prize. The stories behind the nominations and the votes for the
Nobel awards are summarized in Jenkin (2001) and Eckert (2012). By July 1914, the Nobel
Committee had received twenty-three nominations for the 1914 Prize in Physics, including
nominations for Laue, Laue and Bragg jointly, Einstein, and others. The declaration of war,
however, disrupted the usual routine, and Sweden's neutrality called for a postponement.
The Physics Committee proposed Laue, and the Swedish Academy endorsed the choice, but
the decision was not made public. Nominations were then considered for 1915 and again Laue

was suggested. Other nominations were for Laue and the Braggs, for the Braggs only, but also for Planck, Moseley, and others. Planck's quantum theory was considered so difficult that the Committee felt unable to honour him. Moseley's work was too recent and had not yet been properly evaluated. Since Laue had been chosen for the 1914 Prize, the Committee proposed the Braggs, which provided a political balance between Germany and Great Britain. The Academy confirmed the 1914 vote and endorsed the choice of the Braggs for the 1915 prize. Since the hostilities were continuing, it was decided that there would be no further postponement, but that the award ceremonies would wait until after the end of the war.

Laue learned about the prize decisions on 11 November 1915. He asked a member of the Committee to send his congratulations to the Braggs, and he received the reciprocal congratulations from the Braggs through the same Committee member. Laue received the diploma and the medal in 1916 through the German Embassy in Stockholm. The prize money he shared with Friedrich and Knipping.

The Braggs were informed by two telegrams, redirected from Leeds, which arrived on 14 November 1915 at the new family home in London, where W. H. Bragg had just been appointed Professor of Physics at University College London. W. H. Bragg and W. L. Bragg are the only father-and-son joint recipients of the Nobel Prize for Physics, and Lawrence is the youngest ever to receive the Nobel Prize. He was at that time stationed in the village of La Clytte in Belgium. He was of course elated: 'I am the most lucky fellow in the world. It is awfully nice to be coupled to Dad in this way', he wrote to his sister, Gwendoline (quoted by Jenkin 2008). There was not much time for celebration: 'I remember that the friendly priest, in whose house we were quartered, brought up a bottle of Lachrymae Christi from his *cave* to celebrate', and his comrades rejoiced with him briefly (quoted by Jenkin 2001).

M. von Laue received his prize in June 1920 in Stockholm when the award ceremonies were resumed. The Braggs did not come. Bragg the elder, who never recovered from the loss of his younger son, Robert, at Gallipoli, told Rutherford, in way of explanation: 'I believe that several Germans are coming' (Jenkin 2001). Indeed, M. Planck and J. Stark, who were awarded the 1918 and 1919 Nobel Prizes in Physics, and F. Haber, who was awarded the 1918 Nobel Prize in Chemistry, attended. W. H. Bragg never delivered a Nobel lecture, while Lawrence Bragg did not deliver his until September 1922. Lawrence and Alice Bragg's visit to Stockholm on this occasion is related in Glazer and Thomson (2015).

8

THE ROUTE TO CRYSTAL STRUCTURE DETERMINATION

We can obtain enough equations to solve the structure
of any crystal, however complicated, although
the solution is not always easy to find.

Sir W. L. Bragg (1914)

8.1 The beginnings

One of the major applications of X-ray diffraction is the determination of crystal structures. The amount of information available to the early investigators was very limited. Theoretical expressions of the diffracted intensity were given as early as 1914 by Darwin (1914*a*, *b*, Section 8.4), but they did not agree with experimental results, and the interpretation of observed diffracted intensities was difficult. Bragg and Bragg (1915) noted that 'theory does not seem able to give an adequate explanation of the intensity of the reflection'. Nine years later, R. W. G. Wyckoff (1924) would still write: 'Several efforts have been made to obtain theoretically sound intensity expressions. They do not, however, give results in even approximate accord with experiment unless terms are introduced which are dependent upon the electronic arrangement within atoms. At present too little is known of the structures of atoms to permit an evaluation of these terms. For these reasons, theoretically sound equations cannot now be used in the determination of the structures of crystals'.

8.1.1 1912–1920

The first structure determinations were limited to simple structures with few parameters and relied on the capacity of their authors to visualize geometrical constructions in space. The Braggs were masters at that. The starting point was the correct interpretation, by 22-year-old Lawrence Bragg, of the zinc-blende diagrams obtained by Friedrich and Knipping. Laue had considered the zinc-blende molecules, ZnS, to be arranged at the corners of a simple cubic lattice, but W. L. Bragg noted that the indices of the diffraction spots were systematically all even or all odd integers, and rightly deduced from this observation that the lattice of zinc-blende was in fact face-centred cubic. This was the first experimental confirmation of Barlow and Pope's atomic model (see Section 12.13) and of Kepler's close-packed arrangement (see Section 11.4.4).

In the months that followed came the determination of the crystal structures of alkali halides and of the complete structure of zinc-blende by the young Bragg (W. L. Bragg 1913*b*, see Section 7.7). By showing that inorganic crystals were composed of a regular pattern of atoms, these first structure determinations had a profound influence on the chemical ideas of the time, paving the way for the future study of ionic crystals (Section 10.1.2). Even more exciting was the publication by the two Braggs together, father and son, of the structure of diamond (Bragg and Bragg 1913*b*, *c*, see Section 7.7), validating the hypothesis of the tetrahedral character of carbon's four bonds in methane. In late 1913, Bragg the son determined the structures of fluorspar, of the trigonal carbonates and of pyrite (W. L. Bragg 1914*a*, Section 8.3.1), also discussed by Ewald and Friedrich (1914) in Munich. In 1914 the young Bragg determined the structure of a metal, copper (W. L. Bragg 1914*b*). In 1915 his father determined the structures of magnetite and spinel (W. H. Bragg 1915*a* and *b*); these were also determined at about the same time by S. Nishikawa (1915) in Japan (see Section 7.5).

In Norway, back in Oslo after his stays in England and Germany, L. Vegard (Fig. 6.19) took up the new science of X-ray crystallography while simultaneously doing work in auroral research, and had a Bragg spectrometer built. He determined the structures of silver (Vegard 1916*a*), zircon ($ZnSiO_4$), rutile (TiO_2), cassiterite (SnO_2), and anatase, another variety of titanium oxide (Vegard 1916*b*). The structures of rutile and cassiterite were also studied in England by C. M. Williams (1917).

In the United States the first structure to be determined was that of another tetragonal mineral, chalcopyrite ($CuFeS_2$), by C. L. Burdick and J. H. Ellis (1917). Burdick, a former graduate student of A. Noyes at the Massachusetts Institute of Technology (M.I.T.), Cambridge (Massachusetts), had left the United States in 1914 to do doctorate work in Europe, and happened to sail on the last German liner that managed to reach Hamburg.[1] After working in Basel and Berlin for two years, he managed, with great difficulty in this time of war, to join W. H. Bragg at University College in London. There, he learnt the art of structure determination with the ionization spectrometer, and, with E. A. Owen, worked out the structure of carborundum (Burdick and Owen 1918). Back in the United States in 1916, Burdick built improved ionization spectrometers, first at M.I.T., then at the California Institute of Technology (Caltech) in Pasadena, California, where he followed A. Noyes. Caltech is where he determined the structure of chalcopyrite. After the war, with the arrival of R. G. Dickinson, Caltech became an important centre for structure determinations.[2]

R. W. G. Wyckoff (Fig. 8.1) first heard about X-ray diffraction in 1916 when he started his doctoral work at Cornell.[3] The topic he had been given was a study of the cesium polyhalides. His adviser suggested to Wyckoff that he should take advantage of the presence of S. Nishikawa, a visiting scientist there at Cornell, and should learn the Laue method from him so that he could determine the structure of a cesium halide. In reminiscences written for a memorial volume in honour of S. Nishikawa,[4] Wyckoff remembered: 'we made a Laue camera out of a wooden box and constructed a correspondingly primitive apparatus to photograph the spectra from a single crystal. The laboratory's old medical unit provided the X-ray source. This was during the early months of 1918'. Under Nishikawa's guidance, Wyckoff determined the structures of cesium dichloro-iodide ($CsCl_2I$) and sodium nitrate (Wyckoff 1920*a* and *b*, respectively). For details of Wyekoff's life and work, see Drenth and Looyenga-Vos (1995).

[1]C. L. Burdick, in McLachlan, Jr & Glusker (1983).

[2]L. Pauling, in McLachlan, Jr & Glusker (1983).

[3]R. W. G. Wyckoff, in McLachlan, Jr & Glusker (1983), and in Ewald (1962*a*).

[4]The author is grateful to Yuji Ohashi for a copy.

Fig. 8.1   Ralph W. G. Wyckoff in 1956. Photograph by V. E. Taylor, National Institutes of Health. After Drenth and Loogenga-Vos (1995).

Ralph Walter Graystone Wyckoff: born 9 August 1897, in Geneva, NY, USA, died 3 November 1994 in Tucson, AZ, USA, was an American crystallographer. He obtained his BSc in 1916 at Hobart College, NY, and his PhD in 1919 at Cornell University, prepared under the supervision of S. Nishikawa. After his graduation he joined the Geophysical Laboratory in Washington (DC), working on the structure of minerals with a high symmetry. In 1927 Wyckoff moved to the Rockefeller Institute for Medical Research (now Rockefeller University) in New York, where he determined the structure of urea and studied the effect of X-rays on living cells. He left the Rockefeller Institute in 1937, working for the private industry on a virus causing *Western equine encephalitis*, which resulted in the creation of a vaccine. During World War II he produced first a vaccine against epidemic typhus fever and then dried blood plasma for in-field use by the armed forces. In 1943 he joined Michigan University and the Michigan State Department of Health. In Ann Arbor he invented a metal-shadowing technique to take three-dimensional electron microscope images of the influenza virus and of bacteria. Just after the war, he was scientific attaché at the US Embassy in London, where he installed an electron microscope and took spectacular micrographs of virus crystals. In 1946 he obtained a position at the National Institutes of Health (NIH) in Bethesda, where he studied viruses and macromolecules. In 1960 he became unhappy with the growing bureaucracy at the NIH and moved to the University of Arizona in Tucson to become a Professor of Physics and Microbiology. In his final years he continued studies at a mineral company. In 1948 he was involved in the foundation of the International Union of Crystallography, of which he was Vice-president and then President from 1951 to 1957. He was elected a member of the National Academy of Sciences in 1949.

MAIN PUBLICATIONS

1922 *The Analytical Expression of the Results of the Theory of Space Groups.*

1924, 1931, 1935 *The structure of crystals.*

1923 *On the Hypothesis of Constant Atomic Radii.*

1927 *Die Kristallstruktur von Zirkon und die Kriterien für spezielle Lagen in tetragonalen Raumgruppen.*

1932 *The crystal structure of thiourea.* With R. B. Corey.

1935 *X-ray diffraction from hemoglobin.* With R. B. Corey.

The next major step was the observation of diffraction by crystalline powders, independently by P. Debye and P. Scherrer (1916*a*, *b*, 1917) in Göttingen, Germany and by A. Hull (1917*a*, *b*) in Schenectady, NY, United States, see Section 8.5. Applications of M. de Broglie's rotating crystal method were developed shortly afterwards by Schiebold (in Rinne 1919) and Polanyi (1921*a*, *b*), see Section 8.6.1.

By 1920, about fifty elements and inorganic compounds had been studied (W. L. Bragg, 1920*b*), from Bragg spectrometer recordings, Laue diagrams, rotating crystal photographs, and powder diffraction. These structures were usually determined case by case using *ad hoc* intuitive methods.

8.1.2   1920–1925

Nishikawa was the first to use space groups in studies of crystal structure (Section 7.5), but it is Niggli[5] (1919) who really showed, in his seminal book, *Geometrische Kristallographie des Diskontinuums*, how space group theory could be applied to the determination of crystal structures, and that reflection conditions revealed a small number of possible space groups to which a crystal could belong. In that way, a frame-work of symmetry elements to which the atoms must conform is provided. Wyckoff (1921, 1922) published tables of the 230 space groups with positional coordinates, and both general and special positions were later called 'Wyckoff positions' in Wyckoff's honour. In London, two of W. H. Bragg's first students, W. T. Astbury (see Fig. 10.13) and K. Yardley (the future Katherine Lonsdale, see Fig. 8.18) (1924), not knowing about Niggli's book nor about Wyckoff's work (K. Lonsdale, in Ewald 1962*a*), provided tables of the symmetry elements of the 230 space groups, based on Hilton's 1903 book. They used a pictorial representation very different from that of Niggli, intended for immediate practical use. J. D. Bernal (Fig. 8.17), another of Bragg's students at the Royal Institution, remembered that it needed all of Astbury's persuasive powers to convince W. H. Bragg that these tables should be published: 'Sir William (Bragg) did not think they were necessary—any good scientist could arrive at all the results contained in them. Astbury was able to persuade him that because not all people who might be good scientists had his supply of common sense, some of them might find the tables useful'.[6] These tables are the ancestors of the first edition of the International Tables for the Determination of Crystal Structures, by W. T. Astbury, J. D. Bernal, C. Hermann, C. Mauguin, P. Niggli, and R. W. G. Wyckoff (*Internationale Tabellen zur Bestimmung von Kristallstrukturen*), edited by C. Hermann and published in 1935. Updated versions have been used constantly and extensively in nearly all crystallographic laboratories since that time.

For simple structures, exhaustively testing all the possible atomic arrangements provided by the theory of space groups became possible for crystals of given composition, space-group symmetry and unit cell dimensions. One of the first structures determined using space groups was that of ammonium chloroplatinate by Wyckoff and Posnjak (1921). The crystallographer or chemist of the 1920s, wishing to determine crystal structures, had at his disposal four main textbooks: Bragg and Bragg (1915), frequently updated, Ewald (1923), Mauguin (1924*c*), and Wyckoff (1924). The last one was the most practical, with an exhaustive yearly list of all publications in the field. The modern crystallographer wishing to read about the early developments of crystal structure determination may consult Ewald (1962), Bijvoet *et al.* (1969) and Lima de Faria (1990).

The study of complicated crystals with lower symmetry such as organic crystals required the unambiguous measurement of as many reflections as possible. This was achieved by K. Weissenberg (1924) by moving the recording film simultaneously with the rotation of the crystal (Section 8.6.3). Techniques were introduced for the graphic analysis of Laue diagrams using a stereographic (Bragg and Bragg 1915, Haga and Jaeger 1916) or a gnomonic projection (Rinne 1919; Schiebold 1922*a*; Wyckoff 1920*c*, 1924), Debye-Scherrer diagrams (Hull and Davey 1921; Ewald 1923; Schiebold 1924) or rotating crystal photographic diagrams (Schiebold 1924).

---

[5]See footnote 43 in Chapter 6.

[6]J. D. Bernal (1963). William Thomas Astbury. 1898–1961. *Biogr. Mems. Fellows R. Soc.* **9**, 1–35.

The excitement of structure determination by X-ray crystallographers in those days was best described by H. Lipson:[7] 'Then every crystal-structure determination was a personal triumph since there were no standard ways of deriving a result, as there are today. One tried various possibilities of atomic arrangements, most of them wrong of course. It was an exciting moment when one of them turned out to give a diffraction pattern in reasonable agreement with that observed'.[8] These were trial-and-error methods.

8.1.3 1925–1930

By 1925, nearly 600 crystal structures had been analysed: sixty elements (including many of their allotropes), 300 inorganic compounds, 160 organic compounds, essentially aliphatic, including long chain compounds, and thirty alloys; these were all listed by R. W. G. Wyckoff.[9] For each of them, lattice mode, structure type and lattice parameters were reported, as well as their space group, if known. An exhaustive survey of simple structures by V. M. Goldschmidt (1926) and his school established tables of atomic and ionic radii (see Section 10.1.2). As one of the first applications, the structure of beryl was determined by W. L. Bragg and J. West (1926) using nothing else but space groups and known atomic radii. Bragg remembered that they worked out the structure in the record-breaking time of less than an hour (W. L. Bragg in Ewald 1962a).

It is only in 1925 that Fourier methods were introduced by W. Duane[10] and his student R. J. Havighurst at Harvard University, USA, in the determination of the electron distribution of sodium and chlorine in sodium chloride (Duane 1925; Havighurst 1925, 1926, 1927). The idea had been proposed by W. H. Bragg (1915c) and was implemented for crystal structure determinations by W. L. Bragg (W. L. Bragg 1929a; Bragg and West 1930), at his father's suggestion (W. L. Bragg 1961; W. L. Bragg in Ewald 1962a). In their work on minerals they noted: 'The first two-dimensional Fouriers appeared therefore as the three principal plane projections of diopside'. According to a letter from W. L. Bragg to L. Pauling, quoted by Pauling in his reply to Bragg (29 March 1927, Archives of the Deutsches Museum, Munich), Bragg et al. (1922b) had already made use implicitly of the Fourier series method. In his letter, Pauling expressed his conviction that 'the three-dimensional Fourier series method permits the evaluation of the distribution of diffracting power without any hypothesis regarding the distribution'.

In the same year, 1929, two-dimensional Fourier projections were also used for the determination of the structure of potassium chlorate by the Norwegian crystallographer from Oslo,

---

[7]Henry Lipson, born 11 March 1910 into a family of polish Jewish immigrants, died 26 April 1991, was a British physicist. He graduated from Liverpool University in 1930 and stayed on to do X-ray structure determinations. He teamed up with A. Beevers, and together they went to Manchester to get advice from W. L. Bragg. After solving two crystal structures, they tackled triclinic copper sulfate pentahydrate. It was when trying to calculate the necessary Fourier summations that they invented the Beevers Lipson strips. H. Lipson joined W. L. Bragg in 1936, and followed him to the National Physical Laboratory and to Cambridge. He was awarded a Liverpool DSc in 1939 and a Cambridge MA in 1942. In 1945 he became Head of the Physics Department of Manchester College of Technology (later University of Manchester Institute of Science and Technology, UMIST), and, in 1954 he was appointed professor. He was elected a Fellow of the Royal Society in 1957.

[8]H. Lipson (1973). Weissenberg's Influence on Crystallography. In *Karl Weissenberg: The 80th Birthday Celebration Essays*. <http://weissenberg.bsr.org.uk/>.

[9]*X-ray diffraction data from crystals and liquids*, in *International critical tables of numerical data, Physics, Chemistry and Technology*, National Research Council of the USA, 1926. New York: McGraw-Hill, with 400 references.

[10]See footnote 37 in Chapter 7.

William Zachariasen[11] (1929), who was visiting W. L. Bragg in Manchester. The use of Fourier syntheses became rapidly systematic both for inorganic and organic structures. One example is the structure of potassium dihydrogen phosphate (KDP) determined by J. West (1930), a co-worker of W. L. Bragg[12]. Other examples are the structures of hexachlorobenzene by K. Lonsdale (1931), and anthracene and naphthalene by J. M. Robertson (1933a, b, see Fig. 10.4), both former students of W. H. Bragg. Numerical methods were rapidly developed for the summation of Fourier syntheses, optical by W. L. Bragg (1929b), and mechanical with Beevers–Lipson strips (Beevers and Lipson 1934; Lipson and Beevers 1936).

The measurement of absolute integrated intensities was a major step forward in tactics for solving structures (W. L. Bragg et al. 1921a, b, see Section 8.7). Crystallographers had at their disposal a whole battery of experimental methods to record diffracted intensities: the ionization spectrometer and photographic techniques (Laue method, rotating crystal and oscillation methods, Weissenberg camera, powder diffraction). There were those who favoured the exact tool, the spectrometer, such as K. Lonsdale, W. T. Astbury or R. J. James, and those who thought it was too much for them: J. D. Bernal (in Ewald 1962a) decided that 'he was not made for it', and that 'to spend a whole day for only two accurate reflections was quite behind [his] patience'. H. Lipson found the instrument very awkward to use, and 'that it required a great deal of patience and manipulative skill. Photographic methods were much easier, no control of the X-ray intensity was required and all the information was presented on a single film'.[13]

The role of crystal imperfections began to be understood (Darwin 1922; W. L. Bragg et al. 1922a) and, in 1926, appropriate formulæ were published to describe the diffracted intensities (W. L. Bragg et al. 1926)—see Section 8.8. The variations of the atomic scattering factor with angle of diffraction were measured and were calculated for comparison (see Section 8.2.4). The first structure determination based on the quantitative approach was that of baryte ($BaSO_4$) by James and Wood (1925). It is an orthorhombic structure with eleven parameters. A simplification was introduced by assuming the SO_4 group to be tetrahedral, but the assigned structure was rigorously confirmed by comparing calculated and observed absolute values (W. L. Bragg in Ewald 1962a)—the trial-and-error method.

It was, however, quite usual to combine several methods. For instance, the structure of oxyfluoromolybdate was determined by L. Pauling (Fig. 10.3) at Caltech, combining spectrometry, the Laue method, and powder diffraction (Pauling 1924). The structure of zircon, studied successively by Vegard (1916b, 1926) and O. Hassel (1926), also in Oslo, was revisited very carefully by Wyckoff (1927) using Laue diagrams, the ionization spectrometer,

---

[11]William H. Zachariasen, born 5 February 1906 in Langesund, Norway, died 24 December 1979, Santa Fe, New Mexico, USA, was a Norwegian-born American crystallographer, one of the best-known specialists in the determination of the structure of crystals of inorganic substances by X-ray diffraction. He studied at the University of Oslo, where he obtained his PhD under V. M. Goldschmidt in 1928, when he was only 22 (the youngest person ever to receive this degree in Norway). He was appointed Assistant Professor at the University of Oslo in 1928, and spent one year as Postdoctoral Fellow in W. L. Bragg's laboratory, in Manchester. After one year back in Oslo he was invited by A. H. Compton to Chicago, where he was successively Assistant Professor (1930–40), Associate Professor (1940–45), and Professor (1945–74).

[12]KDP is ferroelectric below 135 K. Its ferroelectricity was discovered by G. Busch and P. Scherrer (1935). Eine neue seignette-elektrische Substanz. Naturwiss. **23**, 737.

[13]H. Lipson (1973), see footnote 8.

and powder diagrams. E. Schiebold (1929a, b) in Leipzig combined Laue diagrams and rotating crystal photographs to study the structure of feldspars. Despite the difficulty of its analysis, the Laue method remained useful for many years. Pauling (1937) found that it was 'the most reliable method for determining the size of the true unit and the space-group symmetry of crystals, due to the appearance on Laue photographs of reflections from very complicated planes'. Buerger (1964) noted that this was in the days when reflections were made from extended faces, absorption was neglected and structure factors approximated by integers.

From that time on, it began to be possible to determine structures with a greater number of parameters, as shown by Bragg and West (1929); a good example is given by the structure of diopside, solved by Warren and Bragg (1929), a turning point in understanding silicates, according to W. L. Bragg (in Ewald 1962a), see Section 10.2. The yearly increase of the number of parameters in typical structure determinations is shown in Figure 8.2, as plotted by W. L. Bragg. The curve starts with one parameter in 1913. The structure of corundum, Al_2O_3, by Bragg and Bragg, 1915, and by Burdick and Owen, 1918, had two parameters. W. L. Bragg (1925a) considered that any structure with more than six parameters presented a difficult problem. The structure of diopside (Warren and Bragg 1929) had fourteen parameters. By 1930, the number of parameters was about twenty; it continued to increase slowly until the mid-1950s when it started to rise sharply.

The second half of the 1920s saw two major developments. One was the determination of the structure of the benzene ring in hexamethylbenzene by K. Lonsdale (1928, 1929a, b). She showed that benzene (in molecular formulæ) exists as a planar ring with six equal C-C bonds, instead of alternating single and double bonds, as in the Kekulé model (see Section 8.9.3). The other development involved the study of the silicates (Section 10.2). The number of resolved crystal structures increased rapidly and, in the 1930s, it became necessary to store information about them. This was done by R. W. G. Wyckoff (*The structure of crystals*, Wyckoff 1924, 1931, 1934), and by P. P. Ewald and C. Hermann who were the Editors of the *Strukturbericht* published by *Zeitschrift für Kristallographie*, and continued as 'Structure Reports' when the IUCr was founded in 1948.

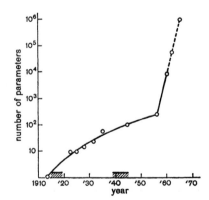

Fig. 8.2   Increase of the number of parameters in typical structures (logarithmic scale) with years. After W. L. Bragg in Ewald (1962a), © International Union of Crystallography.

8.2   Intensity of the X-ray reflections: first considerations

8.2.1   Variation of the intensity with order of the reflection

Early on, the first investigators observed that the intensity of the X-ray reflections decreased with the order of reflection, and, more generally, with increasing diffraction angle, θ. Several factors were identified that act on the intensity, such as the Debye temperature factor, which was discussed in Section 7.9, and, as described below, the polarization, the Lorentz and atomic form factors.

8.2.2   The polarization factor

J. J. Thomson (1903) showed that the amplitude of the wave scattered by an electron is proportional to $\sin \psi$, where ψ is the angle between the electric field vector $\mathbf{E_0}$ of the incident wave and the direction, $\mathbf{s_h}$, of observation of the scattered wave (Fig. 8.3, see equation (5.1), Section 5.5). If the incident wave is polarized in a direction Oz normal to the scattering plane $(\mathbf{s}, \mathbf{s_0})$, $\sin \psi = 1$, and if it is polarized in the scattering plane, then $\sin \psi = \cos 2\theta$ (θ: diffraction angle). If the incident wave is polarized in any direction, it can be decomposed into two components—one normal to the scattering plane, and one parallel to it. If the incident wave is unpolarized, the scattered intensity takes into account the average of the intensities corresponding to the two directions of polarization and is proportional to $(1 + \cos^2 2\theta)$. This effect was first noted in the case of X-rays scattered by an electron by Barkla and Ayres (1911) and by a co-worker of J. J. Thomson, J. A. Crowther (1912), Fellow of St John's College in Cambridge. It was introduced in the case of X-rays diffracted by a crystal by P. P. Ewald (1913a). Darwin (1914a) showed that the expression for the intensity, as given by geometrical theory, should be multiplied by a factor $(1 + \cos^2 2\theta)/2$, see equation (8.14). This factor is the *polarization factor*. Compton (1917a), using long wavelengths, could measure the reflections at sufficiently large glancing angles to check the effect due to this factor. In the dynamical theory, polarization is taken into account by a factor $(1 + |\cos 2\theta|)/2$, see equation (8.18).

8.2.3   The Lorentz factor

In an appendix to his final paper on the influence of thermal agitation, Debye (1913d) reported a calculation made by H. A. Lorentz to take into account the divergence of the incident beam and

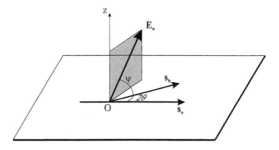

Fig. 8.3   Scattering of an electromagnetic wave. $\mathbf{E_0}$: incident electric field; $\mathbf{s_0}$: direction of the incident wave; $\mathbf{s_h}$: direction of the scattered wave; 2θ: scattering angle.

its spectral width. In order to do this, expression (6.3) of the diffracted intensity given by Laue (Section 6.6.5) should be integrated with $d\Omega \, d\theta$ as variable of integration for monochromatic or polychromatic radiation, where $d\Omega$ is a solid angle element, θ, the angle between the incident beam and the lattice planes. Rewriting (6.3) by replacing A, B, C, by their values, $A = h_1$, $B = h_2$, $C = h_3$, (equation 6.5), which are now to be considered as variables (they are in fact numerical coordinates in reciprocal space, but that was only understood later), the diffracted intensity is of the form, for monochromatic radiation:

$$I \propto \int \int \int \frac{\sin^2 \pi M h_1}{\sin^2 \pi h_1} \frac{\sin^2 \pi N h_2}{\sin^2 \pi h_2} \frac{\sin^2 \pi P h_3}{\sin^2 \pi h_3} d\Omega d\theta \qquad (8.1)$$

where the crystal is assumed to be a parallelepiped, and M, N, P are the numbers of unit cells along the three sides of the parallelepiped. The variable of integration can be rewritten, in terms of $dh_1 dh_2 dh_3$:

$$d\Omega d\theta = \frac{\lambda^3}{V \sin 2\theta} dh_1 dh_2 dh_3 \qquad (8.2)$$

where V is the volume of the unit cell and λ the wavelength.[14] The diffracted intensity is therefore proportional to a factor, $1/\sin 2\theta$, which is now called the Lorentz factor. It is included in Darwin's expressions (8.14) and (8.18) of the diffracted intensity, according to both the geometrical and the dynamical theories.

In his initial calculation, reported by Debye (1913d), Lorentz considered a polychromatic incident beam. The variable of integration $d\theta$ can be expressed in terms of $d\lambda$ by differentiation of Bragg's law:

$$d\theta = \tan \theta \frac{d\lambda}{\lambda}, \qquad (8.3)$$

and the diffracted intensity is proportional to

$$\frac{\lambda^3}{V \sin 2\theta} \times \frac{\tan \theta}{\lambda} = \frac{\lambda^2}{2V \sin^2 \theta} = \frac{d^2}{2V} \qquad (8.4)$$

where d is the interplanar spacing of the family of lattice planes. The fact that the diffracted intensity should decrease with decreasing density of nodes in the reflecting plane was noted by the early investigators, for instance by Friedel (1913b) who thought that the relationship should be direct proportionality. W. H. Bragg (1914c) knew about Lorentz's $(1/\sin^2 \theta)$ factor and about the 'obliquity factor' $(1 + cos^2 2\theta)$ (which he did not call polarization factor), but considered that the intensity measurements were 'hardly accurate enough as yet to bear discussion on these factors'. The Lorentz factor is also mentioned by Ewald (1914b) and Compton (1917a). Webster (1915) knew about it, but not its origin, and gave a different derivation. Surprisingly, Wyckoff (1924) wrote in his textbook that the relation $I \propto d^2$ was an empirical one! Laue (1926), noting that there were still some doubts concerning the Lorentz factor, clarified the situation and rederived it.

---

[14]A derivation can be found, for instance, in Warren (1969) or Authier (2003).

8.2.4   The atomic scattering factor or atomic form factor

During the First World War, W. H. Bragg and the Department of Physics of Leeds University were engaged in some work in connection with the war effort, but otherwise Bragg continued his research and his teaching. On 18 March 1915 he delivered the Bakerian lecture to the Royal Society. After presenting the basic principles of X-ray crystallography and describing a few basic structures, he discussed the decline of intensity with increasing order of reflection (W. H. Bragg 1915*c*). He made at this occasion two hypotheses which were suggested to him by the analogy with diffraction gratings: if the transparent portions are narrow, the different orders of diffracted spectra all have the same intensity. If this is not the case, and the transparent portions are broader, the intensity decreases with increasing order. The two hypotheses, which provide a basis for the development of crystal structure determinations, were:

1) 'The scattering power of the atom is not localized at one central point, but is distributed through the volume of the atom'; furthermore, 'the scattering centres of the atom are expected to be not only diffused through its volume, but also to be less dense at the edges than at the centre, thus producing exactly those conditions which would reduce greatly the intensities of the higher orders of the spectra'.

2) 'If we know the nature of the periodic variation of the density of the medium we can analyse it by Fourier's method into a series of harmonic terms.'

A similar analogy with a diffraction grating was made at the same time by a young American PhD student at Princeton University, Arthur H. Compton (Fig. 8.4). Compton had become interested in X-rays as an undergraduate while working with his brother Karl at the College of Wooster, Ohio. He took up this topic at Princeton, where there was modern X-ray equipment. Compton built an X-ray spectrometer and had been planning to study crystal structures, when the first structure determinations by the Braggs were published. He was strongly encouraged by a visit to Princeton by E. Rutherford, who praised Compton's experimental methods.[15]

From the analogy with the diffraction grating, Compton concluded, in a note submitted to *Nature* on 29 April 1915, that 'the relative intensity of the higher orders of spectra will depend upon the ratio of the effective diameter of the atoms to the distance between the successive layers of atoms', and that 'there are good reasons for believing that it is the electrons in atoms that scatter the X-rays' (Compton 1915). A short approving note was added by W. H. Bragg to Compton's paper, with a remark about his own suggestion of analysis of the periodic distribution of electron density by Fourier's series.

In his thesis, defended on 28 June 1916, Compton developed a theory of X-ray diffraction similar to Darwin's geometrical theory (Section 8.4.1). From W. H. Bragg's measurements on rock-salt and calcite, Compton obtained the variations of the scattering amplitude with diffraction angle, and, from them, deduced some possible arrangements for the electron distributions in atoms, and the distances of the electron rings from the nucleus. It is to be noted that A. W. Hull (1917*a*), quite independently it seems, also concluded that the fall-off of the intensities was due to the distribution of the electrons round the nucleus (see Section 8.5.2).

The angular variations of the atomic scattering factor were measured by W. L. Bragg *et al.* (1921*a* and *b*, see Section 8.7), Havighurst (1926) with powder diffraction, and Bearden (1927) with a double-crystal spectrometer. They were first calculated for sodium and chlorine,

[15]According to Robert S. Shankland, a former pupil of Compton, in *Scientific papers of Arthur Holly Compton: X-rays and Other Studies*. University of Chicago Press (1973).

Fig. 8.4   Arthur Holly Compton. AIP Emilio Segre Visual Archives, with permission.

Arthur Holly Compton: born 10 September 1892 in Wooster, Ohio, USA, the son of a one-time dean of the College of Wooster, and the brother of a future President of M.I.T; died 15 March 1962 in Berkeley, California, USA, was an American experimental physicist. After graduating from the College of Wooster in 1913, Compton went to Princeton where he received a master's degree in 1914 and his PhD in June 1916. From 1916 to 1917 he taught physics at the University of Minnesota, after which he spent two years as research physicist for the Westinghouse Lamp Company in Pittsburgh. In 1919 he became a Fellow of the National Research Council, and spent one year in Rutherford's laboratory at Cambridge, England, where he studied the scattering and absorption of γ-rays. He observed that the scattered radiation was more absorbable than the primary, an observation which led to his discovery of the Compton effect (see Section 9.5). In 1920 he was appointed Professor of Physics at Washington University, St Louis, and in 1923 Professor of Physics at the University of Chicago. About 1930 Compton's interests shifted from X-rays to cosmic rays. During the war he directed the efforts to initiate the chain reaction, and was heavily involved in the Manhattan Project. Compton returned to St Louis as Chancellor in 1945, and from 1954 until his retirement in 1961 he was Professor of Natural Philosophy at Washington University. Compton was awarded the Nobel Prize for Physics in 1927, 'for his discovery of the effect named after him'; the prize was divided between him and C. T. R. Wilson (see footnote 1, Chapter 1). He was elected a member of the National Academy of Sciences in 1927.

MAIN PUBLICATIONS

1917 *The Intensity of X-Ray Reflection, and the Distribution of the Electrons in Atoms*

1923 *A Quantum Theory of the Scattering of X-Rays by Light Elements.*

1926 *X-Rays and Electrons; an Outline of Recent X-Ray Theory.*

1935 *X-Rays in Theory and Experiment.* With S. K. Allison.

1935 *The Freedom of Man.*

1940 *The Human Meaning of Science.*

1956 *Atomic Quest; a Personal Narrative.*

using classical methods by Compton (1917*a*), Hull (1917*a*), Bragg *et al.* (1921*a, b*), and Glocker (1921). These authors assumed that electrons lie on Bohr circular orbits, at certain distances from the nucleus. Bragg *et al.* (1922*a, b*) improved the calculation by assuming elliptical orbits. The angular variations of the atomic scattering factor were calculated by D. R. Hartree,[16] at first using Bohr orbits (Hartree 1925), then using wave mechanics and the 'self-consistent fields'

---

[16]Douglas Rayner Hartree, born 27 March 1897 in Cambridge, England, the son of a lecturer at Cambridge University, died 12 February 1958 in Cambridge, England, was an English mathematician and physicist. He attended St John's College, Cambridge, but the First World War interrupted his studies. He returned to Cambridge after the war and graduated in 1922. Following a visit by N. Bohr to Cambridge, where he gave a course, Hartree chose Bohr's atomic theory as the topic of his thesis, for which he obtained his PhD in 1926. In 1929, Hartree was appointed Professor of Applied Mathematics at Manchester University. He was elected a Fellow of the Royal Society in 1932.

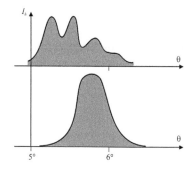

Fig. 8.5   Reflection profiles of the first order spectrum from a (100) face of a large rock-salt crystal (the 200 reflection – that notation was introduced later). The top and bottom curves were recorded at two neighbouring positions on the crystal. I_h: current from the ionization chamber; θ: reflection angle. Redrawn after W. H. Bragg (1914c).

introduced by him (Hartree 1928; Waller and Hartree 1929). The agreement with experimental results was checked by James *et al.* (1928) on rock-salt and by James and Brindley (1928) on sylvine. James *et al.* (1928) noted that the early calculations of the atomic scattering factor had been made at a time when incoherent scattering had not been taken into account, and therefore contained errors.

8.2.5   *W. H. Bragg and the integrated reflection (integrated intensity), 1914*

W. H. Bragg (1914c) made a major contribution by giving a 'definite meaning to the term intensity of reflection'. He rotated the crystal by a succession of small steps through the angle at which the reflection takes place, measuring the ionization current at each step and he plotted the results as a curve (Fig. 8.5). Bragg noted that the observed profile is irregular and depends on the position of the crystal that is being impinged upon by the incident beam of X-rays. This is due to the imperfections in the crystal, as shown by the plots in Fig. 8.5, recorded with a particularly irregular crystal. Bragg concluded that 'clearly no measure of intensity is to be obtained from the maximum ordinate of either of these curves. The *areas* of the curves are nearly the same, however, and can be taken as a measure of the intensity of the reflection'. This area was later called 'integrated reflection' (W. L. Bragg *et al.* 1926). An equivalent way of recording the integrated intensity was 'to observe the ionization current for a given time at each step in the movement of the crystal, and subsequently to add together all the currents observed'. Another way, analogous to de Broglie's (1913c) rotating crystal method, was to leave the X-rays in action, to look at one's watch and to turn the crystal at each beat of the clock through the whole reflection domain.

8.3   The first stages of the analysis of crystal structures

8.3.1   *W. L. Bragg's empirical scale of intensities, 1914*

W. L. Bragg's analysis of crystal structures relied on the use of two empirical factors, 1) the dependence of intensity on the nature of the scattering atoms, and 2) the variation of intensity with angle of diffraction.

The first structures determined by the Braggs were either cubic (alkali halides, diamond, fluorite, pyrite) or derived from the cubic, such as the trigonal carbonates. The structure of trigonal carbonates can, indeed, be described with respect to a deformed sodium chloride structure by flattening the cube into a rhombohedron, with the metal atom (Ca, Mg, Mn or Fe) occupying the position of sodium and the groups CO_3 those of chlorine atoms.

The analysis was based on the comparison of the spectra recorded from the (100), (110) and (111) faces with the ionization spectrometer. The intensities of successive orders of a given spectrum, measured in this manner, were given values with respect to that of the most intense one, which was arbitrarily given the value 100. The spectra thus obtained were compared to an empirical scale, called 'normal spectrum', corresponding to the intensity decrease expected for a simple cubic crystal (W. L. Bragg 1914a, Bragg and Bragg 1915):

$$I_1 : I_2 : I_3 : I_4 : I_5 :: 100 : 20 : 7 : 3 : 1.$$

Any deviation from this 'normal' decline of the intensities gave an indication as to the contents of the planes (100), (110) or (111). As an example, the spectra corresponding to four orders of reflection from the (111) planes of four carbonates and sodium nitrate, all isomorphous with calcite, are given in Table 8.1.

An interesting situation occurs when certain orders are systematically absent, and initially such observations were interpreted in terms of the atomic weights of the constituents, as illustrated here by three examples:

1. In potassium chloride (sylvine), reflections with odd indices (111, 333, ..) are absent. The crystal structure consists of a stacking of (111) planes occupied successively by potassium and chlorine with atomic weights 39 and 35, respectively and for which scattering could be considered constructive or destructive (see Section 7.7).
2. Similarly in fluorite, the 200 and 222 reflections are absent. The structure therefore consists, on the one hand, of a stacking of (100) planes occupied alternately by calcium and fluorine, and, on the other hand, by a similar stacking of calcium and fluorine along the (111) planes. Since the atomic weights of calcium and fluorine are 40 and 19, respectively, it was considered that there are twice as many fluorine atoms as calcium atoms in successive (100) and (111) planes. From this result it was immediately deduced that the structure of fluorite can be derived from that of diamond (see Section 7.7). The atoms of calcium lie on a face-centred cubic lattice and the fluorine atoms occupy the centres of the eight little cubes into which the cubic cell is divided. Fluorite cleaves easily into small octahedra, with faces parallel to 111 planes. The fact that it could not be divided into small parallelepipeds created problems for the early investigators such as Haüy (see Section 12.1.2) and Wollaston (see Section 12.2).

Table 8.1   Relative intensities of the first four orders of the (111) spectra of four carbonates and sodium nitrate

Substance	Observed spectrum			
order	first	second	third	fourth
Calcite ($CaCO_3$)	30	100	0	14
Dolomite ([Ca,Mg]CO_3)	100	100	0	8
Rhodochrosite ($MnCO_3$)	0	100	0	10
Siderite ($FeCO_3$)	0	100	0	–
Sodium nitrate ($NaNO_3$)	100	50	–	–

3. In rhodochrosite ($MnCO_3$) and siderite ($FeCO_3$), the 111 reflection is absent. Their crystal structures are composed of (111) planes occupied alternately by manganese or iron atoms, of atomic weights 55 and 56 and CO_3 of atomic weight 60.

These interpretations of systematically absent reflections led to information on the crystal structure involved. As a result of them Bragg and Bragg (1915) concluded that 'the alternate planes become equal in reflecting power when their masses per unit area are equal', and 'that the amplitude of the diffracted wavelet, as an X-ray wave passes over an atom, is proportional to the atomic weight of that atom'. The observations reported in Table 8.1 can be interpreted in the light of this remark, by noting that the intensities of the first two orders, 111 and 222, are proportional respectively to the differences and the sums of the scattering powers of the metal on the one hand and of the CO_3 or NO_3 groups on the other hand. The atomic weight of sodium is 23, and is much smaller than that of the NO_3 groups, 62, and the intensities are hardly affected by the presence of sodium. In calcite, the atomic weight of calcium is 40 and is quite important relative to that of the CO_3 group, 60. The intensity of the 111 reflection is, for that reason, much smaller than that of 222. Dolomite presents an intermediary case, the average weight of calcium and magnesium being 32.

It was only after Compton's 1917 paper that it was clear that the scattering power of an atom or ion is proportional to the number of electrons in it, as was assumed by Debye and Scherrer (1918) in their discussion of the structure of lithium fluoride (Section 10.1.2).

W. L. Bragg used a similar analysis to determine the relative positions of the carbon and oxygen atoms in the CO_3 groups of calcite. The oxygen atoms lie between two carbon atoms at a fraction x/d of the distance between two carbon atoms (Fig. 8.6). The structure can be described by the stacking, parallel to the (100) planes, of successions of one plane containing calcium, carbon and oxygen and two planes of oxygen atoms. The observed (100) spectra and the corresponding spectra calculated for several values of x/d are compared in Table 8.2. It can be concluded that the value x/d is of the order of 0.30 to 0.33. The structure of the carbonates was revisited by Wyckoff (1920c) using the Laue method and gnomonic projection to analyse the diagrams. He obtained a slightly different value of x/d (0.24 to 0.26) for calcite, and he determined the corresponding value for the other carbonates.

The case of pyrite is more complex because its symmetry is hemihedral (point group $m\bar{3}$). The atoms of sulphur are displaced from the centres of the eight little cubes along the diagonals

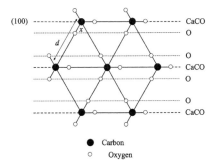

Fig. 8.6   Structure of the CO_3 groups in the (111) planes of calcite. The dashed lines show the traces of the (100) planes. Adapted from W. L. Bragg (1914a).

Table 8.2   Observed and calculated (100) spectra of calcite

Calcite		(100) spectrum		
order	first	second	third	fourth
Observed	100	20	20	9
$x/d = 0.25$	100	6	7	7
$x/d = 0.30$	100	11	18	8
$x/d = 0.33$	100	20	30	3
$x/d = 0.35$	100	30	31	2

by a certain amount, which is an unknown parameter of the structure. W. L. Bragg (1961, 1975) recalled: 'The structure of pyrite provided the greatest thrill. It seemed impossible to explain its queer succession of spectra until I discovered, going through Barlow's geometrical assemblages, that it was possible for a cubic crystal to have non-intersecting trigonal axes. The moment of realization that this explained the iron pyrite result was an occasion I well remember'. The structure of pyrite was revisited one year later by Ewald (1914b) and Ewald and Friedrich (1914) in Munich, who determined the shift of the sulphur atoms by consideration of the structure factor, as described in the next Section, 8.3.2. The space-group for pyrite was first given by Schoenflies (1915).

Bragg's empirical 'normal spectrum' was still referred to in Ewald's and Wyckoff's textbooks that were published much later in 1923 and 1924.

8.3.2   P. P. Ewald and the structure factor, 1914

Laue, in one of the addenda he wrote at the occasion of the reprinting in *Annalen der Physik* of his 1912 article on the interpretation of the diffraction spots (15 March 1913), was the first to note that the contribution of each of the atoms in the unit cell should be taken into account in the expression of the diffracted intensity, but it is Ewald (1914a and b) who introduced the structure factor, as announced by Sommerfeld at the Solvay Congress (equation (7.10), Section 7.10.3), and who coined the term. In his expression of the structure factor, Ewald had, however, taken the atoms to be point scatterers, and had not included a direction-dependent atomic form factor.

Ewald (1914b) made a first use of the structure factor to make a more accurate determination of the structure of pyrite (FeS_2) than W. L. Bragg (1914a) had. In pyrite, the iron atoms occupy the nodes of a face centred-cubic lattice, and there are eight sulphur atoms located along the four diagonals of the cube at a distance xu of each of the summits of the cube, where u is the length of the diagonal. Bragg had assumed that $x = 0.2$. Ewald (1914b) observed on the Laue diagrams that the intensities of the spots $(18\bar{2})$ and $(81\bar{2})$, and $(37\bar{2})$ and $(73\bar{2})$, respectively, were different. By calculating the structure factor, he showed that these intensities would have been equal if x was indeed 0.2. In order to account for the observed intensities, he found that x was 0.224 instead.

8.4   C. G. Darwin and the theory of X-ray diffraction, 1914

C. G. Darwin (Fig. 8.7) joined E. Rutherford's laboratory at Manchester in 1910, after his graduation from Cambridge University. After doing some experimental and theoretical work on α-rays, he turned towards X-ray diffraction. As has been recounted earlier in Section 7.8.2, he and H. G. J. Moseley initially measured X-ray intensities with an ionization spectrometer. In the fall of 1913, H. G. J. Moseley started systematically recording the X-ray spectra of the elements, and C. G. Darwin, who was at heart an 'applied mathematician',[17] decided to derive the expression of the diffracted intensity. Laue's expression (6.3) of the intensity had been derived assuming plane waves, but Darwin felt that this was unsatisfactory and that 'a complete theory regarding the whole problem as one of spherical waves' was necessary (Darwin 1914*a*). The two papers that resulted were considered 'landmarks in the history of X-ray analysis' by W. L. Bragg, who added that 'X-ray crystallographers have always considered this imaginative

Fig. 8.7   Charles Galton Darwin. Photo from George Grantham Bain collection, US Library of Congress.

Charles Galton Darwin: born 18 December 1887 in Cambridge, England, the son of an astronomy professor at Cambridge, and the grandson of Charles Darwin, the father of the theory of evolution; died 31 December 1962 in Cambridge, was an English physicist. After schooling at Marlborough College, Wiltshire, England from 1901 to 1906, he entered Trinity College, Cambridge. He was Fourth Wrangler in the Mathematical Tripos of 1909. In 1910 he joined Rutherford's laboratory at Manchester University as Reader in Mathematical Physics. During World War I, he served in the Royal Engineers and, after one year, joined the sound-ranging unit organized by W. L. Bragg. From 1919 to 1922 he was a Lecturer in Mathematics and a Fellow of Christ's College, Cambridge, where he worked with R. H. Fowler on statistical mechanics. After spending one year at the California Institute of Technology, he was appointed Tait Professor of Natural Philosophy at the University of Edinburgh in 1924, where he worked on quantum optics and magneto-optic effects, being the first in 1928, to calculate the fine structure of the hydrogen atom under P. A. M. Dirac's relativistic theory of the electron. In 1936 Darwin left Edinburgh to take up the position of Master of Christ's College, Cambridge. He held this post for two years, then became Director of the National Physical Laboratory, which he remained until his retirement in 1949. During the war years he worked on the atom bomb Manhattan project. Darwin was elected a Fellow of the Royal Society in 1922.

MAIN PUBLICATIONS

1913 *The reflection of X-rays.* With H. G. J. Moseley.

1914 *The theory of X-ray reflection.*

1922 *The reflection of X-rays from imperfect crystals.*

1926 *The Intensity of Reflection of X-rays by Crystals.* With W. L. Bragg and R. W. James.

1931 *The new conception of matter.*

1937 *The mosaic structure of crystals.*

1952 *The Next Million Years.*

---

[17]G. P. Thomson (1963). Charles Galton Darwin, 1887–1962. *Biogr. Mems. Fellows R. Soc.* **9**, 69–85.

and original work of Darwin, produced at such an early stage of the subject, as one of the finest contributions to science' (quoted in Thomson 1963, see footnote 17).

The theory was worked out by Darwin, taking as model that experimental arrangement which he and Moseley had proved most fruitful, reflection from the planes parallel to an external face of a crystal. He presented the results in two parts. In the first one (Darwin 1914a) he assumed that the scattering by one atom did not affect that by others (the so-called geometrical or kinematical approximation). In the second part (Darwin 1914b) the interaction between the waves scattered by successive planes was taken into account. The same results for the diffracted intensity were obtained independently by Ewald with his 'dynamical theory' (Section 9.2).

8.4.1   The geometrical theory of X-ray diffraction

Scattering by a single plane.   In a first step, Darwin calculated the amplitude reflected by a single plane of atoms, using Fresnel zones. The complex reflection coefficient $-iq$ taking into account the phase shift associated with the scattering is

$$-iq = i \frac{Nd}{k \sin \theta} f(2\theta, k) \tag{8.5}$$

where $k = 1/\lambda$, N is the number of atoms per unit volume of the crystal, d the distance between successive planes (Nd is the number of atoms per unit area of the diffracting plane), θ the glancing angle and $f(2\theta, k)$ is the scattering amplitude of a single atom. Following J. J. Thomson (1903), assuming one atom per unit cell and assuming that the atom can be reduced to a point, $f(2\theta, k)$ is proportional to ne^2/mc^2 where n is the number of electrons of the atom. Actually, the scattering amplitude should take into account the electron distribution around the atoms (W. H. Bragg 1915c; Compton 1915, Section 8.2.1), and all the atoms in the unit cell, through the structure factor introduced by Ewald (1914b, Section 8.3.2):

$$f(2\theta, k) = R F_{hkl} \tag{8.6}$$

where F_{hkl} is the structure factor, and $R = e^2/mc^2$ is the so-called classical radius of the electron.

Refractive index.   According to the dispersion theories available when the first hypotheses concerning the electromagnetic nature of X-rays were suggested, the refractive index was expected to be very close to 1 for radiation in the very short wavelength range (Lodge 1896c). Darwin (1914a) was the first to derive a correct expression. His argument was as follows. If $-iq_o$ is the complex amplitude scattered in the forward direction by the first lattice plane, the amplitude of the spherical wave incident on the second lattice plane is

$$A = (1 - iq_o) \exp 2\pi i\nu t \frac{\exp -2\pi ikr}{r}. \tag{8.7}$$

One has, after crossing s planes:

$$A = (1 - iq_o s) \exp 2\pi i\nu t \frac{\exp -2\pi ikr}{r} \tag{8.8}$$

$$\approx \exp 2\pi i\nu t \frac{\exp[-2\pi ikr - iq_o s]}{r} \tag{8.9}$$

since $|q_o|$ is very small. Darwin remarked that the presence of the term $iq_o s$ implies a refractive index. If θ is the glancing angle and $z = sd = r/\sin\theta$ the thickness of the crystal, taking the origin of the position vector $\mathbf{r}$ at the entrance surface, there comes, for the transmitted amplitude:

$$A \approx \exp 2\pi i\nu t \, \frac{\exp[-2\pi i[k + q_o/d\sin\theta)]r}{r}, \tag{8.10}$$

and, using (8.5):

$$A \approx \exp 2\pi i\nu t \, \frac{\exp[-2\pi inkr]}{r}, \tag{8.11}$$

where:

$$n = 1 - \frac{RN\lambda^2 F_o}{2\pi} = 1 - \delta_o \tag{8.12}$$

is the refractive index, and F_o the amplitude of the wave scattered in the forward direction. This result is obtained assuming that the wavelength λ is far from an absorption edge. When this is not the case, the theory of dispersion shows that corrections have to be applied to δ_o (Lorentz 1916) and:

$$n = 1 - (\delta_o + \delta_o' + i\delta_o'') \tag{8.13}$$

The corrections δ_o' and $i\delta_o''$ are called anomalous dispersion corrections, and the absorption coefficient of the material is proportional to the imaginary part of the refractive index. The amplitude F_h of the wave scattered in the reflected direction is corrected in a similar way. This is called anomalous scattering.

The variations of anomalous dispersion with wavelength have been measured experimentally by Hjalmar and Siegbahn (1925) in an indirect way. They plotted the variations with wavelength in the 0.7–5.2 Å range of the ratio $d_{calcite}/d_{gypsum}$ of the lattice parameters of calcite and gypsum. These parameters had been obtained by applying Bragg's law, $2d\sin\theta = n\lambda$, and depended therefore on the anomalous dispersion corrections for the index of refraction, which vary with wavelength. The curves exhibited two discontinuities, one corresponding to the absorption edge of calcium, the second one to the absorption edge of sulphur.

The refractive index is very close to 1; for instance, for silicon and $CuK\alpha$, $n = 1 - 0.757 \times 10^{-5}$. The theory of anomalous dispersion was extended by I. Waller (1928) within the framework of quantum mechanics, and the real and imaginary parts of δ_o close to an absorption edge were first calculated using quantum mechanics by H. Hönl[18] (1933) in Ewald's Institute at Stuttgart Technical University.

Scattering by a set of lattice planes. In the next step, Darwin summed up the amplitudes diffracted by the successive planes. The incident beam is in practice a divergent beam, approx-

---

[18]Helmut Hönl, born 10 February 1903 in Mannheim, Germany, died 29 March 1981 in Freiburg im Breisgau, Germany, was a German theoretical physicist. He obtained his PhD in Munich under Sommerfeld in 1926. He became Ewald's Assistant in Stuttgart in 1929, Privatdozent in 1933. In 1940 he was appointed Professor at Erlangen University and in 1943 at Freiburg University.

imated by a spherical wave, and the total reflected intensity is the result of an integration over the glancing angle (it is now called the integrated intensity). For a small non-absorbing crystal entirely bathed in the incident beam, and in the reflection geometry, it is given by:

$$I_{geom} = I_o \frac{N^2 \lambda^3 R^2 |F_{hkl}|^2}{\sin 2\theta} \frac{1 + \cos^2 \theta}{2} \exp{-M \Delta v} = I_o Q \Delta v \qquad (8.14)$$

where $(1 + \cos^2 \theta)/2$ is the polarization factor and $1/\sin 2\theta$ the Lorentz factor (see Section 8.2), $\exp{-M}$ the Debye factor taking thermal agitation into account (Debye 1913d), Δv the volume of crystal and I_o the energy incident per unit area in the beam. If the crystal is very large and absorbing, this expression is replaced by:

$$I_{geom} = I \frac{Q}{2\mu} \qquad (8.15)$$

where μ is the linear absorption coefficient and $I = I_o S_o$ is total energy of the incident beam (S_o cross-section of the incident beam). The same results were later rederived by Compton (1917a) using a slightly different method of integration.

Darwin compared the calculated intensities with those measured with rock-salt crystals by Moseley and Darwin (1913b), and found that the calculated values were too large. He supposed that this was due to the assumption made when the interaction between the waves transmitted and reflected at each atomic plane was neglected. He took this interaction into account in the second of his papers (Darwin 1914b), which is devoted to the dynamical theory of diffraction, although Darwin did not use that expression.

8.4.2   Reflection of X-rays by a perfect crystal

In the second paper, published three months after the first one (Darwin 1914b), the mutual influence of the waves scattered by the different atoms is taken into account. For that reason it deserves to be called a dynamical theory, in the same way as Ewald's and Laue's theories are dynamical theories (see Section 9.2). Laue (1931b) acknowledged that Darwin's theory was the first form of the dynamical theory of diffraction.

In view of the greater complexity of the problem, Darwin found that it was more convenient to deal with plane waves, contrary to what had been done in the first paper. For that, it is only necessary to calculate the diffracted intensity at a point sufficiently distant from the crystal. As the incident wave propagates inside the crystal, it generates both a reflected and a transmitted wave at each lattice plane it crosses (Fig. 8.8, *Left*). These in turn generate reflected and transmitted waves whenever they cross an atomic plane, and so on. The problem is thus very similar to that of the Fabry–Pérot interferometer or etalon, but with an infinite number of parallel equidistant plates. The amplitudes and phases of the transmitted and reflected waves are related by a set of recurrent equations derived by Darwin. Let S_n and S_{n+1} be the amplitudes of the waves incident on the nth and $(n + 1)^{th}$ planes, respectively, and T_n and T_{n+1} the amplitudes reflected from these planes. They are related by:

$$S_n = iq T_n + (1 - iq_o) \exp(-i\phi) S_{n+1},$$

$$T_{n+1} \exp(i\phi) = (1 - iq_o) T_n - iq \exp(-i\phi) S_{n+1},$$

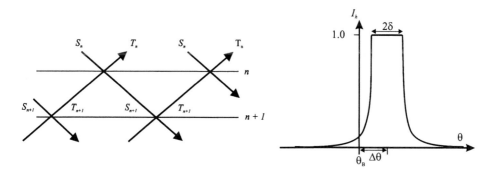

Fig. 8.8 *Left*: Darwin's dynamical theory. S_n: incident amplitude on plane n; T_n: amplitude reflected from plane n. After Authier (2003). *Right*: Total reflection domain. θ_B: Bragg angle; $\Delta\theta$: shift of the middle of the reflection domain with respect to Bragg angle. 2δ: width of the total reflection domain.

where $\phi = 2\pi a \sin\theta / \lambda$ is a phase factor and q_o the amplitude of the wave scattered in the forward direction. By solving this set of equations, it is possible to obtain the expression of the amplitude reflected at the surface of the crystal. Darwin showed that, if absorption is neglected, the amplitude is imaginary within a narrow angular range proportional to $|q|$, and that there is total reflection within that range. He did not, however, plot the variations of the diffracted intensity with diffraction angle. They were first represented, quite independently, by Ewald (1917), who, in his X-ray medical unit on the Russian front during World War I, had no way of knowing about Darwin's works. They are shown in Fig. 8.8, *Right*. The width of total reflection domain is

$$2\delta = \frac{N\lambda^2 R |F_{hkl}|}{\pi \sin 2\theta}. \tag{8.16}$$

It is much narrower than the angular range of reflection given by the geometric theory, and is now called the *Darwin width*. In practice, crystals are always absorbing and the reflectivity is never 100%. The expression of the reflected intensity for absorbing crystals was first given by the Dutch physicist, J. A. Prins (1930), a student of D. Coster in Groningen.

The middle of the total reflection domain is shifted with respect to the Bragg angle by a small angle $\Delta\theta$. In the symmetric reflection geometry, the only case considered by Darwin, it is:

$$\Delta\theta = \frac{N R\lambda^2 F_o}{\pi \sin 2\theta} = \frac{\delta_o}{2} = \frac{1-n}{2}, \tag{8.17}$$

where F_o is the scattered amplitude in the forward direction and n the refractive index (8.12). This shift is due to the effect of the refraction of the X-rays in the medium. The complete calculation with the dynamical theory shows that (8.17) should be multiplied by an asymmetry factor, which is equal to zero for a symmetric transmission geometry (see, for instance, Authier 2003).

For a divergent incident beam, the intensity of the total reflected beam (integrated intensity) is given by:

$$I_{dyn} = I_o \frac{8NR\lambda^2 |F_{hkl}|}{3\pi \sin 2\theta} \exp(-M) \frac{1 + |\cos 2\theta|}{2} \qquad (8.18)$$

This expression is very different from that given by geometrical theory, (8.14). It is proportional to the absolute value of the structure factor and not to its square, it does not depend on the size of the crystal, and it is much smaller.

Darwin found that the reflected intensities calculated with (8.18) are not in better agreement with the experimental ones than those calculated from (8.14) or (8.15). This time, they were too small. This result was very baffling and Darwin attributed it to the presence of imperfections. He supposed that at various depths the crystal was twisted by an amount sufficiently large to allow a new reflection. One would therefore expect a wider angular range of reflection and a larger total reflected intensity. This was a first qualitative attempt at what would be the theory of extinction developed a few years later by Darwin (1922), see Section 8.8.

8.5   Powder diffraction, P. Debye and P. Scherrer, A. W. Hull

8.5.1   Peter Debye and Paul Scherrer

Paul Scherrer (Fig. 8.9) came to the University of Göttingen, Germany, in 1913 to prepare a doctorate under P. Debye (Fig. 7.12). This was at a time when Bohr's atomic theory was becoming known (Section 10.6). Physicists were immediately very excited by this new atomic theory, but they had some difficulty being convinced of the reality of the electron orbits. Scherrer (*Personal reminiscences*, in Ewald 1962a) recounted that 'Debye had come to the conclusion that specific diffraction effects should be produced with X-rays by the regular spacing of electrons on circular orbits'. Debye (1915) showed that this should be the case, even though the atomic structure changes its orientation in space continuously and randomly. He proposed to Scherrer that they should try some X-ray diffraction experiments that would enable them to observe these effects. This is how powder diffraction was discovered!

The first diffraction photographs, with paper and charcoal as the scattering substances, and a gas-filled medical X-ray tube with platinum target as source, showed no diffraction effects. Scherrer then constructed a metal X-ray tube, water-cooled and with copper target, and 'a cylindrical diffraction camera, of 57 mm diameter, with a centring head for the sample, of the type which is being used still nowadays'. For the sample, he used the finest grain powder of lithium fluoride (LiF). Debye and he 'were most surprised to find on the very first photographs the sharp lines of a powder diagram', which they correctly interpreted as crystalline diffraction by the randomly oriented micro-crystals of the powder. Scherrer remembered that they had been lucky to use as first sample a cubic crystal powder with a simple structure, which they determined easily using copper and platinum radiation.

Debye and Scherrer's first work was presented to the Göttingen Science Society by D. Hilbert on 4 December 1915 (Debye and Scherrer 1916a, part I) and was submitted to *Physikalische Zeitschrift* on 28 May 1916 (Debye and Scherrer 1916b). It included descriptions of the set-up, of the theory and of the structures of lithium fluoride and silicon. They showed that the structure of silicon was identical to that of diamond, described by the Braggs, and measured the unit cell parameters of both crystals. Further analysis of the lithium fluoride data led to the conclusion that lithium and fluorine were present as ions in the crystal (Debye and Scherrer 1918, see Section 10.1.2). In the second paper (Debye and Scherrer 1916a, part II), they took diffraction

Fig. 8.9   Paul Scherrer. Photo courtesy Paul Scherrer Institut, Villingen, Switzerland.

Paul Scherrer: born 3 February 1890, in St. Gallen, Switzerland, the son of a painter; died 25 September 1969 in Zürich, Switzerland, was a Swiss physicist. After studying at ETH Zürich and Königsberg University, he joined P. Debye at Göttingen University where he obtained his PhD in 1916. The topic of his doctorate was the Faraday effect in the hydrogen molecule. It is at that time that he developed the powder diffraction method jointly with P. Debye. He was appointed Professor of Experimental Physics at ETH Zürich in 1920, and became Principal of the Physical Institute at ETH in 1927. In 1930 he turned towards ferroelectrics (he discovered the ferroelectricity of KDP) and then to nuclear physics. He became President of the Swiss Study Commission on Atomic Energy in 1946, and President of the Swiss Commission for Atomic Sciences in 1958.

MAIN PUBLICATIONS

1916 *Interferenzen an regellos orientierten Teilchen im Röntgenlicht.*

1916 *Die Rotationsdispersion des Wasserstoffs.*

1918 *Atombau.* With P. Debye.

1942 *Gekürzte Vorlesung über Physik.*

1942 *Neuere Ergebnisse kernphysikalischer Forschung.*

1960 *Beiträge zur Entwicklung der Physik.*

pictures of liquids, in particular benzene (C_6H_6). They observed a diffuse distribution of intensity, with maxima and minima, and made a preliminary interpretation of the diagram, using a one-dimensional interference function. From the position of the first maximum, they could estimate the diameter of a molecule of benzene. The third paper (Debye and Scherrer 1917) was devoted to the structure of graphite (Fig. 8.11, *Right*, see Section 8.9.2).

Scherrer tried many types of crystals and determined the structure of magnesium oxide (MgO) and iron, both cubic, and of a molecular crystal, potassium chloroplatinate (K_2PtCl_6), also cubic (Scherrer and Stoll 1922), but he failed with boron (Scherrer in Ewald 1962a). The case of potassium chloroplatinate was easy because the structure of such coordination complexes had been surmised by the Swiss chemist A. Werner[19] from chemical considerations. Scherrer also took diffraction diagrams of fibres such as ramie[20] or silk fibroin, and observed the regular arrangement of cellulose crystallites along the direction of the fibre axis.[21]

For his habilitation, Scherrer studied the structure and size of colloidal particles by means of X-ray powder diffraction. He used colloidal gold and silver, which could be obtained with a definite particle size. P. Scherrer derived the relation between the half-width h of the diffraction peaks and the particle size, assuming a uniform distribution of small cubic particles of edge t (Scherrer 1918):

---

[19] See footnote 2, Chapter 10.   [20] A woody urticaceous shrub of Asia, used in making fabrics, cord, etc.

[21] P. Scherrer, in R. Zsigmondy (1920). *Über Kolloidchemie.* 3rd Edition, p. 408.

$$h = \frac{K\lambda}{t \cos \theta} \qquad (8.19)$$

where K is a numerical factor, called 'shape factor'. In Scherrer's original work, $K = 2\sqrt{\ln 2/\pi}$. Equation (8.19) is known as 'Scherrer formula'. This expression was slightly modified by N. Seljakov (1925), who gave a different value for K, and was generalized by M. von Laue (1926) to take into account both the shape and the size of the crystallites. The use of that formula in metal physics is discussed in Section 10.1.6.

8.5.2   Alfred Hull

A. W. Hull (Fig. 8.10) had been teaching physics at the Worcester Polytechnic Institute in Massachusetts for a few years when his wife suggested that he should, at last, present his work at a meeting of the American Physical Society (Hull, in Ewald 1962a). This he did in June 1912. He impressed I. Langmuir[22] and W. Coolidge[23] so much that they invited him to the General Electric Research Laboratory in Schenectady for the summer of 1913. After finishing a year of teaching, he returned to Schenectady in 1914, never to leave. Soon after his arrival he invented the dynatron, a vacuum tube having true negative resistance, but his interests shifted abruptly on the occasion of a lecture given early in 1915 at General Electric by Sir William Bragg. At the end of the lecture, Hull asked Bragg whether he had been able to find the structure of iron. Bragg answered very simply, 'No, we have tried, but have not succeeded' (Hull in Ewald 1962a, and in McLachlan and Glusker 1983). Hull saw there a challenge and decided to try to find the crystal structure of iron.

The necessary X-ray equipment was available at General Electric: the Coolidge X-ray tube and Kenotron rectifiers recently developed at the laboratory. Hull filtered the rectified current by a pair of condensers with an inductance between them. This system was patented and was later used in broadcast receiver circuits. Since single crystals of iron were not available, Hull used iron filings, which were rotated continuously, in order to provide randomness. He obtained rapidly good diffraction patterns, but being busy with the dynatron, turned them over to a young assistant, asking her to see, using Bragg's formula, whether the measurements were consistent with any of the three cubic lattices. The young lady, however, made a mistake, and did not find

---

[22]Irving Langmuir, born 31 January 1881 in Brooklyn, NY, USA, died 16 August 1957 in Woods Hole, MA, USA, was an American chemist and physicist. He graduated from the Columbia University School of Mines in 1903, and obtained his PhD in 1906 under W. Nernst in Göttingen. Langmuir then taught at Stevens Institute of Technology in Hoboken, New Jersey, and joined the General Electric Research Laboratory in Schenectady, NY. in 1909. He is best-known as the inventor of the high-vacuum tube, for his work on 'the arrangement of electrons in atoms and molecules (see Section 10.1.2), and for the concept of monolayer thin films he developed with Katharine B. Blodgett (Langmuir-Blodgett films). In 1932 he was awarded the Nobel Prize for Chemistry, 'for his discoveries and investigations in surface chemistry'.

[23]William D. Coolidge, born 23 October 23 1873 in Hudson, MA, USA, died 3 February 1975 in Schenectady, NY, USA, was an American physicist. He graduated from M.I.T. in 1896, and obtained his PhD in 1899 from the University of Leipzig. After being Research Assistant at M.I.T. from 1899 to 1905, he joined the newly established General Electric Research Laboratory in Schenectady, NY, where he conducted experiments that led to the use of tungsten as filaments in light bulbs. In 1913 he invented the Coolidge tube, an X-ray tube with an improved cathode. Coolidge became Director of the General Electric Research Laboratory in 1932, and a vice-president of General Electric in 1940, until his retirement in 1944.

Fig. 8.10   Alfred Wallace Hull. AIP Emilio Segre Visual Archives, with permission. Physics Today Collection.

Alfred Wallace Hull: born 19 April 1880, on a farm, in Southington, Connecticut, USA; died 22 January 1966, was an American physicist and engineer. He studied at Yale University, where he majored in Greek and took one undergraduate course in physics. He taught languages at Albany Academy in Albany, NY, for a time, before returning to Yale, where he obtained a PhD in physics in 1909. He taught physics at the Worcester Polytechnic Institute in Massachusetts for five years, doing some research on photoelectricity. In 1914, he joined Langmuir's group at the General Electric Research Laboratory in Schenectady, NY. Following a visit by W. H. Bragg in 1915, he developed the powder diffraction method for crystal analysis, and was awarded the Potts Medal from the Franklin Institute in 1923 for this work. During World War I, Hull originated the use of piezoelectric Rochelle salt crystals to pick up noise from submarines. After the war he returned to X-ray crystal analysis for a short time, after which he went back to electronics. In the field of electronics he is best known for his invention, in 1920, of the magnetron, a vacuum tube that generates microwaves using the interaction of a stream of electrons with a magnetic field, and which was applied later as a source of microwave power for radar. Hull became Assistant Director of the General Electric Research Laboratory in 1928, and remained at General Electric until his retirement in 1949.

MAIN PUBLICATIONS

1917 *A new method of X-ray crystal analysis.*

1918 *The dynatron: a vacuum tube possessing negative resistance.*

1919 *A new method of chemical analysis.*

1920 *The arrangement of atoms in some common metals.*

1921 *X-ray crystal analysis of thirteen common metals.*

1921 *The magnetron.*

1929 *Hot-cathode thyratrons.*

any agreement. Some time later, Hull had an opportunity of measuring Bragg reflections from a 2 mm thick, 6 × 6 mm square single crystal of 3.5% silicon steel, and found that it was body-centred cubic. He then went back to the powder diffraction data and found that it was indeed compatible with that structure. The results were presented on 27 October 1916 by Hull at a meeting of the American Physical Society in the General Electric National Lamp Works, Cleveland, Ohio, and were published in early 1917 (Hull 1917*a*). It is only after the war that Hull heard of Debye and Scherrer's work, and that they had published it a few months before he had.

In his analysis of the data, Hull discussed the decline of the intensities with increasing diffraction angle. He found that it could not be accounted for if the atoms were point scatterers, and concluded that the scattering was entirely due to the electrons. He calculated the diffracted intensities assuming that the electrons were displaced along the cube diagonals, and found a good agreement with the observed values. For the same reason, he revisited the structure of carborundum determined by Burdick and Owen (1918), taking the electron distribution into account, and found that it had exactly the diamond structure (Hull 1919*b*).

 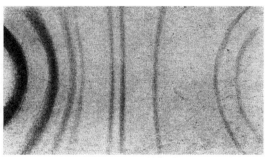

Fig. 8.11   Powder diffraction diagrams. *Left*: aluminium filings (after Hull 1917*b*; ©1917 by The American Physical Society). *Right*: natural graphite (after Debye and Scherrer 1917).

Immediately after determining the structure of iron, Hull started to determine the structure of many elements: aluminium (Fig. 8.11, *Left*) and nickel (fcc = face-centred cubic), lithium and sodium (bcc = body-centred cubic), magnesium (hcp = hexagonal close-packed), silicon (diamond structure) and graphite—see Section 8.9.2—(1917*b*); chromium (bcc) (1919*c*), calcium (fcc) (1921*a*); cobalt*α*, iridium, palladium, platinum, and rhenium (fcc), molybdenum and tantalum (bcc), cobalt*β*, cadmium, ruthenium, and zinc (hcp), and indium (face-centred tetragonal) (1921*b*); thorium (fcc), cerium, titanium, osmium, and zirconium (hcp) (1921*c*); vanadium (bcc) and germanium (diamond structure) (1922).

8.5.3   Applications of powder diffraction

The powder diffraction method developed very rapidly, as it did not require beautiful specimens from a mineral cabinet, and constituted a real breakthrough for the applications of X-ray analysis. The method can be applied to any type of material and is now a widely used technique in chemistry, mineralogy, petrography, metallurgy, and materials science.[24]

The first application, as shown both by Debye and Scherrer, and by Hull, was the determination of crystal structures and lattice parameters, notably of metals and alloys, as will be reported in Sections 10.1.4, 10.1.5 and 10.1.6. W. H. Bragg (1921*a*) showed that a better accuracy could be obtained by using a flat layer of powder and measuring the intensities with the ionization spectrometer. That method was used by Owen and Preston (1922, 1923*a* to *c*, Section 10.1.5) in their studies of alloys. H. Seemann (1919*a*) and H. Bohlin (1920) avoided the absorption problems of the usual Debye–Scherrer method and improved the resolution by using a curved sample and a focusing spectrometer. J. C. M. Brentano[25] (1919, 1924) added improvements

---

[24]For a review of modern applications, see, for instance, Langford, J. I. and Louër, D. (1996). Powder diffraction. *Rep. Prog. Phys.* **59**, 131–234, and Louër, D. (1998). Advances in Powder Diffraction Analysis. *Acta Crystallogr.* **A54**, 922–933.

[25]Johannes C. M. Brentano, born 27 June 1888, died 1969, was an Italian-born crystallographer. After being brought up in Florence, Italy, he studied at Munich University where he obtained his PhD in 1914 under W. C. Röntgen. He then became M. von Laue's assistant in Frankfurt. In 1915 he joined A. Piccard at the Zürich Polytechnic Institute, where he developed an X-ray monochromator. In 1920 he went to Manchester to work with W. L. Bragg. Among other things, he took part in the development of the powder spectrograph. In 1940 he obtained a position at Northwestern University, Illinois, USA, where he remained until his retirement.

with the so-called para-focusing geometry. W. L. Bragg (1961) could rightly say that 'the powder method was employed with a virtuosity which has perhaps never been excelled since'. The determination of complex crystal structures from powder diffraction is possible nowadays using the 'Rietveld refinement method'.[26]

The indexing of powder diffraction patterns of a non cubic material is complicated if no crystallographic data are available. The first to propose a systematic mathematical analysis was C. Runge (1917), but the method was, in general, laborious, and had to wait for computer-based programs. A. W. Hull and his co-worker W. P. Davey at General Electric made the guessing easier by using plots of the d-spacings of all possible planes as a function of the axial ratio. This was done for several point systems of tetragonal, trigonal or hexagonal symmetry (Hull and Davey 1921).

The second major application of powder diffraction is the identification of phases and of unknown substances. It was pioneered by Hull (1919a), who proposed a new method of 'X-ray chemical analysis', besides X-ray emission and absorption spectroscopy. He substantiated his proposal by two examples. In the first one he took the powder diffraction diagram of a sample of sodium fluoride (NaF), 'taken from stock'. It exhibited some extra lines besides those of pure sodium fluoride, indicating the presence of a large amount of an impurity. In order to determine the nature of the impurity, a series of diffraction photographs was taken of substances which were considered the most probable constituents, such as sodium carbonate, sodium chloride, sodium hydrogen fluoride, $etc.$, and their diagrams were compared with the extra lines in the impure sample of NaF. In this way, Hull was able to identify the impurity as sodium hydrogen fluoride (NaHF$_2$).

The second example given by Hull is that of the analysis of two samples of identical chemical content, $viz.$, 33.5% potassium, 19.7% sodium, 16.3% fluorine, and 30.5% chlorine. The powder diagrams of the two samples were totally different. Comparison with diagrams of pure samples of NaF, KF, NaCl, and KCl revealed that the composition of the two samples were mixtures of 50.2% sodium chloride and 49.8% potassium fluoride for one, and 36% sodium fluoride and 64% potassium chloride for the other one.

The X-ray diagram of a crystalline powder is characteristic of the substance. It is therefore possible to identify the substance by comparison of its diffraction pattern with those of other substances sorted in atlases or databases of standard diagrams. In 1936, J. D. Hanawalt and H. W. Rinn of the Dow Chemical Company proposed a search scheme,[27] and in 1938 a new presentation of data, in table form, was introduced, along with a method for searching the file to retrieve data.[28] About 1000 patterns were published in this way by the Dow Chemical Company in 1938.

In 1937, a Joint Committee on Chemical Analysis by X-ray Diffraction Methods, chaired by W. P. Davey,[29] was set up by the American Society for Testing and Materials (ASTM) in

[26]See, for instance, *The Rietveld Method*. R. A. Young, Ed. (1993). Oxford University Press.

[27]Hanawalt, J. D. and Rinn, H. W. (1936). Identification of crystalline materials. *Ind. Eng. Chem. Anal. Ed.* **8**, 244–250.

[28]Hanawalt, J. D. and Rinn, H. W., and Frevel, L. K. (1938). Chemical analysis by X-ray diffraction. *Ind. Eng. Chem. Anal. Ed.* **10**, 457–512.

[29]Wheeler Pedlar Davey, born 19 March 1886, died 12 October 1959, was an American physicist. After earning an MSc at Pennsylvania State University in 1911, he was research fellow at Cornell University for three years. From 1914 to 1926, he served as research physicist with General Electric. He was then appointed Professor of Physical Chemistry at Pennsylvania State University, where he remained until his retirement in 1949.

Fig. 8.12   Growth of the Powder Diffraction File (PDF) since 1941. Courtesy of the International Centre for Diffraction Data.

order to design standards for chemical analysis by X-ray diffraction. The 'Powder Diffraction File' resulted from this effort. For each compound, it includes the d-spacings (related to angle of diffraction) and the relative intensities of observable diffraction peaks. The data of the Dow Chemical Company were reprinted in 1941 under the auspices of the Joint Committee, in the form of files with a 3"×5" card format, known as ASTM files. The format was designed to use the search scheme proposed by Hanawalt and Rinn (1936, see footnote 27). In 1969 the Joint Committee became the 'Joint Committee on Powder Diffraction Standards' (JCPDS) and was incorporated as a separate corporation. JCPDS became the International Centre for Diffraction in 1978. Figure 8.12 shows the growth of the Powder Diffraction File (PDF) since 1941.

Another very important application of powder diffraction is the study of microstructural properties using line-broadening analysis (see Section 10.1.6).

8.6   Application of the rotating crystal method to X-ray analysis

8.6.1   Stationary film: Schiebold 1919, Polanyi 1920

The rotating crystal method, initiated by M. de Broglie (1913c, d, 1914a to d, see Section 7.11) was first used for spectroscopy purposes by E. Wagner[30] (1914, 1915a, b, 1916), in Munich, and by H. Seemann (1919b) in Würzburg. Hugo Seemann (1884–1974) was the first to observe

---

[30]See footnote 4 in Chapter 7.

the out-of-plane reflections due to the Laue cones having as their axis a zone axis normal to the incident beam; he named this the 'complete' (*vollständig*) diagram.

E. Schiebold[31] (in Rinne 1919) and his supervisor, F. Rinne[32] (1921), at the University of Leipzig, were the first to apply the rotating crystal method to crystal analysis. They determined lattice parameters, indexed reflections, and attempted to determine the space groups of some minerals, but failed in the case of tourmaline because of the ambiguities encountered when indexing crystals with low symmetry and large unit cells.

Nishikawa (1914) was the first to note that one could obtain X-ray spectra without rotating the crystal if one used a fibrous substance like asbestos or hemp (see Section 7.5), consisting of elementary crystals of prismatic structure arranged parallel to the axis of the fibres. Similar fibre diagrams of cellulose fibres were observed in 1920, with a Debye-Scherrer camera, independently by P. Scherrer (Section 8.5.1) and by R. O. Herzog's group at the Institute of Fibre Chemistry of the Kaiser-Wilhelms Gesellschaft in Berlin-Dahlem (Herzog and Jancke 1920; Herzog *et al.* 1920; Polanyi 1921*a*, *b*). When M. Polanyi[33] arrived at the Institute of Fibre Chemistry in Berlin-Dahlem in 1922, its Director, R. O. Herzog, had just found that ramie fibres irradiated by an X-ray beam at right angles produced a diffraction pattern (Herzog and Jancke 1920). Polanyi was asked to solve the mystery. He quickly learned the principles of X-ray diffraction and found that the hyperbolic layer lines were due to the intersections of Laue cones with the flat photographic film (Polanyi 1921*a*), an observation already made by Schiebold who pointed out that the rotating crystal photographs and the fibre diagrams gave essentially the same results (Schiebold 1922*b*).

M. Polanyi was joined in 1922 by H. Mark, E. Schmid and K. Weissenberg, from Vienna. R. Brill[34] was also part of the group. Polanyi at first derived the trigonometric relations necessary to interpret the fibre diagrams in all situations: uniquely oriented fibres normal to the incident beam or at an angle, which give a point diagram, or bundles of fibres disoriented

[31]Ernst Schiebold, born 9 June 1894, died 4 June 1963, was a German mineralogist. He studied at the University of Leipzig with the mineralogist F. Rinne, and obtained his PhD in 1919. In 1922 he joined the Applied Physics division of the Kaiser-Wilhelm-Instituts für Physik in Berlin, where he developed X-ray methods for the study of metals. In 1926, he became Professor of Mineralogy at the University of Leipzig, and in 1928 Director of the Institute of Mineralogy. During the Second World War he suggested the possibility of 'X-ray weapons' against enemy aircraft. In 1954 he was appointed Professor of Material Science at the High School for Heavy Machinery in Magdeburg.

[32]Friedrich Rinne, born 16 March 1863, died 12 March 1933, was a German mineralogist. After graduating from Göttingen University in 1883, he was assistant, then Privatdozent in Göttingen. In 1887, he moved to Berlin where he was assistant at the Friedrich-Wilhelms-Universität. He was then successively Professor at the Technische Hochschule Hannover (1894), at the Universities of Giessen (1904), of Kiel and Königsberg (1908), of Leipzig (1909), and, finally, of Freiburg (1928). Rinne (1917) used the term *Leptonenkunde*, literally the science of leptons, or *Feinbaulehre*, to designate the science of the inner structure of crystals.

[33]Michael Polanyi, born 11 March 1891 in Budapest, Hungary, died 22 February 1976 in Northampton, England, was an Hungarian-British physical chemist, economist, and philosopher. After completing a medical degree in 1913, he went to Karlsruhe to study chemistry, but his studies were interrupted by the First World War. After serving as a medical officer from 1914 to 1917 he came back to Budapest, where he obtained a PhD in physical chemistry in 1917. In 1920 he was appointed to the newly founded Kaiser-Wilhelm Institute of Fibre Chemistry in Berlin-Dahlem. In 1923 he moved to the Kaiser-Wilhelm Institute for Physical Chemistry and Electrochemistry, and in 1926 he became Professor. When the Nazi party came into power in 1933, he accepted a chair in Physical Chemistry at the University of Manchester. In 1944, Polanyi was elected a Fellow of the Royal Society. In 1948, Manchester University created a chair of Social Science for him.

[34]See footnote 23 in Chapter 10.

within a narrow range, giving rise to streaks (*Streifendiagramme*). With K. Weissenberg, he showed that it was possible to measure the lattice parameters, index the spots of the diagram, and determine the corresponding values of $1/d_{hkl}^2$ and of the structure factor (Polanyi and Weissenberg 1922*a*, *b*). It is Polanyi who had the idea to use an extended Debye–Scherrer camera and a cylindrical film instead of a flat one (Polanyi in Ewald 1962*a*, Polanyi and Weissenberg 1922*a*).

The same set-up was used by M. Polanyi to study crystal imperfections. With an unworked copper wire, he observed a powder diagram, and, after cold-working, a fibre diagram with streaks (Becker *et al.* 1921, Polanyi 1922). Polanyi and his co-workers studied several other metals such as tungsten and zinc, and showed that the directions of preferred orientation could be determined from the fibre diagrams. Mark and Weissenberg (1923*a*) studied the effect of cold-rolling on metals. Mark and Polanyi (1923) determined the orientation of the glide planes of zinc and Mark, Polanyi and Schmid[35] (1923) elucidated the plastic flow of zinc crystals. Polanyi's thoughts on the mechanism of plastic flow eventually led him to invent the concept of edge dislocation, at the same time as E. Orowan did in Budapest and G. I. Taylor in Cambridge (Polanyi 1934).

The rotating crystal method was used by H. Mark[36] and M. Polanyi (1923*a*) to determine the structure of white tin, already discussed by Bijl and Kolkmeijer (1919) in Utrecht, the Netherlands, from powder diagrams. The method was also used by H. Mark and K. Weissenberg (1923*b*) to determine the structure of urea and of the tetra-iodide of tin (SnI_4). Reviews of the rotating crystal method were published by Schiebold (1924), Polanyi *et al.* (1924), and Bernal (1926). Graphical analyses of rotating crystal diagrams using the reciprocal lattice were proposed by Bernal (1926) and Mauguin (1926*a*).

The rotating crystal method is very useful, but has important drawbacks. Firstly, it provides no information on symmetry (all rotation photographs have the symmetry $2mm$), secondly, as mentioned above, it does not supply unambiguous information about the intensities and indices of the reflections since several reflections may be superimposed. Two types of solutions have been proposed to remedy this situation: J. D. Bernal's oscillation method (1926) and the Weissenberg camera (1924).

---

[35]Erich Schmid, born 4 May 1896, died 22 October 1983, was an Austrian physicist. Following studies at the University of Vienna, he worked from 1922 to 1924 at the Institute of Fibre Chemistry in Berlin-Dahlem with M. Polanyi, from 1924 to 1928 at the Institute of Metal Research, in Frankfurt/Main, and from 1928 to 1932 in the Kaiser-Wilhelms Institute for Metallurgy. He obtained his habilitation from the Technische Hochschule in Berlin-Charlottenburg in 1928, and was Privatdozent there from 1929 to 1932. From 1932 to 1936, he was Professor of Physics in Fribourg, Switzerland, and from 1936 to 1951 was Director of the Metallurgy Department in Frankfurt/Main. He is known for the 'Schmid factor', which is involved in the expression of the critical resolved shear stress.

[36]Hermann Franz Mark, born 3 May 1895 in Wien, Austria, died 6 April 1992 in Austin, Texas, USA, was an Austrian-American chemist and crystallographer. He served as an officer during the First World War, and started chemistry studies at the occasion of a convalescent leave. He obtained his PhD in 1921, and joined the Institute of Fibre Chemistry in Berlin-Dahlem in 1922. In 1926, he became Research Diretor at I. G. Farbenindustrie, later Badische Anilin- und Sodafabrik (BASF) in Ludwigshafen, Germany, where he carried out X-ray studies of polymers. When the Nazi party came to power, Mark went to Vienna, Austria, as Professor of Physical Chemistry. After the *Anschluss*, Mark emigrated to the USA through England and Canada, becoming Assistant Professor, then Professor at the Brooklyn Institute of Technology. In 1944 he founded there the Institute of Polymer Research.

8.6.2   Oscillation method: Bernal 1926

In the course of his study of the structure of graphite (Section 8.9.2) J. D. Bernal (Fig. 8.17) used the rotating crystal method. Here is his description of his first camera (Bernal in Ewald 1962*a*): 'I had to make my own cylindrical camera which I did in the most amateur way out of a piece of brass tubing which I had cut with a hack-saw, bored a hole in it, stuck in with sealing wax a smaller piece of brass tubing with two bits of lead with pin holes through them for the aperture. The film was held together in place with bicycle clips and I used an old alarm-clock and a nail to mount and turn the crystal'.

Bernal (1926), in an 'epoch-making' paper (Buerger 1964), brought three innovations to the rotating crystal method: 1) he interpreted the rotation photographs by referring to the reciprocal lattice, which had been introduced by Ewald (1913*a*) for orthorhombic lattices and by Laue (1914) for any kind of lattice (Section 6.8), and the use of which for analysis of structures had been popularized by Ewald (1921) after the war; 2) by means of formulæ, tables and charts, Bernal showed how the reciprocal lattice cylindrical coordinates could be found for any diffraction spot; 3) he introduced the oscillation method in which the crystal is oscillated backwards and forwards through a small angle. In that way only one spot is recorded at any given location on the film; this removes any ambiguity in indexing. Bernal developed a universal photogoniometer for rotation, oscillation and powder photographs which was commercialized and used by many generations.[37]

8.6.3   Rotating crystal and moving film: Weissenberg camera 1924

K. Weissenberg's[38] answer to the problem of indexing rotating crystal photographs was to screen out all but one layer line. To do this, he let the diffracted rays corresponding to this layer reach the film through an annular slit in a cylinder placed coaxial with the film, thus isolating a single Laue cone. In addition, he arranged for the film to move parallel to the axis of the instrument in coordination with the rotational orientation of the crystal (Weissenberg 1924). As a result, the data corresponding to the selected layer line was spread out in two dimensions, avoiding any overlap of diffraction spots on the film. Weissenberg's design was improved by J. Böhm (1926), also in Berlin-Dahlem, by arranging for the translational movement of the cylindrical camera to be horizontal.

The Weissenberg camera was a real breakthrough, but it was not immediately adopted. Bernal's oscillation method prevailed in England in the second half of the 1920s (Buerger 1973[39]). H. Lipson (1973, see footnote 8) remembered that he had not heard of the Weissenberg

[37]J. D. Bernal. *J. Sci. Instrum.* (1927). **4**, 273–284. (1928). **5**, 241–250, 281–290. (1929). **6**, 314–318, 343–353.

[38]Karl Weissenberg, born 11 June 1893 in Vienna, Austria, died 6 April 1976, in The Hague, The Netherlands, was an Austrian physicist. He studied at the Universities of Vienna, Berlin and Jena, where he obtained his PhD in 1916. In 1922 he joined Polanyi's research team at the Institute of Fibre Chemistry in Berlin-Dahlem. In 1929 his interests turned towards rheology. He was at that time Professor of Physics at the University of Berlin, but, when Hitler came into power in 1933, he emigrated to England where he was appointed Professor of Physics at the University of Southampton.

[39]M. J. Buerger (1973). Karl Weissenberg and the development of X-ray crystallography. In *Karl Weissenberg: The 80th Birthday Celebration Essays.* <http://weissenberg.bsr.org.uk/>.

camera in 1930, when he started research. It then became progressively widespread. There was, however, some frustration with the Weissenberg photographs, in that they looked so different from what they in fact were, images of the reciprocal lattice. In 1933, W. A. Wooster and his wife Nora in Cambridge, England, proposed a graphical method for transforming the diffraction spots on a Weissenberg photograph to a map of a reciprocal lattice plane (Wooster and Wooster 1933). Later, the retigraph, invented by de Jong and Bouman (1938) in The Netherlands, and the precession camera, invented by Buerger (1942) at M.I.T. in Cambridge, USA, provided direct images of the reciprocal lattice.

8.7   Absolute intensities: W. L. Bragg 1921

Major Lawrence Bragg, O. B. E., M. C., was demobilized in January 1919. He returned to Trinity College, Cambridge, to take up the fellowship and lectureship to which he had been elected just before the war. That same year, J. J. Thomson resigned his Cavendish Professorship and E. Rutherford was appointed in his place, leaving the position at Manchester University vacant. W. L. Bragg was appointed, largely due to the influence of the mineralogist Sir H. Miers,[40] who was Professor of Crystallography at Victoria University. W. L. Bragg always received ample support from H. Miers, who became a family friend. Lawrence took up his job in September 1919. He asked R. W. James (Fig. 8.13), who had been his assistant at the sound-ranging school during the war, to join him, and James quickly became his right-hand man: 'My main helper in getting the laboratory into full activity again was R. W. James' (Bragg in Ewald 1962a). W. L. Bragg (1949), in his speech of acceptance of the Roebling Medal of the Mineralogical Society of America, remembered that 'when we all returned to our laboratories in 1919, scientific friends advised me to drop the study of crystals. They pointed out that all crystals would soon be worked out, and that I would find myself out of a job. I did not take their advice and I also have no regrets'.

W. L. Bragg soon had a spectrometer in working order, and in 1920 was offered a Coolidge X-ray tube by the General Electric Laboratory at Schenectady (Phillips 1979). His first work was to complete the determination of the structure of zinc oxide, zincite (ZnO), that he had started before joining the army (W. L. Bragg, 1920a). Zincite is hexagonal and was the first non-cubic structure determined by W. L. Bragg. While he did not have good enough crystals to determine the structure of wurtzite, the hexagonal variety of zinc sulfide, he correctly guessed that it had the same crystal structure as zincite.

By that time, 1920, many molecular and ionic crystal structures had already been determined all over the world, and new ideas had emerged about the structure of atoms and chemical bonding (Lewis 1916, Langmuir 1919). Bragg's work on zinc oxide led him to the idea that interatomic distances in ionic compounds obey an additive law. He assigned sizes to the most common ions and showed that the sums of their radii agreed with the measured interatomic distances. He presented these ideas on 28 May 1920 at the Friday Evening Discourse at the Royal Institution, and published them in *Philosophical Magazine* (W. L. Bragg 1920b). They are discussed in Section 10.1.2.

Some information about the distribution of electrons in atoms can be obtained from the decrease in the diffracted intensities with glacing angle, as had been shown by Compton (1917a)

---

[40]See footnote 66 in Chapter 12.

Fig. 8.13   Reginald James. Source: University of Cape Town, South
Africa.

Reginald William James: born 9 January 1891, in London, England;
died 7 July 1964 in Cape Town, South Africa, was an English physicist.
After receiving his education at the City of London School, he obtained
a scholarship at St John's College, Cambridge, in 1909. He read for
the Natural Sciences Tripos, where he met W. L. Bragg, and graduated
in 1912. He then started work in the Cavendish Laboratory under J. J.
Thomson, but, in the summer of 1914, just before the outbreak of war,
he joined Sir E. Shackleton's expedition to the Antarctic as physicist
and sailed on the *Endurance*; he played a significant role in the survival
of the crew. On his return to England at the end of 1916 he enlisted in
the army where he did sound-ranging with W. L. Bragg. In 1919 he joined W. L. Bragg's laboratory
at Manchester University as Lecturer, then Reader (1934). He was appointed Professor of Physics at
Cape Town University in 1937 and spent the rest of his life there. Between 1953 and 1957, James was
Vice-Chancellor and Acting Principal of Cape Town University. He was elected a Fellow of the Royal
Society in 1955.

MAIN PUBLICATIONS

1921 *The intensity of reflexion of X-rays by rock-salt.* With W. L. Bragg and C. H. Bosanquet.

1922 *The distribution of electrons around the nucleus in the sodium and chlorine atoms.* With W. L. Bragg and C. H. Bosanquet.

1925 *The influence of temperature on the intensity of reflexion of X-rays from rocksalt.*

1926 *The intensity of reflexion of X-rays by crystals.* With W. L. Bragg and C. G. Darwin.

1928 *An investigation into the existence of zero-point energy in the rock-salt lattice by an X-ray diffraction method.* With D. R. Hartree and I. Waller.

1948 *The optical principles of the diffraction of X-rays.*

1950 *X-ray crystallography.*

1963 *The dynamical theory of X-ray diffraction.*

and by Debye and Scherrer (1918), but Lawrence Bragg realized that very accurate measurements of these intensities were necessary in order to be able to make a significant comparison of
measured scattering factors with values calculated for various models of the atomic structure.
He set out to do this with the aid of R. W. James and C. H. Bosanquet, from Balliol College,
Oxford, who had also been a sound-ranger, like James and himself. W. L. Bragg *et al.* (1921*a, b*)
improved on W. H. Bragg's seminal paper (W. H. Bragg 1914*c*, see Section 8.2.5), by extending
the measurements over a larger range of glancing angles and by making direct comparisons
between the reflected intensities and the energy of an incident 'homogeneous' (monochromatic)
beam (absolute intensities). They considered two cases. In the first one the diffraction is
observed in reflection geometry (1921*a*), and in the second one the crystal is set in transmission
geometry (1921*b*).

   The procedure defined by W. L. Bragg *et al.* (1921*a*) to measure absolute intensities has
been the standard procedure ever since. An ionization spectrometer was used, with a Coolidge
X-ray tube equipped with a rhodium anticathode. The geometry was that described in Bragg and
Bragg (1913*a*, see Fig. 7.4). The intensity of the incident beam was maintained at a constant
value. The reflected intensities were measured by sweeping the crystal with a uniform angular

velocity through the entire reflection range and by counting the total ionization produced during this process (the integrated intensity). The background, which can be very important for higher orders was carefully removed. The crystal faces were prepared by grinding, since it was observed that the intensity reflected from cleaved faces was smaller. The faces were cut as nearly parallel to the lattice planes as possible, since it was noticed that asymmetric reflections produce different values. The range of angles over which the crystal was swept must be sufficiently large; the width of the ionization chamber slit must be adjusted and must be neither too large nor too small.

In order to be able to compare the energies of the incident and reflected beams directly, a homogeneous beam must be obtained by reflection from a first crystal. It was the reflection from a second crystal that was measured, in the reflection geometry in W. L. Bragg *et al.* (1921*a*) and in the transmission geometry in W. L. Bragg *et al.* (1921*b*). The beam incident on the second crystal was not limited by a slit but by pressing a lead wedge against the surface of the first crystal, as was first done by Seemann (1914, 1916)—the so-called 'slitless spectrograph'.

The absolute intensity is defined by the ratio $E\omega/I$, called the *reflecting power*, where E is the total amount of energy reflected when the crystal is rotated at constant angular velocity, ω, expressed in radians per second, and I is the total amount of energy of the incident beam passing into the ionization chamber during one second. Using a simplified derivation with respect to Darwin's (1914*a*) and Compton's (1917*a*), W. L. Bragg *et al.* showed that:

1. **Reflection geometry** (W. L. Bragg *et al.* 1921*a*):

$$\frac{E\omega}{I} = \frac{Q}{2\mu},$$

(8.20)

where Q is given by (8.14), and μ is the linear absorption coefficient.

2. **Transmission geometry** (W. L. Bragg *et al.* 1921*b*):

$$\frac{E\omega}{I} = \frac{Qt}{\cos\theta}\exp-[\mu t/\cos\theta],$$

(8.21)

where t is the thickness of a crystal plate cut perpendicular to the reflecting planes.

W. L. Bragg *et al.* (1921*a*) studied the reflections from a rock-salt crystal, which was fortunate, since the samples they used were imperfect enough for geometrical theory to apply. They measured $E\omega/I$ very carefully, using the double-crystal arrangement and cross-checking their results. The intensity of all the other reflections were measured in the single-crystal mode and their reflecting powers were determined by reference to that of 200.

The values of the scattering factors for chlorine and sodium were obtained by comparing the measured absolute intensities with the values calculated from (8.20). This required knowledge of the Debye temperature factor, $\exp-M$, which they took from W. H. Bragg (1914*c*), and of the absorption coefficient, which they measured. They had, however, some problems because the intensity of the 200 reflection was much lower than expected. As a result, they had to use an abnormally high absorption coefficient, as W. L. Bragg wrote to his father on 5 October 1920.[41] W. H. Bragg replied, telling about similar problems he had been having with diamond.

---

[41]Bragg archives at the Royal Institution (RI MS WHB 28A/11). The quotations from the Bragg archives are by courtesy of the Royal Institution, London, Great Britain.

In a particularly interesting work, W. H. Bragg (1921*b*) had observed the forbidden 222 reflection from diamond and had deduced from the analysis of the reflection data that carbon atoms were not spherical but had a tetrahedral form (see Section 10.1.2). He had also found that he had to take into account a higher absorption coefficient, which he wrote as $\rho + r$, where ρ is the normal absorption coefficient and r a 'special absorption coefficient' varying from plane to plane.

W. L. Bragg found his father's results on diamond 'tremendously' interesting because they showed how important it was to have a correct value for the low order reflections which strongly affect the atomic scattering curves.[42] In a later letter, dated 25 February 1921,[43] W. L. Bragg came back to this matter, writing the abnormal absorption coefficient $\mu + \epsilon$, where ϵ is the *extinction* coefficient, the first mention of the term. In the first of their two papers, W. L. Bragg *et al.* (1921*a*) did not make any allowance for the correction to the absorption coefficient, but they did in the second one (W. L. Bragg *et al.* 1921*b*), in which they introduced the term 'extinction'. Following Darwin (1914*a*, *b*), they attributed this effect to crystal imperfections and rightly suspected it should be smaller for higher orders of reflection.

Lawrence Bragg acknowledged that exchanges between him and his father were always rewarding for him, as he wrote to his father on 5 October 1920: 'Whenever I have a good talk about work with you it bucks me up like anything and gives me lots of new ideas for the work'.[44]

W. L. Bragg *et al.* (1921*a*) compared the results they obtained for the angular variations of the atomic scattering factors of chlorine and sodium with those deduced from three atomic models: 1) one in which the electrons are supposed to be distributed uniformly throughout a sphere, 2) a second one in which the electrons are arranged in a series of spherical shells with two, eight, and eight electrons respectively for chlorine and two and eight electrons respectively for sodium, 3) a third one, where the electrons are assumed to be similarly arranged in successive shells but also in oscillation about their mean positions. The best agreement was obtained with the third model. The most important result from these analyses, however, is their conclusion that an electron has been transferred from sodium to chlorine in NaCl.

R. Glocker,[45] at the Technische Hochschule Stuttgart, then made new calculations of the scattering factor with a slightly different model, which fitted better with the data of W. L. Bragg *et al.* (1921*a*, *b*). This prompted W. L. Bragg *et al.* (1922*a*) to do calculations with a fourth model in which the electrons were arranged in elliptical shells. After Darwin's (1922) paper was published, they reevaluated their work, this time with seventeen electrons for a chlorine ion (Cl^-) and eleven electrons for a sodium ion (Na^+), and had to conclude that, due to the uncertainties related to the extinction effect, it was impossible to tell which were the real configurations of the electrons, seventeen and eleven, or eighteen and ten (Bragg *et al.* 1922*b*). They could only say that electron distributions extended right through the volume of the crystal. After progress had been made in the calculation of the atomic scattering factor, Hartree (1928) obtained a very good agreement between theory and experiment assuming eighteen electrons for chlorine (Cl^-) and ten for sodium (Na^+).

---

[42]Letter from W. L. B. to W. H. B. dated 21 October 1920. Bragg archives at the Royal Institution (RI MS WHB 28A/9).

[43]Bragg archives at the Royal Institution (RI MS WHB 28A/20).

[44]Bragg archives at the Royal Institution (RI MS WHB 28A/9).   [45]See footnote 5 in Chapter 7.

8.8 Extinction: C. G. Darwin 1922

Bragg, James and Bosanquet's papers led Darwin to re-examine diffraction from crystals in a theoretical way (Darwin 1922). Darwin in 1922 was a Fellow at Christ's College, Cambridge, and a lecturer in mathematics. His research topic at that time involved statistical mechanics, but he went back to his theory of diffraction to try to explain extinction. He was able to relate the expressions of the intensity diffracted according to the geometrical theory (8.14) and according to the dynamical theory (8.18), and he showed that, if the thickness of the crystal corresponds to m lattice planes:

$$I_{dyn} = I_{geom} \frac{\tanh mq}{mq} \leq I_{geom}. \tag{8.22}$$

where q is defined by (8.5).

For a thick perfect crystal (m large), (8.18) holds, while for a very thin perfect crystal (m small), it is (8.14) that is valid. This suggested to Darwin a way out of the difficulty due to the disagreement between the experimental reflectivities and those calculated with either (8.14) or (8.18). He imagined the crystal to be a conglomerate of small blocks of perfect crystal, now described as the mosaic model. The term 'mosaic crystal' was in fact introduced by Ewald (Mark 1925, James in Ewald 1962a). Two situations may arise: 1) if the small blocks are not thin enough, their reflectivity becomes closer to that predicted by the dynamical theory; it must therefore be multiplied by a correction factor, which was called *primary extinction* correction by Darwin. 2) If the lattice planes in two successive blocks, however, are nearly parallel, part of the incident intensity will be reflected off by the first block before it reaches the second one. This requires a correction, named *secondary extinction* correction, which is taken into account by an artificial absorption coefficient. The corrections to the absorption coefficient introduced by W. H. Bragg (1921b) and by W. L. Bragg *et al.* (1921b) were in fact due to this secondary extinction.

W. L. Bragg (1924), when determining the structure of aragonite (the orthorhombic variety of calcium carbonate, $CaCO_3$), observed that, for the stronger reflections, the diffracted intensities appeared to be 'proportional to the amplitude-factor (structure factor) and not to its square'. He attributed this effect to 'an extinction factor', and noted that his father had made a similar observation for calcite (W. H. Bragg 1915c). In a note on Hartree's (1925) paper on the calculation of atomic scattering factors, W. L. Bragg (1925b) stressed 'the unsatisfactory nature of the assumptions usually made in calculating the intensity of reflection to be expected from a given atomic arrangement, and the uncertainty of the results due to an ignorance of the laws governing the intensity'.

Ewald's dynamical theory of diffraction (see Section 9.2) had been published during the war and was not well known about in England. For that reason, Ewald (1925), who since 1921 was Assistant Professor at the Technische Hochschule Stuttgart, summarized briefly the results of his theory and pointed out that a scattered amplitude was proportional to the structure factor for a perfect crystal. As examples, he referred to W. H. Bragg's paper on diamond and W. L. Bragg's 1924 paper on aragonite. The confusion about which intensity formulæ to use for real crystals led Ewald to organize a meeting in Holzhausen on the Ammersee, in Bavaria, to discuss this difficulty. The meeting took place in September 1925 at the house of his mother, who was a painter and had a big studio suitable for the purpose. It was the first international conference on

X-ray analysis. The participants included most of the pioneers of X-ray diffraction. They were: W. L. Bragg, L. Brillouin, C. G. Darwin, P. Debye, A. D. Fokker, K. Herzfeld, R. W. James, M. von Laue, H. Mark, H. Ott, I. Waller, and R. W. G. Wyckoff (Bragg *et al.* 1926). Figure 8.14 shows some of the participants. W. L. Bragg later remembered that, to the dismay of the other English participants (James and himself), Darwin had 'forgotten the logic of his own papers and was unable to present his theory'! (W. L. Bragg 1968a).

The Ammersee meeting resulted in a much better understanding of the respective roles of crystal perfection and crystal defects in the process of diffraction. The conclusions of the discussions, which led to appropriate formulæ to use for the reflected intensities were reviewed in Bragg *et al.* (1926), where extinction was discussed in detail. The new confidence that this generated in the interpretation of the intensities and the quality of the experimental data marked a cornerstone in the development of crystal structure determinations. W. L. Bragg wrote (in Ewald 1962a): 'This led to our examining more complex crystal structures, in which the values of a number of parameters had to be determined in order to fix the positions of the atoms. Some of our more ambitious crystal analyses were at first viewed with deep suspicion by our colleagues in other laboratories, but I think that one of the chief contributions of the 'Manchester School' was the demonstration that these quantitative measurements enabled one to tackle such structures with confidence'.

Darwin's model of extinction was in reality too crude, and not applicable to most crystals. The problem of extinction has remained acute till today, since it affects mainly the low-order reflections, which are particularly important for the determination of accurate electronic distributions. More and more sophisticated theories have been elaborated over the years, by such crystallographers as W. C. Hamilton, W. H. Zachariasen, P. Coppens, P. J. Becker, and N. Kato.

H. Mark (1925) tried to evaluate what quantitative information could be obtained about the Darwin's mosaic structure from X-ray measurements. His views were discussed by A. Smekal (1927) and F. Zwicky (1929), who proposed other models of mosaic crystals. These models

Fig. 8.14   Photograph taken after the 1925 Ammersee conference. Left to right: P. P. Ewald, C. G. Darwin, H. Ott, W. L. Bragg, R. W. James. After (W. L. Bragg 1968a).

were discussed from the structural viewpoint and compared to Darwin's model in the reviews by M. J. Buerger (1934), H. E. Buckley (1934), and A. Smekal (1934).

8.9   A few landmark structures

8.9.1   Hexamethylene tetramine

The first complete determination of the structure of an organic crystal is that of hexamethylene tetramine, $(CH_2)_6N_4$, by R. G. Dickinson and A. L. Raymond (1923) at the California Institute of Technology, USA, using data obtained from Laue diagrams, and, independently, by H. J. Gonell and H. Mark (1923) in Berlin-Dahlem, Germany, using the rotating crystal method. The fact that this crystal is cubic—a rare occurrence for organic crystals—made it much easier to determine its crystal structure.

R. Dickinson[46] began his X-ray work with the spectrometer built by C. L. Burdick at Caltech (Section 8.1.1) in 1918. His thesis bore on the structure of wulfenite and scheelite (Dickinson 1920). In 1921 he paid a short visit to Wyckoff at the Geophysical Laboratory of the Carnegie Institution in Washington, DC, and, in 1922–23 Wyckoff spent a year at Pasadena. Under Wyckoff's influence, Dickinson gave up the spectrometer and adopted the Laue technique used by Nishikawa and Wyckoff. Pauling, who was a student of Dickinson, explained the advantages of the Laue method:[47] it made it possible to study crystallographic planes that had quite small spacings. The idea was to determine which multiples of the smallest lengths of the edges of the unit cell given by the rotation photographs were required in order to account for the existence of all of the observed Laue spots. This was a powerful technique, and it prevented errors that were sometimes made by other investigators. For instance, Dickinson determined the structure of tin tetra-iodide, SnI_4 in that way (Dickinson 1923); this structure was also studied by Mark and Weissenberg (1923b). Dickinson found the cubic unit of tin tetra-iodide to have edge 12.23 A, with 32 tin atoms and 128 iodine atoms in the unit cube, whereas Mark and Weissenberg reported a unit with edge only half so great, containing only four $SnI_4$4 molecules. The reason for that was that, with the rotating crystal method used by them, the weaker reflections requiring a larger unit cell were overlooked.

For the same reasons, Dickinson and Raymond's determination of the structure of hexamethylene tetramine was more complete than that of Gonell and Mark. The structure of hexamethylene tetramine was rather exceptional for organic crystals in that the atomic positions were closely limited by space group considerations. From the extinctions observed on the Laue diagrams, the space group could be restricted to two or three possibilities: $I23$, $I\bar{4}3m$ or $P\bar{4}3n$. In fact, the structure is cubic-centred, with space group $I\bar{4}3m$. Most importantly, there was strong evidence of the real existence of the molecule postulated by the organic chemist, with two molecules $(CH_2)_6N_4$ per cubic cell. The structure is represented in Fig. 8.15. The atoms of carbon are at the summits of an octahedron, and the atoms of nitrogen at the summits of

---

[46]Roscoe Gilkey Dickinson, born 3 May 1894 in Brewer, Maine, USA, died 13 July 1945 in Pasadena, California, USA, was an American chemist and X-ray crystallographer. He received his undergraduate education at M.I.T., Cambridge, USA, and followed A. Noyes to the California Institute of Technology at Pasadena, where he obtained his PhD in 1920, the first person to receive a PhD at Caltech. He spent his career as Professor of Chemistry at Caltech. His most famous student was Linus Pauling, Nobel Laureate in 1954 and 1962.

[47]L. Pauling in McLachlan, Jr and Glusker (1983).

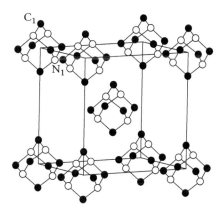

Fig. 8.15   Structure of hexamethylene tetramine, $(CH_2)_6N_4$, according to Dickinson and Raymond (1923) and Gonell and Mark (1923). Black circles: carbon atoms; white circles: nitrogen atoms; the hydrogen atoms are not represented.

a tetrahedron. There are two parameters: the coordinates $0, 0, u$ of the carbon atom C_1 and v, v, v of the nitrogen atom N_1 (Fig. 8.15). Dickinson and Raymond determined u and v to be approximately 0.235 and 0.12, respectively. The nitrogen to carbon distance was found to be 1.44 Å. Gonell and Mark (1923) obtained similar results, with $I\bar{4}3m$ as its highly probable space group, and a nitrogen–carbon distance of 1.48 Å. The structure was later confirmed using Fourier syntheses by R. W. G. Wyckoff and R. B. Corey (1934). Accurate electron density maps were established by Brill *et al.* (1939).

8.9.2   Graphite

The structure of diamond was very important in our understanding the structure of aliphatic compounds. Similarly, that of graphite was important for aromatic compounds. The first to obtain quantitative information on graphite was Ewald (1914c) who took Laue diagrams of a small crystal (2 × 2 mm, 0.1 mm thick), found that it had hexagonal symmetry and measured the axis ratio of the hexagonal cell c/a to be 1.63. Bragg and Bragg (1915) made one measurement and found that graphite was pseudo-hexagonal, with a lattice distance 3.42 Å for the cleavage basal plane.

The structure was then determined using powder diffraction, independently, by P. Debye and P. Scherrer (1917) in Göttingen, Germany, and by Hull (1917b) in Schenectady, USA. Debye and Scherrer found the structure to be trigonal. It consisted of a stacking of molecular planes parallel to the cleavage plane, with an equidistance of 3.41 Å. In each plane, the atoms of carbon occupied the nodes of a honeycomb lattice with C-C distances 1.45 Å (Fig. 8.16, *Left*).

Hull (1917b) concluded that graphite had hexagonal symmetry, and that its structure consisted of a stacking of pairs of planes in which the carbon atoms occupied respectively positions A and B, and A and C (Fig. 8.16, *Right*); the distance between neighbouring planes was 3.40 Å and 1.50 Å between neighbouring carbon atoms.

The structure of graphite was revisited using the rotating crystal method, independently, by O. Hassel and H. Mark (1924) at the Kaiser-Wilhelms Institute of Fibre Chemistry in Berlin-Dahlem, and by J. D. Bernal (1924) at the Royal Institution in London. They concurred with Hull's description of the structure. Hassel and Mark found the distance between two neighbouring planes to be 3.39 Å and the distance between two neighbouring carbon atoms

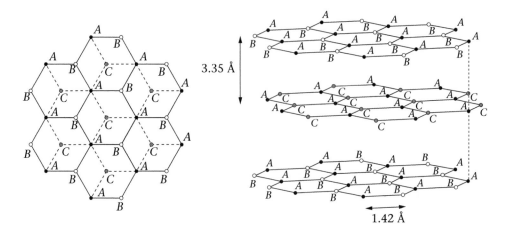

Fig. 8.16   Structure of graphite, according to Hull (1917*b*), Hassel and Mark (1924), and Bernal (1924). *Left*: *A*, *B*,
C, projections of the atoms of carbon on the basal plane; broken lines indicate a molecular plane below that of the
structure illustrated with solid lines. *Right*: three-dimensional view.

in the basal plane to be 1.44 Å. C. Mauguin[48] (1926*b*) rederived the structure completely with
Laue and rotation crystal diagrams at the Sorbonne in Paris, unaware of Bernal's and of Hassel's
and Mark's papers when he started his study. His results are essentially in agreement with
Bernal's, differing only in the value of the distance between two successive planes, which he
finds equal to 3.37 Å.

The most detailed and complete analysis of graphite was, however, Bernal's. In particular, he
showed that Debye and Scherrer's analysis was incorrect, and that the values they had obtained
for the dimensions of the unit cell were wrong. The reason was that they often mistook α and β
lines and this resulted in indexing errors. In addition to his rotating crystal photographs, Bernal
used measurements made with the ionization spectrometer by W. T. Astbury and by K. Yardley
(the future Kathleen Lonsdale), who was able to measure five orders of the cleavage plane
reflections. Laue photographs taken by W. T. Astbury showed hexagonal symmetry and the
presence of twinning in some samples. Bernal found the distance between two neighbouring
planes to be 3.41 Å and the distance between two neighbouring carbon atoms in the basal plane
to be 1.42 Å. The structure of graphite was confirmed by H. Ott, E. Wagner's successor in
Würzburg, for his habilitation, from rotating crystal data (Ott 1928).

It is interesting to tell how Bernal came to study the structure of graphite and, firstly, how he
came to work with W. H. Bragg.[49] John Desmond Bernal was nicknamed 'Sage' by a friend,
Dora Grey, a bookshop attendant. His first contact with crystallography was when he read, as an
undergraduate at Cambridge, W. L. Bragg's 1920 paper on the arrangement of atoms in crystals.
He was particularly fascinated with Langmuir's theory. Soon after, he became interested in

---

[48]See footnote 6 in Chapter 7.

[49]The following account is taken from D. C. Hodgkin (1980). John Desmond Bernal. 10 May 1901–15 September
1971. *Biogr. Mems. Fellows R. Soc.* **26**, 16–84, and A. Brown (2005). *J. D. Bernal, the Sage of Science*. Oxford University
Press.

Fig. 8.17   John Desmond Bernal. After W. W. F. Taylor (1972). Photo Henry Grant.

John Desmond Bernal: born 10 May 1901 in a farm in Nenagh, Co. Tipperary, Ireland; died 15 September 1971 in London, England, was an Irish crystallographer. He was educated at Stonyhurst College and Bedford School in England. In 1919 he went to Emmanuel College at Cambridge University with a scholarship, and in 1922 obtained a BA. After graduation, Bernal joined W. H. Bragg at the Royal Institution. In 1927 he was appointed as the first lecturer in structural crystallography at Cambridge, and in 1934 Assistant Director of the Cavendish Laboratory. In 1937 he became Professor of Physics at Birkbeck College, University of London. From 1963 until his retirement in 1968 he was Director of the newly established Department of Crystallography in the same college. He was elected a Fellow of the Royal Society in 1937. During the Second World War he served as scientific adviser to Lord Louis Mountbatten. Two of Bernal's many students, D. C. Hodgkin and M. Perutz, were awarded the Nobel Prize.

MAIN PUBLICATIONS

1927 *A universal X-ray goniometer.*
1929 *The world, the flesh and the devil.*
1939 *The social function of science.*
1951 *The physical basis of life.*
1953 *The use of Fourier transforms in protein crystal analysis.*

1954 *Science in history.*
1958 *World without war.*
1967 *The origin of life.*
1972 *The extension of man: Physics before 1900.*

vector analysis. He first wrote a paper on 'the vectorial geometry of space lattices', which led him to rederive the 230 space-groups! He showed the manuscript of his work 'On the analytic theory of point group systems' to A. Hutchinson, then lecturer in crystallography, who recommended him to W. H. Bragg in June 1923, describing him 'as a shy, diffident, retiring kind of creature, but something of a genius'. The problem of the structure of graphite was suggested to Bernal by Bragg who had recently revised the book he had written with his son (Bragg and Bragg 1915), and had found 'disheartening' the disagreement between the structures proposed by Hull and by Debye and Scherrer.

The atmosphere at the Royal Institution, with its bright young men and women, was described very vividly by J. D. Bernal in his biography of A. T. Astbury:[50] 'the people who had gathered there were diverse but all shared in common a lively enthusiasm for the discovery of the new world of crystal structure which they were privileged to share. It was a very happy time: there was no real rivalry because that world was quite big enough for all their work. They were effectively and actually a band of research workers, dropping into each other's rooms discussing informally over lunch and ping-pong and formally at Bragg's colloquia every week'. W. T. Astbury, in his reminiscences (in Ewald 1962*a*, p. 356), wrote: 'Sir William—the Old Man, or Bill Bragg, as we called him behind his back—never "led" any of us, in the technical sense. That was not his way—if he had a way—and if you were stupid enough you might even claim

[50]Bernal, J. D. (1963). William Thomas Astbury. *Biogr. Mems. Fellows R. Soc.* **9**, 1–35.

that you "led" him, since, especially after we migrated in 1923 to the Davy–Faraday Laboratory of the Royal Institution, as often as not when he popped into your room (not terribly often) it was to ask you a question in connection with some lecture that he was due to give. Or you might meet him on the stairs and he would say: "Hello! How is the family", or some such. He turned up at tea as often as possible, where we rarely talked "shop"in any case'.

8.9.3   The benzene ring: K. Lonsdale 1928

The determination of the structure of the benzene ring by K. Lonsdale (1928, 1929a, b) was a major event, important to chemistry in general. By establishing its strictly planar nature, it 'provided the key to the structure of all aromatic structures' (Robertson 1964). Up to then, it had been a matter of controversy. Barlow and Pope (1906, Section 12.13) had proposed a structure in which the carbon atoms were arranged in a puckered hexagon. Several formulæ were discussed at the 1913 meeting of the British Association for the Advancement of Science for the molecule of benzene: 1) that of Kekulé,[51] with three double linkings (1865), 2) that of Armstrong in which each individual carbon atom exercises an influence upon every other carbon atom (the 'centric formula', 1897), and 3) that of Thiele, with six 'inactive double linkings' (1899). No conclusion was reached. Debye and Scherrer (1916a, II) observed a diffraction pattern of liquid benzene and gave an estimate of the diameter of the molecule (Section 8.5.1).

In a lecture delivered before the Chemical Society on 26 October 1922, W. H. Bragg indicated that solid benzene was orthorhombic and had a non-planar 'boat' shape (W. H. Bragg 1922a); he thought that benzene could not have a trigonal axis, still less a hexagonal axis. In his book, Wyckoff (1924) also stated that benzene crystals being orthorhombic, the carbon atoms of its molecule could not be equivalent.

L. Pauling (1926) applied to benzene the dynamic model of the chemical bond, based on the 'Electron-Orbit-Sharing Theory of Chemical Bonding', and concluded that 'all six carbon atoms and all six hydrogen atoms are in the same plane, at the corners of two similar regular hexagons. Each carbon atom is connected by pairs of electron orbits to the two adjacent carbon atoms and to one proton, the angles between pairs of orbits being 120°'.

In 1928, E. G. Cox (1906–1996), in W. H. Bragg's laboratory, determined, using the rotating crystal method, that crystalline benzene was simple orthorhombic, with $a = 7.44$ Å, $b = 9.65$ Å and $c = 6.81$ Å at –22°, and contained four molecules per unit cell (Cox 1928). The molecule itself had a centre of symmetry, and the evidence favoured a flat ring with 1.42 Å carbon-carbon distances.

Kathleen Lonsdale was a small, high-principled and very determined woman, studying at Bedford College for Women. She headed the university list in physics at the age of 18, with the highest marks in ten years, and was noticed by W. H. Bragg, one of the examiners.[52] Although scientific jobs were scarce, he offered her a place in his research team at University College with a government grant. Lonsdale remembered, like Astbury, that W. H. Bragg 'did not appear to lead at all', 'he inspired me with his own love of pure science and with his enthusiastic spirit of enquiry and at the same time left me entirely free to follow my own line of research'. He

---

[51]See footnote 15 in Chapter 7.

[52]The following account is taken from K. Lonsdale in Ewald 1962a and D. C. Hodgkin (1975). Kathleen Lonsdale. 28 January 1903–1 April 1971. *Biogr. Mems Fellows R. Soc.* **21**, 447–484.

advised her that it would take some three months to collect together the apparatus she would need, ionization spectrometer and gold-leaf electroscope, of which she became rapidly a skilled user. At the same time, W. H. Bragg suggested to his students that they should read Hilton's 1903 book on space groups. Kathleen had been Hilton's student at Bedford College. Together with Astbury, they produced the space-group tables mentioned in Section 8.1.2.

It was understood that her problem would be the structure analysis of some organic compound. Inorganic compounds were studied by W. L. Bragg's group at Manchester. Kathleen asked W. T. Astbury for suggestions. Astbury, who was studying tartaric acid, suggested succinic acid. She produced a structure which gained her her MSc in 1924, but was later proven to be 'wholly wrong'. In those early days, the structure of organic compounds had often to be reached by guess-work, and errors were common. Lonsdale was more cautious in the future as she studied several aliphatic compounds. Her major success, however, was the solution of the

Fig. 8.18   Dame Kathleen Lonsdale. Source AIP Emilio Segre Visual Archives, with permission.

Dame Kathleen Lonsdale: born 28 January 1903 in Newbridge, County Kildare, Ireland, *née* Yardley, the daughter of Harry Yardley, the town postmaster, married to Thomas Jackson Lonsdale in 1927; died 1 April 1971 in London, England, was an Irish crystallographer. Her parents moved to England when she was five. After schooling at Woodford County High School and Ilford County High School, she obtained her BSc from Bedford College for Women in 1922, and her MSc from University College London in 1924. She then followed W. H. Bragg to the Royal Institution. In the late 1920s she moved to the University of Leeds, where she studied the structure of benzene. She returned to work with Bragg at the Royal Institution as a researcher in 1934, and was awarded a DSc from University College London in 1936. In 1949 Lonsdale became Professor of Chemistry and head of the Department of Crystallography at University College London, the first woman tenured professor. She was given the title Dame Commander of the Order of the British Empire in 1956, and was one of the first two women elected a Fellow of the Royal Society, in 1945. She was elected Vice-President of the International Union of Crystallography in 1963, and assumed the office of President when she replaced J. D. Bernal, who had to resign just before the Seventh General Assembly in 1966 for reasons of health. She and her husband were pacifists and Quakers. Lonsdale served a month in Holloway prison during the Second World War because she refused to register for civil defence duties, and later became a tireless campaigner in the cause of international peace.

MAIN PUBLICATIONS

1924 *Tabulated data for the examination of the 230 space-groups by homogeneous X-rays. With W. T. Astbury.*

1929 *The structure of the benzene ring.*

1931 *Analysis of the Structure of Hexachlorobenzene, Using the Fourier Method.*

1935 *Structure Factor Formulae for the 230 Space Groups of Mathematical Crystallography* (Vol. I of International Tables of Crystallography).

1936 *Simplified Structure Factor and Electron Density Formulae for the 230 Space Groups of Mathematical Crystallography.*

1943 *Experimental study of X-ray scattering in relation to crystal dynamics.*

1949 *Crystals and X-rays.*

structure of hexamethylbenzene in 1928, which she found her 'most fundamental and satisfying piece of research'.

Kathleen Lonsdale left the Royal Institution in 1927 when she married Thomas Lonsdale, whom she followed to Leeds, where he worked in the Silk Research Association, housed in the Textile Department of the University of Leeds. She had intended to give up scientific research work, but her husband would have none of it: 'he had not married to get a free housekeeper'. So she continued research in the Physics Department at the University of Leeds, where W. H. Bragg and his son had carried out their early studies of crystal structures. She built up once more an X-ray equipment. C. K. Ingold, Professor of Chemistry at Leeds, offered her some large crystals of hexamethylbenzene. Their shape was that of flat prisms. The crystals were triclinic, pseudo-monoclinic and contained one molecule per unit cell. There was an excellent cleavage plane parallel to (001), and the intensities from that plane fell off in almost the same proportion as those from the (001) plane of graphite observed by Bernal (Lonsdale 1928). This implied a planar arrangement of the carbon atoms. Furthermore, the close correspondence between the reflections from the $(hk0)$, $(\overline{h + k}h0)$ and $(k\overline{h + k}0)$ planes proved the existence of a hexagonal symmetry of the benzene ring. The dimensions of the hexagon were estimated by trial structure factor determinations to be 1.42–1.45 Å, in agreement with Bernal's value of 1.42 Å. She tried to explain the triclinic symmetry by packing problems related to the methyl groups. These problems were elucidated much later with a Fourier analysis by Brockway and Robertson (1939) who, otherwise, fully confirmed Lonsdale's analysis. W. H. Bragg was delighted and wrote to her on 30 October 1928: 'I think your new result is perfectly delightful: many compliments upon it! I like to see the benzene ring *emerging*.'

The study of hexamethylbenzene was completed by that of hexachlorobenzene, which was more difficult. For the first time, the parameters of an organic molecule were determined by Fourier analysis (Lonsdale 1931).

The structure of the benzene ring was calculated quantum-mechanically by L. Pauling and his student G. M. Wheland according to Pauling's concept of resonance.[53]

---

[53]Pauling, L. and Wheland, G. W. (1933). The nature of the chemical bond. V. The quantum-mechanical calculation of the resonance energy of benzene and naphthalene and the hydrocarbon free radicals. *J. Chem. Phys.* **1**, 362–374.

9

X-RAYS AS A BRANCH OF OPTICS

[Laue's] discovery was primarily a contribution to optics.

Sir C. W. Raman (1937)

9.1 Optical properties of X-rays

The title of this chapter is taken from A. H. Compton's Nobel lecture (1927), in which he pointed out that 'it has not always been recognized that X-rays is a branch of optics', but that 'there is hardly a phenomenon in the realm of light whose parallel is not found in the realm of X-rays'. All the properties characteristic of light have been found also to be characteristic of X-rays:

1. **Specular reflection.** Röntgen and his contemporaries tried in vain to observe specular reflection of X-rays (Sections 4.3 and 4.4). This is not surprising, since the refractive index of matter for X-rays is only very slightly less than 1, as shown by Darwin (1914a, Section 8.4.1) and Lorentz (1916). The existence of the refractive index for X-rays was first detected by the observation of deviations from Bragg's law (Section 9.3). Total reflection can only be observed if the glancing angle is smaller than a critical angle, $\omega = \sqrt{2\delta_o}$, where $\delta_o = 1 - n$, and the refractive index, n, is given by (8.12). A. H. Compton (1923a) was the first to observe the specular reflection of X-rays from plate glass and silver mirrors. The work was presented at the meeting of the American Physical Society in Washington on 21/22 April 1922.[1] A beam collimated by two slits, and of effective width 2 minutes of arc fell on the mirror, and the totally reflected beam was detected with an ionization spectrometer. For an X-ray wavelength of 1.279 Å from a tungsten anticathode, Compton measured the critical angle to be 10 minutes of arc for crown glass and 22.5 minutes of arc for silver. The corresponding experimental values of δ_o (the deviation of the refractive index from 1) were 4.2×10^{-6} and 21.5×10^{-6}, respectively, to be compared with the theoretical values 5.2×10^{-6} and 19.8×10^{-6}, respectively. Specular reflection was also observed, one year later, by M. Siegbahn and O. Lundquist, from a silver mirror.[2]

2. **Refraction.** For the same reasons, refraction of X-rays was not observed by the early investigators (Röntgen 1895, Perrin 1896, Voller and Walter 1897, and others). Barkla (1916) made an attempt to observe the refraction by putting two crystals of potassium bromide one above the other, so as to form two refracting prisms end to end, and letting an X-ray beam pass between the two crystals; but he observed no effect. The first successful observation was by Larsson *et al.* (1924) in Siegbahn's laboratory with a glass prism, using a photographic method to observe the deviation of the X-ray beam. They used an X-ray tube with an anticathode of mixed copper and iron, and separated the deviations due to Cu $K\alpha$ and Fe $K\alpha$ X-radiation. The next observation was by B. Davis and C. M. Slack (1925) with a prism of copper and an ionization chamber and by the

---

[1]Compton, A. H. (1922). *Phys. Rev.* **20**, 84.

[2]Siegbahn, M. and Lundquist, O. (1923). Röntgenstrålarnas total reflexion. *Fysisk Tidsskr.* **21**, 170–171.

same authors with a prism of aluminium inserted in the path of the X-ray beam between the two calcite crystals of a double-crystal spectrometer (Davis and Slack 1926). More sensitive versions of the latter experiment were developed much later, when highly perfect crystals became available, for instance by Okkerse (1963) and by Malgrange *et al.* (1968), and with the X-ray interferometer developed by Bonse and Hart (1965).

3. **Diffraction by a slit.** The experiments by Haga and Wind, and by Walter and Pohl, have been discussed in Section 5.9. Diffraction by a ruled reflection grating was first observed by Compton and Doan (1925), at a small glancing angle, within the critical angle, with Cu $K\alpha$ and Mo $K\alpha$ radiation. A beautiful picture of X-ray diffraction fringes by a slit (Bäcklin 1935) is reproduced in Laue (1960). For a modern study, see Le Bolloch *et al.* (2002).

4. **Diffuse scattering** was observed by Röntgen himself and by the early investigators (Section 5.4). Diffuse scattering by crystals is superimposed on the Bragg peaks and is due to thermal vibrations and static disorder in the structure, but there is, in fact, no clear-cut separation between the two.

5. **Polarization** of X-rays was discovered by Barkla (1905)—see Section 5.6.

6. **Emission and absorption spectra**, see Section 10.6.

7. **Photoelectric effect.** The emission of electrons following the absorption of X-rays was first observed by Perrin (1897), Curie and Sagnac (1900), and by Dorn (1900), and was studied further by Innes (1907), see Section 5.10 and A. H. Compton (1926). The photoelectric effect is now used, for instance, in X-ray photoelectron spectroscopy.

Nowadays, many optical systems involving X-ray optics are used at synchrotron facilities for focusing or imaging purposes—curved mirrors, wave-guides, Fresnel zone plates, and diffractive and refractive lenses; for a review, see, for instance, Authier 2003 or <http://www.x-ray-optics.com>. Furthermore, when X-rays are diffracted at the Bragg angle by perfect crystals, entirely different optical properties appear, and these are summarized in Section 9.7.

9.2   Ewald's dynamical theory of X-ray diffraction, 1917

9.2.1   The dispersion surface

Ewald's dynamical theory of diffraction (Ewald 1917) was developed by him during World War I while he was stationed on the Russian front, where he was servicing a mobile medical X-ray unit (Ewald 1962a). The theory is derived from his thesis, and it is similarly based on Maxwell's equations. Its name can be traced back to Maxwell's paper, *A dynamical theory of the electromagnetic field*,[3] in which Maxwell explains that 'it may be called a Dynamical Theory, because it assumes that in space there is matter in motion, by which the observed electromagnetic phenomena are produced'.

The most important result of Ewald's dispersion theory (Section 6.2) is that the optical field inside the crystal can be expanded as a sum of plane waves. The corresponding total electric field $\mathbf{E}$ is:

$$\mathbf{E} = \sum_{\mathbf{h}} \mathbf{E_h} \exp(-2\pi i\, \mathbf{K_h} \cdot \mathbf{r}) \exp(2\pi i\, \nu t) \tag{9.1}$$

where $\mathbf{E_h}$ and $\mathbf{K_h}$ are the amplitudes and the wave vectors of the diffracted waves propagating in the crystal, respectively, ν is their frequency and $\mathbf{r}$ a position vector. The vectors $\mathbf{K_h} = \mathbf{HP}$ are vectors in reciprocal space, called *Anregungsvektoren* by Ewald, where H is a reciprocal

---

[3]Maxwell, J. C. (1865). *Phil. Trans. Roy. Soc.* **155**, 459–512.

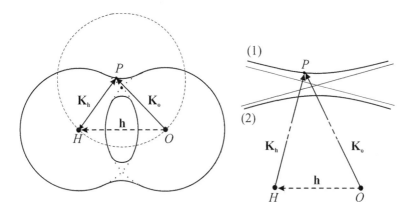

Fig. 9.1   *Left*: Full line: Dispersion surface in the reciprocal lattice. O: origin of the reciprocal lattice; H: node of the
reciprocal lattice. **h**: reciprocal lattice vector; P: a tiepoint; **OP**: incident wave; **HP**: reflected wave; dashed circle:
trace of the Ewald sphere. *Right*: Enlarged view of the centre of the diagram on the left.

lattice node (Fig. 9.1). The summation is over all the reciprocal lattice vectors, $\mathbf{h} = \mathbf{OH}$, where
O is the origin of the reciprocal lattice. Expansion (9.1) was called wavefield by Laue (1931*a*)
and is sometimes called *Ewald wave*. Solid-state physicists call it a Bloch wave, after Bloch
(1928), who derived the theory of electrons in solids, and introduced a similar expression for
the wave function of the electrons.

The most frequent case is that when there are only two nodes of the reciprocal lattice that
lie on the Ewald sphere (the 'two-beam case'). Two waves only propagate inside the crystal,
the reflected wave of wave vector $\mathbf{K_h} = \mathbf{HP}$, and the refracted wave, $\mathbf{K_o} = \mathbf{OP}$ (Fig. 9.1, *Left*).
The extremity, P, of the wave vectors was called *Anregungspunkt*, excitation point, by Ewald
(1917), but he later called it 'tiepoint' to emphasize the connection between the two waves.
Indeed, the two waves propagate together in the crystal; they interfere to generate standing
waves, and they both undergo the same anomalous absorption properties.

The relation between the amplitudes of the waves are deduced from Maxwell's equations. It
is, for a plane-polarized wave:

$$\mathbf{E_h} = \frac{K_h^2}{K_h^2 - k^2} \sum_{h'} \chi_h' \mathbf{E_{h'}}, \tag{9.2}$$

which expresses the self-consistency of the problem. The summation is over all the reciprocal
lattice nodes and χ_h' is the h' coefficient of the Fourier expansion of the polarizability of the
medium. For the set of linear equations (9.2) to have a non-trivial solution, its determinant
must be equal to zero. The corresponding secular equation is the 'dispersion equation'. It is the
equation of the surface on which the tiepoint P must lie—the 'dispersion surface'.

The only terms in expansion (9.1) that have a non-negligible amplitude are those for which
the resonance factors $1/(K_h^2 - k^2)$ are very large, namely those for which K_h is not very
different from the wave number in a vacuum, k. They correspond to those reciprocal lattice
points that are close to the Ewald sphere, O and H in the two-beam case. Far from the Bragg

condition for any reflection, there is only one such term, and one wave only propagates inside the crystal. In the two-beam case, the dispersion surface is composed of two sheets connecting the two spheres centred at O and H and of radii n/λ, where n is the refractive index, branch (1) and branch (2) (Fig. 9.1). All the properties of the diffracted waves are deduced from the analysis of the dispersion surface.

9.2.2 *Wavefields excited in the crystal by the incident wave: reflection profiles*

The next step is to introduce the boundary conditions and then to find out which are the waves actually excited in the crystal. The dispersion surface is the equivalent of the surface of indices in optics and one simply applies Huygens's construction. Two geometrical situations are to be distinguished: transmission, or Laue geometry, and reflection, or Bragg geometry.

- **Transmission geometry**: the normal, **n**, to the entrance surface of the crystal cuts across both branches of the dispersion surface (Fig. 9.2, *Left*). There are two tiepoints, P_1 and P_2, and two wavefields propagating inside the crystal. The boundary conditions are applied in the same way at the exit surface of the crystal to determine the reflected wave. The variations of its intensity with the glancing angle, θ, of the incident wave are shown in Fig. 9.2, *Right*.
- **Reflection geometry**: the normal, **n**, to the entrance surface intersects one branch only of the dispersion surface, at P' and P'' (Fig. 9.3, *Left*). The wavefield corresponding to P'' would propagate towards the outside of the crystal and is not excited. The wavefield corresponding to P' propagates towards the inside of the crystal and is the only one excited. If the normal to the entrance surface lies within the shaded region in the figure, the intersection points are imaginary and there is total reflection, as shown in Fig. 9.3, *Right*, representing the variations of the reflected intensity with glancing angle of the incident wave. It is the famous 'top-hat' curve.

Twenty years after his main article, P. P. Ewald published a development of his theory for any kind of lattice and taking the full structure factor into account (Ewald 1937). Ewald's dynamical

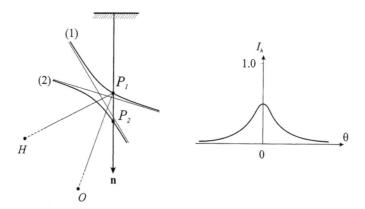

Fig. 9.2   Transmission geometry. *Left*: Dispersion surface; P_1, P_2: tiepoints of the waves excited in the crystal by the incident wave; **OP$_1$**: refracted wave; **HP$_1$**: reflected wave. *Right*: Reflection profile; horizontal axis: glancing angle of the incident beam.

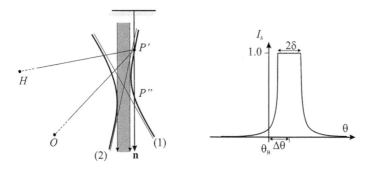

Fig. 9.3   Reflection geometry. *Left*: Construction in reciprocal space. P': tiepoint of the wavefield propagating inside the crystal; **OP′**: refracted wave; **HP′**: reflected wave. *Right*: Reflection profile; horizontal axis: glancing angle of the incident beam.

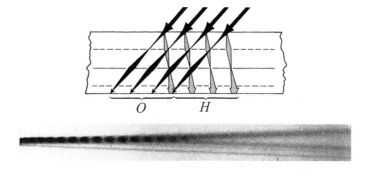

Fig. 9.4   Plane-wave Pendellösung. *Top*: As predicted by Ewald (1927). After Ewald (1927), with kind permission of Springer Science and Business Media. *Bottom*: As observed by Malgrange and Authier (1965) with a wedge-shaped silicon crystal and MoKα radiation.

theory was extended to the case of electrons by H. Bethe,[4] soon after the discovery of electron diffraction (Bethe 1928).

9.2.3 Pendellösung

In the transmission geometry, the diffracted waves overlap as they propagate inside the crystal. The waves associated with the two branches of the dispersion surface interfere, and Ewald (1917) predicted that a periodic transfer of energy should occur between the waves diffracted in the incident and reflected directions, as shown in Fig. 9.4 reproduced from a later publication

[4]Hans Bethe, born 2 July 1906 in Strasbourg, then Germany, died 6 March 2005 in Ithaca, NY, USA, was a German-American nuclear physicist. He studied physics at the Goethe University in Frankfurt, and at the University of Munich where he obtained his PhD prepared under the supervision of A. Sommerfeld. He became Privatdozent at the University of Munich in 1930, and Assistant Professor at the University of Tübingen in 1933, but soon after emigrated, first to England, then to the USA. He was appointed Assistant Professor at Cornell University, Ithaca, NY, in 1935, and Professor in 1937. He was married to Ewald's daughter, Rose, in 1939. Bethe was awarded the 1967 Nobel Prize in Physics for 'his contributions to the theory of nuclear reactions, especially his discoveries concerning the energy production in stars'.

(Ewald, 1927). He called this effect *Pendellösung*, after the German verb, *pendeln*, to oscillate. It took more than forty years until it was observed in real life, by Kato and Lang (1959) for spherical waves and by Malgrange and Authier (1965) in the plane wave case. It is now commonplace and is also observed in the reflection geometry. The early developments of Ewald's dynamical theory of diffraction have been recounted by him in Ewald (1962*a*, 1969*b*, 1979*a*) and in an Annex to Authier (2003).

9.3   Deviation from Bragg's law

It has been shown in Section 8.4.2 that the reflection peak is shifted in reflection geometry with respect to Bragg's angle by a small angle, $\Delta\theta$, given by (8.17); see Figs. 8.8 and, particularly, 9.3. This shift was first observed by students of M. Siegbahn in Lund University, Sweden. Very careful measurements of the Bragg angle were needed for the determination of X-ray wavelengths. The first to detect the deviation from Bragg's angle was W. Stenström (1919), who wrote about it in his thesis. He compared the values of $\sin\theta/n = \lambda/2d$ with increasing values of n and found that they were not constant. The work was pursued in more detail by E. Hjalmar (1920, 1923), who observed discrepancies in the values of the wavelength of $CuL\beta_1$ for several orders of n in gypsum, and by A. Larsson, who measured the Bragg angle for the reflection of copper $K\alpha_1$ up to the 11th order (Siegbahn 1925). They used a very accurate X-ray spectrograph developed by Siegbahn (1919*a, b*) for the measurement of X-ray spectral lines. The deviation from Bragg's law was suspected by W. Duane and R. A. Patterson (1920) with measurements of tungsten L lines with high order reflections on calcite and an ionization spectrometer, but it lay within experimental errors. It was studied more fully in B. Davis's laboratory in Columbia University, New York, USA, with the double-crystal spectrometer, first by Davis and Terrill (1922) with calcite and $MoK\alpha_1$, and then, at B. Davis's suggestion, with crystals ground so as to give an asymmetric reflection, which increases the deviation, by Hatley (1924) with calcite, also with $MoK\alpha_1$, and by Nardroff (1924) with pyrite and $MoK\beta_1$, $CuK\alpha_1$ and $CuK\beta$.

It was Ewald (1920, 1925) who explained the shift by the effect of refraction and calculated it with his dynamical theory. Siegbahn (1925) checked the explanation with crystals of gypsum and calcite and agreed with Ewald's explanation.

9.4   The double-crystal spectrometer

Double-crystal settings using twice the same crystal with the same reflection are used to produce monochromatic radiation (Fig. 9.5). They were first used by Wulff and Uspenski (1913*b*) and Wagner and Glocker (1913) to confirm that a Laue spot corresponded to a given wavelength (Section 7.2). They were used for the measurement of absolute integrated intensities (Compton 1917*b*; W. L. Bragg *et al.* 1921*a, b*; Wagner and Kulenkampf 1922; Bearden 1927), and for the accurate measurement of X-ray wavelengths (Compton 1931; Bearden 1931).

B. Davis (Fig. 9.6) was one of the main developers of the double-crystal technique in the United States, at Columbia University in New York. He and W. M. Stemple were the first to record rocking curves and to show the value of the double-crystal spectrometer as an instrument of high resolving power (Davis and Stemple 1921). The source of X-rays was a Coolidge tube with a tungsten anticathode, and the two crystals were calcite crystals in the parallel setting (Fig. 9.5, *Left*). Davis and Stemple compared crystals that had various origins and degrees of

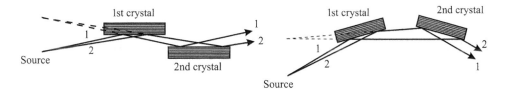

Fig. 9.5   Double-crystal settings. *Left*: Parallel, (+n,-n) non-dispersive; if rays 1 and 2 satisfy Bragg's condition on
the first crystal, both also satisfy Bragg's condition on the second crystal. *Right*: Antiparallel, (+n,+n) dispersive;
if rays 1 and 2 both satisfy Bragg's condition on the first crystal, and one of them satisfies Bragg's condition on the
second crystal, the other one does not. After Authier (2003).

Fig. 9.6   Bergen Davis. AIP Emilio Segre Visual Archives, with permission.

Bergen Davis: born on a farm 31 March 1869; died 30 June 1958, was an American physicist. He obtained his bachelor's degree in 1896 at Rutgers University, and his PhD in 1900 at Columbia University. He then did postgraduate work for one year at Göttingen, Germany, and for one year at the Cavendish Laboratory, in Cambridge, with J. J. Thomson. In 1903 he returned to New York to be tutor at Columbia University, where he successively held appointments as Instructor from 1907 to 1909, as Adjunct Professor from 1909 to 1913, as Associate Professor from 1913 to 1919, and as Professor of Physics from 1919 until his retirement in 1939. He was elected a member of the National Academy of Sciences in 1929.

MAIN PUBLICATIONS

1921 *An Experimental Study of the Reflection of X-Rays from Calcite*. With W. M. Stempel.

1922 *Reflection of X-rays from Rock Salt*. With W. M. Stempel.

1922 *The Refraction of X-Rays in Calcite*. With H. M. Terrill.

1924 *Refraction of X-Rays in Pyrites*. With R. von Nardroff.

1926 *Measurement of the Refraction of X-Rays in a Prism by Means of the Double X-Ray Spectrometer*. With C. M. Slack.

1927 *Measurement of the MoK Doublet Distances by Means of the Double X-Ray Spectrometer*. With H. Purks.

perfection, and had cleaved or polished surfaces. The reflected intensity was measured with an ionization chamber when rocking the second crystal. The half-widths and the maximum reflectivities depended on the degree of perfection of the crystals and on the state of the surfaces. With two freshly cleaved very clear Iceland spar crystals (not polished), they obtained a beautiful rocking curve, with a half-width of 16 seconds of arc and a maximum reflectivity nearly 50%, which showed that, indeed, their crystals were highly perfect. With a less perfect crystal, the half-width was 57 seconds of arc. They also made measurements with different wavelengths and found that the wavelength dependence of the maximum reflectivity decreased with increasing perfection of the crystals. Allison (1932) and Tu (1932) with the double-crystal spectrometer and highly perfect calcite crystals obtained very good agreement with the results of dynamical theory, and found no trace of mosaicity. It was the first time that it was clearly proved that there could be crystals perfect enough to diffract according to the perfect-crystal theory.

Wagner and Kulenkampf (1922) in Munich were the first to distinguish the two possible settings for the two crystals. Consider, in the 'parallel' setting (Fig. 9.5, *Left*), two rays, 1 and 2, with slightly different glancing angles on the reflecting planes, and which satisfy Bragg's condition on the first crystal for wavelengths, λ_1 and λ_2, respectively. Both satisfy Bragg's angle on the second crystal because their glancing angles on the second crystal are equal to their glancing angles on the first crystal. The setting is *non-dispersive*. In the 'anti-parallel' setting (Fig. 9.5, *Right*), on the contrary, if ray 1 satisfies Bragg's condition on the second crystal, ray 2 does not, because its glancing angle on the second crystal is different from its glancing angle on the first crystal. Only monochromatic rays of wavelength λ_1 are reflected by the second crystal. The setting is *dispersive*. The two settings are denoted (+n,–n) and (+n,+n), respectively.

Ehrenberg and Mark (1927), and Ehrenberg and von Süsich (1927) in Berlin-Dahlem, recorded very narrow rocking curves with diamond and calcite, respectively, with Mo$K\alpha$ radiation. An important progress in the theory of double-crystal settings was made by Ehrenberg, Ewald and Mark (1928) who showed that the reflectivity observed when rocking the second crystal is the convolution of the intrinsic reflectivities of the two crystals. They measured the half-widths of the rocking curves and compared them with the values calculated with Ewald's dynamical theory. They found a rather good agreement for the 111 reflections of zinc-blende and diamond. Reasonable agreement between measured and calculated rocking curve widths was also observed by Davis and Purks (1929) for highly perfect crystals of calcite.

The optical and geometrical properties of the double-crystal spectrometer were discussed in detail by M. Schwarzschild (1928) in B. Davis's laboratory. A detailed account of the theory is given in Compton and Allison's textbook.[5] The double-crystal settings are the archetypes of the multiple optical devices used for beam conditioning with synchrotron radiation (Authier 2003).

9.5 The Compton effect and the corpuscular nature of X-rays, 1923

The discovery of the Compton effect by A. H. Compton (1923b) was a major event in the history of physics. It was the outcome of five years of work, both experimental and theoretical. Compton gave his own account of how he arrived at this result,[6] and the historians of science have commented in detail the successive steps of his thought.[7]

After his thesis, Compton's interests shifted toward scattering phenomena. He tried to find classical explanations for anomalies unaccounted for by Thomson's scattering theory. For instance, the intensity of γ-rays scattered by a plate should be the same on both sides of the plate, but, in fact, it is higher on the emergent side. To explain this effect, Compton suggested the existence of electrons of various sizes and shapes. The work was presented at the 1917

---

[5]Compton, A. H. and Allison, S. K. (1935). *X-rays in Theory and Experiment*. Van Nostrand, New York.

[6]Compton, A. H. (1961). The scattering of X-rays as particles. *Am. J. Phys.* **29**, 817–820.

[7]See, for instance, R. H. Stuewer (1975). *The Compton Effect. Turning Point in Physics*. Science History Publications. New York, and B. R. Wheaton (1983).

December Meeting of the American Physical Society, while he was with Westinghouse, and published in 1919, but was not very convincing. Compton then went to Rutherford's laboratory in Cambridge, England, on a National Research Council Fellowship. There, he studied the progressive softening of γ-rays with angle of scattering (Compton 1921), which had first been observed by Sadler and Mesham (1915). This effect was not new but Compton's experiments were quantitative and far more accurate than those of his predecessors.

In the ship, on his way back to the USA, Compton (Fig. 8.4) drew the plans of the experiments he intended to undertake at Washington University in St Louis, Missouri, where he had been appointed Professor of Physics.[8] There, he carefully measured the increase in wavelength of the scattered X-rays with an ionization spectrometer and a special X-ray tube designed by him, and blown by him. The increase in wavelength is a maximum at a scattering angle of 90°. Compton's results indicated a kind of fluorescence whose wavelength was independent of the nature of the substance used as a scatterer, and depended on that of the incident rays, which he called 'general' fluorescence, as opposed to the usual 'characteristic' fluorescence. Furthermore, with C. F. Hagenow, he showed that this new fluorescence was polarized. He suggested that it was 'emitted at the instant of liberation of the secondary cathode rays from the atoms', by a mechanism distinct from that of the 'true' scattering without change of wavelength (Thomson scattering). These results were presented at the Meeting of the American Physical Society in Washington on 22 April 1921. In a short note published in March 1922, Compton suggested a mechanism involving a Doppler shift of the X-rays radiated by the recoil electron.[9]

In a paper that constituted the third part of a report by a Committee on X-ray spectra of the National Research Council,[10] Compton summarized the discussion on the softening of secondary radiation in a very clear way.[11] Part of the secondary rays is of the same nature as the primary radiation, but with an increased wavelength. It looks like scattered radiation, but the increase in wavelength cannot be explained by classical theory. It could be explained by a Doppler shift on the basis of a quantum effect. The scattering electron receives a momentum $h\nu/c$, where ν is the frequency of the primary radiation, and the conservation of energy demands that the electron shall recoil with a momentum $h\nu'/c$, where ν' is the frequency of the secondary radiation. But, Compton concluded, 'it has not been possible to account on either basis [classical or quantum] for all the observed phenomena'. The paper was published in October 1922. Then, in a communication to the American Physical Society on 1 December 1922, Compton gave the explanation of the effect in terms of light quanta, whereby an X-ray quantum transfers part of its momentum, $\mathbf{q_1}$, to an electron which acquires a momentum $\mathbf{q_e}$ (Fig. 9.7):

$$\mathbf{q_e} = \mathbf{q_1} - \mathbf{q_2} \tag{9.3}$$

where $\mathbf{q_2}$ is the momentum of the scattered X-ray quantum.

The change in wavelength is $\Delta\lambda = (2h/mc)\sin^2\theta$, where 2θ is the angle between the incident and the scattered X-ray quanta (Fig. 9.7).[12] The recoil electrons thus predicted by

[8]R. S. Shankland (1973). *Scientific Papers of Arthur Holly Compton*. University of Chicago Press. Introduction.

[9]Compton, A. H. (1922). The spectrum of secondary rays. *Phys. Rev.* **19**, 267–268.

[10]The Committee consisted of W. Duane, A. H. Compton, B. Davis, A. W. Hull, and D. L. Webster.

[11]A. H. Compton (1922). Secondary radiations produced by X-rays. *Bull. Nat. Res. Council.* **4**, number 20, 1–56.

[12]A modern treatment of Compton scattering is given in Cooper, M., Mijnarends, P., Shiotani, N., Sakai, N., and Bansil, A. (2004). *X-ray Compton scattering*. Oxford Series on Synchrotron Radiation. Oxford University Press.

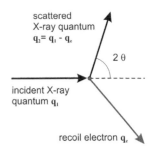

Fig. 9.7   Compton scattering. An incident X-ray quantum transfers part of its momentum $\mathbf{q}_1$ to the electron.

Fig. 9.8   Fifth Solvay Conference at Institut International de Physique Solvay, Brussels, 1927. *Top row*: A. Piccard, E. Henriot, P. Ehrenfest, E. Herzen, Th. de Donder, E. Schrödinger, J. E. Verschaffelt, W. Pauli, W. Heisenberg, R. H. Fowler, L. Brillouin. *Middle row*: P. Debye, M. Knudsen, W. L. Bragg, H. A. Kramers, P. A. M. Dirac, A. H. Compton, L. de Broglie, M. Born, N. Bohr. *Bottom row*: I. Langmuir, M. Planck, M. Curie, H. A. Lorentz, A. Einstein, P. Langevin, Ch.-E. Guye, C. T. R. Wilson, O. W. Richardson. Photo Benjamin Couprie.

Compton were observed a few months later by C. T. R. Wilson[13] who was to share the 1927 Nobel Prize for Physics with Compton, and by W. Bothe.[14] Figure 9.8 shows A. H. Compton

---

[13]C. T. R. Wilson (1923). Investigations on X-rays and β-rays by the cloud method. *Proc. Roy. Soc. A*. **104**, 1–24.

[14]W. Bothe (1923). Über eine neue Sekundärstrahlung der Röntgenstrahlen. *Z. Phys.* **16**, 319–320.

and C. T. R. Wilson at the Fifth Solvay Conference in Brussels, in 1927, where the newly formulated quantum theory was discussed by the leading physicists of the time.

The idea of light quanta, although more than ten years old, was not very familiar to the physicists of Anglo-Saxon countries. According to Carl H. Eckart (1902–1973), a graduate student who had a desk in the office of the Australian-born physicist, G. E. M. Jauncey (1888–1947), next door to Compton's, there was, some time in the autumn of 1922, a discussion between Jauncey and Compton on a series of papers by Schrödinger on light quanta that may have triggered Compton's train of thought[15]—a view shared by J. Jenkin.[16] In Europe, the idea of light quanta was quite familiar. P. Debye (1923), on the basis of Compton's report in *Bull. Nat. Res. Council*, proposed on 14 March 1923 a mechanism similar to Compton's.

Compton's discovery was the source of a lively controversy between him and W. Duane.[17] Experiments in Duane's laboratory had not confirmed Compton's measurements, and Duane did not accept Compton's proposed mechanism with X-ray quanta. The controversy lasted until, finally, Compton's experimental results were confirmed in Duane's laboratory.

9.6   Laue's dynamical theory of X-ray diffraction, 1931

Laue admired Ewald's thesis (Fig. 9.9) and considered it one of the all-time *masterpieces* in mathematical physics (Laue 1931*a*). He noted, however, that in it crystals were represented by

Fig. 9.9   *Left*: Paul P. Ewald in 1950, after Authier (2009). *Right*: Max von Laue in 1950. Photo Deutsches Museum Munich.

[15]Interview of Dr Carl Eckart by John L. Heilbron on 31 May 1962, Niels Bohr Library & Archives, American Institute of Physics, College Park, MD, USA, <http://www.aip.org/history/ohilist/4586.html>.

[16]J. Jenkin (2002). G. E. M. Jauncey and the Compton Effect. *Phys. Perspect.* **4**, 320–332.

[17]See, for instance, S. K. Allison (1965). Arthur Holly Compton, 1892–1962. *Bibliographical Memoirs of the National Academy of Science.* Washington, DC, USA.

a discrete distribution of single-point dipoles, while the recent theories about the structure of the atom pointed to a continuous distribution of electronic charge, a view confirmed by the studies of W. L. Bragg *et al.* (1922) and James *et al.* (1928) on rock-salt. This led him to reformulate Ewald's dynamical theory on an entirely different basis. He assumed a continuous distribution of the dielectric susceptibility of the medium for X-rays and considered it to be proportional to the electron density. A theory of the diffraction of X-rays by a medium with a continuous dielectric susceptibility had already been developed in Vienna by Lohr (1924), but it did not have any practical applicability.

Laue's theory is, in fact, simpler than Ewald's and is the more widely used of the two. It consists in looking for solutions of Maxwell's equations in a medium with a triply periodic dielectric susceptibility. The electric negative and positive charges are distributed in a continuous way throughout the whole volume of the crystal, and cancel out so as to ensure the neutrality of the crystal. The local electric charge and density of current may therefore be put equal to zero in Maxwell's equations.

Laue found it more convenient to represent the electromagnetic field through the electric displacement $\mathbf{D}$ because div $\mathbf{D} = 0$. By elimination of the electric and magnetic fields in Maxwell's equations, one obtains the propagation equation:

$$\Delta \mathbf{D} + \mathrm{curl\ curl\ } \chi \mathbf{D} + 4\pi^2 k^2 \mathbf{D} = 0 \qquad (9.4)$$

The dielectric susceptibility χ can be expanded in a Fourier series:

$$\chi = \sum_{\mathbf{h}} \chi_h \exp\ (2\pi \mathrm{i}\ \mathbf{h} \cdot \mathbf{r}), \qquad (9.5)$$

where the coefficients χ_h are proportional to the structure factors F_h.

The electric displacement is therefore also triply periodic and can be expanded in a Fourier series analogous to (9.1),

$$\mathbf{D} = \sum_{\mathbf{h}} \mathbf{D_h} \exp(-2\pi \mathrm{i}\ \mathbf{K_h} \cdot \mathbf{r}) \exp(2\pi \mathrm{i}\ \nu t), \qquad (9.6)$$

which expresses the wavefield propagating in the crystal.

By substitution of the expansions of χ and $\mathbf{D}$ in the propagation equation (9.4), one finds that the amplitudes D_h satisfy a set of equations similar to (9.2), from which the dispersion surface can be deduced in the same way.

In his 1931 article, Laue discussed the properties of the dispersion surface and derived expressions for the reflected intensity in both the transmission and reflection geometries, which are more convenient than Ewald's, but, at the time, he did not go further into the study of dynamical diffraction. A first extension of the theory included Laue's explanation of the contrast of Kossel lines (Laue 1941). A second edition of that book was published in 1945 with only small changes. The third edition, which included discussions of the properties of wavefields, such as anomalous absorption and propagation direction, appeared on the year of Laue's accidental death (Laue 1960). An account of Laue's theory covering the progresses that took place during the forty years that followed can be found in Authier (2003).

Laue's theory of a continuous distribution of dielectric susceptibility was later justified quantum-mechanically by G. Moliere (1939). The correspondence between Ewald's and Laue's dynamical theories was worked out by H. Wagenfeld (1968). The dynamical theory had been developed by Ewald and Laue for an incident plane wave. It was extended to the case of an incident spherical wave by N. Kato (1960) and to that of any type of wave by S. Takagi (1962).

9.7   Optical properties of wavefields

The notion of wavefield was introduced initially as a purely mathematical entity. For Ewald (1913*a*, 1917), expression (9.1) described the optical field in the crystal. For Laue (1931*a*), equation (9.6) expressed the solution of the propagation equation (9.4). That notion did not appear in Darwin's theory. In fact, wavefields have a physical reality; wavefields belonging to different branches of the dispersion surface undergo different anomalous absorption (Borrmann 1941, 1950), propagate along different directions inside the crystal (Borrmann 1954; Borrmann *et al.* 1955), interfere to produce *Pendellösung* fringes (Kato and Lang 1959, Malgrange and Authier 1965), and the path of individual wavefields can even be isolated (Authier 1960, 1961). The first evidence of the physical existence of the wavefields came from Laue's interpretation of the contrast of Kossel lines, based on the standing waves generated by the interference of the waves which constitute a wavefield.

9.7.1   *Kossel lines*

Kossel lines occur when the fluorescent radiation from one type of atom in a crystal is Bragg-reflected by the lattice planes of that same crystal. One speaks then of lattice sources. The primary radiation maybe either electrons or X-rays. These lines lie at the intersections of cones having as axes the normals to each family of lattice planes with the photographic plate. As an example, the lines due to the reflections on $(\bar{1}11)$, $(11\bar{1})$ and $(1\bar{1}1)$ can be observed in Fig. 9.10, *Left*. These line are in general dark or light or have a double contrast, dark–light or light–dark.

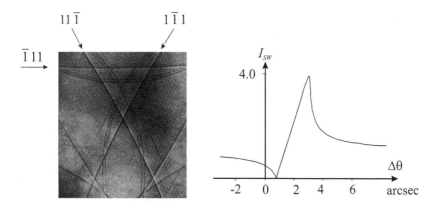

Fig. 9.10   *Left*: Kossel lines in a copper crystal with Cu$K\alpha$ radiation. Photo H. Voges, reproduced in Laue (1960). *Right*: variation of the intensity of the wavefield $|D|^2$ across the reflection domain, recalculated after Laue (1935).

This effect was surmised by Mie (1923), Clark and Duane (1923) and by Kossel (1924), but was not explained by them. It was clearly observed for the first time in 1935, by Kossel[18] and his co-workers using electrons as the primary source (Kossel 1935; Kossel and Voges 1935; Kossel *et al.* 1935). The Kossel lines, which are analogous to the Kikuchi lines observed in electron diffraction, were also studied in detail by G. Borrmann for his PhD thesis, prepared under the supervision of Kossel in Danzig (now Gdansk); see Authier and Klapper (2007). Borrmann used X-rays as the primary source, both in the reflection setting (Borrmann 1935, 1936) and in the transmission setting (Borrmann 1938), and, like Kossel and his co-workers, a copper crystal as the source of the diffracted secondary radiation. As mentioned in Section 6.5, W. Friedrich and P. Knipping had chosen a copper salt for their first attempts at observing diffraction of X-rays by a crystal, because they thought it would have something to do with fluorescence radiation, and copper was a suitable element for that.

Borrmann observed that, in the transmission geometry, the double contrast of the lines is inverse for thick crystals (Borrmann 1938). This was not explained at the time but was, in fact, the first indication of anomalous absorption (Schülke and Brümmer 1962).

It was Laue (1935) who explained the fine structure of the Kossel lines in the reflection geometry, using the properties of wavefields and the reciprocity theorem. The intensity of the wavefield excited by an incident plane wave is, after equation (9.6), in the two-beam case:

$$|D(r)|^2 = |\mathbf{D_o} \exp(-2\pi i \, \mathbf{K_o} \cdot \mathbf{r}) + \mathbf{D_h} \exp(-2\pi i \, \mathbf{K_h} \cdot \mathbf{r})|^2 \tag{9.7}$$

$$= |D_o|^2 + |D_h|^2 + |D_o D_h| \cos(2\pi \mathbf{h} \cdot \mathbf{r} + \psi), \tag{9.8}$$

where ψ is the phase of D_h/D_o. Fig. 9.10, *Right*, shows its variations across the reflection domain, in the reflection geometry. Laue (1935) argued that, according to the reciprocity theorem, the intensity distribution in space of the beams resulting from the reflection of the spherical waves emitted by the lattice sources should be identical. The fact that it was, indeed, what is observed was considered by Laue, at the time he was writing his 1960 book (1959), as the only direct evidence of the physical existence of the wavefields.

9.7.2 Standing waves

Expression (9.8) of the wavefields shows that the interference of the waves D_o and D_h generates a set of standing waves in the crystal (Laue 1941). The term $\cos 2\pi(\mathbf{h} \cdot \mathbf{r} + \psi)$ indicates that the nodes lie on planes parallel to the lattice planes and that their periodicity is equal to $1/h = d_{hkl}/n$ where d_{hkl} is the periodicity of the $hk\ell$ family of lattice planes and n is the order of the reflection.

1. **In transmission geometry**, phase ψ is equal to π for wavefields associated with branch (1) of the dispersion surface and to 0 for wavefields associated with branch (2). The nodes of standing waves therefore lie on the lattice planes (planes of maximum electronic density) for wavefields associated with branch (1) of the dispersion surface (Fig. 9.1). These wavefields undergo anomalously low absorption. The antinodes (maxima of electric field) of the wavefields associated with branch (2) lie on the lattice planes. They undergo anomalously high absorption.

---

[18]See footnote 23 in Chapter 7.

2. **In reflection geometry**, the phase ψ varies from π to 0 across the total reflection domain. For an incidence on the low-angle side of the reflection domain, the nodes of standing waves lie on the lattice planes. The absorption is anomalously low. As the incidence sweeps the reflection domain, the nodes are progressively shifted until they lie at mid-distance between the reflecting planes for an incidence on the high-angle side of the reflection domain. It is then the antinodes which lie on the reflecting planes, and the absorption is anomalously high.

This shift of the system of nodes and antinodes in the reflection geometry when one rocks the crystal through the reflection domain can be made use of to localize the position of atoms in the crystal. If the incident radiation excites the emission of secondary radiation, either fluorescent X-rays or photoelectrons, by atoms of the crystal or impurity atoms, this emission will be maximum when the atom lies at an antinode of electric field. The position of these atoms can therefore be localized by detecting this secondary radiation with an appropriate detector synchronously with the recording of the intensity reflected by the crystal.[19]

9.7.3   Anomalous absorption

Anomalous absorption of X-rays, which is one of the most remarkable properties of wavefields, was discovered by G. Borrmann[20] (1941) and bears his name (*Borrmann effect*). G. Borrmann and his students played a decisive role in the first revival of the dynamical theory. When Ewald submitted his habilitation thesis in 1917, Sommerfeld had found it a beautiful mathematical construction but predicted that it would never have any practical applications. These came more than twenty years later, with Borrmann's investigations.

The discovery of anomalous absorption came from the observation by Borrmann of the forward-diffracted beams transmitted through good-quality crystals of calcite and quartz of various thicknesses, but only the quartz results were published at the time (Hildebrandt 1995, 2002; Authier and Klapper 2007). His experimental set-up was the same as that already used by Rutherford and Andrade (1914*b*) to measure the wavelength of γ-rays diffracted by a rock-salt crystal: a point source and a very divergent beam—the wide-angle method (see Section 7.10.3). The trace of the forward-diffracted beam was expected to show a deficit of intensity against the back-ground because of the intensity drawn out of the incident beam by the reflected beam. It was the contrary that was observed, which baffled Laue considerably. It could only mean an anomalously low absorption. Laue (1949) accounted for the effect by calculating the intensities of the reflected and forward-diffracted beams taking absorption into account. Borrmann (1950, 1954) made very careful measurements of the anomalous absorption with calcite crystals and explained it by the relative positions of the nodes and antinodes of the standing wavefields with respect to the lattice planes (Section 9.7.2). Wavefields associated with branch (2) are, in practice, completely absorbed out in thick crystals.

[19]See, for instance, *The X-ray Standing Wave Technique: Principles and Applications.* J. Zegenhagen and A. Kazimirov, editors (2013). World Scientific, Singapore.

[20]Gerhard Borrmann, born 30 April 1908, died 12 April 2006, was a German physicist. He received his higher education at the Technische Universität München and the Technische Hochschule Danzig (now Gdansk, Poland) where he was awarded the title Diplom-Ingenieur in 1930, and where he obtained his PhD in 1936. He became then Kossel's assistant. In 1938, he was called by M. von Laue to the Kaiser-Wilhelm-Institut für Physikalische Chemie und Elektrochemie in Berlin-Dahlem (now Fritz-Haber-Institut der Max-Planck-Gesellschaft), where he turned to the study of reflection by perfect crystals. When Laue was appointed Director of the Fritz-Haber-Institut, Borrmann became head of a department of his own (Kristalloptik der Röntgenstrahlen).

Anomalous absorption takes place in a similar way in reflection geometry and is exhibited by the reflection profiles, which become asymmetric. This effect was first observed by Renninger (1955).

9.7.4 Path of the wavefields: Borrmann triangle, or fan

A surprising result of Borrmann's 1950 article on anomalous absorption in calcite had been that the propagation of X-rays in thick crystals was neither along the incident nor the reflected directions, but in between, along the lattice planes. This had already been guessed at earlier, in a very qualitative way, by Murdock (1934), who had observed 'triple Laue spots' in quartz crystals. Laue had at first not been convinced by Borrmann's observations. But, from Maxwell's theory of electromagnetism, it is known that the direction of propagation of the energy of an electromagnetic wave is along the Poynting vector, $\mathbf{S} = \mathbf{Re}(\mathbf{E} \wedge \mathbf{H}^*)$, where $\mathbf{Re}(\mathbf{E})$ is the real part of $\mathbf{E}$ and $\mathbf{H}^*$ the complex conjugate of $\mathbf{H}$. Laue (1952c) calculated the Poynting vector by means of the dynamical theory and showed that it is normal to the dispersion surface (Fig. 9.11, *Left*). A natural incident beam is divergent and should therefore excite tie points along the whole dispersion surface. It is therefore to be expected that there should be wavefields propagating inside the crystal along all the directions lying between the incident and reflected directions Borrmann (1954, 1959b). They fill out what is now called the *Borrmann triangle*, or fan (Fig. 9.11, *Right*). The absorption is minimum for waves propagating along the lattice planes (maximum anomalous absorption effect), and, for thick crystals, are the only ones observed, as Borrmann (1950) had shown. The path of wavefields in a calcite crystal was then studied carefully by Borrmann *et al.* (1955), confirming Laue's calculations. That calculation was later generalized by Kato (1958) to the *n*-beam case, but Ewald (1958) pointed out that Laue's and Kato's calculations implied incident plane waves, which was not the case in practice. He then proposed a very simple physical proof by substituting wave bundles for plane waves and showing that their group velocity is along the normal to the dispersion surface.

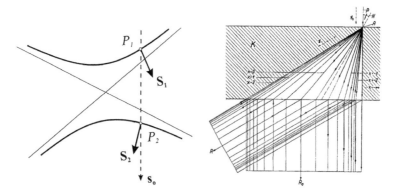

Fig. 9.11   Propagation of wavefields in a crystal. *Left*: reciprocal space. $\mathbf{s_0}$: normal to the crystal surface; P_1, P_2: tiepoints of the two waves excited in the crystal; $\mathbf{S_1}$, $\mathbf{S_2}$: Poynting vectors. *Right*: Borrmann fan in direct space. After Borrmann (1959b).

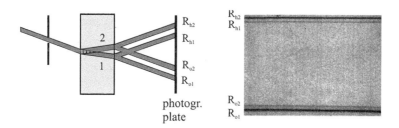

Fig. 9.12   Experimental proof of the double refraction of X-rays in a silicon crystal. *Left*: experimental setup. *Right*: traces of the reflected and forward diffracted beams on the photographic plate. After Authier (2003).

The perturbations in the propagation of wavefields by crystal defects such as dislocations, stacking faults, twins, inclusions, *etc.* can be taken advantage of to visualize these defects by topographic imaging techniques (Lang 1958, Borrmann *et al.* 1958).

9.7.5   Double refraction

An incident plane wave excites two wavefields inside the crystal in the transmission geometry, with tiepoints P_1 and P_2 (Fig. 9.11, *Left*), and with Poynting vectors $\mathbf{S_1}$, $\mathbf{S_2}$. The two wavefields therefore propagate along separate paths inside the crystal. In the general case of unpolarized radiation, there are in fact four wavefields, two for each direction of polarization. This is why Borrmann (1955) spoke of quadruple refraction (*Vierfachbrechung*) of X-rays. In practice, the paths corresponding to the two directions of polarization are so close to each other that it is hopeless to try to observe their separation. The separation of the paths of wavefields 1 and 2 is in principle also impossible to observe, since either the incident wave is a spherical wave and all the possible directions of propagation within the Borrmann fan are excited, or it is a plane wave and its lateral expansion is, by definition, infinite. The paths of the two wavefields then overlap and cannot be separated.

A way to go round this difficulty was found by isolating from the Borrmann fan a wave packet that is narrow in both direct and reciprocal space (Authier 1960). The paths of the two packets of wavefields, 1 and 2, can then be separated (Fig. 9.12, *Left*). The result is shown in Fig. 9.12, *Right* (Authier 1961). It provides the most direct experimental proof of the physical existence of the wavefields.

EARLY APPLICATIONS OF X-RAY CRYSTALLOGRAPHY

X-ray analysis ranks as one of the most powerful weapons of modern science.

Sir W. L. Bragg (1943)

10.1 Chemical sciences

10.1.1 Introduction

In the mind of R.-J. Haüy, each chemical entity had a characteristic crystalline form, while E. Mitscherlich (1819) considered that different chemical substances that possessed analogous composition could have the same crystalline form (isomorphism), and, in contrast, that the same chemical substance could have different crystalline forms (polymorphism). L. Pasteur (1848*a*, *c*) found that geometrically enantiomorphic or chiral molecules of tartaric acid crystallized with enantiomorphic crystalline forms. P. Curie (1894) related physical properties and crystal symmetry. P. von Groth had treasured a very extensive wealth of observations and measurements on crystals, but in 1913 the chemist still did not know the atomic structure of matter. W. Barlow and W. H. Pope's concept of close-packed arrangements and theory of valency-volume (proportionality between the volumes of the spheres of influence of atoms and their valencies; see Section 12.13) had little impact, and the American physical chemist H. C. Jones, already mentioned in Section 1.2, could write in June 1913 in his book, *A New Era in Chemistry*: 'We know comparatively little about matter in the solid state. We know the geometrical forms in which it crystallizes; we do not know what is the formula of solid sodium chloride or rock-salt, or of solid water or ice; and we have no reliable means at present of finding out these simplest matters about solids. Our ignorance of solids is very nearly complete'[1] (Jones 1913, p. 145). This was not true any longer when it was written. By that date the Braggs had already determined the structures of rock-salt and other alkali halides, zinc-blende, fluorite, calcite and isomorphous carbonates, sodium nitrate.

The foremost result, in confirmation of Barlow and Pope's predictions and Groth's ideas, and in contradiction to Sohncke's belief (Sohncke 1884, see Section 12.13), was that there was no discrete molecule of sodium chloride in the solid state. In its crystalline state, each sodium is surrounded by six chlorines and each chlorine by six sodiums. There was no connection between the commonly accepted valency of atoms and the structure of crystals containing them: NaCl (monovalent atoms) and MgO (divalent atoms) have the same crystal structure; CsCl and CsF (both with monovalent atoms) have different structures. This was difficult to understand for chemists. The German chemist P. Pfeiffer (1915) noted that 'the ordinary notion

---

[1]Quoted by L. Pauling in Ewald (1962*a*), p. 136.

of valency didn't seem to apply' and that 'this bond structure was remindful of the octahedral arrangement of the six chlorine around the platinum in the Werner[2] coordination complexes such as chloroplatinate $PtCl_6K_2$'. The structure of such complexes were confirmed later by the crystal structure determination from Laue diagrams of ammonium chloroplatinate by Wyckoff and Posnjak (1921) and of potassium chloroplatinate by Scherrer and Stoll (1922).

The number of structure determinations increased very rapidly, particularly after the development of powder diffraction methods which greatly increased the variety of crystals that could be studied. Very soon a new science was born, *structural chemistry* (Pauling 1937). It provided a quantitative evaluation of the size and shape of ions, atoms and organic molecules, and showed that there are different structure types depending on whether the bonding is between ions, atoms or molecules. The first structures to be solved were those of simple inorganic compounds or of elements.

10.1.2   Inorganic crystals

Information about the nature of bonds is one of the major contributions of X-ray crystallography to chemistry. R. W. G. Wyckoff (1924) in his book on the structure of crystals noted that the crystal structure results were capable of supplying information concerning the equivalence of the bonds joining an atom. For instance, three of iodine atoms of SnI_4 are identically related to their surrounding atoms and different from the fourth. In calcite, $CaCO_3$, the three oxygen atoms are alike, and such a formula as $CaO \cdot CO_2$ is without justification.

The chemists Lewis[3] (1916) and Langmuir[4] (1919) and the physicist Kossel[5] (1916a, 1919) developed a theory of the atomic structure in which the electrons are distributed in concentric shells. The valency properties are related to the number and properties of the electrons in the outer shell. Certain arrangements of electrons such as those in inert gases are very stable and the electrons in a compound tend to assume the most stable arrangement. Two types of binary compounds AB may for instance be expected, 1) those in which one or more electrons are transferred from atoms A to atoms B, so that both have complete outer shells of eight electrons (polar crystals with electronic charge), 2) those in which one or more electrons are held in common or shared between atoms A and B (non-polar crystals).

The scattering factor of an atom is altered on ionization, especially at low values of $\sin\theta/\lambda$. Debye and Scherrer (1918) deduced from estimations of the intensities diffracted by powdered lithium fluoride crystals that a valency electron had been transferred from the lithium to the fluorine atom. This they did by comparing the intensities of X-ray reflections for which the scattered amplitude was respectively proportional to the sum and to the difference of the contributions by fluorine and lithium. They also analysed diamond, but became convinced (surprisingly) that this crystal structure did not contain a shared-electron bond between the two types of carbon atoms in it. D. Coster (1920) in Delft, The Netherlands, discussed this point and showed that the presence of directional bonds would entail the presence of reflections such as 222; but he estimated that they would have been too weak to be observed by Debye and

---

[2]Alfred Werner, born 12 December 1866, died 15 December 1919, was a Swiss chemist. He studied at the ETH Zürich, where he obtained his PhD in 1890, and where he became Professor in 1895. He was awarded the Nobel Prize in Chemistry in 1913 for proposing the octahedral configuration of transition metal complexes.

[3]See footnote 58 in Chapter 3.   [4]See footnote 22 in Chapter 8.   [5]See footnote 23 in Chapter 7.

Scherrer, and that no conclusion could be drawn. At the same time, N. H. Kolkmeijer (1920) in Utrecht, also using powder diffraction, confirmed Debye Scherrer's results. The first evidence for a shared-electron C–C bond came from W. H. Bragg (1921b) who did observe a weak forbidden 222 reflection from a small diamond bathed in the incident X-ray beam. This led him to write: 'It is necessary, therefore, to suppose that the attachment of one atom to the next is due to some directed property, and that the carbon atom has four such special directions: as indeed the tetra-valency of the atom might suggest'. A full experimental proof of the covalent nature of the C–C bond had to wait for the Fourier electron density maps of R. Brill[6] et al. (1939).

W. L. Bragg (1920b) attacked the same problem and interpreted the polar structure of potassium chloride as a stacking of K^+ and Cl^- ions in which both ions have a complete stable argon shell with 8 electrons, chlorine having extracted an electron from potassium. W. L. Bragg et al. (1921a and b, 1922) tried to evaluate the electron distribution in sodium chloride by measuring absolute intensities diffracted by single crystals of rock-salt. They concluded that because of the uncertainties affecting the measurements of the low-order reflections, in particular extinction, it was impossible to ascertain whether the number of electrons around sodium and chlorine was respectively 10 and 18 or 11 and 17. Furthermore, they estimated that the electron distribution was not concentrated around the centres of the atoms, as Debye and Scherrer (1918) had assumed, but distributed throughout the volume of the crystal (See Section 8.2.4). It is only when the Fourier syntheses became usual that derivation of accurate electronic densities became possible. The first electron-density maps of sodium chloride, diamond and hexamethylene-tetramine showed the nature of the bonds in these compounds (Brill et al. 1939), opening up a very important research topic. For a review, see, for instance, Coppens (1992).

The increasingly high number of structure determinations led to the establishment of elementary rules and of classification schemes. The notion of atomic and ionic radii began to emerge. W. L. Bragg (1920b) showed that the arrangement of the atoms in a crystalline structure may be pictured as that of an assemblage of spheres of appropriate diameters, each sphere being held in place by contact with its neighbours. The distance between the centres of two neighbouring atoms can be expressed as the sum of two constants characteristic of the component atoms: this is the empirical rule of additivity of radii. In the same paper, W. L. Bragg gave a list of ionic radii deduced from the diffraction data for more than twenty crystals (W. L. Bragg 1920b). In a letter to his father, dated 21 October 1920,[7] W. L. Bragg mentions values of atomic radii obtained by the kinetic theory of gases, which are larger than the values given by crystals, and writes: 'I cannot see how all these quantitative relationships can be explained by anything so indefinite as a planet system of electrons round the nucleus'!

P. Niggli (1921) considered a large number of compounds of known crystal structure and related their composition to the various crystal systems in which they are observed. He also compiled a list of atomic radii, which he found to be in agreement with Bragg's values and noted that they vary periodically with atomic number.

The Swede Jarl A. Wasastjerna (1923) calculated atomic and ionic radii from optical refractometric observations, and published more realistic values than Bragg's or Niggli's, in particular for fluorine F^- and oxygen O^{--}. W. L. Bragg (1925a) admitted that in his 1920 paper he had

---

[6]See footnote 23.   [7]Bragg archives at the Royal Institution (RI MS WHB 28A/11).

'made the domains of the positive ions too large and those of the negative ions too small', although the sums were correct. Taking the new results into account, W. L. Bragg (1926) accordingly modified his values of atomic and ionic radii. 'It is very striking', he notes in that paper, 'in working out a new structure, to observe how closely the first approximation obtained by packing together an assemblage with the dimensions of the atomic domains, corresponds with the final structure attained by a careful consideration of the intensities of X-ray reflection. One cannot help being convinced that the idea is useful, and gives a helpful conception of the reasons for the stability of the structure'. A very large number of articles on atomic radii were published in the 1920–25 period, often inspired by Bragg's 1920 article. There was a certain confusion as the atomic radius, or 'sphere of influence' of a given element was sometimes measured from data in crystals where its state of ionization took different values. For Wyckoff (1923), there were many disagreements between calculated and measured atomic radii and 'it must be concluded that the hypothesis of constant atomic radii is not in agreement with existing information'. In a letter to W. H. Bragg, dated 27 March 1923, and congratulating him on the structure of naphthalene and anthracene, P. P. Ewald wrote: 'I cannot help distrusting atom radii in inorganic crystals except as a very crude rule'. To which W. H. Bragg answered, on 22 April 1923: 'I think with you that the principle of ionic radii is to be taken cautiously. It is quite clear that the atom must be given a radius appropriate to its surroundings'[8].

Atomic radii were also considered by the Norwegian L. Vegard in the course of his study of solid solutions (*Mischkristalle*) of inorganic crystals. In a first study (Vegard and Schjelderup 1917), he had analysed solid solutions of potassium chloride and bromide with the ionization spectrometer. This was followed a few years later by a systematic analysis of the KCl-KBr, KCl-NH$_4$Cl, K$_2$SO$_4$-(NH$_4$)$_2$SO$_4$, NH$_4$Cl-NH$_4$Br, NH$_4$Cl-NH$_4$I systems by powder diffraction (Vegard 1921, 1928). He found that a linear relation exists, at constant temperature, between the crystal lattice parameter and the composition for a continuous substitutional solid solution in which atoms or ions that substitute for each other are randomly distributed. This empirical rule known as 'Vegard's law' also applies to alloys and is used in mineralogy (see Section 10.2), materials science and metallurgy (see Section 10.1.5). It has, however, many exceptions and its deviations are of special interest. V. M. Goldschmidt investigated the variations in the sizes of the atoms or ions for which substitution can still occur, and this gave rise to the Goldschmidt rules. The variations of the atomic radius with the atomic number of the elements were found by L. Vegard to be similar to those described by W. L. Bragg (1920*b*).

It was V. M. Goldschmidt, a Swiss-born Norwegian geochemist and mineralogist, considered one of the founders of crystallo-chemistry (Fig. 10.1), who, with his collaborators, made the major contribution to the study of atomic or ionic radii. He established an extensive list of them, deduced from the interatomic distances in crystals. For this work he and the members of his laboratory synthesized more than 200 compounds, from the simpler AX and AX$_2$ to the more complex AX$_3$, A$_2$X$_3$, AXY, ABX, ABX$_4$, made up of 75 elements, substituting other atoms for the various constituents (A,X,Y,B). By use of powder patterns and the Weissenberg camera, they determined their crystal structure. In this way X-ray techniques brought to light the extreme complexity and variety in the structures of chemical substances.

---

[8]Bragg archives at the Royal Institution.

Fig. 10.1   Victor Goldschmidt and Albert Einstein in 1920 during a boat trip near Oslo. Photo: Halvor Rosendahl. By permission of University of Oslo history photobase.

Victor Moritz Goldschmidt: born 27 January, 1888 in Zürich, Switzerland, the son of a Professor of Physical Chemistry; died 20 March 1947 in Oslo, Norway. His family came to Norway in 1901 when his father became professor of Chemistry in Kristiania University in Oslo. He prepared his doctorate obtained in 1911, under the Norwegian geologist W. C. Brøgger, at the University of Oslo. The following year, he became Docent of Mineralogy and Petrography at the same University, and in 1914 was promoted to full Professor. In 1917 he was appointed Chair of the Government Commission for Raw Materials and head of the Raw Materials Laboratory, at a time when there was a shortage of raw materials due to the First World War. This led him to the study of the geochemical distribution of elements in the Earth's crust and to his major work published in nine volumes from 1923 to 1938. In 1929 he was called to the chair of Mineralogy in Göttingen, but had to return to Oslo in 1935 because of the growing antisemitism in Germany. The persecution of Jews made the period of the Second World War very difficult for Goldschmidt. He managed to escape to Sweden from where he was able to go to England in 1943. He came back to Oslo after the war. He had many students, among whom was the Norwegian-born American crystallographer, W. Zachariasen.

MAIN PUBLICATIONS

1911	*Die Kontaktmetamorphose im Kristianiagebiet.* (Doctorate thesis.)	1925	*Band V: Isomorphie und Polymorphie des Sesquioxyde.* With G. Lunde.
1912–1921	*Geologisch-petrographische Studien im Hochgebirge des südlichen Norwegens.*	1926	*Band VI: Crystal structure of the rutile type with remarks of the geochemistry of the bivalent and quadrivalent elements.* With T. Barth, D. Holmsen, G. Lund and W. Zachariasen.
1923	*Geochemische Verteilungsgesetze der Elemente, Band I.*		
1924	*Band II: Beziehungen zwischenden geochemischen Verteilungsgesetzen und den Bau der Aome.*	1926	*Band VII: Die Gesetze der Krystallochemie.* With T. Barth and W. Zachariasen.
1924	*Band III: Röntgenspektrographische Untersuchungen über die Verteilung der seltenen Erdmetalle in Mineralen.* With L. Thomassen.	1927	*Band VIII: Untersuchungen über Bau und Eigenschaften von Krystallen.*
		1938	*Band IX: Die Mengenverhältnisse der Elemente und der Atomarten.*
1925	*Band IV: Zur Krystallstruktur der Oxyde der seltenen Erdmetalle.* With F. Ulrich and T. Barth.	1954	*Geochemistry.* (posthumous)

Goldschmidt divided substances into two categories, ionic and atomic (covalent), and formulated two sets of radii, atomic and ionic, deduced from the interatomic distances provided by the crystallographic data. In particular, he showed that ionic radii do not account for the observed inter-atomic distances in atomic crystals. Furthermore, he gave some rules for the possibility of substituting an atom by another in a given structure: ionic radii must differ by less than 15%, the charge can differ by one unit if electrical neutrality is maintained by coupled substitution, etc. ('Goldschmidt rules').

Crystal structures were classified by Goldschmidt into various types according to their types of coordination and in agreement with Werner's structural chemistry. The principle of classification was based on 'the number and arrangement of neighbours around any single atom in the crystal lattice' (Goldschmidt 1926, 1929). As an example, Fig. 10.2 shows the various structure types of compounds AX and AX_2. The number of nearest neighbours around any atom is called the co-ordination number for that atom. In binary compounds there are in general four, six, or eight non-metal atoms around a metal atom, and the corresponding most common coordination polyhedra are the tetrahedron, the octahedron, and the cube. 'What are the causes which determine the type of structure of any given substance?' asks Goldschmidt (1929); 'why, for instance, has magnesium fluoride the structure of rutile, and strontium fluoride the structure of fluorite?'. Assuming that anions and cations are in mutual contact, this is determined by the ratio between their radii, and it is easy to calculate the limiting ratios for each coordination number. This had in fact already been done by Hüttig (1920) and Magnus (1922) for the coordination complexes, but Goldschmidt only refers to Magnus. The limiting

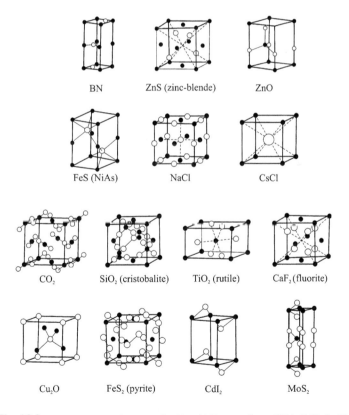

Fig. 10.2   Structure types of compounds AX and AX_2 according to V. M. Goldschmidt.

ratio between the rutile structure (coordination 6) and that of fluorite (coordination 8) is 0.73. Lower values of the radius ratio result in the rutile structure (as in magnesium fluoride, radius ratio 0.59) and higher values in the fluorite structure (as in strontium fluoride, radius ratio 0.95).

For Goldschmidt (1929), 'the task of crystal chemistry is to find systematic relationships between chemical composition and physical properties of crystalline substances, especially to find how crystal structure, the arrangement of atoms in crystals, depends on chemical composition'. In effect, he showed how the physical properties are modified by substitutions of one ion by another and their dependence on the distances between ionic centres. As a matter of conclusion, Goldschmidt defined what he called the *fundamental law of crystal chemistry*: 'The structure of a crystal is determined by the ratio of numbers, the ratio of sizes and the properties of polarization of its building stones. As the building stones of crystals we visualise atoms (or ions) or group of atoms'.

The empirical radii of ions determined by Goldschmidt were confirmed with a most remarkable agreement by the values calculated by L. Pauling (1927) using wave mechanics. Pauling (Fig. 10.3) had started his research career at Caltech by solving several inorganic structures. After his PhD obtained in 1925, he visited Europe for more than one year, where he studied with Arnold Sommerfeld in Munich, Niels Bohr in Copenhagen, and Erwin Schrödinger in Zürich—all three experts in quantum mechanics. In Munich, he learned about the new approach proposed by one of Sommerfeld's former students, Werner Heisenberg. Pauling developed for many-electron atoms a model similar to that of Schrödinger for hydrogen by considering each electron to be hydrogen-like, and used it to calculate ionic radii. He noted that, 'in the past, ionic radii have often been compared with observed interatomic distances without much regard to the nature of the crystal from which they were derived' (Pauling 1927).

In a 'landmark' study (W. L. Bragg in Ewald 1962), Pauling (1928a, 1929) introduced a set of simple rules governing the fundamental principles underlying all inorganic crystals, which have become known as 'Pauling's rules'. W. L. Bragg and his co-workers undertook in the years 1926–28 a systematic study of the crystal structure of silicates (see Section 10.2). They had assumed that, in a crystal composed of large and small ions, the structure can be approximated by a close-packed arrangement of the large ions with the small atoms located at the interstices and linked to four or six large atoms. Pauling noted that this is in fact not always true and introduced instead a theory based on the consideration of coordination polyhedra. He assumed that each cation is located at the centre of a polyhedron, the corners of which are occupied by anions, the cation–anion distance being determined by the radius sum and the coordination number of the cation by the radius ratio. This is the first of Pauling's five rules. The polyhedra may be linked by a corner, an edge, or a face. This method leads, for a given substance, to a small number of possible structures. Pauling applied it to the determination of the structure of brookite (TiO_2), the orthorhombic variety of tetragonal rutile and anatase, and gave an interpretation for the sequence of the three varieties: rutile, brookite, and anatase. He showed how the known structures of silicates could be interpreted with his rules and he determined further structures of silicates himself, such as topaz (Pauling 1928c), micas (Pauling 1930a), and zeolites (Pauling 1930b). B. E. Warren's and W. L. Bragg's (1929) determination of the structure of diopside was an example of the application of Pauling's rules.

Pauling's coordination theory is only applicable to ionic or partially ionic crystals and not to non-polar crystals or crystals with shared-electron pair bonds. At the same time as he worked

Fig. 10.3 Linus Pauling in 1950. Photo: Maryland Studio, Pasadena, CA, USA. Courtesy Special Collections and Archives Research Center, Oregon State University Libraries.

Linus Pauling: born 28 February 1901 in Portland, Oregon, USA, the son of a travelling salesman, later owner of his own drugstore; died 19 August 1994 in Big Sur, California, USA. He received his bachelor's degree from Oregon Agricultural College, now Oregon State University, in 1922. During his studies he had the occasion to read the papers of Lewis and Langmuir on valency theory and the book by W. H. and W. L. Bragg, 'X-rays and Crystal Structure'. In September 1922 he joined the California Institute of Technology, where he worked under Roscoe Dickinson on the X-ray determination of crystal structures. He received a PhD in 1925, after which he visited Europe with the help of a National Research Council Fellowship (1925–26) and a Guggenheim Foundation Fellowship (1926–27). When he came back he was appointed Assistant Professor at Caltech where he was on the professorial staff until 1964. From 1963 to 1967, Pauling was Research Professor at the Center for the Study of Democratic Institutions at Santa Barbara, California, from 1967 to 1969 Professor of Chemistry at the University of California at San Diego, and from 1969 Professor at Stanford University. His research bore initially on X-ray diffraction, quantum mechanics, and the nature of the chemical bond, and then turned towards biochemistry. His landmark book on the *Nature of the Chemical Bond* was widely read. With R. B. Corey and H. Branson he proposed a satisfactory model for components of a protein structure (before any complete protein structures had been determined)—the α-helix and the β-sheet (1951). In his later years he became a peace activist. In 1931 he was awarded the first Langmuir Prize, in 1954 the Nobel Prize in Chemistry 'for his research into the nature of the chemical bond and its application to the elucidation of the structure of complex substances', and in 1962 the Nobel Peace Prize, becoming the only person to have received two unshared Nobel Prizes. He had many students, among whom was the Nobel Prize winner, W. N. Lipscomb.

MAIN PUBLICATIONS

1929 *The principles determining the structure of complex ionic crystals.*

1935 *Introduction to Quantum Mechanics, with Applications to Chemistry.* With E. Bright Wilson, Jr.

1939 *The Nature of the Chemical Bond and the Structure of Molecules and Crystals: An Introduction to Modern Structural Chemistry.*

1941 *General Chemistry.*

1947 *General Chemistry: An Introduction to Descriptive Chemistry and Modern Chemical Theory.*

1950 *College Chemistry: An Introductory Textbook of General Chemistry.*

1958 *No More War!*

1964 *The Architecture of Molecules.* With Roger Hayward.

1967 *The Chemical Bond.*

1970 *Vitamin C and the Common Cold.*

on ionic structures, Pauling studied the nature of the chemical bond, starting with the shared-electron chemical bond (Pauling 1928*b*). This was his major work for the next ten years. It was published in a series of memorable papers and in his 1939 book, which contains also tables of covalent, van der Waals, ionic, and other radii.

Reviews of the early work on inorganic compounds can be found for instance in W. L. Bragg and L. Pauling in Ewald (1962a), L. Pauling (1937), F. Laves (1964), G. Menzer (1964), and J. Glusker in Lima di Faria (1990).[9]

10.1.3 Organic crystals

The determination of the structure of organic crystals started more slowly than for inorganic ones, because of their greater complexity. The first important result was the determination of the structure of diamond by W. H. and W. L. Bragg (1913b). The chemists were surprised that each carbon atom was linked to four other carbon atoms and that there were no C_n complexes (Pfeiffer 1920). As in the case of sodium chloride, the notion of an isolated molecule had lost its meaning in the diamond structure. If the atoms in it are linked by shared-electron bonds, every atom shares an electron with each of its four neighbours, and this is continued indefinitely, making the entire crystal a single molecule.

The structure of diamond validated Barlow's description (see Section 12.13); it corroborated both Kekulé's[10] theory of the tetravalency of carbon and van't Hoff's[11] and Le Bel's[12] assumption of the tetrahedral character of carbon's four bonds in aliphatic compounds (see Section 12.13). In fact, the angle of 109°5 between the tetrahedral bonds of carbon in diamond is equal to that observed in chair-shaped cyclohexane. They were explained by Pauling (1928b) who found that, 'as a result of the resonance phenomenon, a tetrahedral arrangement of the four bonds of the quadrivalent carbon atom is the stable one'.

The next step was the determination of the structure of graphite (Section 8.9.2), which gave a clue to the bindings in aromatic compounds (Pfeiffer 1920). The first measurements of X-ray diffraction lines of organic crystals were made by W. H. Bragg (1921c) at University College London, on several aromatic compounds, naphthalene, anthracene, benzoic acid, naphtol, and acenaphtene. These investigations were not full determinations of the structure of these compounds, which would have been far too complex at that time. Bragg recorded the dimensions of their unit cells and the possible orientations of the molecules in the unit cell by use of an ionization spectrometer and powder diffraction. Most importantly, he showed that these crystals were molecular, and that the molecules, of the type predicted by organic chemistry, have a real existence in the crystal. The structures of anthracene and naphthalene were later fully determined by J. M. Robertson (1933a and b, respectively), also at the Royal Institution, using Fourier syntheses. The molecules were found to be planar and not boat- or chair-shaped. Figure 10.4, *Left* shows the unit cell of naphthalene, as determined by W. H. Bragg (1921c); the middle figure shows the result of the two-dimensional 1933 investigation by Robertson, and, for comparison, Fig. 10.4, *Right* shows a section of the three-dimensional electron distribution obtained in 1949 after refinement of improved experimental data of Robertson.

Only a few structures of aliphatic compounds were determined in the early 1920s, but they confirmed classic organic stereochemistry: the tetrahedral symmetry of bonding for a four-coordinated carbon atom, the possibility of free rotation of a single C-C linkage, and the immobility of C=C double bonds.

---

[9]About 116 000 structure types of inorganic crystals are by now deposited in the Inorganic Crystal Structure Database (ICSD). *Fachinformationszentrum Karlsruhe* and National Institute of Standards and Technology, <http://www.fiz-karlsruhe.de/icsd.htm>.

[10]See footnote 15 in Chapter 7.   [11]See footnote 16 in Chapter 7.   [12]See footnote 17 in Chapter 7.

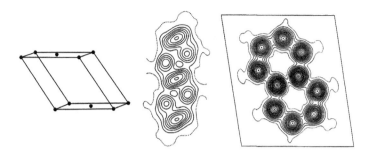

Fig. 10.4   Structure of naphthalene. *Left*: Unit cell (monoclinic), after W. H. Bragg (1921*c*). *Middle*: Projection along
c axis. After Roberston (1933*b*). *Right*: Section through the plane of the molecule, showing the electron density
distribution and the presence (low peaks) of hydrogen atoms. After Abrahams *et al.* (1949).

The first complete determination of the structure of an organic crystal is that of hexamethylene tetramine (Section 8.9.1). Another example is that of urea, which is also tetragonal,
determined by H. Mark and K. Weissenberg (1923*b*) by use of the Laue and rotating crystal
methods. The structures of these two compounds was confirmed using Fourier syntheses by
R. W. G. Wyckoff and R. B. Corey (1934). Accurate electron-density maps were reported by
Brill *et al.* (1939).

Many interesting studies were made on long-chain compounds and fatty acids (Müller and
Shearer 1923; Müller 1927; Piper 1922, 1929). Although they were not complete structure
determinations, for the lack of good crystals, they gave useful information about the geometry
and dimensions of the zig-zag carbon chains. The smectic phase of potassium oleate was shown
by de Broglie and Friedel (1923) to give a good diffraction pattern, which is not the case for
nematic phases. This supported the view that smectic phases, which are formed of layers of
molecules oriented along the normal to the layers, are ordered structures.

A very important event took place at the end of the 1920s: the first determination of the
structure of an aromatic compound, that of hexamethylbenzene by K. Lonsdale (1928, 1929*a*
and *b*) who showed that the structure of benzene consists of a strictly planar ring with six
equal bonds, instead of alternating single and double bonds, as in the Kekulé model (see
Section 8.9.3). This was a key to the structure of all aromatic compounds. W. H. Bragg (1922*a*)
had thought that benzene could not have a trigonal axis, still less a hexagonal axis!!

The number of structure determinations of organic crystals increased rapidly in the 1930s
with the development of Fourier syntheses and the Patterson vector method (Patterson 1934,
1935). The structures essentially confirmed the predictions of the chemists and provided
accurate bond-angles and bond-lengths, and a three-dimensionality that was important for
understanding the nature of the electronic distribution in bonds. It was only after the war (1918)
that the structures determined by the crystallographer began to show to the organic chemist
details of chemical structure that he did not already know (W. L. Bragg in Ewald 1962). By the
1960s the number of known crystal structures became so large that it was necessary to catalogue
them in a database. For this purpose, Olga Kennard established the Cambridge Crystallographic
Database[13] in the (then) Department of Organic, Inorganic, and Theoretical Chemistry of the

[13]<http://www.ccdc.cam.ac.uk/>.

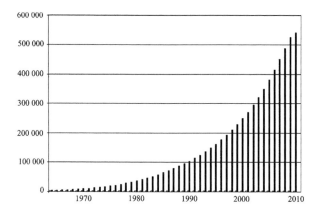

Fig. 10.5   Growth of the Cambridge Structural Database. Plotted according to the data in <http://www.ccdc.cam.ac.
uk/products/csd/statistics/stats_growth11.pdf>.

University of Cambridge (Allen 2002). The information is currently available as a computer database, the Cambridge Structural Database. It contains structural data concerning small-molecule organic or metallo-organic crystals. The number of entries reached 100 000 in 1990, 250 000 in 2000, and more than 500 000 in 2010. Its growth is represented in Fig. 10.5. The database is of primary importance to the structure chemist for studies of structure correlation and intermolecular interactions, conformational analysis, and combined crystallographic/quantum mechanical researches (Allen and Motherwell 2002).

Reviews of early work on organic crystals can be found for instance in W. H. Bragg (1929), J. M. Robertson (in Ewald 1962), R. Brill (1964), J. M. Robertson (1964), and J. Glusker in Lima di Faria (1990).

10.1.4 Metallurgy

'Prior to the age of X-ray crystallography, the nature of metals was very poorly understood' (Kasper 1964); 'it may be claimed fairly that the results of X-ray diffraction have revolu-tionized metallurgical science' (Hume-Rothery in Ewald 1962); 'it is no exaggeration to say that the principles of metal chemistry for the first time began to emerge' (with the first X-ray determinations of the structure of metals and alloys) (Bragg in Ewald 1962).

The status of physical metallurgy before X-ray diffraction can be found, for instance, in the book by the metallurgist Walter Rosenhain (1875–1934), Head of the Department of Metallurgy and Metallurgical Chemistry at the National Physical Laboratory, UK, from 1906 to 1931 (Rosenhain 1915), the last book that did not take X-ray diffraction into account (Guinier 1964). The knowledge of metals at that time was limited to what could be deduced from the observation with the optical microscope. Chemical attack of cut sections revealed a structure in grains of polygonal shape, which were 'recognized to be true crystals, the metal being thus an agglomerate of crystals, much as a mass of rock-salt or of granite is an agglomerate of minute crystals' (Rosenhain 1915). These grains are chemically equivalent but differ in orientation. The inner constituents of metals were thought to be 'molecules or groups of molecules'. For lack of single crystals there was no morphological study of metal crystals, as is routinely done with minerals; their symmetry had to be determined from that of etch pits. The different phases of an alloy were identified by the examination of polished and etched sections under the microscope, and the general principles of phase diagrams ('constitutional diagrams') describing the transformations of alloys and solid solutions with composition and temperature had been well established.

The action of strain on the polygonal grains was not very well understood. How could they become elongated while remaining crystals? At the yield point, when plastic deformation occurs, slip bands appear in ductile materials. They are 'the result of a process of sliding or slip which occurs on certain of its crystallographic planes, which is well illustrated by the behaviour of a pack of cards, or of a pile of books, when the pile is distorted the shape of each individual card or book remains unchanged but the shape of the pile is changed by the sliding of the individual cards or books over one another' (Rosenhain 1915).

The first observation of Laue diffraction spots by metals was reported by E. A. Owen and G. G. Blake (1914) at the National Physical Laboratory in the UK, and the first crystal structure, that of copper, was determined the same year by W. L. Bragg (1914*b*). The next metal crystal structures were those of silver and gold, by Vegard (1916*a* and *b*) at the University of Christiania in Norway, with a Bragg ionization spectrometer. But good single metal crystals are rare, and limited essentially to those of noble metals. It is therefore only after the development of powder diffraction that the structure determination of metallic elements and alloys became widespread. The researches on metals and alloys constituted in fact a significant fraction of the structures determined at the beginnings of X-ray crystallography. Many groups were involved, the main ones being in the USA General Electric (A. W. Hull and E. C. Bain), in Sweden the Metallographic Research Institute (A. Westgren and G. Phragmén), and in the UK, the National Physical Laboratory (E. A. Owen and G. D. Preston) and Manchester University (A. J. Bradley and J. Thewlis). A non-exhaustive list of groups active before 1930 follows:

1. North America

   - *General Electric Research Laboratory*, Schenectady, N.Y., USA: A. W. Hull, Fe (1917*a*); Al, Ni, Li, Mg, Na (1917*b*); Cr (1919*c*); Ca (1921*a*); Co, Cd, In, Ir, Mo, Pd, Pt, Rh, Ru, Ta, Zn (1921*b*); Ti, Os, Ce, Th, Zr (1921*c*); V, Ge (1922), powder diffraction. See Section 8.5.2.
   - *General Electric National Lamp Works*, Cleveland, Ohio, USA: E. C. Bain, Mn, Ag–Au, Ag–Cd, Ag–Zn, Cu–Al, Cu–Mn, Cu–Ni, Cu–Zn, Fe–Cr, Fe–Mn, Fe Mo, Fe W, Mo–W (1921*a*, *b*, 1923*a*, *b*), austenite and martensite (1921, 1922), powder diffraction; M. R. Andrews, Cu–Zn, Ni–Fe, Co–Fe (1921), powder diffraction.
   - *Research Laboratory of the American Telephone and Telegraph Company*, New York, USA: L. W. McKeehan, crystallized mercury, K, Be, Ag–Pd, Ag–Au (1922), powder diffraction.
   - *Vassar College*, Poughkeepsie, NY, USA: Slattery M. K., Te and Se (1925), powder diffraction.
   - *California Institute of Technology*, Pasadena, USA: L. Pauling, $MgSn_2$, Laue and rotation photographs (1923); J. B. Friauf, $MgZn_2$ (1926, 1927), powder diffraction and rotation photographs.
   - *University of Toronto*, Toronto, Canada: J. F. T. Young; Heusler alloys, Cu–Mn–Al (1923), powder diffraction.

2. Europe

   - *Laboratory of Physics and Physical Chemistry*, Veterinary College, Utrecht, The Netherlands: J. Bijl and Kolkmejer N. H., white and grey tin (1919), powder diffraction.

- *Kaiser-Wilhelm Institut für Faserstoffchemie*, Berlin-Dahlem, Germany: H. Mark and M. Polanyi, white tin (1923), rotating crystal method.
- *Mineralogy Institute*, University of Stockholm, Sweden: G. Aminoff and G. Phragmén, IrOs (1922), Laue diagrams; N. Alsen and G. Aminoff: crystallized mercury (1922), powder diffraction.
- *Metallographic Research Institute*, Stockholm, Sweden: A. Westgren and G. Phragmén, steel and cementite (1921, 1922, 1928), α-, β- and γ-Mn, Cu–Zn, Ag–Zn and Au–Zn Alloys (1925), Cu–Sn (1928); G. Phragmén, iron–silicon alloys (1926); with A. J. Bradley: Ag–Al (1928); with T. Negresco, iron–chromium-carbon system (1928); Ag–Cd (1928); with G. Hägg and S. Eriksson, Cu–Sb and Ag–Sb (1929), powder diffraction, Seeman-Bohlin diffractometer.
- *National Physical Laboratory*, Teddington, UK: E. A. Owen and G. D. Preston, Pb, Al–Cu, Al–Mg, Mg_2Si, AlSb, Cu–Zn alloys, α-brass, β-brass (1923*a* to *c*), ionization spectrometer; Ag–Mg, AuZn (1926), powder diffraction; Au-Sn (1927), rotating crystal; G. D. Preston: α- and β-Mn (1928), powder diffraction.
- *Physical Laboratories*, University of Manchester, UK: A. J. Bradley, As, Se, Te (1924), α-, β- and γ-Mn (1925), powder diffraction; with E. F. Ollard, Cr (1926), powder diffraction; Cu_9Al_4 (δ Cu–Al) (1928); with J. Thewlis: γ-brass, Ag–Zn, Au–Zn, Laue diagrams and rotating crystal method (1926); α-Mn (1927), Seeman-Bohlin powder diffractometer; with C. H. Gregory, Cu–Zn–Al ternary alloys (1928), powder diffraction.

The impact of X-ray diffraction on the metallurgical sciences has been essentially along two directions: atomic structure and metal chemistry on the one hand, and crystal imperfections and metal physics on the other hand.

Reviews of early work on metals and alloys can be found for instance in A. Westgren and G. Phragmén (1929), A. Westgren (1932), W. L. Bragg (1948), W. Hume-Rothery (in Ewald 1962), A. Guinier (1959, 1964), and J. S. Kasper (1964).

10.1.5 *Atomic structure and metal chemistry*

It is X-ray diffraction that showed that the constituting elements of metals are atoms, and not molecules. To quote, for instance, one of the pioneers of X-ray metallurgy, the American metallurgist Edgar C. Bain[14] (1921*a*), 'The fact is that we must regard atoms—not molecules—as the basis of the crystal structure: the bricks, if you will, that build the crystal walls of the crystal. The bricks are not alike, but fit together to form a stable whole.'

The atoms are generally of spherical shape. The cohesion between these spheres is due to the valency electrons and there are no oriented bonds. The crystal structure is therefore expected to be close packed, as predicted by Barlow and Pope (see Section 12.13). For instance, copper, silver, gold, nickel, γ-iron, and α-brass are face-centred cubic, while zinc, magnesium,

---

[14]Edgar C. Bain, born 14 September 1891 in La Rue, Ohio, USA, died 27 November 1971 in Edgeworth, PA, USA. He received his higher education at Ohio State University. He was first a chemical engineer (Goodrich Rubber Company, 1917–18), and then a metallurgist, at the General Electric National Lamp Works in Cleveland, Ohio (1918–23), where he started his X-ray work. From there he moved successively to Atlas Steel Corporation (1923–24), Union Carbide (1924–28), and finally to United States Steel Corporation, to which he remained attached all his life, becoming its Vice-President from 1943 to 1957. He was elected a member of the National Academy of Sciences in 1954. For details about Bain's scientific contributions and his life story, see H. W. Paxton and J. B. Austin, *Metall. Trans.* (1972). **3**, 1035–1042, and his Biographical Memoir by J. B. Austin, National Academy of Sciences of the USA (1978).

cobalt, zirconium, and beryllium are hexagonal close-packed and α-iron, chromium, molybdenum, vanadium, and β-brass are body-centred cubic. In some cases the structure is much more complex, such as α-manganese, with 58 atoms per cubic cell (Bradley and Thewlis 1927; Preston 1928) and γ-brass, Cu_5Zn_8, with 52 atoms per unit cell (Bradley and Thewlis 1926).

Solid solutions.   Solid solutions obtained by dissolving one element in another constitute an important class of alloys. X-ray diffraction showed that they can be either substitutional or interstitial. In the former case (substitutional) solute atoms replace the atoms of the solvent so that both occupy equivalent positions in the structure, such as copper in nickel or silicon in iron and copper. In the latter case (interstitial), smaller atoms can fit in the interstices left by the packing of the bigger spheres, for instance carbon in iron or silicon in niobium. These solid solutions follow approximately Vegard's law (see above, Section 10.1.2) and are governed by the so-called Hume-Rothery empirical rules deduced from X-ray diffraction studies on many metals and alloys in the 1920–35 period (Hume-Rothery *et al.* 1934, Hume-Rothery 1936), echoing Goldschmidt rules for inorganic mixed crystals (see Section 10.1):

1. For substitutional solid solutions, the atomic radii of the solute and solvent atoms must differ by no more than 15%, the crystal structures of solute and solvent must match, and complete solubility occurs when the solvent and solute have the same valency.
2. For interstitial solid solutions, solute atoms must be smaller than the interstitial sites in the solvent lattice, and the solute and solvent should have similar electronegativity.

E. C. Bain (1923*a*) was one of the first to propose the notion of solid solution. But that was not readily accepted. At a meeting of the American Mining and Metallurgical Engineers (AIME) in February 1924 in New York, Bain (1924) explained: 'In general, it appears that a solid solvent metal dissolves a metal solute by substituting foreign atoms in its own space lattice. Thus iron may dissolve nickel in either of its crystalline forms by permitting an atom of nickel here and there to occupy the normal position of an iron atom in the pure-iron crystal arrangement'. In the discussion that followed, one participant, Jerome Alexander, from New York, objected: 'Another difficulty that confronts us is the difficulty in deciding when we have a solid solution and when we have chemical combination. With all due respect to the X-ray spectrographic evidence, I hardly see how that is going to be conclusive, for it is pretty generally conceded that sodium chloride is a chemical compound and yet the X-ray spectrometer shows an absolutely definite crystal lattice. Are you going to consider sodium chloride a solution of chlorine and sodium?' For good measure, he added: 'Somebody went so far as to say that there is no such thing as a molecule of sodium chloride, that the whole crystal of sodium chloride is the molecule with billions upon billions of atoms of each kind in it'. This shows that, twelve years after the discovery of X-ray diffraction, minds had not yet changed!

The mechanical properties of alloys are usually better than those of pure metals, for instance, brass (Cu-Zn) and bronze (Cu-Sn) are harder than copper. For that reason, as shown by the list given above, the structures of many different binary systems were determined in the early days of X-ray crystallography. When an alloy is obtained by dissolving one element in another, several phases are usually formed. The phase diagram of a binary alloy has regions of single phase of which the composition can be varied continuously and regions where two phases coexist. W. L. Bragg and E. J. Williams (1934) explained: 'This feature of an alloy is in contrast to the constant atomic ratio of a chemical compound, and is explained by the nature of the binding forces in an alloy which are predominantly those between the metal atoms of both kinds on the one hand and the common electronic system on the other hand, as opposed to the binding forces between atom and atom which predominate in other chemical compounds'.

Thanks to X-ray diffraction, these phase diagrams could now be established with a much higher accuracy than could be done with the optical microscope, which did not provide the structure of each phase. One of the first systems to be studied was the iron-carbon system. Two groups, led respectively by A. Westgren[15] at the Metallographic Research Institute in Stockholm, Sweden, and by E. C. Bain in the United States, independently built high-temperature stages for *in situ* recording of powder diffraction diagrams. Arne Westgren had contacted M. Siegbahn in 1919, asking whether the structures of the phases of iron could be determined by means of powder diffraction. He was invited to Lund, where he made his first investigations with a student of Siegbahn, A. E. Lindh (Westgren and Lindh 1921). In this work an electric current was used to heat an iron wire in an ordinary Debye–Scherrer camera. Westgren then built a high-temperature camera in Stockholm, where he worked in collaboration with Gösta Phragmén who had learned X-ray crystallography with Gregori Aminoff.[16] They made observations at 800° (β-iron), 1100° (γ-iron) and 1425° (γ-iron + δ-iron). On the other side of the Atlantic, Edgar Bain visited A. Hull in Schenectady in 1920 to explore the possible usefulness of X-rays in solving metallographic problems. He subsequently built his own powder diffraction apparatus in Cleveland. The two groups determined the structures of α- and β-iron (body-centred) and of austenite, γ-iron, in the 910°–1390° temperature range, but Westgren and Phragmén (1922a, b) beat Bain (1922, 1924) to publication by a few months. These authors recognized the role of interstitial carbon for the properties of steel, and understood the nature of the martensitic transformation. Carbon is much more soluble in γ-iron than in α- or δ-iron. On slow cooling, austenite transforms into α-iron and the carbon atoms precipitate to form cementite, Fe_3C, while on rapid cooling martensitic transformation takes place. This transformation is too rapid for cementite to be formed, and a metastable phase, martensite, appears, which is a supersaturated solution of carbon in α-iron, with the carbon atoms in octahedral interstices.

Another important phase diagram is that of the copper–zinc system, brass. A preliminary study of the phases of brass was made by E. C. Bain (1921a, 1923b) and his co-worker M. R. Andrews (1921), identifying the Bravais lattices of α- and β-brass. Bain surmised that brass was a substitutional solid solution rather than an interstitial one. This was proved by E. A. Owen and G. D. Preston (1923b) at the National Physical Laboratory, UK, who determined

[15]Arne Westgren, born 11 July 1889 in Årjäng, Sweden, died 07 March 1975 in Stockholm, Sweden, was a Swedish physicist and metallurgist. After his studies at Uppsala University, he obtained his doctorate's degree in 1915 with a thesis on Brownian motion. In 1918, he became a physical metallurgist with the SKF ball-bearing factory in Gothenburg. Following his first successes in determining the structure of the high-temperature phases of iron, he was appointed in 1921 Assistant Professor of metallurgy at the University of Stockholm, and, from 1927 until his retirement in 1943, full Professor. He was Secretary of the Nobel Committee for physics and chemistry from 1926 to 1943, and Vice President of the International Union of Crystallography during the period 1948–51. For details about his life and his career, see his personal reminiscences in Ewald (1962) and his obituary by G. Hägg (*Acta Crystallogr.* (1976) **A32**, 172–173).

[16]Gregori Aminoff, born 8 February 1883 in Stockholm, Sweden, died 11 February 1947 in Stockholm, Sweden, was a Swedish artist and mineralogist. After receiving his higher education in Uppsala, he turned towards the fine arts at the age of 22 and studied music and painting in Paris and Italy, but turned back to science after ten years. He obtained his PhD in 1918 and became Lecturer in Mineralogy and Crystallography at the University of Stockholm. In 1923 he became Director of the Mineralogy Department of the Museum of Natural History in Stockholm.

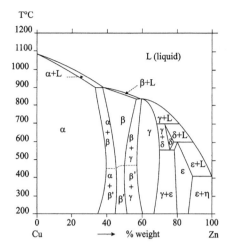

T°C

Fig. 10.6 Phase diagram of brass.

the complete phase diagram of brass, represented in Fig. 10.6. Up to a composition of 30–35% zinc, the material is a substitutional solid solution, α-brass, which is face-centred cubic and fairly malleable. At higher zinc concentrations several types of crystals are found. Initially, as the percentage of zinc in the brass increases, the material is the β-phase, which has the CsCl structure and is harder and stronger; β-brass is disordered at high temperatures, ordered at lower temperatures. Then, as the zinc concentration is further increased, the complex cubic γ phase and the hexagonal ϵ phase which differs from the hexagonal zinc structure appears. E. A. Owen and G. D. Preston (1923a) also showed the substitutional nature of the solid solutions copper–aluminium, aluminium–magnesium, and copper–nickel.

The structure of γ-brass, Cu_5Zn_8, was determined by A. J. Bradley and J. Thewlis (1926). Albert James Bradley[17] was W. L. Bragg's first research student at Victoria University in Manchester, and he followed him all his life in all his successive appointments. Although Bragg, personally, was of course a specialist of single crystal work, he recognized the virtue of powder diffraction, and he encouraged Bradley to work with powder diffraction. Bradley gave once a talk at the Royal Institution in London on the structure of Cu_5Zn_8, which stimulated Bernal to try his hand with metal structures. Bernal thought that the rotating crystal method would provide better results than the powder method, and set out to determine the structure of δ-bronze, Cu_4Sn. In the course of his study an incident occurred that he always remembered. He 'was trying, by a flotation method, to measure to several points of decimals the density of a very small single crystal of delta bronze. It involved a great number of weighings and measurings and on the very last weighing on which all the rest depended, the little crystal slipped out of my fingers

---

[17]Albert James Bradley, born 5 January 1899 in Chesterfield, UK, died 4 September 1972 in the same locality. He started his higher education in Manchester University in 1916, but left it after one year to serve during the war. He returned after the war and graduated in 1921 in chemistry. Having already learned crystallography from H. Miers' lectures, he joined W. L. Bragg's laboratory in 1922 as a research assistant, and was awarded his PhD in 1924. He then had successive research fellowships, always in Bragg's laboratory, in Manchester University, and in 1937 at the National Physical Laboratory. In 1938 he was appointed Assistant Director of Research at the Cavendish Laboratory in Cambridge. He was elected a Fellow of the Royal Society in 1939.

and disappeared. I never found it again' (Bernal, in Ewald 1962a). For K. Lonsdale, 'her most joyous recollection of that period was of J. D. Bernal on his knees looking for his one and only crystal' (Lonsdale, in Ewald 1962a). Bernal did nevertheless determine the structure, and declared that it was much more complicated than metallurgists thought. He showed that δ-bronze had a very large unit cell with 328 atoms of copper and 88 atoms of tin; he suggested that its formula was in fact $Cu_{41}Sn_{11}$ (Bernal 1928). This work led Bernal to discuss the notion of the 'metallic state' and to conclude that this term covered substances bound together not only by metallic but also by homopolar and ionic forces (Bernal 1929).

Bradley[18] determined the structures of selenium, tellurium and arsenic for his doctorate. He next studied the phases of manganese, also by powder diffraction, identifying three allotropes, α, obtained electrolytically, and β and γ present in commercial manganese (Bradley 1925). Similar results were obtained at the same time by A. Westgren and G. Phragmén (1925b). Bradley eventually solved the structure of γ-manganese (Bradley and Thewlis 1927), with fifty-eight atoms in the unit cell, a remarkable achievement for a structure determination with powder diffraction. In 1926, Bragg sent Bradley for one year to study in Westgren's laboratory in Sweden in order to learn their techniques. There, he analysed the silver-aluminium alloys, identifying two phases, one cubic, Ag_3Al, with twenty atoms, isomorphous with β-manganese, and another one, hexagonal close-packed containing from twenty-seven to forty atomic % aluminium (Westgren and Bradley 1928). Westgren and Phragmén (1925a) had, at that time, analysed the Cu–Zn, Ag–Zn and Au–Zn alloys and had found five different types of crystal structure in the Cu–Zn alloys. They had solved four of them but not that of γ-brass. They had obtained a few single crystals of this form of brass and worked out the unit cell, using Laue diagrams and rotation photographs. They allowed Bradley to take the data back to Manchester with him. There, he solved the structure. The crystal is cubic, space group $I\bar{4}3m$ and fifty-two atoms per unit cell. Bradley also showed that the exact formulae of the alloys studied by Westgren and Phragmén were Cu_5Zn_8, Ag_5Zn_8, and Au_5Zn_8, and that the crystal structures of the γ-phases of these alloys were isomorphous.

The existence of phases of definite composition such as Cu–Au and Cu_3–Au was explained by assuming that the free energy of each phase varies continuously and presents a minimum for the particular composition. It was quite intriguing to observe that these compositions were in general very simple, for instance the β-solid solutions are body-centred cubic for the alloys CuZn, Cu_3Al, Cu_5Sn. Hume-Rothery (1926) noted that they all correspond to an electron/atom ratio, or 'electron concentration', of three valency electrons to two atoms (3/2). This was generalized by Westgren and Bradley. The γ-alloy structure type (Cu_5Zn_8, Cu_9Al_4, $Cu_{31}Sn_5$) occurs for a ratio 21/13 and hexagonal structures occur for a ratio 7/4. The corresponding compounds are called 'electron compounds' (Hume-Rothery in Ewald 1962). This so-called 'Hume-Rothery law' (see earlier) suffers many approximations. It was later justified quantum-mechanically.[19]

---

[18]For details on Bradley's life and contributions, see his obituary by H. Lipson, *Bibliographical Memoirs of Fellows of the Royal Society* (1973), 116–128.

[19]See, for instance, N. F. Mott and H. Jones (1936). *Theory of the Properties of Metals and Alloys*. Clarendon Press, Oxford.

Order–disorder transformations.   It was at first believed that in a binary alloy the two kinds of atoms were distributed at random according to the stoichiometry amongst the sites of the structure. This would be normal if the two types of atoms were identical, but if they are not, one should expect that they segregate in particular sites to form an orderly arrangement. The German chemist G. Tammann (1918) predicted that this should happen in alloys such as Cu–Au, Ag–Au, Pd–Au, and Pd–Ag subjected to long annealing. If the alloy is ordered, the unit cell is larger than in the disordered phase, and extra lines should appear in the diffraction diagram, as is described in the next paragraph. This phenomenon was first observed by Bain (1923*b*) with Cu_3Au, C. H. Johansson and J. O. Linde (1925) with Au–Cu and Pd–Cu, and G. Phragmén (1925) with Fe_3Si. Very soon many other cases were found.

Let us take the order–disorder transformations of the alloys Cu–Zn (β-brass), Cu_3Au, and Cu–Au as examples. In the disordered phase, where the copper and zinc atoms are randomly distributed, Cu–Zn is body-centred cubic. In the ordered phase, the atoms of one type (say, zinc) occupy the nodes of a simple cubic lattice and atoms of the other type (say, copper) lie at the centres of the unit cells, as in the CsCl structure (Fig. 10.7, *Left*). The common alloy Fe–Al presents the same transition as β-brass. The disordered phases of Cu_3Au and Cu–Au are both face-centred cubic. In the ordered structures, the copper and gold atoms occupy respectively the positions represented by black and open circles in Fig. 10.7. In Cu_3Au the gold atoms occupy the nodes of a simple cubic lattice, while the copper atoms occupy sites at the centre of the faces (Fig. 10.7, *Middle*). In Cu–Au the gold atoms also occupy the nodes of a simple cubic lattice, but the copper atoms occupy one third only of the positions at the centre of the faces, so that the structure is constituted of a succession of layers of gold and copper atoms and is tetragonal (Fig. 10.7, *Right*).

In disordered phases the scattering factor to be taken into account to calculate the diffracted intensities is an average over the scattering factors of the two components of the alloy (Laue 1918). In the ordered phase of the three examples given above, reflections such as 100, which are forbidden in the disordered phase, appear in the diffraction diagram and are called 'super-structure' lines. The structure of the ordered phase is therefore referred to as a 'superstructure' or 'superlattice'. The intensity of such reflections is the strongest, the larger the difference between the scattering factors of the two atoms.

The first theory of an order–disorder transition is due to Bragg and Williams (1934). They postulate that an alloy is in a state of dynamic equilibrium. On one hand, the atoms tend to segregate in the positions corresponding to a minimum potential energy while, on the other hand, thermal agitation has an opposing effect, creating a random arrangement. At a temperature

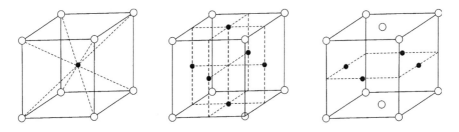

Fig. 10.7   Ordered structures. *Left*: Cu–Zn. *Middle*: Cu_3Au. *Right*: CuAu. Black circles: Cu. Open circles: Zn or Au.

higher than a certain critical value, $460°$ for CuZn, $390°$ for Cu_3Au and $410°$ for CuAu, the structure is disordered, and it is ordered below that temperature. The transition is a second-order transition, according to modern classification. The disordered phase can usually be frozen in by quenching rapidly, and the ordered structure will be formed after a long annealing of the quenched alloy.

Reviews of the order–disorder transformations and of their influence on the properties of alloys are given, for instance, in Nix and Shockley (1938) and Lipson (1950).

Intermetallic compounds.   Intermetallic compounds are another class of alloys. Their study is, in general, rather difficult. The compounds Mg_2Si and AlSb are among the first ones to have been studied. E. A. Owen and G. D. Preston (1923c) showed that Mg_2Si consists of a face-centred cubic lattice of silicon atoms intermeshed with a simple cubic lattice of magnesium atoms, and AlSb of a face-centred cubic lattice of antimony intermeshed with an identical lattice of aluminium atoms. More complex structures such as the so-called 'Friauf phases', Cu_2Mg and Zn_2Mg, are particularly interesting (Friauf 1927a and b). In cubic Cu_2Mg, the larger magnesium atoms have the diamond structure and the smaller copper atoms occupy the summits of corner-sharing tetrahedra. The Zn_2Mg structure is a hexagonal variant.

The study of age-hardening of some aluminium alloys furnishes one of the most striking examples of the power of X-rays in providing insight into industrial processes. These alloys, such as aluminium-copper, are relatively soft when rapidly cooled, but gradually harden on standing at room temperature. This is due to the precipitation of an intermetallic compound, $CuAl_2$, but the interesting fact is that hardening begins before any visible precipitates can be detected. Rosenhain's successor at the head of the Metallurgy Department of the National Physics Laboratory in England, C. Desch, predicted that this should be due to the formation of 'minute nuclei of only a few molecules, producing local distortion of the space lattice, followed by the growth of these nuclei'.[20] These 'minute nuclei' were detected at the same time, independently, by G. D. Preston[21] (1938a) at NPL, and by A. Guinier[22] (1938a) in the research laboratory of the Ecole Normale Supérieure in Paris. Their material was a 4% copper aluminium alloy for Preston and a 5% one for Guinier. Preston observed streaks accompanying diffraction spots on both oscillation and Laue photographs, which he interpreted as due to 'segregation of copper atoms on (100) planes' (Preston 1938a), while Guinier observed diffuse streaks radiating

[20]Quoted by Hardouin-Duparc (2010).

[21]George Dawson Preston, born 8 August 1896 in England, died 22 June 1972 in Scotland, was a British physicist. He received his secondary education at Oundle School until 1914 when he joined the army. He was discharged after a severe leg injury. He then studied at Gonville and Caius College in Cambridge, UK. After his graduation in 1921 he joined the National Physical Laboratory in Teddington, UK, in Rosenhain's Department of Metallurgy and Metallurgical Chemistry, where he worked with E. Owen on X-ray diffraction. In 1943 he was appointed Professor of Physics at the University College, Dundee. For details on Preston's life and work, see Hardouin-Duparc (2010).

[22]André Guinier, born 1 August 1911 in Nancy, France, died 3 July 2000 in Paris, was a French physicist. After studying at the Ecole Normale Supérieure from 1930 to 1934, he prepared his thesis with C. Mauguin at the University of Paris. He obtained his doctorate in 1939 on X-ray diffraction at very low angles. He was then appointed at the Conservatoire National des Arts et Métiers (CNAM). In 1949 he became Professor at the University of Paris and in 1959 founded the Solid State Physics Laboratory with Jaques Friedel and Raymond Castaing. He was elected a member of the French Academy of Sciences in 1971 and was President of the International Union of Cristallography from 1969 to 1972. For details about his life and work, see Hardouin-Duparc (2010) and M. Lambert (2001). *Acta Crystallogr.* **A57**, 1–3.

from the centre of the Laue diagram. He could analyse them and proposed that they were due to 'groups' of copper atoms localized in disk-shaped regions about 100–400 Å in diameter and 3–4 Å thick, parallel to the (100) planes of the aluminium (Guinier 1938a). Eventually, they both published a short paper, on successive pages of the 24 September 1938 issue of *Nature* (Guinier 1938b, Preston 1938b). These regions are now known as *Guinier–Preston zones* or G–P zones. They are also found in many other systems, such as aluminium–silver, aluminium–zinc, aluminium–magnesium, and others. The inside story of the discovery is told by Hardouin-Duparc (2010). Guinier's observation was the first application of the small-angle X-ray scattering method, which he had invented, and which was to have world-wide success.

10.1.6   Crystal imperfections and metal physics

The very first observations of the influence of cold-rolling and annealing on the diffraction of X-rays by metal sheets were made very shortly after the discovery of X-ray diffraction, revealed by anomalies such as halos and star-shaped figures (asterisms) in Laue diagrams (Hupka 1913a; Keene 1913b, c; Knipping 1913, de Broglie 1914b) but it is only after the development of the powder diffraction techniques that such studies became systematic. Some of the first studies of plastic deformation with powder diffraction, were done by Bain and Jeffries (1921), and Mark *et al.* (1923).

Metal crystals are usually highly imperfect. The actions of mechanical and heat treatments result in modifications of the size of the grains, in strain or faulting within the grains, and of their reorientation along preferred directions. Powder diffraction has proven to be the most suitable technique of analysis for the study of the deformations of metals and alloys. The information available from the powder diagrams is the width, the shape, and the intensity of the diffraction lines, as well as their variation with the indices of the reflection.

The width of the diffraction peaks is directly related to the particle size by Scherrer's formula (8.19), modified by N. Seljakov (1925), and generalized by M. von Laue (1926) to take into account both the shape and the size of the crystallites (see Section 8.5.1). R. Brill[23] (1928) discussed Scherrer's and Laue's formulae and applied them to the determination of the particle size of iron powders obtained by various techniques. Laue's expression was further generalized by A. L. Patterson (1928) who considered a Maxwellian distribution of sizes for the particles. These derivations were based on a number of simplifying hypotheses, and an exact derivation of the Scherrer formula was given later by Patterson (1939) for particles of spherical shape, using Fourier analysis. A more general treatment, taking distortions within the grains into account, was given by A. R. Stokes and A. J. C. Wilson (1942, 1944).

Many studies have been devoted to the deformation of metals and alloys, and only the first detailed analyses performed in the 1920s will be mentioned here. Pioneer work was performed by M. Polanyi and his co-workers at the Institute of Fibre Chemistry in Berlin-Dahlem (Section 8.6.1). They observed that the Debye–Scherrer diagram of a copper wire

---

[23]Rudolf Brill, born 7 September 1899, died in 1989, was a German chemist. After graduating from the University of Berlin under R. O. Herzog with a work on the constitution of silk fibroin, Brill joined the Forschungslaboratorium Oppau (Ludwigshafen) der I.G. Farbenindustrie in 1923. In 1941, he was appointed Professor of Inorganic and Physical Chemistry at the Technische Hochschule in Darmstadt. After the war, he spent several years in the United States. He came back to Germany in 1958 as Director of the Fritz-Haber-Institut in Berlin-Dahlem, where he remained until his retirement in 1969.

transformed into a fiber diagram after cold-drawing (Becker *et al.* 1921, Polanyi 1922). By interpreting the fiber diagram of zinc samples, they could determine the orientation of the glide planes of zinc crystals (Mark and Polanyi 1923), and make a complete analysis of its plastic flow (Mark *et al.* 1923). Most of the first studies were carried out in industrial research laboratories, in particular on the properties of tungsten wires: Philips in Eindhoven (A. E. Van Arkel 1925), The Netherlands; General Elelectric, Wembley in the UK (F. S. Goucher 1926); Osram-Konzern in Berlin (now Osram GmbH in Munich), Germany (K. Becker 1927); and I. G. Farbenindustrie Ludwigshaffen, Germany (R. Brill 1928).

In general, the experimental set-ups for these studies were designed so that there could be a high enough angular resolution for the $K\alpha_1$ and $K\alpha_2$ lines of copper radiation, so that they could be well separated. Van Arkel (1925) observed that the widths of these lines increased with cold work of tungsten wires up to the point where the doublet could not be resolved any longer. On annealing, the lines become narrower and can be resolved again. Van Arkel estimated that the merging of the two lines of the $K\alpha$ doublet corresponded to a lattice deformation of about 0.2 %. He observed comparable effects with iron and platinum, but not with soft metals such as lead or copper. Similar studies were made by Becker (1927), a former co-worker of Polanyi at the Institute of Fibre Chemistry in Berlin-Dahlem, who also observed discontinuous Debye–Scherrer lines in recrystallized tungsten wires after annealing. Figure 10.8 compares the powder diagrams of a tungsten wire, a) after cold drawing, b) after annealing, c) after recrystallization. In Fig. 10.8 a), the two $K\alpha_1$ and $K\alpha_2$ lines of the 400 line are merged, while they are resolved in Fig. 10.8 b). In both cases, distinct signs of preferred orientation can be observed, revealed by the presence of maxima and minima in the intensity of each line. In Fig. 10.8 c) the lines are broken up in separate dots corresponding to larger individual crystallites.

U. Dehlinger (1927) at the Technical University in Stuttgart, Germany, studied the effect of successive cold-rolling and annealing on copper, silver, aluminium, tantalum, zinc, and brass. He used photometric analysis of the blackening of the $K\alpha_1$, $K\alpha_2$ doublet to study the variation

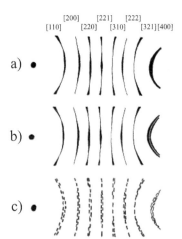

Fig. 10.8   Powder diagrams of tungsten wires. CuKα. a) Cold-drawn. b) Annealed. c) Recrystallized. Sketches redrawn after Becker (1927).

of the line broadening with degree of cold-rolling. He made various hypotheses concerning the structure of the grains in the cold-rolled material and, using Laue's formula, was able to distinguish the broadening due to the small size of the grains from that due to their distortions. He showed that the relative importance of the latter increases with indices of the reflection. Van Arkel and Burgers (1928) also used a photometric technique to measure the variations of the broadening of the 321 and 400 reflections in terms of the temperature and duration of the annealing, and concluded that the origin of the broadening is essentially due to the lattice distortions.

The quantitative proof of the overwhelming importance of broadening of reflections as a result of distortions had to wait for the treatment by B. E. Warren and B. L. Averbach (1950, 1952). They showed that the shape of the diffraction can be expressed as a Fourier series, the coefficients of which are the products of two terms, one due to distortion broadening and the other to particle size broadening. Further analysis (Patterson 1952; Warren and Warekois 1955) showed that if the distortions due to cold work of face-centred cubic metals are stacking faults, there is not only a broadening, but also a shift of the diffraction peaks.

A number of authors have also tried to correlate the intensity of the diffraction peaks with the mechanical and heat treatments of metals. For instance, J. Hengstenberg and H. Mark (1929, 1930) at I. G. Farbenindustrie in Ludwigshafen, Germany, measured the ratio of the intensities of the 200 and 400 reflections from tantalum, tungsten, and molybdenum after cold-rolling and after annealing, and found that it was systematically smaller after annealing. This is to be associated with a higher extinction effect in the annealed crystals, but no serious conclusions could be deduced about the inner structure of the metals.

10.2 Mineralogical sciences

Up to 1912, crystallography had been derived from mineralogy. It is the study of minerals that led Haüy, Mitscherlich, and Groth to formulate progressively the relationships between chemical composition, crystal habit, and crystal structure. The first structures to be determined were those of minerals. After the first few years, X-ray crystallography grew independently of mineralogy. But X-ray diffraction, in particular powder diffraction, made major contributions to our knowledge of the structure of minerals and to our understanding of the natural processes of their formation. W. L. Bragg (1949) noted: 'One interesting feature of this new work was the complete justification of the mineralogist in basing his classification on crystalline form and not on chemical constitution'.

X-ray diffraction is an ideal tool for the identification of phases, even if they are micro-crystalline or badly crystallized. For example, P. F. Kerr at Stanford, California, wrote a thesis in 1924 on *The identification of opaque ore minerals by X-ray powder diffraction patterns*.[24] With X-ray diffraction, it is possible to show that two substances which were thought to be different have, in fact, the same structure and, conversely, that two substances which were thought to be identical are, in fact, different from the structural viewpoint. Many minerals form solid solutions, for example, at high temperatures, Na^+ and K^+ readily substitute for each other in the alkali feldspar minerals. The composition of any feldspar then lies between that of

[24]*Econ. Geol.* (1924). **19**, 1–34.

each of the end members of the series, albite ($NaAlSi_3O_8$) and microcline ($KAlSi_3O_8$). Their lattice parameters vary approximately according to Vegard's law (see Section 10.1.2), and the composition of a given mineral can be deduced from the measurement of its lattice parameters by X-ray diffraction.

Historically, one may distinguish two stages in the development of structural mineralogy. In the first stage, which culminated in the classification of feldspars by W. H. Taylor (1933), the main structure types were determined. Minerals are, in general, far from perfect, and after 1930 the emphasis shifted to a different stage, the study of crystal imperfections, which are usually related to the growth conditions of the minerals.

Lattice defects such as inclusions, point defects, growth horizons, stacking faults, or twins are present in most minerals. An early example is given by the study of iron sulphides, FeS. Laves, in his thesis prepared under the supervision of P. Niggli in Zürich (1929), showed that the iron sulphides FeS have a composition $Fe_{n-1}S_n$, as a result of the presence of vacancies (Laves in Ewald 1962a). Mineral structures are often disordered. For instance, in the structure of spinel ($MgAl_2O_4$), determined by W. H. Bragg (1915a, b) and S. Nishikawa (1915), there are eight equivalent positions occupied by magnesium atoms and sixteen occupied by aluminium atoms. F. Machatschki in Tübingen, however, using powder diffraction data, showed that the twenty-four atoms are distributed with some randomness over the two sets of positions (Machatschki 1931). Diffuse scattering is a very useful tool for the study of defects, but it was not used much in the first years. An early example was given by Mauguin (1928b), who observed diffuse streaks in rotation-crystal photographs of chlorites (phyllosilicate minerals), corresponding to missing rows in the reciprocal lattice.

The first structures of minerals to be determined were:

- **1913**: rock-salt, sylvine, zinc-blende by W. L. Bragg (1913b) and of diamond by W. H. and W. L. Bragg (1913a, b) in Cambridge and Leeds, with the ionization spectrometer and Laue diagrams.
- **1914**: fluorite, pyrite, and the trigonal carbonates by W. L. Bragg (1914a) in Cambridge with the ionization spectrometer.
- **1915**: magnetite (Fe_3O_4) and spinel ($MgAl_2O_4$) by W. H. Bragg (1915a, b) in Leeds with the ionization spectrometer, and, independently, by S. Nishikawa (1915) in Tokyo with Laue diagrams.
- **1916**: rutile and anatase (TiO_2), cassiterite (SnO_2), zircon ($ZrSiO_4$), and xenotime (YPO_4) by L. Vegard (1916b) with the ionization spectrometer in Oslo.
- **1917**: graphite by P. Debye and P. Scherrer (1917) with powder diffraction in Zürich and by A. W. Hull (1917b) with powder diffraction in Schnectady, USA; chalcopyrite by C. L. Burdick and J. H. Ellis (1917) with the ionization spectrometer in Pasadena, USA; rutile (TiO_2) and cassiterite (SnO_2) by C. M. Williams (1917), independently of L. Vegard, with the ionization spectrometer in Cardiff, Wales.
- **1919**: isostructural brucite ($Mg(OH)_2$) and pyrochroite ($Mn(OH)_2$) by G. Aminoff in Stockholm, by means of Laue diagrams.[25]
- **1920**: zincite (ZnO) by W. L. Bragg (1920a) in Cambridge with the ionization spectrometer; wulfenite ($PbMoO_4$) and scheelite ($CaWO_4$) by R. G. Dickinson (1920) in Pasadena, USA, with the ionization spectrometer.
- **1922**: zincite (ZnO) by G. Aminoff (1922) in Stockholm with Laue diagrams; ice, by W. H. Bragg (1922b) at University College, London, with the ionization spectrometer.

[25]In *Geol. För. Förh.* (1919), **41**.

- **1923**: wurtzite (ZnS) and nickeline (NiAs) by G. Aminoff (1923) in Stockholm with powder diffraction.
- **1924**: aragonite ($CaCO_3$), by W. L. Bragg; corundum (Al_2O_3), hematite (Fe_2O_3) and calomel (Hg_2Cl_2) by C. Mauguin (1924a and b, respectively) in Paris with the ionization spectrometer.
- **1925**: preliminary structure of garnet by G. Menzer in Berlin with powder diffraction.[26]
- **1926**: chrysoberyl ($BeAl_2O_4$) by W. L. Bragg and G. B. Brown (1926a) in Manchester with the ionization spectrometer.
- **1928**: brookite (TiO_2) by L. Pauling and J. H. Sturdivant (1928) in Pasadena, USA, with the rotating crystal method.

The major contribution of X-ray diffraction to structural mineralogy was, however, the determination of the crystal structure of silicates. Initially, the structures of quartz and its polymorphs were determined. A preliminary study of quartz had been made by W. H. Bragg (1914d, Bragg and Bragg 1915), as shown in Fig. 10.12, *Left*, but the structures of α- and β-quartz were fully determined by W. H. Bragg and R. E. Gibbs (1925) and Gibbs (1926a). The structure of cristobalite was determined by Wyckoff (1925), and that of tridymite by Gibbs (1926b).

The main actors in the determination of silicate structures were:

- W. L. Bragg and his co-workers: olivine (W. L. Bragg and G. B. Brown 1926b), beryl (W. L. Bragg and J. West 1926), zircon (W. Binks 1926), phenacite (W. L. Bragg 1927), monticellite (G. B. Brown and J. West 1927), topaz (N. A. Alston and J. West 1928), cyanite (W. H. Taylor and W. W. Jackson 1928), sillimanite (W. H. Taylor 1928), andalusite (W. H. Taylor 1929), amphibole (B. E. Warren 1929), diopside (B. E. Warren and W. L. Bragg 1929), tremolite (B. E. Warren 1929), benitoite and titanite (W. H. Zachariasen 1930), chrysotile (B. E. Warren and W. L. Bragg 1930), feldspars (W. H. Taylor 1933), melillite (Warren 1930), muscovite (Jackson W. W. and J. West 1930), willemite (W. L. Bragg and W. H. Zachariasen 1930).
- F. Machatschki (1928), feldspars.
- C. Mauguin (1927, 1928a), micas; (1928b), chlorites.
- G. Menzer (1925, 1926, 1929), garnets.
- L. Pauling (1928c) topaz; (1930a) mica, talc and pyrophillite; (1930b) zeolites.
- E. Schiebold (1929a, b), feldspars.

Chemists and mineralogists had postulated the existence of a number of silicic acid radicals with different Si/O ratios in the silicates, but X-ray diffraction showed that their structure could be described in terms of groupings of SiO_4 tetrahedra: isolated as in olivine (Bragg and Brown 1926b), in chains as in pyroxenes such as diopside (Warren and Bragg 1929) and amphiboles (Warren 1929), in rings as in beryl (Bragg and West 1926, see Fig.10.9), in sheets as in mica (Mauguin 1927, 1928a, Jackson and West 1928, Pauling 1930a) or in frameworks as in feldspars (Machatschki 1928, Schiebold 1929a, b, Taylor 1933), zeolites (Pauling 1930b), and quartz. W. L. Bragg (1943) commented: 'The general scheme of all the families of silicates turns out to be beautifully simple, and at the same time quite different from anything pictured in textbooks before the advent of X-ray analysis'. The classification of silicates was established by W. L. Bragg (1930, 1937) on the basis of these results; it was also discussed by F. Laves (1932).

---

[26]G. Menzer (1925). Die Kristallstruktur von Granat. *Zentralbl. Miner. A*, 344–345.

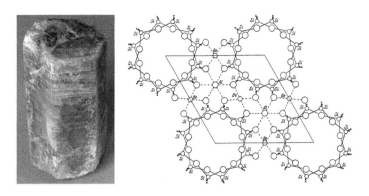

Fig. 10.9   Beryl. *Left*: crystal of beryl. Photo by the author. *Right*: Structure of beryl. Beryl is a cyclosilicate, of space-group $P6/mmc$. After W. L. Bragg (1930), reproduced with permission from Oldenburg Wissenschaftsverlag.

Many groups took part in the study of silicates. F. Machatschki in Tübingen, Germany, attached a great importance to the linking of the tetrahedral groups, and showed that aluminium can replace silicon in silicate structures, thus opening the way for the determination of the structures of feldspars (Machatschki 1928). Pauling's rules about the balancing of electrostatic valences played an important role in the description of ionic compounds in general, and silicates in particular, and the structures of the silicates provide a striking illustration of the application of Goldschmidt's and Pauling's rules. The main contribution to our understanding of the structure of silicates is, however, due to the Manchester group who studied more than fifteen different minerals.

W. L. Bragg (Fig. 10.10) has told how he came to study silicates: 'After the 1914–18 war we developed more powerful forms of X-ray analysis in my research school at Manchester, and cast round for subjects on which to try our new methods. I chose the silicates because fine crystals of them were available and because they were a good deal more complex than anything we had yet attempted' ... 'I always regard this as one of the most exciting and aesthetically satisfying researches with which I have been associated' (Bragg 1949).

The successive steps of the study can be found in Bragg and West (1927), Bragg (1929c), and Bragg (1930). But the turning point was, according to W. L. Bragg (in Ewald 1962), the solution of diopside $CaMg(SiO_3)_2$ by Warren and Bragg (1929). Diopside is monoclinic with space group $C2/c$ and four molecular groups per unit cell. The tetrahedral SiO_4 groups are linked by the corners to form endless chains of composition SiO_4 parallel to the c axis. They lay side by side and are held together by the calcium and magnesium atoms. The structure, which is characterized by fourteen parameters, was solved using the analysis scheme introduced by Bragg and West (1929) to determine structures with many parameters.

B. E. Warren's[27] first contact with crystal physics was when M. Born gave a set of lectures on lattice dynamics during a visit to M.I.T. (Warren in Ewald 1962a). Some time later, Warren had

---

[27]Bertram Eugene Warren, born 28 June 1902 in Waltham, MA, USA, died 27 June 1991 in Arlington, MA, USA, was an American crystallographer whose X-ray studies contributed to an understanding of both crystalline and non-crystalline materials. He studied at M.I.T. from 1919 to 1925 and 1927 to 1929. In 1926 he spent one semester at the Technische Hochschule in Stuttgart. He received his ScD in 1929, after which he spent one year with W. L. Bragg.

Fig. 10.10   Sir William Lawrence Bragg. After Bragg (1970). © International Union of Crystallography.

the opportunity of spending a semester in Stuttgart, Germany, where he attended R. Glocker's course on X-ray diffraction and did some theoretical work with P. P. Ewald. Back home in the USA, Warren had the chance that W. L. Bragg spent four months at M.I.T. in 1928 as visiting Professor. Bragg asked Warren to build models of crystal structures for the course he was giving: 'I was lucky enough to win the opportunity to be Professor Bragg's chore boy in charge of building models in a hurry' (Warren in Ewald 1962*a*). In addition to building models for Bragg, Warren had another opportunity. Bragg had brought with him the data that he and J. West had measured on diopside, and he asked Warren to solve the structure. Bragg remembered later, 'I was fortunate to find a clever research student willing to apply himself to the problem and he found the solution. I feel proud that I enticed Professor B. E. Warren to take up X ray analysis' (Bragg 1949). The chain structure of diopside was 'completely unexpected' (Warren in Ewald 1962). Warren then spent a year at Manchester where he solved silicate structures, in particular those of amphibole and chrysotile. After his stay in England, he continued for a while to work on complex silicate structures, but soon shifted to the study of glasses and amorphous substances. Warren was a remarkable teacher besides being an excellent researcher, and he soon attracted many students. His teaching had a profound impact on his students who had both affection and respect for him. Former members of the 'Warren School' had the feeling of belonging to a common fraternity who stuck together wherever they met,[28] a feeling shared by the present author who had the privilege of spending one year in B. E. Warren's laboratory (1955–56).

In 1930 he returned to M.I.T., where he served successively as Assistant, Associate, and, finally, full Professor of Physics. He was Vice-President of the International Union of Crystallography from 1966 to 1972.

[28]L. Muldawer, *The Warren School of X-ray Diffraction at M.I.T.*, in McLachlan Jr and Glusker (1983).

L. Pauling also made a significant contribution to the study of silicates, whose structures were determined by applying his rules. He worked out the structures of muscovite and biotite micas, already studied by Mauguin (1927, 1928*a*, *b*), and those of pyrophillite and talc, which also have layered structures, but are held together so weakly in two directions that they are soft but do not split like micas (Pauling 1930*a*). Pauling also studied another type of aluminosilicate, the zeolites, such as sodalite and natrolite, and scapolite, which are tectosilicates. He explained the porosity properties of zeolites by showing that they are honeycombed, with passages so tiny that they form molecular sieves, letting in only molecules small enough to squeeze through and keeping out others (Pauling 1930*b*).

10.3 Materials science

Many important physical properties of crystals depend on their structure. An extreme example is given by the polymorphs of carbon: graphite, soft and conducting, on one hand, and diamond, hard, and semiconducting on the other hand. V. M. Goldschmidt (1926) showed the relationships between physical properties and crystallographic structure (Section 10.1.2). Atoms of similar size can be exchanged in a crystal structure no matter what site they occupy in the Periodic Table and what their valency is. This can cause changes in the physical properties of the compound; for instance, the hardness of a material increases when the valency increases. Goldschmidt's work laid the basis of crystal engineering: which elements should be changed in a given material in order to obtain the required properties.

Applications came, however, much later. The term 'crystal engineering' was coined by R. Pepinsky (1955) as a new concept in crystallography. He showed that by co-crystallizing organic ions one could control packing in the crystal lattice. There are many examples of materials for which properties can be changed by exchanging ions or atoms. Laves (1964) quoted the cases of ferrites and garnets. Ferrites are compounds with the formula AB_2O_4. W. H. Bragg (1915*a*, *b*) and S. Nishikawa (1915) determined the structures of magnetite (Fe_3O_4) and spinel ($MgAl_2O_4$). By substituting magnesium and aluminium by barium, iron, manganese, nickel or zinc, one obtains a variety of materials with interesting magnetic properties. Garnets are silicates of formula $X_3Y_2(SiO_4)_3$. For instance, almandine has composition $Fe_3Al_2(SiO_4)_3$. By substituting iron and aluminium by yttrium and iron, one obtains a ferrimagnetic material, yttrium iron garnet $Y_3Fe_2(FeO_4)_3$ (YIG). One could also mention the family of zeolites such as natrolite $Na_2Al_2Si_3O_{10},2H_2O$, in which a large variety of cations can be substituted (Fig. 10.11), used widely as molecular sieves, catalysts, and high-temperature superconductors such as $YBa_2Cu_3O_{7-x}$ (YBaCuO).

The concept of crystal engineering, in the sense suggested by Pepinsky, was first applied by G. M. J. Schmidt (1971) in his studies of photodimerization in the solid state and synthetic photochemistry. For a review of modern applications, see, for instance, G. Desiraju (1989, 2010) and D. Braga (2003).

Drug design is another type of crystal engineering, which has taken a great importance in recent years. Macromolecular crystallography is now a routine tool for determining how a pharmaceutical drug interacts with its protein target, and what changes in the chemical formula of the drug might improve its required action as a drug that can control a specific health problem.

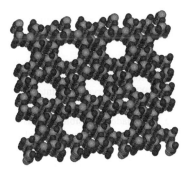

Fig. 10.11   Synthetic zeolite. Source: Wikimedia commons.

10.4   Physical sciences

10.4.1   Structure of the atom

The study of the structure of the atom was certainly the foremost application of X-ray diffraction to physics in the early years. This is discussed in Section 10.6.

10.4.2   Structural interpretation of physical properties

Physical properties of solids can be divided into two categories. 1) Properties depending on the chemical composition, the nature of the chemical bonds, the three-dimensional arrangements of ions, atoms or molecules, the symmetry, or the lack of it (optically active, pyroelectric, ferroelectric or piezoelectric materials); they are sometimes called 'intrinsic' properties. 2) Properties related to the real structure of the materials and their defects, disorder, point defects, dislocations, stacking faults, sub-grain and grain boundaries, twins, *etc.*; they are sometimes called 'extrinsic' properties. They were called 'structure-sensitive' (*strukturempfindliche*) by A. Smekal[29] (1933, 1934). Thermal and electrical conductivity, diffusion phenomena, plastic deformation are examples of such properties.

1. **Intrinsic properties**. During the first years following Laue's discovery in 1912, the results of X-ray diffraction were applied to an improvement in our understanding of chemical bonds and to the derivation of the main structure types of solids. The application to the relationships between intrinsic physical properties and structure came later. Three early examples of such investigations are given here.

   (a) **Zero-point energy**. Debye (1913*d*) and Waller (1923, 1927) showed that the presence of lattice vibrations affect the X-ray diffracted intensities (Section 7.9).While the phonon dispersion curves can be deduced from measurements of thermal diffuse scattering (TDS), one was, in the 1920s, very far from doing so. One could, however, estimate the Debye–Waller factor from elastic constants. As an application, R. W. James, I. Waller, and D. R. Hartree (1928) joined forces to prove the existence of a zero-point energy.

   As soon as physicists realized the possibilities offered by X-ray diffraction, they wondered whether it was possible to detect the presence of zero-point energy from X-ray diffraction

---

[29]Adolf Smekal, born 12 September 1895, died 7 March 1959, was an Austrian solid-state physicist. He studied at the Universities of Vienna, Graz, and Berlin, and was Professor at the Universities of Vienna, Halle, Darmstadt, and Graz.

measurements (Section 7.10). Appropriate conditions were met in 1928 with the improved accuracy of intensity measurements and quantum-mechanical computation of atomic scattering factors.

Two points were decisive. 1) I. Waller (1923, 1927), in his thesis, showed that the temperature factor to be applied to experimental data was $\exp -2M$ for diffracted intensities, and $\exp -M$ for amplitudes. 2) R. W. James and Miss E. Firth (1927) observed that intensities diffracted by a rock-salt crystal, when measured at the temperature of liquid air, were considerably increased above those measured at room temperature. This made it possible to measure the scattering factor at much larger angles of scattering. They compared reflections such as 200 for which the scattering amplitude was proportional to $f_{Cl} + f_{Na}$, and reflections such as 111 for which the scattering amplitude was proportional to $f_{Cl} - f_{Na}$. They concluded that the scattering amplitude should include a different Debye–Waller factor for each type of atom in the crystal structure.

Waller and James (1927) calculated the Debye–Waller factors for sodium and chlorine in the presence or absence of zero-point energy. James et al. (1928) then compared the theoretical angular variations of the atomic scattering amplitudes of sodium and chlorine with the measured values, corrected for the appropriate Debye–Waller factors in the presence or absence of zero-point energy. They concluded to the presence of zero-point energy of amount half a quantum per degree of freedom, as proposed by Planck. As a result, they could estimate root-mean square amplitudes of vibration at absolute zero to be 0.12 Å for sodium and 0.11 Å for chlorine.

(b) **Piezoelectricity of quartz.** Piezoelectricity was discovered in 1880 by Pierre Curie and his brother Jacques (Curie and Curie 1880), in C. Friedel's laboratory, at the Sorbonne in Paris. It was widely utilized during the First World War for the sonar detection of submarines, and quartz is still the most widely used piezoelectric material. The first attempt to give a structural interpretation of the piezoelectricity of α-quartz was by A. Meissner[30] (1927). The crystal structure of α-quartz was determined by W. H. Bragg and Gibbs (1925), and was assumed at the time to be purely ionic—the ions of silicon and oxygen were considered to lie on helices (Fig. 10.12 a). As an approximation, Meissner assimilated each pair of oxygen ions to a single ion of charge $-4e$, Figures 10.12 b) and c) show the projection of the ions on a plane normal to the trigonal axis. Meissner suggested that under an applied stress represented by arrows on the figures, the ions are slightly displaced and the electrostatic balance is disturbed; electric charges appear on the electrodes A and B. Actually, according to modern views, quartz is only partially ionic. Pietsch et al. (2001), however, proved that Meissner's model is incorrect and suggested that the piezoelectric effect is due to a change in the bonding angle between oxygen and silicon. In a detailed study by X-ray diffraction, Guillot et al. (2004) have shown that the mechanism is more complex and involves variations of both the intra-tetrahedral angles and bridging angles.

(c) **Phase transformations.** X-ray diffraction, in particular powder diffraction, is an ideal tool for the study of phase transformations, and this has been the subject of innumerable studies. The first example may be the study of the dimorphism of boracite by H. Haga and F. M. Jaeger (1914) in Groningen, The Netherlands. Boracite is a colourless, glassy mineral, $Mg_3B_7O_{13}Cl$,

[30]Alexander Meissner, born 14 September 1883 in Vienna, Austria, died 3 January 1958 in Berlin, Germany, was an Austrian engineer and physicist. After studying at the Technical University of Vienna, he obtained his ScD in 1902. In 1907 he joined the Telefunken company in Berlin, where he worked on antenna design, and devised amplification systems and new vacuum-tube circuits—in particular the inductively-coupled oscillator which bears his name. He is at the origin of the heterodyne principle for radio reception.

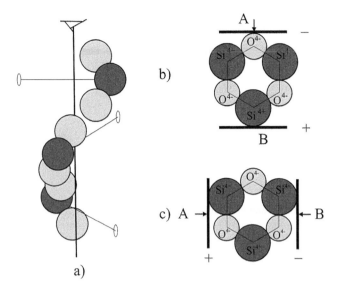

Fig. 10.12   Meissner's explanation of the piezoelectricity of α-quartz. a): structure of quartz, according to W. H. Bragg; dark circles: silicon ions, light circles: oxygen ions. b) stress applied along a binary (electric) axis, electrodes *A* and *B* are charged negatively and positively, respectively; c) stress applied normal to a binary (electric) axis, electrodes *A* and *B* are charged positively and negatively, respectively.

usually found in sedimentary deposits. Its space group at room temperature is $Pca2_1$, with parameters $a = 8.54$Å, $b = 8.54$Å, $c = 12.07$Å. On heating, it undergoes a first order transition at 265°C to a cubic phase of space group $F\bar{4}3c$ and of parameter $a = 12.01$Å. Its habit at room temperature is the same as at high temperature, which baffled mineralogists, since the mineral was birefringent with a cubic habit (see Section 12.10.2). The mineral is also highly pyroelectric and piezoelectric. Haga and Jaeger studied a crystal of boracite at 18°C and at 300°C, both with a polarizing microscope and with Laue photographs. They ascertained the phase transformation by observing the symmetry change from orthorhombic to cubic, as well optically as on the Laue diagrams. The structure of boracite was determined by Ito *et al.* (1951). Synthetic boracites are used as an ionic conductor.

2. **X-ray diffraction studies of crystal imperfections**. Most crystals contain imperfections. Polanyi (1922) noted that ten years after the discovery of X-ray diffraction, hardly any article was devoted to the study of crystalline imperfections. One of the first observations of the effect of crystal imperfections on a diffraction diagram was by W. Friedrich himself, in a footnote added to the reprinting in *Annalen der Physik* of the original paper describing the discovery of X-ray diffraction (Friedrich *et al.* 1913, p. 984). Mandelstam and Rohman (1913, 17 February) observed the effect of scratches on the surface of a sheet of mica, leading to a locally increased diffraction intensity— the ancestor of X-ray topography (see Section 7.1). Polanyi (1922) suggested some of the possible inner arrangements which one could infer from diffraction diagrams of highly imperfect materials such as fibres or cold-drawn metallic wires. In the 1920s, most studies of crystalline imperfections by X-ray diffraction were carried out on fibres or metals. The study of fibres is briefly reviewed in Section 10.5, and the use of X-ray diffraction in metal physics is described in Section 10.1.6. As one of the only examples of the study of a non-metallic material, Ioffe and Kirpitcheva (1922) at

Table 10.1   Planck constant *h* measured by X-ray diffraction

Year	Authors	Planck constant
1915	Duane and Hunt	6.51×10^{-34} J.s
1916	Hull	6.59
1916	Webster	6.53
1917	Blake and Duane	6.555
1918	Wagner	6.49
1921	Duane *et al.*	6.556

the Röntgenological and Radiological Institute in Petrograd, USSR, studied the Laue diagrams of a rock-salt crystal at various degrees of homogeneous strain (compression) by loading it by means of an electromagnet. They observed that, beyond the elastic limit, the separate diffraction spots became elongated, namely when plastic flow started. They also observed the reflections from the (110) planes with a spectrometer, before and after strain, and showed that the atomic structure had not changed, thus answering some of the questions that had been worrying the physicists of the time about the deformation process (see the remark at the beginning of Section 10.1.4: how can crystalline grains be elongated and still remain crystals?).

10.4.3   *Evaluation of the Planck constant* h

R. T. Birge (1919) enumerated no less than seven different methods to measure the Planck constant,[31] with values ranging from 6.542×10^{-27} to 6.579×10^{-34} J.s, the average being 6.554×10^{-34} J.s. The first measurement deduced from the smallest wavelength of the white spectrum was by Duane and Hunt (1915); see Section 7.13. The measurement was repeated successively by A. W. Hull (1916), D. L. Webster 1916, F. C. Blake and W. Duane (1917), E. Wagner (1918), and W. Duane *et al.* (1921). The values they obtained are shown in Table 10.1. The more recent are based on Millikan's (1917) value of *e*, the electric charge of the electron. Their accuracy increased as the lattice constant of the analyser crystal became better known. Duane and Webster used calcite and Wagner sylvine. The values obtained by the Duane group are the most accurate and the closest to the average of the values obtained by other methods.

10.5   Biological sciences

The first diffraction diagrams of biological materials were diagrams of fibres. Nishikawa and Ono (1913) obtained Laue diagrams of fibres of silk, wool, bamboo, and *asa* (a kind of hemp). These diagrams presented sets of straight bands passing through the centre of the diagrams. In 1920, diagrams of cellulose fibres were obtained independently by Scherrer (see footnote 21, Chapter 8) and by Herzog and Jahncke (1920), using a Debye–Scherrer camera (see Section 8.6.1). Scherrer observed ramie (see footnote 20, Chapter 8) and silk fibroin, Herzog and Jahncke observed cotton, ramie, silk, and wood cells. Fibre diagrams were further studied

---

[31]Birge, R. T. (1919). The most probable value of the Planck constant *h*. *Phys. Rev.* **14**, 361–368.

by R. O. Herzog's group and M. Polanyi at the Institute of Fibre Chemistry of the Kaiser-Wilhelms Gesellschaft in Berlin-Dahlem (Herzog *et al.* 1920; Polanyi 1921*a*, 1922; Polanyi and Weissenberg 1922*a*, *b*) who showed that the fibres of cellulose were nearly 'ideal' fibres in that the crystallites have a perfectly random distribution about the axis of the fibres. Gonell (1924) from the same Institute showed that the fibres of dyer's broom (*genista tinctoria*) and white mulberry (*Morus alba*) presented some deviations from the ideal fibre diagrams of cellulose. Hengstenberg and Mark (1928), at I. G. Farbenindustrie in Ludwigshafen, Germany, estimated the size of the crystallites of cellulose, and Meyer and Mark (1928) proposed a structural model for cellulose, built up in a rather imperfect way of molecular chains (*Hauptvalenzketten*).[32]

The studies of fibres that have had the most important impact on our knowledge of the structure of biological substances are those of W. T. Astbury (Fig. 10.13), who is considered as the 'father' of molecular biology. Astbury learned crystallography at Cambridge, with Professor A. Hutchinson, who recommended him to W. H. Bragg at University College, London. The first structures discussed by Astbury were those of tartaric acid, one of the first monoclinic crystals to be studied (Astbury 1923*a*), and of anhydrous racemic tartaric acid, the first triclinic crystal to be studied (Astbury 1923*b*). The very interesting point made by Astbury in the second structure was that there is no such thing as a racemic molecule in the crystal: there are two molecules per unit cell in the crystal, one molecule of right-handed tartaric acid and one molecule of left-handed tartaric acid.

Bernal, in his obituary of Astbury,[33] has told how it came about that Astbury started working on fibres. W. H. Bragg used to ask the members of his research group to help him prepare material for his popular lectures. In 1926 he asked Astbury to find pictures of fibres such as wool for a lecture 'The imperfect crystallization of common things'. Astbury looked at the diffraction photographs obtained by the group at Berlin-Dahlem and obtained some diffraction photograph of human hair himself (Astbury and Street 1932). He was immediately 'fascinated by the biological implications of the work' (Bernal, see footnote 33).

Leeds was the capital of the textile industry. J. B. Speakman, a textile chemist at the Textile Department of the University of Leeds, had started physico-chemical investigations of the wool fibre and had done some preliminary X-ray analyses. W. H. Bragg was asked 'to supply someone to carry out complementary investigations in textile physics' (Astbury, quoted by Bernal). Astbury was appointed Lecturer in Textile Physics in 1928 and set about building up a Textile Physics Research Laboratory.

Astbury showed that the wool fibre diagrams are due to diffraction by a crystalline, or pseudo-crystalline protein, keratin (Astbury and Street 1932). Wool fibre has remarkable elastic properties and can be stretched up to 30% without rupture. Astbury showed that the X-ray diagram of unstretched hair is changed into quite a different diagram when the hair is stretched, and that the change is reversible. Astbury called the two forms of keratin α- and β-keratin where α- is for the unstretched form and β- for the stretched form (Astbury and Street 1932, Astbury and Woods 1934). The crystallites of β-keratin are better aligned than those of α-keratin, and have undergone a definite elastic transformation. Astbury found keratin to be formed by

[32]Meyer, K. H. and Mark, H. (1928). Uber den Bau des Krystallisierten Anteils der Cellulose *Berl. Deutsche Chem. Gesellschaft.* **61**, 593–614.

[33]J. D. Bernal (1963). William Thomas Astbury. 1898–1961. *Biogr. Mems Fell. R. Soc.* **9**, 1–35.

Fig. 10.13   William T. Astbury. © National Portrait Gallery, UK.

William Thomas Astbury: born 25 February 1898 in Longton, Staffordshire, England, the son of a potter; died 4 June 1961 in Leeds, was an English physicist and molecular biologist. After schooling in Longton High School he won a scholarship to Jesus College, Cambridge, in chemistry, physics and mathematics. After two terms in Cambridge, his studies were interrupted by the war, during which he served in the Royal Army Medical Corps in Cork, Ireland. He returned to Cambridge after the war, and learnt crystallography with A. Hutchinson. After his graduation in 1921, Astbury went to work with William H. Bragg as Assistant at University College London. In 1923 he followed W. H. Bragg at the Davy–Faraday Laboratory at the Royal Institution. In 1928 he became Lecturer in Textile Physics at the University of Leeds, where he studied the properties of fibrous substances such as keratin and collagen. In 1937 he became Reader, and, in 1945, Professor of a newly established Department and Laboratory of Biomolecular Structure. He was elected a Fellow of the Royal Society in 1940.

MAIN PUBLICATIONS

1932 *X-Ray Studies of the Structure of Hair, Wool, and Related Fibres. I. General.* With A. Street.

1933 *Fundamentals of fibre texture.*

1934 *X-Ray Studies of the Structure of Hair, Wool, and Related Fibres. II. The molecular structure and elastic properties of hair keratin.* With H. J. Woods.

1935 *X-Ray Studies of the Structure of Hair, Wool, and Related Fibres. III. The configuration of the keratin molecule and its orientation in the biological cell.* With W. A. Sisson.

1938 *Recent developments in the X-ray study of proteins and related structures.* With F. O. Bell.

1940 *Textile fibres under the X-rays.*

a molecular chain of repeating unit 10.3 Å, the same as had been observed in cellulose by the German groups. He suggested that the structure of β-keratin was based on fully extended polypeptide chains, linked side to side by molecules of cystine or cysteine, and that α-keratin was constructed out of the same chains in a shorter, folded form, his celebrated α-fold. The properties of hair were summarized in a jingle by A. L. Patterson, a former colleague of Astbury at the Royal Institution:

> 'Amino acids in chains
> Are the cause, so the X-ray explains,
> Of the stretching of wool
> And its strength when you pull,
> And show why it shrinks when it rains'[34]

[34]Quoted in W. T. Astbury (1938). X-ray adventures among the proteins (4th Spiers Memorial Lecture 1937). *Trans. Faraday Soc.* **34**, 378–388.

Astbury observed also that the diagram of β-keratin was very similar to that of silk fibroin, which shows that the fibre substance of silk is for the most part built of fully-extended polypeptide chains. This explains why silk does not have the elastic properties of wool, because it is already in the extended state.

The next topics tackled by Astbury were the keratins of feathers and reptile skin (a kind of β-keratin), myosin, and muscle, which gives a characteristic α-protein photograph, and then nucleic acids of which he started to unravel the structures. In 1938 he obtained the first X-ray diffraction patterns of DNAs showing their regular structure.[35] Noting the strong birefringence of the nucleic acid fibres, and, knowing that they included flat purine and pyrimidine bases, he concluded that these must be at right angle to the axis of the fibre, with spacing 3.4 Å. It is true that W. T. Astbury missed the actual structure of DNA, but he opened the road in a decisive way for Pauling's α-helix (1951), and to the later work on DNA by R. Franklin, M. Wilkins, J. Watson and F. Crick (1953). For an account of Astbury's studies on DNA, see Hall (2014).

In fact, the prodigious development of the applications of X-ray diffraction to the study of biological substances only took off in the 1930s, with W. T. Astbury, J. D. Bernal, D. Crowfoot (the future Dorothy Hodgkin), and M. Perutz, among others.

10.6 X-ray spectroscopy

10.6.1 Birth of X-ray spectroscopy

X-ray spectroscopy started with C. G. Barkla's identification of the harder K series and the softer L series in the emission spectra of the elements. This he did by measuring the absorption through a series of aluminium foils of the fluorescent radiation of twenty elements from calcium to barium for the K series and of nine elements from silver to bismuth for the L series (Barkla 1911). Both of these radiations were found to become stepwise harder (shorter wavelength) if the atomic weight of the element that emitted them is higher. K-radiation was also found to be roughly 300 times more penetrating than L-radiation.

In April 1913 the two Braggs, father and son, obtained the first X-ray spectrum with their ionization spectrometer, corresponding to three characteristic lines of platinum, $L\alpha$, $L\beta$, $L\gamma$ (W. H. Bragg and W. L. Bragg 1913a); see Fig. 7.5 and Section 7.6. In June of the same year, W. H. Bragg (1913b) observed the K characteristic line of nickel and the L line of tungsten, and in July G. J. Moseley and C. G. Darwin (1913b), working in E. Rutherford's laboratory at Manchester Victoria University, found two more platinum lines, $L\delta$ and $L\epsilon$, with very fine measurements on crystals of rock-salt, potassium ferrocyanide, $K_3[Fe(CN)_6]$, $3H_2O$, and selenite (gypsum); see Section 7.8. These early experiments were repeated by J. Herweg in Greifswald University, Germany (Herweg 1913b, 1914).

The first photographic recordings of X-ray spectra were made by M. de Broglie (1913c and d, November, see Section 7.11) in Paris and by H. Moseley (1913, December), but the most spectacular contribution is that of Moseley 'whose work will always stand as a milestone in the struggle to penetrate the secrets of the atom' (Siegbahn 1937). In a series of brilliant experiments described in Section 7.8, Moseley (1913, 1914) recorded the K-lines of twenty-one

---

[35]W. T. Astbury and F. O. Bell (1938), *Some recent developments in the X-ray study of proteins and related structures.* Cold-Spring Harbor Symposia en Quantitative Biology. **6**, 109–118.

lighter elements, from aluminium to silver, and the *L*-lines of twenty-four elements, from zirconium to gold.

10.6.2 X-ray spectroscopy and the structure of the atom

J. J. Thomson (1910) and E. Rutherford (1911) both proposed models of the atom. In Thomson's model, the atom consisted of a number of *N* negatively charged corpuscles uniformly distributed through the atom, accompanied by an equal quantity of positive electricity. In Rutherford's model, a minute positively charged nucleus is surrounded by a cloud of negatively charged electrons that have a total charge equal to that of the nucleus. Bohr (1913) applied his quantum theory to Rutherford's model. In Bohr's quantum theory of the atom, an atomic system possesses a number of stationary states, and any emission or absorption of energy corresponds to a transition between two stationary states. The radiation emitted during such a transition has a frequency determined by the relation $h\nu = A_1 - A_2$, where A_1 and A_2 are the energies of the system in the two stationary states. The electrons are arranged in coaxial rings and rotate in circular orbits around the nucleus. In agreement with an earlier proposal by J. J. Thomson (1912), Bohr (1913) supposed that 'the characteristic Röntgen radiation is emitted during the settling down of the system if electrons in inner rings are removed by some agency, *e.g.* by impact of cathode particles'. Furthermore, 'the *permanent* state of any atomic system, *id est* the state in which the energy emitted is maximum, is determined by the condition that the angular momentum of every electron round the centre of its orbit is equal to $h/2\pi$'. Bohr's hypotheses of a single quantum and electron rings were to be modified in the later theories.

Moseley's results were in agreement with Rutherford's nucleus theory and Bohr's quantum theory. They showed that the sequence of frequencies of X-ray spectra follows that of the positive charges of the nucleus, *N*, or that of the position, *Z*, of the element in the Periodic Table, which van den Broek (1913) had suggested were equal, and not that of the atomic weights. Moseley further showed that the square root of the frequencies varies linearly with this number $N = Z$, called the 'atomic number'[36] (Moseley's law, equation 7.7), thus proving that *Z* was not simply an ordinal number but also had a physical significance. Moseley's pioneer investigations showed that the X-ray spectra were a complement to the optical spectra and opened the way for the experimental researches in X-ray spectroscopy by de Broglie, E. Wagner, and M. Siegbahn's school, and to the theoretical studies by W. Kossel and A. Sommerfeld.

For details about the early history of X-ray spectroscopy, the reader may consult, for instance, Siegbahn (1924, 1937, in Ewald 1962), Sommerfeld (1937), Cauchois (1964), Heilbron (1967).

10.6.3 Experimental studies: the Siegbahn school

Absorption spectra.   The absorption measurements by Barkla and Sadler (1907, 1908, 1909) and Barkla (1909) revealed the existence of absorption edges. These were first observed by de Broglie (1914*c, d*) on his rotating crystal photographs, and the first accurate measurement of absorption spectra were made by Wagner (1914, 1915*b*), using de Broglie's rotat-

---

[36]See footnote 22 in Chapter 7.

ing crystal method, and rock-salt as analyser crystal. Kossel (1914 I.) compared the mass absorption coefficients of Al, Fe, Ni, Cu, Zn and Ag measured by Barkla and noted 1) that their values decreased with increasing atomic number, 2) that when plotted on a logarithmic scale against the wavelengths measured by Moseley, their variations were linear, 3) the plots presented a discontinuity due to the existence of absorption edges. Siegbahn (1914) also plotted the variations of the mass absorption coefficient of aluminium, as measured by Barkla, against wavelength on a logarithmic scale. He also obtained a straight line, thus showing that the wavelength dependence of the mass absorption coefficient μ/ρ was of the form:

$$\mu/\rho = A\lambda^x, \tag{10.1}$$

where A and x are constants. Siegbahn further noted discontinuities (absorption edges) in the plots for Zn, Cu, Ni, Mg and Fe. W. H. Bragg (1915d) measured the wavelengths and absorption coefficients of the K emission lines of palladium, rhodium and silver, using a calcite analyser. Like Siegbahn and Wagner, and independently of them, he noted that the variations of the absorption coefficients with wavelength were represented by a straight line, if plotted with a logarithmic scale. Summarizing Moseley's results and his own, he concluded that a condition for the fluorescent radiation to be emitted is that the hardness of the primary radiation should be greater than that of the characteristic radiation of the fluorescent element.

De Broglie later recorded the K absorption spectra for the elements copper and bromine to bismuth (Broglie 1916a), and the L absorption spectra for platinum, gold, lead, thorium, and uranium (Broglie 1916b).

Thanks to the increased resolution and precision of the new spectrometers, a fine structure was observed in the absorption spectra:

• Siegbahn's student Stenström (1919) detected a fine structure in the M-series of uranium and thorium, which Kossel (1920) attributed to the degree of ionization of the atoms, and which Sommerfeld (1920) discussed from the quantum mechanical viewpoint.
• The Dane H. Fricke (1920) observed a fine structure in the neighbourhood of the K absorption edges of the elements Mg to Cr, during a stay in Siegbahn's laboratory in 1918.
• The Swede J. Bergengren (1920), also during a stay in Siegbahn's laboratory, observed a different fine structure of the K absorption edge in orthorhombic black phosphorus, monoclinic violet phosphorus, and in chemically bound phosphorus (ammonium phosphate and P_2O_5).
• A fine structure was observed in the neighbourhood of the L absorption edges of the elements Cs to Nd by G. Hertz (1920) in Berlin with a Seemann-type slitless spectrograph.

The emission and absorption spectra of elements are an atomic property, but measurements of X-ray spectra were usually made with solids in which the atoms were linked to their neighbours by chemical binding. If the outermost shells or the valence electrons are involved in the emission or absorption process, it is to be expected that the neighbouring atoms should influence the X-ray spectra. The relation between the fine structure and the degree of ionization was first observed by Siegbahn's student, A. E. Lindh (1921), who compared the fine structure of the K absorption spectra of chlorine in various compounds in which it took valencies 1-, 5-, or 7-. He observed a shift of the wavelengths toward shorter wavelengths for higher valencies. Lindh (1922) observed similar effects with sulfur in compounds where it had valencies 2-, 4-, or 6-.

The fine structure of absorption edges is now widely applied for analysis in the 'Extended X-Ray Absorption Fine Structure' (EXAFS)[37] and 'X-ray Absorption Near Edge Structure' (XANES) techniques.

Emission lines.　During the First World War, little research work could be done in England, France, or Germany, leaving the door open for the Swedish school. During the ten to fifteen years that followed the beginnings of X-ray spectroscopy, M. Siegbahn (Fig. 10.14) had many students: I. Malmer, E. Friman, W. Stentström, E. Hjalmar, A. Leide, A. E. Lindh, N. Stensson, D. Coster, O. Lundquist, A. Larsson, I. Waller. Siegbahn soon realized that in order to improve our understanding of the structure of the atom it would be necessary to increase the precision of measurements of the wavelengths of the emission lines and of the absorption edges. This required, Siegbahn wrote in Ewald (1962*a*), 'an increase also of the resolution in the registered spectra. Only by complete and exact measurements of the X-ray spectra could a reliable picture of the atomic structure be obtained'. Siegbahn, who was very skilful himself, devised ever more accurate spectrometers covering a range of wavelengths from 0.1 Å to 20 Å, corresponding to voltages from 130 000 to 600. Special spectrometers, including vacuum spectrometers for long wavelengths, and special X-ray tubes had to be constructed for each part of the X-ray range, short, medium, and long waves (Siegbahn and Friman 1916; Siegbahn 1919*a*, *b*; Siegbahn and Leide 1919). In the next ten years, Siegbahn achieved a resolution of wavelengths surpassing Moseley's by more than a hundred-fold. The vast amount of results obtained and their remarkable quality made Siegbahn's laboratory, first in Lund, then in Uppsala, the world's

Fig. 10.14　Manne Siegbahn. © The Nobel Foundation.

Karl Manne Georg Siegbahn: born 3 December 1886, in Örebro, Sweden, the son of a stationmaster of the Swedish State Railways; died 26 September 1978 in Stockholm, Sweden, was a Swedish physicist. He entered the University of Lund in 1906, where he obtained his doctor's degree in 1911. He was appointed Assistant at the University of Lund in 1910, and Lecturer in 1911. In 1919 he became Rydberg's successor as Professor of Physics at the University of Lund, and, in 1923, Professor of Physics at the University of Uppsala. In 1937, Siegbahn was appointed Director of the newly founded Physics Department of the Nobel Institute of the Royal Swedish Academy of Sciences. In 1924 he was awarded the Nobel Prize for Physics 'for his discoveries and research in the field of X-ray spectroscopy', and was awarded the Rumford Medal by the Royal Society in 1940.

Main Publications

1921 *Elektricitet.*

1919 *Precision-measurements in the X-ray spectra.*

1923 *Spektroskopie der Röntgenstrahlen.*

1924 *Spectroscopy of X-rays.*

1935 *Zur Spektroskopie der ultraweichen Röntgenstrahlung.* With T. Magnusson.

1950 *Nobel, the man and the prizes.*

---

[37]A detailed history of the development of EXAFS is given, for instance, in Stumm von Bordwehr, R. (1989). A history of X-ray absorption fine structure. *Annales Physique Fr.* **14**, 377–466.

centre for X-ray spectroscopy, and earned Siegbahn the 1924 Nobel Prize for Physics. No better tribute can be paid to Siegbahn's accomplishments than Niels Bohr's in his 1922 Nobel lecture: 'Our picture would, however, be incomplete without some reference to the confirmation of the theory afforded by the study of X-ray spectra. Since the interruption of Moseley's fundamental researches by his untimely death, the study of these spectra has been continued in a most admirable way by Prof. Siegbahn in Lund. On the basis of the large amount of experimental evidence adduced by him and his collaborators, it has been possible recently to give a classification of X-ray spectra that allows an immediate interpretation of the quantum theory'.

From the start, M. Siegbahn employed de Broglie's rotating crystal method, and his double-angle technique, namely measuring the angle between the lines at $-\theta$ and $+\theta$ (Broglie 1913e). The accuracy of the wavelength measurements depend on the quality of the value of the lattice parameter adopted for the analyser crystal. Wagner (1916) made a careful comparison of palladium and platinum K lines with several rock-salt and sylvine crystals, obtaining a good agreement between the results. W. H. Bragg (1915d) had recommended using calcite instead of rock-salt, because of the poor quality of many rock-salt crystals. During the first years, Siegbahn did use rock-salt as analyser crystal. He later explained that he had obtained several hundred very good spectrograms with selected rock-salt crystals from the salt mines of Stassfurt, Sachsen-Anhalt, Germany, but that at large angles somewhat better results came out with calcite (Siegbahn 1919b-I). In the 1920s, most investigators in the world preferred calcite, which became the standard, for instance in Duane's and Davis's schools in the United States. Careful measurements of the calcite lattice parameter were reported by Bearden (1931).

M. Siegbahn's first student, I. Malmer, published measurements of the wavelengths of the $K\alpha$ and $K\beta$ lines of fourteen elements, 39 Y to 58 Ce, in 1914 (Malmer 1914). In the further development of his work for his thesis, defended in 1915, he discovered that the $K\alpha$ and $K\beta$ lines were in fact doublets ($K\alpha_1$, $K\alpha_2$ and $K\beta_1$, $K\beta_2$). Siegbahn made this very important result known by a letter to Nature (Siegbahn 1916a, 17 February). In the same publication he announced that the $L\alpha$ lines are also doublets and that Moseley's $L\beta$ and $L\gamma$ lines seemed to be doublets, which he denoted $L\beta_1$, $L\beta_2$. In the course of 1916 Siegbahn and his students, E. Friman and W. Stentström, published the wavelengths of the $K\alpha$ lines of the elements 11 Na to 47 Ag, and 79 Au to 92 U, the L lines of the elements from 59 Pr to 84 Po, 88 Ra, 90 Th, 92 U. In 1916 also, Siegbahn (1916b) announced that he had observed emission lines corresponding to the M series in the heavier elements, 79 Au, 81 Tl, 82 Pb, 83 Bi, 90 Th, and 92 U.

After the end of the First World War (1918), many studies were devoted to X-ray spectroscopy and spectrometry. The wavelengths of materials such as copper, molybdenum, platinum, and tungsten used in anticathodes were remeasured many times. A comparison of the results for the Mo$K\alpha$ and $K\beta$ lines by various authors is given in Larsson (1927), who also reported on the latest version of the Siegbahn spectrometer. Two methods were used in general for measuring the wavelengths, the rotating crystal method, used by Siegbahn, and the double-crystal spectrometer, used by Duane, Davis, and Compton. By the end of the 1920s the latter method was considered the most precise one (Bearden 1932).

10.6.4   *Theoretical studies: Kossel and Sommerfeld*

H. Moseley, in agreement with Bohr's treatment, interpreted the K-lines as due to the radiation emitted during a transition between two states of the electrons in the innermost ring, but his law, (7.7), relating wavelengths of the emission lines with atomic number was a purely empirical one. Walther Kossel[38] (1914 II) went a step further. He was at that time assistant to A. Sommerfeld in Munich. He first heard of Bohr's theory when Ewald reported about it in the Sommerfeld Physics colloquium in the autumn of 1913. He also met Bohr when both gave a talk at that same colloquium on 15 July 1914 (Heilbron 1967). Kossel considered the atom to consist, as in the Bohr model, of a nucleus and electrons arranged in concentric rings,[39] and explained Moseley's result qualitatively by assuming that the emitted X-rays came from atoms that had lost an electron from an inner ring. The hole is filled by an electron coming from an outer ring. The line denoted $K\alpha$ by Moseley corresponds to the transition from ring 2 to the innermost ring, 1. It follows that the energy necessary to remove an electron from the innermost ring by incident cathode rays or primary X-radiation, $h\nu_K$, where ν_K is the frequency of the K absorption edge, should be equal to the sum of the energy necessary to remove an electron from ring 2, $h\nu_L$, and of the energy released by the transition of an electron from ring 2 to ring 1, $h\nu_{K\alpha}$:

$$h\nu_K = h\nu_L + h\nu_{K\alpha}, \tag{10.2}$$

a relation suggested independently by Wagner (1915b). The frequency of the emitted X-rays will be the longer, the larger the distance of the donor ring from the K ring.

Kossel checked approximately relation (10.2) using the values of the absorption edges he deduced from Barkla and Sadler's (1909) measurements (Kossel was the first to theorize the absorption edges) and the values of the frequencies $\nu_{K\alpha}$ from Moseley's results.

In a similar way, the energy associated with the transition from ring 3 to ring 1, $\nu_{K\beta}$, is equal to the sum of the energies associated to the transitions from ring 3 to ring 2 and from ring 2 to ring 1:

$$h\nu_{K\beta} = h\nu_{L\alpha} + h\nu_{K\alpha} \tag{10.3}$$

This expression is represented schematically by the diagram in Fig. 10.15, *Left*. Bohr (1915) checked it for Zr, Mo, Ru, Pd, Ag, Sn, Sb, and La, using Moseley's (1914) values for $\nu_{L\alpha}$ and Malmer's (1914) values for $\nu_{K\alpha}$ and $\nu_{K\beta}$. Kossel (1916b) also checked this relation for Mo, Pd, Sb, La, Ce, and Nd.

When Malmer found in his thesis that the $K\alpha$ and $K\beta$ lines were in fact doublets, Siegbahn (1916a) noticed immediately that relation 10.3 should be replaced by

$$h\nu_{K\beta_1} = h\nu_{L\alpha} + h\nu_{K\alpha_2}, \tag{10.4}$$

which he checked with antimony, using data obtained by himself and his student, E. Friman. Sommerfeld (1916 I.) developed a quantum-mechanical theory of Bohr's atomic model, which he applied to the hydrogen atom, taking relativity into account. He then considered the case of X-rays. The expression of Moseley's law (7.7) is, for $K\alpha$:

$$\nu_{K\alpha} = \nu_o[\frac{1}{1^2} - \frac{1}{2^2}](Z - 1)^2 \tag{10.5}$$

[38]See footnote 23, Chapter 9.

[39]Bohr's concept of rings was later replaced by that of electronic shells.

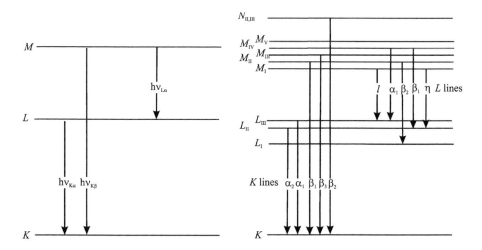

Fig. 10.15   *Left*: Energy levels of atoms according to the early views of Wagner and Kossel. *Right*: Energy levels and emission lines of zinc, according to the data in Bearden (1967).

where the atomic number is here denoted by Z. In order to satisfy Kossel's equation (10.3), Sommerfeld (1916 III.) sets

$$v_{K\beta} = v_o[\frac{(Z-1)^2}{1^2} - M] \qquad (10.6)$$

$$v_{L\alpha} = v_o[\frac{(Z-1)^2}{2^2} - M], \qquad (10.7)$$

expressions which he justified later theoretically, and where M is a constant associated with the energy level M discovered by Siegbahn (1916*b*).

The existence of the $K\alpha$ doublet indicates a splitting of the L level into two levels, and Sommerfeld replaced expression (10.7) by separate expressions for each of the two levels, L and L'.

The existence of the various lines of the K, L and M series implies, of course, a more complex structure of the energy levels, and this was determined by the measurements of the wavelengths of the relevant X-ray emission lines. As an example, Fig. 10.15 *Right* shows schematically the positions of the energy levels of zinc. The further developments of the theory of X-ray spectral lines by Sommerfeld (1919), Pauli, Heisenberg, Pauling, and others are beyond the scope of the present book.

11

UNRAVELLING THE MYSTERY OF CRYSTALS:

THE FORERUNNERS

It was the appearance of regular and beautifully shaped snowflakes rather than the
appearance of the crystals of the mineral world that inspired Kepler with the idea that this
regularity might be due to the regular geometrical arrangement of minute and equal brick-like
units. Thus he was led to think of close-packed spheres, and, although he did not coin
the expression 'space-lattice' and although his development of these ideas is not
always correct, we can find among his illustrations the first pictures of space-lattices.

M. von Laue (1952*a*)

11.1   The terms 'crystal' and 'crystallography'

'Crystal' is the name that was given to quartz, or rock-crystal (Fig. 11.1), by the Greeks because, according to Aristotle, it is produced by the congelation of water.[1] Let us quote Pliny the Elder (23–79 AD, see Section 11.2.2):

> 'Crystal is formed by a very strong congelation and can only be found in regions where the winter snow freezes with the greatest intensity. It is with certainty an ice, hence the name the Greek gave it, $\kappa\rho\upsilon\sigma\tau\alpha\lambda\lambda o\varsigma$, ice'.[2]

The same conception of the formation of minerals still prevailed throughout the Middle Ages. To give just a few examples, we may mention:

> Isidore of Seville (*ca* 560–636), Bishop of Seville, Spain, at the time of the Visigoth King Reccared, considered 'the last scholar of the ancient world', propagated the works of Aristotle in the West well before the Arabs. In his encyclopedia, which compiled the sum of the universal knowledge of his time, he wrote:

> 'Crystal shines brightly and is clear as water. It occurs when snow remains frosted for many years, hence the name given to it by the Greeks. It is also to be found in Asia and Cyprus, but chiefly in the Northern Alps. Exposed to the rays of the sun, it throws flames, to the point of igniting dry moss or leaves.'[3]

---

[1]The term 'quartz' seems to have first been used by Saxon miners (S. I. Tomkeieff, *Miner. Mag.*, 1942, **26**, 172–178. *On the origin of the name 'quartz'*).

[2]*Natural History*, Book XXXVII, The natural story of precious stones, Chapter 9.

[3]Book XVI, *De lapidibus et metallis*, Chapter 13, *de Crystallinis* of his encyclopedia, *The Etymologies (or Origins)*.

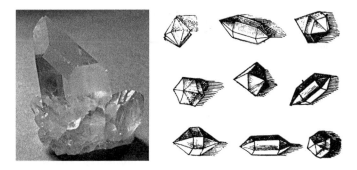

Fig. 11.1   *Left*: A crystal of quartz. Photo by the author. *Right*: Drawings of various quartz crystals, after the French translation of Boece de Boot (1644).

Albertus Magnus (*ca* 1200–1280, see Section 11.2.4):

'Crystal and beryl are formed by the congelation of water under a deep frost, as said Aristotle'.[4]

Girolamo Cardano (1501–1576, see Section 11.2.6) has a slightly different view:

'Crystal consists in a watery material and for that reason melts easily in the fire and transforms into glass. For the same reason, one shouldn't say it is formed from ice, although it is usually found in snow-covered areas, but from a liquid of the same kind. Indeed, the ice from the mountains melts when it is heated by fire, while a crystal does not unless it is surrounded by flames'.[5]

William Gilbert (1544–1603), an English physician and natural philosopher, known for his work on electrical and magnetic phenomena and for using the word *electricus* to describe static electricity:

'Lucid gems are made of water, just as crystal which has been concreted from clear water, not always by a very great cold . . . but sometimes by a less severe one'[6] (quoted by Burke, 1966).

The first uses of the term crystal for other substances than quartz are to be found in the second half of the seventeenth century:

'Crystalline salts' and 'saline crystals' as the results of chemical reactions, in *The Sceptical Chemyst* by the British philosopher, chemist, and physicist Robert Boyle[7] (Boyle 1661).

'Crystals of alum, nitre, or saltpeter, potassium nitrate, and salt' in *De figura nivis dissertatio* by Rasmus Bartholin (Bartholin 1661, see Section 11.6).

'Crystals of alum and saltpeter' in *Micrographia* by Robert Hooke (Hooke 1665, see Section 11.5).

---

[4]*De Mineralibus*, Book I, Chapter 1: *De lapidibus in communi.*   [5]*De Subtilitate*, Book VII: *De Lapidibus.*

[6]*De Magnete, Magneticisque Corporibus, et de Magno Magnete Tellure*, published in 1600.

[7]See footnote 20, Chapter 3.

In *De solido intra solidum naturaliter contento dissertationis prodromus*, Steno notes the analogy between the mountain crystals (quartz) and the crystals of nitre obtained from aqueous solutions (Steno, 1669, see Section 11.7), but in general denotes crystals by the term *angulata corpora* (angular bodies).

In *Traité de la lumière*, written in 1678, Huygens speaks of 'crystals of Iceland spar', while, for him, quartz is 'ordinary crystal' (see Section 3.4.2).

It seems that the term 'crystallography' was introduced for the first time by the Swiss physician and naturalist Moritz Anton Cappeller (1685–1769), in his work *Prodromus crystallographiae de crystallis improprie sic dictis commentarium*, published in Lucerne in 1723, where he gave a classification of crystals (actually, 'crystallized bodies improperly called crystals': *Recensio crystallisatorum corporum quae improprie crystalli vocantur*) according to their shape: spherical, conical, cylindrical, pyramidal, prismatic, polyhedral, dendritic, lamellar, crystals with a transparency comparable to that of quartz but with an uncertain form (Fig. 11.2). For each class, he distinguishes between 'stones', 'metals', and 'salts'.

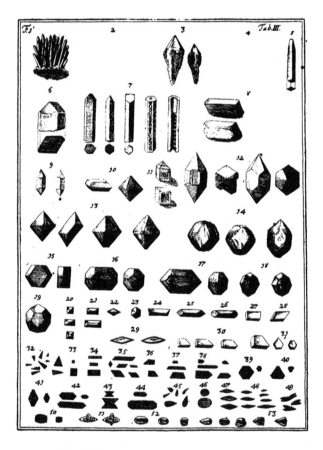

Fig. 11.2   Crystal shapes. After Cappeller (1723).

11.2 From ancient times to the sixteenth century

11.2.1 Marcus Terentius Varro and de Re Rustica, ca 36 BC

Varro (116–27 BC) was a Roman scholar and writer, considered the most learned of his contemporaries. Of the many works he wrote, one of the two that remain is his treatise on agriculture, *De Re Rustica*.[8] Of interest to us here is his mention of the six-fold symmetry of the honeycomb. After asking (Book III, chapters 16, 5), 'Does not the honeycomb cell have six angles, the same number as the bee has feet?', an explanation still given by Pliny the Elder in Book XI, chapter XII-1 of his Natural History, he immediately adds 'The geometricians show that this hexagon inscribed in a circular figure encloses the greatest amount of space'. This important property was also discussed later by Pappus of Alexandria (see Section 11.2.3) and many others. It is sometimes called the *honeycomb conjecture* and was finally solved by A. Thue in 1892.[9] For a history of the honeycomb conjecture, see Hales (2000, 2001). It is interesting to note, *en passant*, that the Latin word for the honeycomb cell, *favus*, was also used by Romans to designate a hexagonal tile.

11.2.2 Pliny the Elder and 'Natural History' (Naturalis Historia), 77 AD

Pliny the Elder (Gaius Plinius Secundus), born 23 AD, was a famous Roman author, naturalist, and natural philosopher as well as a naval and army commander of the early Roman Empire, who died on 25 August, 79 AD, during the eruption of Mount Vesuvius that destroyed the cities of Pompeii and Herculaneum. In Books XXXVI on natural stones and XXXVII on precious stones of his Natural History (*Naturalis Historia*),[10] he described the properties of many minerals. For instance, 'crystal [quartz] is always formed with six faces; it is not easy to understand why, all the more so that the terminations always look different'.[11] Iris (rainbow quartz) casts a rainbow on a wall (Chapter 52). He notes the polyhedric form of diamond and its 'unbelievable' hardness.[12] He also mentions the property of cleavage of the specular stone (*lapis specularis*): 'The specular stone, which deserves also to be called a stone, is sectile (*sectilis*) and can be cut into slices as thin as can be desired'.[13] Pliny adds that it appears that, like crystal, it is formed by congelation.

11.2.3 Pappus of Alexandria and the 'Mathematical Collections', 340 AD

Pappus of Alexandria, born *ca* 290 AD in Alexandria, Egypt, died *ca* 350 AD, was the last of the great Greek mathematicians. He interests us here for his discussion of the space-filling problem and because his writings were quoted by many authors of the seventeenth century, such as Kepler and Bartholin. Little is known of his life, but his main work, the *Mathematical*

---

[8]Available online at <http://penelope.uchicago.edu/Thayer/E/Roman/Texts/Varro/de_Re_Rustica/home.html>.

[9]A. Thue (1892), *Forandlingerneved de Skandinaviske Naturforskeres*, **14**, 352–353. *nogle geometrisk taltheoretiske Theoremer*; A. Thue (1910), *Christinia Vid. Selsk. Skr.* **1**, 1–9. *Über die dichteste Zusammenstellung von kongruenten Kreisen in der Ebene*.

[10]Available online at <http://penelope.uchicago.edu/Thayer/E/Roman/Texts/Pliny_the_Elder/home.html>.

[11]*Natural History*, Book XXXVII, Chapter 9.   [12]*Natural History*, Chapter 15.

[13]*Natural History*, Book XXXVI, Chapter 45. Specular stones were minerals such as selenite (gypsum) or micas that could be cut into thin and transparent sheets, which Roman craftsmen used to seal windows with or in greenhouses.

Fig. 11.3   Hexagonal structure of the honeycomb. Photo
courtesy J. M. Naillon.

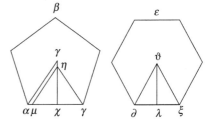

Fig. 11.4   Geometrical comparison of the pentagon and the
hexagon. After Pappus, Mathematical collections, 1588,
reprinted in 1877 by F. Hultsch editor, Berlin: Weidmann.

Collections (*Collectio* or *Synagoge*), written around 340 AD, gives an account of the most important results obtained by his Greek predecessors, for which it constitutes the main source. The *Synagoge* was first translated into Latin in 1588 by Federicus Commandinus in Pesaro and was immediately very popular. It was soon reprinted in Venice in 1589. It often gives new proofs; for instance, in Book IV, Pappus gives a proof of the squaring of the circle that is different from that of Archimedes. In the first part of Book V he discusses the old problem, dating back to Aristotle, of the shape of the regular polygons that can pave the plane without leaving any gap, and that of which polygon has the greatest area. It seems that he was following a work on *isometric figures* written *ca* 180 BC by the Athenian mathematician Zenodorus (Boyer 1991). Pappus starts by admiring the sagacity of the bees. The honeycomb built by bees for storing honey consists of adjacent hexagonal cells (Figure 11.3). He states that bees prefer regular shapes (*nam inaequalia apibus displicebant*) and that their juxtaposition must not leave any voids where foreign substances could fall. He then notes that only three regular polygons meet these requirements: triangles, squares, and hexagons, and, in particular, he specifies that it is impossible to fill out space with pentagons and heptagons, because of the value of the angle between neighbouring sides of these polygons (Figure 11.4).

He goes on to say that it is in hexagonal cells that one can store the maximum quantity of honey. This statement was well known in the ancient times, as already mentioned by Varro (see Section 11.2.1).

The second part of Pappus' Book V is dedicated to Archimedes' theory of solids (the thirteen Archimedean semiregular convex polyhedra, all bounded by equilateral and equiangular but not similar polygons).[14] It is in fact through Pappus that they became known, Archimedes' original

---

[14]For instance, the cuboctahedron, the truncated cube, octahedron, and tetrahedron, the truncated icosahedron or buckyball (the fullerene molecule), *etc.*

Fig. 11.5   Albertus Magnus (1200–1280). Source: gallica.bnf.fr.

treatise having been lost (Boyer 1991); they are discussed by Kepler in his *Harmonice Mundi*. In the same book, Pappus compares Plato's five regular polyhedra and discusses the same problem as for polygons, namely finding among all the polyhedra having the same superficial area, which one has the maximum volume. He also derives the area and the volume of a sphere.

11.2.4   Albertus Magnus and de Mineralibus, 1256

Albertus Magnus (Saint Albert the Great, born *ca* 1200, died 1280), a Dominican friar, bishop of Ratisbon (Regensburg) in 1260, beatified in 1622, canonized in 1931, is one of the most famous scholars, theologians, and philosophers of the Middle Ages (Fig. 11.5). He studied sciences in Padova and taught theology, and philosophy at the universities of Paris, Ratisbon, Strasbourg, and Cologne; in particular, he was the teacher of Saint Thomas Aquinas. He studied the works and the philosophy of Aristotle at great length and did much to make them known in Europe, as well as the works of the Arabic scientists such as Al-Farabi, Algazel, and Avicenne. He was very interested in natural science and stressed the importance of personal observation. He had a vast knowledge of encyclopedic nature and his collected works cover a wide range of topics; they make up 38 volumes (*Opera Omnia*),[15] including a treatise on mineralogy, *de Mineralibus*[16], for which he gathered material from the Goslar mines, in Lower Saxony. It is considered one of the most interesting medieval treatises on mineralogy, and was reprinted several times in the sixteenth century; it constitutes the basis of the knowledge that was typically available to people like Kepler. It is worthwhile to tell here the ideas Albert had of the origin of minerals so that we can better appreciate what a big step forward Kepler's work represents (see Section 11.4). In *de Mineralibus*, Albert discusses the origin, the nature and the properties of stones, precious stones and metals. In Book I, he analyses the 'causes' (in the Aristotelian sense of the word) of minerals according to Aristotle's classification.[17] The 'material' cause of stones lies in a combination of Earth and dryness on the one hand and Water and humidity

[15] Available online in Latin at the web page <http://albertusmagnus.uwaterloo.ca/>.

[16] English translation: D. Wyckoff (1967). *Albertus Magnus, 'Book of Minerals'*. Clarendon Press, Oxford; French translation: M. Angel (1995). *Le Monde Minéral*. Editions du Cerf, Paris.

[17] Aristotle distinguished four types of causes: *material, efficient, formal*, and *final*.

Fig. 11.6   Georgius Agricola. Unknown painter. Source: Wikimedia commons.

on the other hand, under the action of heat and cold. The true 'efficient' cause is to be found in a mineral formative power of the stones (*virtus mineralis lapidis formativa*),[18] which is analogous to the animal formative power that lies in the seminal vessels of animals: the stone is formed by the action of moisture on dry matter or by that of earthen dryness on humid matter, out of the power of the stars (the 'formal' cause). It is the formative power that determines the 'substantial' form of the stone and its specific properties: 'it is folly to doubt that the forms of stones are substantial since sight assures us that they are obtained by coagulation and that their matter has a determined and specific form'.[19] Albert explains thus the formation of quartz: 'in the high mountains where temperatures are very low, the cold squeezes out the moisture and induces the dryness, turning ice into [rock] crystal'.[20] The same process applies to beryl and iris (rainbow quartz), which he says is often hexagonal. There is no 'final' cause because the minerals are inanimate. The second part of Book I describes the properties of minerals in general and Book II those of nearly one hundred precious stones. Book III concerns metals in general and, in particular, bronze, gold, iron, lead, silver and tin.

11.2.5   Georgius Agricola and De natura fossilium, 1546

Georg Pawer (Bauer, farmer), born 24 March 1494 in Grauchau, Saxony, Germany, the son of a cloth manufacturer, died 21 November 1555 in Chemnitz, Germany, was a German scholar, physician, metallurgist, and mineralogist. After doing his humanities in Leipzig (1514–1518), he latinized his name as Agricola and became teacher of Greek in Zwickau until 1522. He then went to Italy to study medicine, in Bologna where he obtained his Doctor's degree in 1524, and Venice. In 1527 he returned to Germany where he was chosen as town physician at Sankt Joachimsthal in Bohemia (now the Czech Republic), a centre of mining and smelting works, where he became interested in mineralogy and metallurgy. In 1531 he returned to Saxony, in Chemnitz, where he remained till the end of his life. His acquired knowledge was the topic of

---

[18]For a discussion of the formative power in Albert's philosophy, see, for instance, Takahashi (2008).

[19]*De Mineralibus*, Chapter I-I-VI.   [20]*De Mineralibus*, Chapter I-I-VIII.

two important books, *De natura fossilium*,[21] a treatise on mineralogy, published in 1546, and *De re metallica*, an exhaustive treatise on mining and extractive metallurgy, published in 1556, which remained unsurpassed for many years.

In the treatise on mineralogy, minerals are sorted in six categories: earths, salts, bitumen, precious stones, metals, and others. Minerals are characterized by their colour, transparency, glitter and brilliance, taste, smell, the various features of their structure, their hardness, their shape, and their external constitution. Agricola is maybe the first to have distinguished the various typical shapes of minerals: sphere or half-sphere, cylinder like beryl and syenite, cone, spinning top. Many have an angular shape, with six or twelve faces like rock-crystal or a six-sided point-like rock-crystal or diamond; others seem made of hair like asbestos or are dendritic like native silver.

11.2.6   *Girolamo Cardano and de Subtilitate, 1550*

Girolamo Cardano (1501–1576), Italian physician, mathematician, and astrologer, in his encyclopedic book *De Subtilitate*, published in 1550 in Nüremberg, mentions the hexagonal form of quartz. He makes a comparison with the six-sided honeycomb cell and notes that quartz crystals always have six faces, never one less and never one more. He considers briefly the stacking of fourteen identical spherical particles (corresponding to the fourteen summits of a rhomb-dodecahedron, the shape of the honeycomb cell) around a central one but does not clearly relate it to the internal structure of quartz (*Liber VII, De Lapidibus*). This may nevertheless be a first attempt at explaining the geometrical form of crystals by close packing.

11.3   Thomas Harriot and the stacking of cannon balls, 1591

11.3.1   *Close packing of cannon balls*

The first person to consider seriously close packing in early modern times is Thomas Harriot. His interest was not in the structure of matter, but in the stacking of cannon balls.

Thomas Harriot (born *ca* 1560, died 1621) was an English mathematician, astronomer, and ethnographer who lived at the time of Queen Elizabeth I. He was the accountant of Sir Walter Raleigh (or Ralegh) and he took part, in 1585–86, in an expedition to Roanoke Island, off the coast of North Carolina, as Raleigh's scientific advisor. Like most scientists of the time, Harriot had many interests and was also a philosopher. Unfortunately, his works were mostly unpublished, but many of his manuscripts are deposited at the British Library.[22]

Harriot became interested in the stacking of spheres in 1591, when Raleigh wanted information about the ground space required for the storage of cannon balls in his vessels. The story is told by one of Harriot's biographers, J. W. Shirley (1983), taken from Harriot's manuscripts. Thomas Harriot used the laws of mathematical progressions to prepare charts in the form of triangular diagrams which enabled him to read directly the number of cannon balls on the ground or in a pyramid pile with triangular (Figure 11.7), square or oblong base. Such a triangular diagram was already known in the Middle East in the eleventh and in China in the

---

[21]Translated into German by G. Fraustadt, (2006), Wiesbaden: Marix Verlag.

[22]For further reading on Thomas Harriot, see, for instance, Lohne 1959; Jacquot 1952; Shirley 1974, 1983.

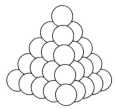

Fig. 11.7   Stacking of cannon balls (a face-centred cubic structure). After Hales (2000).

thirteenth centuries and was discussed by several authors in the sixteenth century; it is now known as Pascal's triangle because the French philosopher and mathematician Blaise Pascal (1623–1662) disclosed some new properties that it had and devoted a small treatise to it in 1654 (*Traité du triangle arithmétique*), published after his death in 1665.

The stacking in Fig. 11.7 is face-centred cubic and Harriot seems to have been the first to have distinguished between the face-centred and the hexagonal close packings. There are of course many other close packings, such as those exhibited by the polytypes. Harriot was an atomist (Kargon 1966) and a follower of Democritus (*ca* 460–*ca* 370 BC) and Lucretius (*ca* 99–*ca* 55 BC); like them, he considered the universe and matter to be made up of indivisible bodies (atoms) and vacuum (void) interposed. Harriot's views caused him some trouble with the Church (Jacquot, 1952). His studies on the stacking of identical spheres certainly encouraged him in his thoughts about the theory of refraction.

11.3.2   Thomas Harriot's atomistic theory of refraction

Thomas Harriot was very interested in optics and established the law of refraction around 1601 (see Section 3.1). In 1609 he acquired a 'Dutch truncke' (telescope) and, on 26 July of that year, drew a sketch of the surface of the moon, a few months before Galileo (Chapman 2009; Lohne 1959).[23] Harriot is also well known for his studies on algebraic theory.

Between 1606 and 1609, Harriot entered into a correspondence with Johannes Kepler. Kepler had unsuccessfully attempted to establish a universal law of diffraction in his *Appendix to Witelo's optics and the Optical Part of Astronomy* (1604), and having heard about Harriot's researches, wrote to him on 2 October 1606, telling him that he ignored the physical nature of light and asking him for his views concerning the refraction of light (Jacquot 1952). In a letter to Kepler dated 2 December 1606, Harriot explains why, according to him, when a ray falls on the surface of a transparent body, part is reflected and part is refracted:

> 'Since by the principle of uniformity a single point cannot at the same time allow a ray to penetrate and reflect it, we must suppose that the ray meets a resistance in one point and not in another. ... Therefore a dense and transparent body which to the sense appears to be continuous in all its parts, is not so in reality, but has corporeal parts, which resist the rays, and incorporeal parts penetrable by the rays; so that refraction is nothing else than an

[23]Harriot only saw blotches and did not understand that the moon's 'strange spottednesse', as he called it, was due to the presence of mountains. According to the art historian S. Edgerton (*Art Journal* (1984). **44**, 225–232) it is because Galileo was skilled in perspective drawing and in Florentine *disegno* that he recognized the moon's relief. The author is grateful to Marjorie Senechal for pointing this out to him.

internal reflexion, and the part of the rays which is received inside, though to the sense it seems straight, is in reality composed of a great number. [*i.e.* a great number of segments forming a broken line] ... I have led you', he adds, 'to the door of nature's mansion, where her secrets are hidden. If you cannot enter on account of their narrowness, abstract yourself mathematically, and contract yourself into an atom, and you will enter easily. And after you have come out, you will tell me what wonders you have seen'. (From Jacquot's translation of the original letters in Latin; Jacquot 1952.)

Kepler held Aristotelian views and did not believe in atoms. In his answer, on 2 August 1607, he declined to follow Harriot *ad atomos et vacua* and said that although understanding very well where Harriot wanted to lead him, he preferred to suppose the union of two opposite qualities, opacity and transparency, in bodies that reflect and refract light at the same time; 'Light ... is of a kind which participates in a surface, and therefore it is influenced by surfaces, and not by thicknesses, with which it does not participate'.[24] In a second letter, dated 13 July 1608, Harriot conceded that his views were based on the doctrine of vacuum and that he did not accept Kepler's explanation 'If such assumptions and reasonings don't satisfy you, I wonder'. Kepler seems to have to a certain extent modified his views about the atomic theory when he wrote his pamphlet about snow crystals (see Section 11.4).

It is interesting to note that until the notion of electron-density distribution was an accepted fact, the problem of vacuum was a big difficulty for all those who described the structure of matter in terms of close packings of spheres or stacking of polyhedra (see for instance, Haüy, Section 12.1.2, and Wollaston, Section 12.2).

11.4 Kepler and the six-cornered snowflake (*Strena Seu de Nive Sexangula*), 1611

Kepler's little 24-page pamphlet, *Strena Seu de Nive Sexangula*, 'A New Year's Gift of Hexagonal Snow' (Kepler 1611)[25] represents only a very small portion of Kepler's works; his main biographer, Caspar (1993), only devotes a few lines to it. However, as pointed out by Laue (1952a), one can find there the first pictures of space lattices, even if he did not coin the word and failed to explain the shape of the snowflakes. It is important for us because the pamphlet has strong crystallographic implications, which it did not have for Kepler himself. He was struck by the beauty and the regularity of the snowflakes and attempted to understand the reason for it. This led him to think about the genesis of matter and its structure, and he expounded his thoughts in a wittily written 'New Year Gift' to his friend and patron, Wackher von Wackhenfels. It inspired many of the scientists of the seventeenth century, Gassendi, Descartes, Bartholin, Steno, Huygens, but was forgotten soon afterwards; Kepler himself does not seem to have paid too much attention to it, calling it in 1622 a 'physics joke'—*jocus physiologicus* (Halleux 1975). It is not mentioned in Metzger (1918) nor in Groth (1926), and it is really thanks to Laue's remark that interest in it has been revived in recent times (Schneer 1960; Halleux 1975). Kepler's main works were on astronomy, and it is interesting to try to understand why he made

---

[24]*Johannes Keplers Gesammelte Werke* (1954), **16**, Munich: C. H. Beck, quoted by A. R. Alexander (1995). The imperialist space of Elizabethan Mathematics. *Stud. Hist. Phil. Sci.*, **26**, 559–591.

[25]It has been commented in particular by its translators, Hardie (1966) and Halleux (1975), and by Schneer (1960) and Shafranovskii (1975).

this seemingly atomistic incursion in the inner structure of matter. The answer is to be found in Kepler's passion for harmony and in his deep conviction that 'mathematics were behind all physical things because God used mathematics as a prototype for the creation of material quantities' (Caspar 1993). This continuous search for harmony culminated in the publication of *Harmonice Mundi* (World Harmony) in 1619. It covered all the aspects of nature: the geometrical ones with the exploration of regular polygons, the tessellation of the plane, the Archimedean solids and the regular stellated polyhedra (now known as Kepler's solids), music and the causes of the harmonics, astrology and, finally, cosmology, the movements of the planets and his third law.

11.4.1   Kepler's early years

Kepler, a Lutheran, studied theology at the University of Tübingen as well as mathematics, in which he was very gifted. In 1594 he became Professor of Mathematics and Astronomy at the protestant school in Graz, Austria. Kepler was immediately convinced by Copernicus's picture of the universe. For an answer to his many questions regarding that structure, he turned towards geometry. On 19 July 1595,[26] when he was not yet 24, he had a revelation: he saw a way to relate the orbits of the planets to the five Platonic regular solids (see Section 2.1). His idea was that the planets Mercury, Venus, Mars, Jupiter, and Saturn could each be associated with one of the solids. Taking the sphere containing the orbit of the Earth, one can circumscribe a dodecahedron around it, the sphere stretched around it contains the orbit of Mars. Repeating the operation successively with a tetrahedron and a cube, one obtains the orbits of Jupiter and Saturn. If one now places an icosahedron inside the orbit of the Earth, the sphere inside it defines the orbit of Venus. Placing an octahedron inside it provides the orbit of Mercury. From the geometrical properties of the five polyhedra, Kepler could get an estimate of the ratios between the distances of the planets from the sun, which were in rather rough agreement with the values given by Copernicus. That was the starting point for months of calculations which resulted in the publication by Kepler in 1596 of his treatise *Mysterium Cosmographicum* (Mystery of the Universe). In order to improve his model, he needed accurate observations. For that he turned towards the Danish astronomer, Tycho Brahe (1546–1601), who was the only one who could provide them. He sent him his book and received an encouraging answer in 1598. Eventually, in January 1600, after having had to leave Graz when the Protestants were chased out of the city by the Counter Reformation, he became one of Tycho Brahe's assistants in Prague. On 26 October 1601, Tycho Brahe died, leaving his scholarly estate to Kepler. Two days later, Emperor Rudolph II transferred the care of Tycho Brahe's instruments and scientific heritage to Kepler and appointed him as Imperial Mathematician. With the wealth of Tycho Brahe's observations at his disposal, he could now set to work in his quest for world harmony, and he was to be the architect of the resultant new structure of the universe. After lengthy calculations and overcoming difficulties with the orbit of planet Mars, he came to two important conclusions that broke away from the lines of thought of Ptolemy, Copernicus and Tycho Brahe. The first one is

---

[26]He made a note of it (Caspar 1993).

Fig. 11.8   Johannes Kepler *ca* 1610—Portrait by an unknown artist. Source: Wikimedia commons.

Johannes Kepler: born 27 December 1571 in the imperial city of Weil der Stadt in Swabia, in south-west Germany; died 15 November 1630 in Regensburg, Bavaria, Germany. His father was a mercenary soldier and his mother the daughter of an innkeeper. In 1577 his mother took him up a hill to observe the Great Comet, and in 1580 his father took him out one night to observe a lunar eclipse. After schooling at a Latin school and a seminary, he studied theology and classics at the University of Tübingen. There, he became interested in mathematics and astronomy and became an adherent of the heliocentric theory first developed by the Polish priest and astronomer Nicolaus Copernicus.

Kepler's theological studies ended in 1594 when his masters recommended him for a position as teacher of mathematics and astronomy at the Protestant school in Graz, the capital of Styria, Austria. He took his position in April 1594, and held that chair until 1600, when he had to leave because Protestants were forced to convert to Catholicism. He then became assistant to the Danish astronomer Tycho Brahe in the latter's observatory near Prague. On the death of Brahe in 1601, Kepler assumed his position as imperial mathematician and astronomer at the court of the Holy Roman Emperor Rudolph II. In March 1610, Kepler learned of Galileo's discoveries using a telescope. He repeated Galileo's observations early September, concluding that Galileo's planets were satellites (a name he coined) of Jupiter. In 1611, trouble erupted in Prague due to dissensions between the Emperor and his brother, Matthias, resulting in the abdication of Rudolph II in May of that year. Kepler had to find another position and was appointed district mathematician in Linz, upper Austria, in June 1611. There he wrote his *World Harmony* and the *Epitome of Copernican Astronomy*, and worked on the *Rudolphine Tables*, which he finished in Ulm after having been chased from Linz by the Thirty Years War in 1626.

MAIN PUBLICATIONS

1596–1597	*Mysterium Cosmographicum* (Mystery of the Universe), first comprehensive and cogent account of the geometrical advantages of Copernican theory.	1610	*Narratio de Observatis Quatuor Jovis Satellibus* (Observation of the four satellites of Jupiter).
1604	*Ad Vitellionem Paralipomena quibus Astronomia pars Optica traditur* (Appendix to Witelo's optics and the Optical Part of Astronomy), devoted to optics and refraction by the atmosphere.	1611	*Strena Seu de Nive Sexangula* (A New Year's Gift of Hexagonal Snow).
		1611	*Dioptrice* on the optics of the telescope.
1606	*De Stella nova in pede Serpentarii* (On the New Star in Ophiuchus's Foot), concerning the observation of a supernova.	1619	*Harmonice Mundi* (World Harmony), which contains his third law on the movement of planets.
1609	*Astronomia Nova* (New Astronomy), which contains his first two laws on the movement of planets.	1618–1621	*Epitome Astronomiae Copernicanae* (Epitome of Copernican Astronomy, which brought all of Kepler's discoveries together in a single volume).
		1627	*Tabulae Rudolphinae* (Rudolphine Tables), new astronomical tables based on Tycho Brahe's observations and Kepler's calculations.

that the centre of the universe is not the centre of the Earth's orbit, as was their belief, but the sun itself. The fact that the positions of the sun and of the centre of the Earth's orbit occupy slightly different positions led to small errors. The second conclusion is that the Earth does not move uniformly along its orbit: Kepler saw no reason for the Earth to behave differently from the other planets. These two considerations led him to realize that the sun is the centre of a moving force and that the planetary system must be regulated by physical forces and governed by inner laws. In this way, he paved the way for Newton and his findings in the late 1660s. In doing so, he was breaking away from Aristotelian physics. By 1605, Kepler had derived his first two laws of planetary motion, but printing of his work was delayed for financial reasons until 1609, when *Astronomia Nova* (New Astronomy) was finally published. Meanwhile, he had investigated the optical problems related to the observation of the moon and the planets and the refraction of light by the atmosphere. He also made an important contribution to the study of conics.[27] All of this resulted in his book published in 1604, *Ad Vitellionem Paralipomena quibus Astronomia pars Optica traditur* (Appendix to Witelo's optics and the Optical Part of Astronomy).

11.4.2   A New Year's gift: a milestone for the concept of space-lattice

After the publication of his *Astronomia Nova*, Kepler could at last have a period of quiet and happiness. As the Court Astronomer he was well considered and had every facility in the beautiful and flourishing imperial city of Prague, which, under the reign of Rudolph II, was playing an important role in European politics. By taste, the shy Ruldolph II encouraged crafts, arts, and sciences; a general intellectual climate of mixed humanism, science and mysticism reigned. There was a well-developed court life, and Kepler had several patrons and friends among the intellectuals, court officials and nobility. The most prominent and loyal one of these was the imperial councillor, Baron Johannes Matthäus Wackher von Wackhenfels (1550–1619), humanist and diplomat. They had many scholarly conversations together and were wont to exchange remarks full of humour. At New Year 1611, Kepler presented his friend with the *Strena Seu de Nive Sexangula* (Fig. 11.9), 'A New Year's Gift of Hexagonal Snow'. Behind the banter there is a deep mathematical and geometrical cogitation in this little essay. It deals with the symmetry of the snowflakes and attempts to explain their regularity by the regular geometrical arrangement of identical minute pellets. It also shows how Kepler's mind balanced between the old Aristotelian view of nature and the new physical description and explanation of phenomena. We have seen in Section 11.3.2 how Kepler disagreed with Harriot's atomistic theory of the refraction of light. In order to recapture the spirit of the time and to better follow the successive steps of Kepler's reasoning, many quotations of Kepler's original text will next be given.[28]

Kepler starts by saying that he knows 'how much his friend loves *Nothing*' (*quam tu ames Nihil*). Laue (1952a) sees here nothing more than a play on words, *Nix* meaning 'snow' in Latin and 'nothing' in German dialect. According to Halleux (1975), it is in fact a reference to a poem

---

[27]It is Kepler who introduced the term *focus* (Boyer 1991), e.g. in his first law: 'the planets move around the sun in elliptical orbits with the sun at one focus'.

[28]Free translation from the Latin by the author; the Latin version can be downloaded from the web page <http://www.thelatinlibrary.com/kepler/strena.html>.

IO ANNIS KE-
PLERI S.C. MAIEST.
MATHEMATICI
STRENA

Seu

De Niue Sexangula.

Cum Priuilegio S. Cæf Maieft. ad annos x v.

FRANCOFVRTI AD MOENVM,
apud Godefridum Tampach.

Anno M. DC. XI.

Fig. 11.9   First page of *Strena Seu de Nive Sexangula*. After Kepler (1611).

Fig. 11.10   Snowflake. Photo courtesy SnowCrystals.com.

by J. Passerat both Wackher and Kepler liked, entitled *Nihil*, precisely about a poet who does not know what to offer as a New Year gift. He reflects that 'nothing' is more precious than gold, 'nothing' is harder than iron, etc. and concludes that nothing is better than 'nothing' as a gift! In my view, it may also be because Kepler did not succeed in finding a satisfactory explanation to the six-fold symmetry of snowflakes (Fig. 11.10), and concludes that all his efforts led to 'Nothing'. For instance, for an idea which he knows will lead nowhere, he speaks about an 'idea for nothing' (*Nihili opinio*), and when he discusses an opinion seriously, he says it is not for 'nothing' (*non de Nihilo*).

Kepler lists things that could be taken for Nothing. 'Whatever will evoke for you *Nothing* will be very small, cheap and worthless ... Maybe you will think of one of Epicurus' atoms; in truth, it is nothing'. Grains of sand, sparks of fire, chips of pyrite, particles of dust, etc., even the smallest animal known at the time, the subcutaneous mite, all of them are too small.

Kepler writes: 'I was pondering on the minutest things I could offer, and, as I was crossing the bridge over the Moldau in Prague, snowflakes started falling sparsely on my dress, all of them

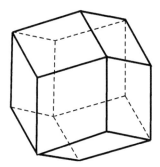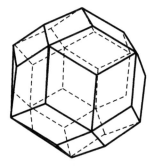

Fig. 11.11   *Left*: Rhomb-dodecahedron, oriented with a three-fold axis vertical. *Right*: rhombic triacontahedron.

six-cornered, with shaggy radii. Here is the desired gift', he exclaims, 'worth being offered by a mathematician to the lover of Nothing, since it falls from the sky and is similar to the stars. What an appropriate name since, if you ask a German, what is *nix*? he will answer *nothing*, unless he knows Latin' (and he will say 'snow').

'Why', asks Kepler, 'are the snowflakes always hexagonal, with hairy radii like little feathers, when they start falling, before they intertwine in bigger flakes? ... There must be a reason, for why do they not fall with five or seven corners?' Kepler considers the possible causes from an Aristotelian viewpoint.[29] 'The matter of snow is vapour', he notes, following Aristotle, 'in which you cannot distinguish small stars. Vapour is fluid and has no boundary and no shape until it condenses into snow or in water droplets. The cause is therefore not material and it must be due to the action of a formative agent; is it an external efficient cause or is the six-cornered figure a material necessity? is the cause in its nature itself, in the archetype of beauty lying in the hexagon or in its finality? To distinguish between these points, examples must be taken, but described geometrically'.

11.4.3   Examples of hexagonal structures given by Kepler

The first example of a hexagonal structure is that of the honeycomb (Fig. 11.3). 'If you ask the geometricians what is the arrangement of the structure of the cells of the bees, he will tell you an hexagonal arrangement'. Kepler notes that each cell has nine neighbours, six in the plane and three below, and that it shares a wall with each of them. He further notes that the three faces bounding the bottom of the cell are lozenge-shaped (*rhombs*) and he looks for polyhedra with rhomb-shaped faces. He finds two, one with twelve faces, the rhomb-dodecahedron (Fig. 11.11, *Left*), and one with thirty faces, the rhombic triacontahedron (Fig. 11.11, *Right*) and concludes that the honeycomb cell is a rhomb-dodecahedron. Most importantly, he further concludes that 'one can fill space completely with this geometrical figure, which is closest to a regular solid, in the same way as one does the plane with hexagons, squares and triangles'.

The second example taken by Kepler is that of the seeds of the pomegranate fruit, which are shaped like rhomb-dodecahedrons. 'What is it', he asks, 'that makes both the bee cell and the

[29]See Section 11.2.4, footnote 17.

pomegranate seed take that geometrical form? In the case of the pomegranate', he argues, 'it cannot be a formal property and it must be a material necessity since when the seeds are small they have enough space and are round-shaped; they can move freely as does a liquid. As the skin of the fruit hardens and the seeds continue growing, they become compressed like peas in their pod'. Kepler then imagines: 'Let us put identical spherules in a round vessel and let us compress the vessel from all sides with brass rings. Many particles will take a rhombic shape',[30] 'particularly if by shaking the vessel you help them to occupy a smaller space. If the spherules lie in unperturbed straight rows, they can even become cubes. In general, there are two modes for identical spherules to arrange themselves in a vessel, corresponding to the two possible modes of ordering in a plane. If free particles lying in a plane are pushed together in a restricted space so that they touch each other, they will form either a triangular or a quadrangular pattern. In the former case, one spherule is surrounded by six others, in the latter by four. ... The pattern formed by identical spheres is never pentagonal.'

11.4.4 Stacking of identical spheres: cubic lattices P and F, hexagonal lattice H

Kepler now looks for various possibilities of stacking identical small spheres, which he calls spherules: 'Let us now superpose layers of spherules on top of each other so that they are as closely packed as possible. [In a given layer], they will form either squares, A, or triangles, B (Fig. 11.12 (1)). If the arrangement of the spherules in their plane is square (A), either a spherule of a given layer lies on top of one spherule of the layer below, or it lies on top of four of them. In the first case, each spherule is in contact with four others in the same plane, one in the layer above it and one in the layer below, namely six altogether, forming a cubic arrangement; under compression it will become a cube'. Kepler has described here a simple cubic lattice P. He continues by noting: 'But this is not the most compact arrangement: in the second case, each spherule is in contact with four others in the same plane, four above and four below, namely twelve altogether, and it becomes a rhomb-dodecahedron when compressed'. This corresponds to the face-centred cubic lattice F, Kepler's layers being parallel to the cubic planes. The arrangement is the tightest, no more spherules can be confined in the same vessel.[31]

'We now come back to the triangular pattern in the plane (Fig. 11.12 (1), B). 'In the structure of the solid, either a spherule lies on top of a spherule of the layer below and the arrangement is again loose or it lies on top of three others of the layer below. In the former case, one spherule touches six others in the same plane, one above and one below, namely eight in total. This structure corresponds to that of prisms. Under compression, the spherules will become columnar with six quadrangular lateral sides and two hexagonal faces'.[32] 'In the latter case, one gets the same result as in the second case described above for the square pattern'. The first case corresponds the hexagonal lattice H and the second one again to the face-centred cubic lattice F.

'Let now B be a group of three spherules (Fig. 11.12 (2)), and A a single spherule that we put on top of them as an apex. Consider in the same way groups of six (C), 10 (D) and

---

[30]*Schema rhombicum*—that is, a rhomb-dodecahedron.

[31]This has been called the Kepler conjecture, see Section 11.4.8.

[32]That is a hexagonal prism terminated by a pinacoid.

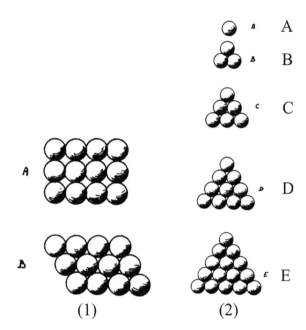

Fig. 11.12   Stacking of identical spherules. (1) The spheres are arranged in a given layer according to a square (A) or a triangular (B) pattern. (2) A, B, C, D, E are successive layers to be stacked on one another. After Kepler (1611).

fifteen (E) spherules. Superposing these groups on top of each other, the smaller ones above the larger ones, one obtains a pyramid' (see Fig. 11.7). 'In this construction, each spherule lies on top of three others. If one then turns the pyramid round and looks at a lateral side, each spherule sits on top of four others and one can see that it has twelve neighbours. Thus, when the structure is close-packed, it is impossible to have triangular patterns without square ones, and vice-versa'. The planes of Fig. 11.12 (2) that are superposed to form a pyramid are {111} planes and the lateral faces of this pyramid are cubic {100} planes. It should be noted that, if Kepler correctly described the face-centred cubic close-packing, he failed to find the hexagonal close-packing.

Kepler concludes that in the case of the pomegranate seeds it is a material cause that leads to the rhomb-dodecahedron, the compression on the seeds as they grow, while in the case of the bees it is through their instinct, which they receive from God, that they find the shape storing the biggest quantity of honey. 'In the plane, only triangles, squares and hexagons fill space completely,[33] without leaving a gap, and it is the hexagon which is the most spacious ... The solids that can fill space without leaving a hiatus are the cube and the rhomb-dodecahedron and it is the latter that has the biggest volume'.[34]

[33]See Fig. 11.20.

[34]This important property of the rhomb-dodecahedron was to be discussed by several other authors, in particular Lord Kelvin (1887).

11.4.5   Considerations as to the origin of the shape of the snowflakes

Kepler then makes a digression in which he asks why many plants have pentagonal-shaped leaves or flowers (Fig. 11.13) and speculates on the beauty and the properties of this shape, which characterizes the 'soul' of these plants. After discussing the golden ratio[35] and the properties of the pentagon-dodecahedron and the icosahedron, he comes back to the causes of the shape of the snowflakes: 'there are extrinsic and intrinsic causes, cold being the first external cause'. Kepler compares the formation of the snowflakes to that of frost on window panes, but asks 'why does the vapour condensate in a figure with six points connected by six radii? Cold, as an external cause, cannot produce such a figure; there must be an internal cause or a cause proper to it'.

Kepler has 'often observed with admiration that, when they start falling, the small stars are not planar, a few of them staying in the air, until, finally, they become planar'. This gives him an idea, to which he is going to devote a long development, but which he warns does not describe the truth (*Nihili opinio*—an idea for nothing). He assumes that the six shaggy radii (*Why is it not five or seven?*) constitute three orthogonal diameters. 'If you ask geometricians which geometrical figure presents three diameters at right angle forming a double cross, they will tell you it is an octahedron' (see Fig. 11.14). 'Indeed, the octahedron has six summits. Why is it that the snowflakes, before they become planar, imitate the skeleton of an octahedron with their three orthogonal diameters? Why is it that the condensation forms three shaggy radii, rather than a whole sphere?'

Kepler is now trying to explain the structure of the three-dimensional snowflakes with three orthogonal diameters. He follows successively various tacks. At first, he considers that vapour condenses into small drops analogous to water drops and that, in order to better resist cold, these drops touch each other, forming regular arrangements similar to the packings of spheres he described before. The drops align themselves either along three orthogonal directions one can associate with the diameters of an octahedron (Fig. 11.14) or along orthogonal and oblique directions one can associate with a cubooctahedron (Fig. 12.51 (5)); the first case corresponds to a simple cubic lattice P and the second one to a face-centred cubic lattice F, but Kepler of

Fig. 11.13   Pentagonal-shaped flowers. *Left*: Borage (*Borago*). *Right*: Balloon flower (*Platycodon*). Photos by the author.

---

[35]Two quantities, a and b, are in the golden ratio if the ratio $\phi = (a+b)/a$ of the sum of the quantities to the larger quantity a is equal to the ratio a/b of the larger quantity to the smaller one. The golden ratio is equal to $\phi = (1+\sqrt{5})/2 = 1.61803$.

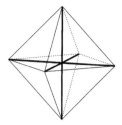

Fig. 11.14   Octahedron with Kepler's three diameters along which are arranged the
        drops into which the vapour condenses.

course does not say so. This explanation implies that 'the internal arrangement of the spheres
is the consequence of their external appearance', but Kepler does not see why the octahedron
should be preferred to the cube. Furthermore, if that explanation was correct, the snowflakes
would all be identical, which is observed not to be the case.

   Another possible explanation is that 'the three hairy diameters hover individually and snap
together accidentally in a cross during their fall', but that is highly improbable; however, 'there
is in the center of the nest a formative force which radiates equally in every direction'. Kepler
makes also an analogy with animals which have three diameters as well: 'upper and lower parts,
forward and hind parts, left and right parts'.

   Finally, Kepler concludes that all the previous reasonings about the origin of the snowflake
have been for Nothing and, 'after having taken everything in consideration', his 'feeling as to
the origin of the hexangular shape of the snowflakes is that it is not different from that of the
regular shapes and constant numbers found in plants ... The shape of the snowflake does not
occur without reason. It is due to the creative skill of the Universe'.

11.4.6   Why are snowflakes hexagonal?

After having 'spoken of who is at the origin of the shape of the snowflake', Kepler discusses the
shape itself: 'does it result from the intersection of the three diameters, which is so far one of
the hypotheses, or is it hexagonal from the beginning, as we shall discuss later?' At first, Kepler
sticks to the idea of a three-dimensional snowflake. For him, the soul of things is the image of
the God Creator and is related to regular and geometric figures. 'Among the regular solids, the
cube is the first-born, father of all others. The octahedron is its corresponding female, having as
many summits as the cube has faces'.[36,37] Kepler asks himself why the snowflake adopts the
shape of the octahedron rather than that of the cube and notes that *the jewellers*[38] 'say that it
is possible to find natural highly perfect and well polished octahedra of diamond'. But, finally,
he does not find any conclusive argument for the snowflake to adopt one shape rather than an
other one. 'Where am I drawn, fool that I am; while I am trying to give near Nothing, in fact,

---

[36]Kepler develops the relations between the regular solids in *Harmonice Mundi*.

[37]Romé de l'Isle (1783), in his *Cristallographie*, still notes that the octahedron is the inverse form of the cube.

[38]Kepler refers to his colleague, Boece de Boot (1550–1632), born in Bruges, gem-cutter, mineralogist, and physician
to Emperor Rudolph II, author of a treatise on mineralogy entitled *Gemmarum et lapidum Historia*, published in 1609,
just before Kepler wrote the *Strena*.

I get to near Nothing because from this near Nothing I nearly formed the world itself, which contains everything: while above I fled from the small soul of a minute animalcule, from an atom of snow I produce the soul of a triply big animal, the globe of the Earth'.

Kepler then says that he is 'finally going to speak seriously about the little stars of snow'. As he was writing, it started snowing again. Observing carefully, he distinguished two kinds of flakes. Both kinds have radii. The most frequent ones are round and have radii converging on a small spherule. 'Scattered in between were small hexagonal stars of a second type, much rarer, which did not hover nor fall in any other way than flat. The hairs were together in the same plane with the stem. Below, there was a seventh radius, like a root, on which they leaned when falling and on which they rested in an elevated position for a little while. This hadn't escaped me before, but had been wrongly interpreted, as if the three diameters were not in the same plane. What I have said so far is thus as close to Nothing as the topic about which I spoke. In the second kind, which is that of the little stars, there is no point in considering cubes and octahedra and neither the contact between drops, since they occur planar and not with three intersecting diameters, as discussed above'.

'First, why is the shape of the snowflakes planar?' In plants too there are planar shapes: in flowers you have a planar pentagon and not a three-dimensional dodecahedron. 'Why is the shape chiefly hexagonal? Is it because, among the regular polygons, this is the first one that is not also the face of a polyhedron? Is it because the hexagon fills the plane without leaving gaps? Is it because the hexagon is the figure closest to the circle'?[39] 'Is there a discrimination between the ability of sterile shapes such as triangles and hexagons and that of fertile shapes such as the pentagon'?[40] 'Or is the principle of the hexagon by essence in the nature of the formative strength[41] (of nature)? 'In favour of the fifth cause are other works of the generating strength, such as crystal'(quartz) 'always hexagonal while octahedral diamonds are very rare'. Examples of various shapes presented by crystals follow.

In the end, Kepler declines to conclude and tells his friend and patron that 'seeing all I still have to discover, I prefer to listen, oh most clever man, to what you have to say and not to tire myself discussing any longer'.

11.4.7 Kepler's contributions and his limitations

Kepler, like others before him, noted that one can fill a plane with triangles, squares and hexagons, but not with either pentagons nor heptagons. He showed that one can fill space not only with cubes, but also with rhomb-dodecahedra. He derived the simple cubic, face-centred cubic and hexagonal lattices, but not the body-centred cubic lattice. He did not use the term 'lattice' nor did he introduce the concept of lattice as we know it; for instance, he did not consider that regular assemblies could be extended to infinity; he correctly explained that one can describe the face-centred cubic lattice as well with a stacking of {100} planes parallel to the faces of the cube as with a stacking of {111} planes parallel to the faces of the octahedron. He showed also that, by compression of an assembly of spheres packed according to

---

[39]Explanation given by Boece de Boot for the hexagonal shape of quartz.

[40]Reference to the importance of the pentagon in plants and in the golden ratio.

[41]Boece de Boot also speaks of a 'formative or seminal strength' as the origin of the formation of precious stones, as did already Albertus Magnus, see Section 11.2.4.

either of these three lattices, these spheres become respectively cubes, rhomb-dodecahedra and hexagonal prisms[42] and thus obtained three polyhedra that entirely fill space; that is, three of the five Fedorov parallelohedra[43] (Shafranovskii 1975; Galiulin 2003). Moreover, he declared that it is face-centred close packing which occupies maximum space. On the other hand, he did not find the hexagonal close packing, which Harriot may have known about (see Section 11.3.1) but was first described by W. Barlow in 1883 (see Section 12.13).

In summary, Kepler gave a correct interpretation of the hexagonal structures of the honey-comb and of the pomegranate fruit—the former because it is that which provides the biggest amount of space to store honey, the latter because it is the one obtained by compressing an assembly of close-packed spheres. In the case of the honeycomb, he went one step further than his predecessors, Varro, Pappus, *etc.* when he considered the honeycomb cell in three dimensions. Robert Hooke also limited himself to packings in the plane (see Section 11.5). Both structures, the honeycomb and the pomegranate fruit, are macroscopic. Kepler, however, failed in his attempt to explain the hexagonal shape of the snowflakes because he did not consider the microscopic inner structure of the snowflakes. Schneer (1960) refers to Kepler as the initiator of 'crystallographic atomism'; Kepler himself, however, did not really hold atomistic views. He related the hexagonal symmetry of the snowflakes to that of the honeycomb cells and that of the pomegranate seeds. These, however, can be explained by the macroscopic assemblies constructed with them, while that of the snowflake must be explained by their inner structure; this is what Kepler could not imagine. He did observe the hairy diameters of the flakes, but could not give a satisfactory explanation for them. However, he correctly guessed that they grow from a nucleus in the centre. Once, he mentioned very briefly the possibility for the structure of an individual spherule to reflect the arrangement of an assembly of close-packed spherules, but he dismissed the idea immediately. According to the views of the time, snowflakes, like plants, have souls; as noted by Schneer (1960), Kepler thought of the Earth as alive. Whether snowflakes are spherical or planar star-shaped, Kepler concluded that their shape lies in their nature and is due to the formative strength of the Universe, in line with his concept of the Harmony of the World.

11.4.8 Kepler's conjecture

Kepler's postulate that the most efficient arrangement of equally-sized spheres is the face centred close packing has been called the 'Kepler conjecture'. The average density of the spheres in close packed structures (face-centred cubic, hexagonal or the others derived from them) is $\pi/(3\sqrt{2}) = 0.74048$. Kepler gave no proof for it, and it turned out to be very difficult to do so. The problem was tackled by many mathematicians. C. F. Gauss[44] is usually credited with having shown that the densest lattice packing is the face-centred cubic lattice (Gauss 1831). In fact, he did not state it that way, it is simply implied by a general property derived by him concerning the volume of the unit cell of a lattice as a consequence of his demonstration of Seeber's theorem (see Section 12.7.2). In 1901, D. Hilbert (1862–1943) included this problem

---

[42]W. Barlow was to use the same image independently, with rubber balls; see Section 12.13.

[43]Parallelohedra are convex bodies which allow tilings of Euclidian space by translations only. They were classified by Fedorov (1885); see Section 12.14.

[44]See footnote 44 in Section 12.7.2.

in his list of twenty three unsolved problems of mathematics;[45] it is part of his eighteenth problem. In 1953, the Hungarian mathematician, L. Fejes Tóth, proposed a program to solve the packing problem.[46] Following this approach, the American mathematician Thomas Hales gave in 1998 a computer-based proof that was finally published in 2005 (Hales 2005).[47]

11.4.9 Other observations of snowflakes in the sixteenth and seventeenth centuries

Aristotle, in his *Meteorologia*, described the formation of snow. This subject was taken up in detail by Albertus Magnus in his treatise on meteorology (*de Meteoris*), where he discusses the formation of hoarfrost and snow from vapour and describes snow falling in the shape of stars (Book II-1-X). The first sketches of snow crystals which were published in Europe were by Olaus Magnus, Archbishop of Uppsala, in 1555 (Frank 1982; Mason 1992). Kepler had studied Aristotle in Tübingen, and probably knew Albertus Magnus' works, but it is unlikely that he knew about Olaus Magnus' sketches. The Danish scientist and physician, Rasmus Bartholin (see Section 11.6.1), however, does mention Olaus Magnus (Bartholin 1661). Kepler's essay prompted other scientists of the time to look at snowflakes. Most of them more or less explained their hexagonal shape by the agglomeration of six small units around a central one.

The French astronomers Nicolas-Claude Fabri de Peiresc (1580–1637) and Pierre Gassendi (1592–1655), who had read the *Strena*, made observations of hexangular snowflakes on 12 February 1623 and 29 January 1629, respectively (Halleux 1975). The latter even observed the snowflakes under the microscope. Gassendi,[48] who was also a philosopher and an Epicurean, put forward several possible explanations for the origin of the hexagonal form: the soul of the World; the eternal wisdom of the Earth; growth from star-shaped seeds according to his theory of seminal strength; coalescence of six droplets around the first one formed, which are then surrounded by others so as to generate the six radii of the snowflakes.

Gassendi discarded the Aristotelian view that matter is constituted of the elements water and earth, with heat as the formative agent, but considered that minerals result from seeds created by God. According to him,[49] a mineral is built by a regular distribution of small particles, which he calls seeds (*semina*), under the action of a formative or seminal strength (*vis seminalis*), these seeds having a shape characteristic of that mineral. Gassendi seems to have been the one who coined the term 'molecule', as a diminutive of the Latin *mole*, mass, in the first version of Book XIII, *De atomis*, of his treatise, *De vita e doctrina Epicuri*, written between 1633 and 1645 (Bloch 1971;[50] Kubbinga 2002).

René Descartes,[51] the French philosopher, was aware of Kepler's and Gassendi's observations. While he was in Amsterdam, in February 1635, he recorded very accurately on successive

---

[45]D. Hilbert, *Mathematische Probleme, Archiv Math. Physik.* (1901), **1**, 44–63.

[46]L. Fejes Tóth (1953), *Lagerungen in der Ebene, auf der Kugel und im Raum*, Berlin: Springer.

[47]For an overview of the Kepler conjecture by T. C. Hales, see the web page <http://arxiv.org/PS_cache/math/pdf/9811/9811071v2.pdf>.

[48]In his work, *Syntagma philosophicum* (Philosophical Treatise), published posthumously in 1658.

[49]In Chapter 3, *De Lapidibus, ac Metallis*, of his book *De Rebus Terrenis Inanimus*, which forms part II of the Philosophical Treatise.

[50]Bloch, O. R. (1971). *La philosophie de Gassendi*, The Hague: Martinus Nijhoff.

[51]See footnote 4, Section 3.1.

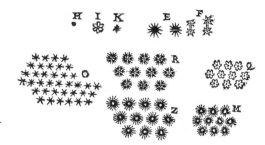

Fig. 11.15   Descartes' sketches of snowflakes.
   After Descartes (1637).

days the various shapes of the snowflakes that had just fallen, which he compared to little stars and roses (Fig. 11.15). He described them in his book, 'Meteorology', *Météores*, included in the 'Discourse on Method'.[52] Descartes had a mechanistic approach to nature; he did not believe in atoms nor in vacuum but supposed 'that water, earth, air, and all other such bodies that surround us are composed of many small parts of various shapes and sizes, which are never so well arranged nor so exactly joined together that there do not remain many intervals around them; and that these intervals are not empty but are filled with that extremely subtle matter[53] through which the action of light is transmitted ... In solid bodies these small parts are linked together, but not simply stacked on each other'.[54] For the snowflakes, he thinks that their hexagonal shape is due to the agglomeration of six small grains around a central one: 'I had some difficulty imagining what could have so accurately assembled six little teeth around each grain, in the middle of free air and by a strong wind, until I considered that this very wind could have flown away some of these little grains outside a cloud and held them up, since they were so small. There, they must have organized themselves in such a way that each of them was surrounded by six others, *in the usual way of nature*'.[55] He also observed, however, some flakes with twelve branches and a few with five.

Snowflakes were also sketched or described by Robert Hooke (see Section 11.5.3), and by the Danish scientists, Rasmus Bartholin (see Section 11.6.1), and his student, Nicolas Steno (see Section 11.7.1). In his treatise on the *Solar corona and parhelia* (Volume XVII, pp. 364–516 of his Complete Works, 1932. La Haye: Martinus Nijhoff), Christiaan Huygens also refers to the shape of snowflakes as described in Kepler's *Strena* and Descartes' *Météores*.

11.5   Robert Hooke and *Micrographia*, 1665

11.5.1   The Curator of Experiments of the Royal Society

Robert Hooke (Fig. 11.16) was a British natural philosopher and architect, considered a master of experiment and observation, with a very wide range of interests and a great inventiveness of

---

[52]The part concerning the observation of snowflakes was translated by F. C. Frank in *J. Glaciology*, 1974, **13**, 535–539. *Descartes' observations on the Amsterdam snowfalls of 4, 5, 6 and 9 February 1635.*

[53]Descartes' *subtle matter*, or *matière subtile*, is not very different from other authors' *ether* or *aether*.

[54]In the first discourse, *On the nature of terrestrial bodies*, of the 'Meteorology'.

[55]In the sixth discourse, *About snow, rain and hail*, of the 'Meteorology'.

Fig. 11.16   Robert Hooke, Portrait by Rita Greer, 2004 (Source: Wikicommons).

Robert Hooke: born 18 July 1635 in the Isle of Wight, England, the son of the curate of the parish; died 3 March 1703 in London, England. He received his early education from his father and served as an apprentice to a painter. After his father's death in 1648 he attended Westminster School in London. In 1653 he entered Christ Church College in Oxford as a chorister and in 1655 became Boyle's assistant. He was interested in astronomy and in the problem of the accurate determination of the longitude while at sea. This led him to design a clock or watch using a circular spring to control its time-keeping movement and to study the conditions for harmonic motion. By hanging weights to a vertical spring, a spiral spring placed vertically or at the end of a horizontal cantilever, he established around 1660 the law of springs (Hooke's law) and devised the anchor escapement mechanism for watches, which was rediscovered by Huygens in 1675. He did not publish his discovery at the time, but much later, in 1676, defining the law by an anagram, *ceiiinosssttuv*, of which he gave the solution, *ut tensio, sic vis*, ('as the extension, so is the force', stress is proportional to strain) in 1678. From 1662 to 1677 he was Curator of Experiments at the Royal Society. In 1663 he was elected a Fellow of the Royal Society and, from 1677 to 1682, Secretary of the Society. In 1664 he was appointed Professor of Geometry at Gresham College in London. Around that time, after the Great Fire of London, he made plans for the reconstruction of London and was made City Surveyor. He built telescopes, invented the iris diaphragm, observed the rotation of Jupiter (1664) and made drawings of Mars (1666). He was always interested in the notion of gravitation. On 21 March 1666, he gave a lecture *On Gravity* before the Royal Society, published in 1674 as an addition to *Attempt to prove the Motion of the Earth*. He was one of those who began to think of an inverse square law for the distance dependence of the force of gravitation, which he mentioned in a letter to Newton (1679). He made preliminary studies of diffraction, presented in a lecture at the Royal Society on 18 March 1675 (Hall, 1990), ten years after Grimaldi, and compared it to the similar effect in sound. His contributions to geology are very important. His descriptions of fossils such as ammonites are very accurate. He compared fossils to living organisms, and, to him, they were a representation of past life; he had very modern views as to the process of petrification and, as an early evolutionist, was well ahead of his time. His geological views are expounded in his posthumous *Discourse of Earthquakes*. In 1691 he was made Doctor of Physics, and in 1700 he presented his last invention, a marine telescope, to the Royal Society.

MAIN PUBLICATIONS

1665 *Micrographia, or Some Physiological Descriptions of Minute Bodies Made by Magnifying Glasses with Observations and Inquiries thereupon.*

1674 *An Attempt to prove the Motion of the Earth.*

1674 *Animadversions on the first part of the Machina Coelestis.*

1676 *A Description of Helioscopes and some other Instruments.*

1677 *Lampas: or, Descriptions of some Mechanical Improvements of Lamps & Waterpoises. Together with some other Physical and Mechanical Discoveries.*

1678 *De Potentia Restitutiva, or of Spring Explaining the Power of Springing Bodies.*

1678 *Cometa, or Remarks about Comets.*

1705 *Posthumous works*, including *Discourse of Earthquakes.*

mind, who played an important role in the Scientific Revolution.[56] He is best known for his law of elasticity, Hooke's law, and for his book of observations with the microscope, *Micrographia*, where he introduced the term 'cell' to describe the basic unit of living organisms. He made also important contributions in meteorology, astronomy, and geology, and worked on combustion and respiration. He was the contemporary of Christiaan Huygens (1629–1695), Antoni van Leewenhoeck (1632–1723), Sir Christopher Wren (1632–1723) and Isaac Newton (1642–1727), and had correspondence and interactions with all of them, as well as quarrels with Huygens and Newton over matters of priority. He was also the contemporary of Rasmus Bartholin (1625–1698) and Nicolas Steno (1638–1686). From 1655 to 1662, he was an assistant to Robert Boyle,[57] for whom he constructed the air pump that was used for Boyle's gas law experiments. These were very exciting times, when the new experimental philosophy was just developing. In 1661 Hooke published a small tract on capillary action and, in 1662, was appointed Curator of Experiments of the newly formed Royal Society—a position which he retained for fifteen years. In this capacity, he was asked to demonstrate a few experiments at each of the Society's weekly meetings. He meticulously collected detailed records of his observations, made with a home-crafted microscope. They are published in his book, *Micrographia* (1665), which contains sixty observations, illustrated by thirty-eight engravings, many of great beauty, of minute bodies such as, among others, fibres of silk, sparks of flint, colours exhibited by thin films (see Section 3.1), snowflakes (see Section 11.5.3), gravel in urine, crystallites of alum (see Section 11.5.2), cells of cork, mosses, sponges, the beard of a wild oat, which can be used as an hygrometer, seeds, peacock feathers, and insects such as lice and fleas. It also includes drawings of the moon and the Pleiades. The book was an immediate success and rapidly became very popular. Newton often referred to it and even acknowledged having made use of Hooke's observations (Andrade 1950). Above all, Hooke favoured the art of observation, as he explained in the Preface to *Micrographia*: 'The truth is, the Science of Nature has been already too long made only a work of the Brain and the Fancy: It is now high time that it should return to the plainness and soundness of Observations on material and obvious things'.

Hooke's *Micrographia* interests us here because of his view of the construction of crystal shapes by the close packing of identical spheres (see Section 11.5.2) and because of his theory of light as a vibrative motion propagated by some fluid in the pores of transparent bodies (see Section 3.1).

11.5.2   *Micrographia, Observation XIII. Of the small diamants, or sparks in flint. Hooke's theory of the structure of crystals*

Hooke explained the regular shape of crystals by close packing of identical spheres. The idea first arose in his mind from his admiration of the play of light in the small crystallites at the bottom of a cavity in a broken flint stone or in small quartz crystals coming from Cornwall (*Cornish diamants*). He notes that this regularity of figures can be found 'in all kinds of minerals, most precious stones and all kinds of salts'. He finds it 'the most worthy, and next in

[56]For further reading on Robert Hooke, see, for instance: Whitrow (1938), Da C. Andrade (1950), 'Espinasse (1956), Chapman (2005).

[57]See footnote 20, Chapter 3.

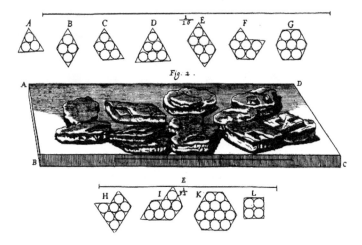

Fig. 11.17 Hooke's sketches showing his reconstruction of the external shape of alum crystals by close packing of identical spheres. After Hooke (1665).

order to be consider'd after the contemplation of the Globular Figure'. In his study of colours, Hooke had considered fluid bodies to be made up of globular particles and, by extension, thinks that 'had I time and opportunity, I could make probable that all these regular Figures [of crystals] that are so conspicuously various and curious ... arise onely from three or four several positions or postures of Globular particles'. He added that 'coagulating particles must necessarily compose a body of such a determinate regular Figure, and no other'. This, he demonstrated *ad oculum* with a 'company of bullets', which, 'if put on an inclined plain, so that they may run together, naturally run into a triangular order, composing all the variety of figures that can be imagin'd to be made out of aequilateral triangles'. Hooke then proceeds to reconstruct the various shapes of surfaces of alum crystals, as shown in Fig. 11.17, A to K, by close-packed spheres. By adding a fourth globule on a group of three such as A, he obtained a tetrahedron and concluded that by continuing this operation one may reconstruct the three-dimensional shape of any alum crystal. He went on to say that you can do the same for crystals of rock-salt by putting the globules in a *cubical form*, such as L in Fig. 11.17. He concluded that one can reconstruct the shape of all crystals, such as vitriol, saltpeter (or nitre), quartz, hoar-frost, etc. by combinations of these textures (the equilateral triangle and the square), but that he had not had the possibility, for a lack of time, to 'prosecute the inquiry so farr as I design'd'. He then listed all the experiments that should be done to check his ideas.

Hooke has thus implicitly shown that the lateral development of crystal faces depends on the way the crystal has grown, by the addition of shorter or longer rows of particles. It is this principle of conservation of interfacial angles that Steno stated in the case of quartz (see Section 11.7.2) and was later generalized by Carangeot and Romé de l'Isle (see Section 11.11).

Hooke's atomistic views, which were probably influenced by Boyle's ideas, are confirmed by his crude kinetic theory of matter. His study of springs, developed in *De Potentia Restitutiva*, led him to discuss harmonic motion and to show that vibrations in which the restoring force

Fig. 11.18   *Left*: Hooke's sketches of snowflakes as he observed them. *Right*: Details of a snowflake seen through a microscope. After Hooke (1665).

is proportional to the displacement are isochronous.[58] He supposed that all the particles of the universe are in perpetual motion: 'Two or more of these particles joyned immediately together, and coalescing into one become of another nature, and receptive of another degree of motion and vibration, and make a compounded particle differing in nature from each of the other particles' (quoted by Andrade, 1950).

Fig. 11.19   Rasmus Bartholin. After P. Hansen (1886) *Illustreret Dansk Litteraturhistorie*.

Rasmus Bartholin: born 13 August 1625 in Roskilde, Denmark, the son of Caspar Bartholin, a Danish physician and theologian; died 4 November 1698 in Copenhagen, Denmark. His father died when Rasmus was four. His elder brother, Thomas (1616–1680), was also a famous physician and anatomist. Rasmus received his first education from private tutors and then went to a Latin School. He entered the University of Copenhagen in 1642. In 1645 he left Copenhagen and travelled in Europe for about ten years, staying successively in Leyden where he met Christiaan Huygens, Paris, Padua, where he received a medical degree, and in England. In 1656 he became Professor of Geometry at the University of Copenhagen and, in 1657, extraordinary Professor of Medicine. In 1671 he was appointed ordinary Professor of Medicine and, in 1667, Royal Mathematician, but his works concern mathematics rather than medicine. He did also some work in astronomy, observing comets and collaborating with his-son-in law, the Danish astronomer O. Rømer.

MAIN PUBLICATIONS

1657   *Dissertatio mathematica qua proponitur analytica ratio inveniendi omnia problemata proportionalium.*

1661   *De figura nivis dissertatio.*

1663   *Auctarium trigonometriae ad triangulorum sphaericorum et rectilineorum solutiones.*

1663   *De poris corporum*

1664   *De problematibus mathematicis tractatus.*

1665   *De cometis anni 1664 et 1665 opusculum.*

1669   *Experimenta crystalli islandici disdiaclastici quibus mira & insolita refractio detegitur.*

1674   *De naturae mirabilibus. Questiones academicae.*

---

[58]Isochronous: an oscillation is isochronous if its frequency is independent of its amplitude. Galileo observed that the oscillation period of a pendulum is constant, regardless of the angle of the swing.

11.5.3   Micrographia, Observation XIV. Of several kindes of frozen Figures. Hooke's description of snowflakes

In Observation XIV, *Of several kindes of frozen Figures*, Hooke at first notes that hoar-frost sometimes forms crystals in the shape of 'hexangular prismatical bodies', which he compares to crystals of nitre. He then describes the sexangular arborescence of frozen urine, which is very similar to that of ferns. In the second part, he recalls his observations of snowflakes:

> 'Exposing a piece of black Cloth, or a black Hatt to the falling Snow, I have often with great pleasure, observ'd such an infinite variety of curiously figur'd Snow, that it would be as impossible to draw the Figure and shape of every one of them, as to imitate exactly the curious and Geometrical Mechanisme of Nature in any one ... In all which I observ'd, that if they were of any regular Figures, they were always branched out with six principal branches, all of equal length, shape and make, from the center, being each of them inclin'd to either of the next branches on either side of it, by an angle of sixty degrees' (see Fig. 11.18, *Left*).

Brian J. Ford[59] thinks that some of Hooke's sketches look very much like some of Bartholin's (see Fig. 11.20) and that he may have copied them. However, Hooke was a very careful observer and he did look at the snowflakes through his microscope. Fig. 11.18, *Right* is a drawing of what he saw.

11.6   Rasmus Bartholin and the double refraction of calcite, 1669

Rasmus Bartholin was a Danish physician and mathematician, coming from a family of well-known physicians. His son-in-law was the Danish astronomer O. Rømer (1644–1710) – see Section 3.4.1. He is himself best known for his discovery of the double refraction of calcite, Iceland spar. He was also the first, well before T. Bergman and R.-J. Haüy, to notice the cleavage properties of calcite and the fact that it can be broken into small rhombohedra all having the same shape, but this is usually ignored. He was an admirer of Descartes and, during his stay in Leyden from 1646 to 1656, collaborated on F. van Schooten's edition of Descartes' geometrical works.[60]

11.6.1   Bartholin's sketches of snowflakes

In 1660 Rasmus Bartholin wrote a small essay on the shapes of snowflakes (*De figura nivis dissertatio*) that was published in 1661 as an annex to a book on the medical uses of snow (*De nivis usu medico*; Copenhagen: Hafniae) by his brother Thomas Bartholin. Rasmus Bartholin's essay was directly inspired by Kepler's *Strena*.[61] After recalling the observations by Olaus Magnus, Kepler, and Descartes, he added his own sketches (Fig. 11.20, *Left*), made from observations he made first in Belgium, then in Denmark on 3 January 1660. Besides the hexagonal stars with feathered, hairy or wool-like radii, Bartholin also observed a few pentagonal and octagonal ones. He compared these structures with the regular shapes, such as cubes and dodecahedra,

---

[59] *The Microscope* (2010), **58**, pp. 21–32.

[60] F. van Schooten (1615–1660), a student of Descartes', was a Professor of Mathematics at the University of Leyden and did much for the rapid development of Cartesian geometry.

[61] It is briefly summarized in Halleux, 1975.

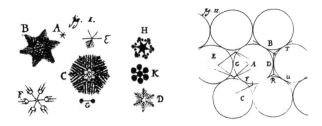

Fig. 11.20   *Left*: Rasmus Bartholin's sketches of snowflakes. *Right*: Circular honeycomb cells before they are compressed into hexagons; see the formation of a hexagon at *R*. After Bartholin (1661).

shown by crystals of alum, nitre, and salt, and, as Kepler did, with the honeycomb and the pomegranate seeds, which form rhomb-dodecahedra. He even constructed a beehive with glass panels and observed that the cells built by the bees are at first circular and close-packed, then become hexagonal due to the compression they undergo as they grow, as seen at point *R* in Fig. 11.20, *Right*; from that observation he deduced that the shape of the honeycomb is not due to the instinct of the bees but to a material necessity. He also notes the hexagonal shape of the cobwebs spun by spiders. In order to understand the star shape of snowflakes, he first refers to the dendritic growth of salts and to the crystallization of starry regulus of antimony[62] by reduction of stibnite in a crucible. To explain this example more specifically, he makes use of the same figure (11.20, *Right*) as for the formation of the honeycomb. He imagines that six small hairy balls of ice agglomerate around a seventh one, *A*. Due to the successive action of cold and heat the matter melts at their contact points, such as *G* and *F*, inducing the star shape around *A*. As in Kepler's *Strena*, the explanation does not make reference to the inner structure of the snowflakes.

11.6.2   Bartholin's observation of double refraction

In 1668 a geological expedition brought back a large quantity of high-quality transparent Iceland spar (calcite) from Helgustadir quarry, central-eastern Iceland (Fig. 11.21, *Left*). Ramus Bartholin got hold of some of it and made a thorough physical and chemical examination of the crystals. He published the results of his experiments in 1669 in his book, *Experimenta crystalli islandici disdiaclastici quibus mira & insolita refractio detegitur*.[63] He admired the perfection of the cleavage faces and noted that the crystals only break easily along the direction of the cleavage planes. He also noticed that breaking a larger crystal produces smaller rhombohedra which all have the same shape and he measured the interfacial angles (101° and 79°). He found that Iceland spar, like amber, can attract very light objects, such as feathers, after being rubbed, and showed that it dissolved into nitric acid.

More importantly, he observed that objects seen through a rhombohedron of Iceland spar give two refraction images, (1) the ordinary one, *solita*, and (2) an extraordinary one, *insolita*,

---

[62]A term used by alchemists and meaning metallic antimony; the regulus of antimony was investigated, among others, by Isaac Newton and Robert Boyle.

[63]Translated into French by Cuvelier (1977) and into English by Lohne (1977).

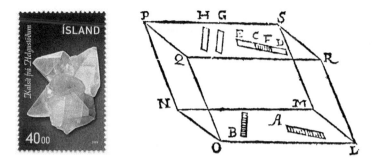

Fig. 11.21   *Left*: Calcite scalenohedra from Helgustadir quarry, Iceland. Stamp from the author's private collection. *Right*: Rasmus Bartholin's sketch of the double refraction of calcite. EF, CD: ordinary and extraordinary images of A. H, G: ordinary and extraordinary images of B. After Bartholin (1669).

that moves when you rotate the crystal; for one position of the calcite crystal, he noted, the two images are superposed. Fig. 11.21, *Right* shows Bartholin's drawing of the double image. B and A are two objects and H, G and EF and CL their ordinary and extraordinary images, respectively. By that time, the refraction law of sines was well established; Bartholin measured the angles of refraction of the two rays and checked that the ordinary ray exactly follows that law, but that the extraordinary did not obey this law and therefore he formulated his own for it. To try to explain the phenomenon of double refraction, Bartholin refers to the notions, accepted by many at the time, of corpuscles and of the porosity of materials. These pores are assumed to be filled with some subtle matter (Descartes) or some fluid (Hooke) which provides the means by which light is propagated. Bartholin himself had published a pamphlet about the pores of matter (*De poris corporum*, 1663). He considered, wrongly, that the extraordinary ray follows the interstices between the particles aligned along the lateral faces of the crystal.

11.7   Nicolas Steno and 'a solid body enclosed by process of nature within a solid', 1669

Nicolas Steno (Niels Stensen, also known under his Latinized name, Nicolaus Stenonis) was a Danish anatomist and geologist, at the end of his life a priest and then a bishop. He made important discoveries in anatomy, and he is considered, along with Robert Hooke, as one of the founders of modern geology and paleontology. While Hooke merely touched on every topic, Steno went much more in depth in those he studied. He related the external shape of a crystal to its growth conditions and was the first to state the constancy of interfacial angles of quartz crystals. He had a very interesting and engaging personality and was one of the great scientists of his time. His doctrine, which is expounded in his diary, entitled *Chaos* and written when he was 21, was to place observation above everything, to be rigorous and to reject any preconceived idea; he never hesitated to criticize ancient views, as well as those of the prominent people of his time, for instance Descartes. His writings are concise, clear, and to the point. During his numerous travels he had the opportunity of meeting some of the brightest intellectuals of his time: in Leyden the Dutch philosopher Baruch Spinoza (1632–1677) whose ideas Steno found after his own conversion to be a danger to the Christian faith,[64] in Paris where scientists were about to found the French Académie Royale des Sciences, in Montpellier the English physician Martin Lister (1638–1712) who would make Steno's work known by the Royal Society, in Rome the Italian anatomist Marcello Malpighi (1628–1694), and, in Florence the members of the Accademia del Cimento, founded in 1657.[65]

11.7.1 Steno and the observation of snowflakes: Chaos

In 1658, Copenhagen was besieged by the Swedes, and studies at the University were inter-rupted. During the respite that followed, Steno took some time to look back on his life and on what he had learned so far. It was the opportunity for him to write, between March and July 1659, a kind of diary, called *Chaos*,[66] rediscovered in 1946 in a Florence library. It is very useful for understanding Steno's motivations and his way of thinking. He recorded there his detailed observations of snowflakes (March 16 and 17 1659), with some sketches.[67] Steno, who knew of Kepler's work, noted that most had six radii and were rose- or fern shaped. In his entry for 28 June, he imagined that 'a part of sea-water evaporates and grows into snow and that it partly solidifies into salt, the latter being formed in hexagonal cubes, the former into stars with six rays' (Ziggelaar 1997), but he did not commit himself as to the cause of the shape of the snowflakes.

In the entry dated 23 March 1659 in *Chaos*, Steno gives his reflections on the cause of figures in stones and why certain fossil stones have a geometrical figure; they prefigure the thoughts he will develop later in the *Prodromus*: 'at the centre of every body in nature a force is hidden that nature has laid in it so it can preserve and propagate itself; this force is by certain rays propagated towards the circumference, not in all bodies spherically but sometimes by longer rays in one direction than in another ... Thus at the centre of the nature of certain fossil stones a force resides by which rays that are not always all equal, but some sometimes the longer the stronger, whereas the others decrease in proportion to their weakness, through the addition of particles to particles attracts particles like themselves, smallest and homogeneous particles' (Ziggelaar 1997). One can see here the appearance of the notion of anisotropy and the idea that minerals grow by the aggregation of small identical particles. It is in this entry that Steno mentions Kepler's *Strena*, but without comment.

11.7.2 Steno and the constancy of interfacial angles: De solido intra solidum ... prodromus

Steno's crystallographic ideas are developed in his small book (78 pages), *De solido intra solidum naturaliter contento dissertationis prodromus* ('Preliminary discourse to a dissertation concerning a solid body enclosed by process of nature within a solid').[68] In Steno's mind it was to be only a preliminary account of his thoughts on the subject (hence *prodromus*, forerunner),

---

[64]In a letter to the Congregazione per la dottrina della fede, 4 September 1677.

[65]For more details on Steno's life and work, see, for instance: Winter (1916), Burke (1971), Shafranovskii (1972), Poulsen and Snorrason (1986), Scherz (1987, 1988), Pedersen (1991), Abbona (2002, 2004), and Van Besien (2009).

[66]English translation by Ziggelaar (1997).

[67]The author is grateful to Francisco Abbona, Turin, Italy, for the relevant excerpts from *Chaos*.

[68]Translated into English by Winter (1916).

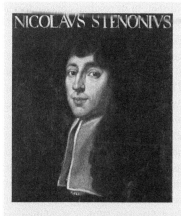

Fig. 11.22   Nicolas Steno painted in 1669, at the time of his first stay in Florence. Galléria Uffizi, Florence, Italy, with permission.

Nicolas Steno (Niels Stensen): born 11 January 1638 in Copenhagen, the son of a Lutheran goldsmith who worked for King Christian IV of Denmark; died 25 November 1686 in Schwerin, Germany. As a child he suffered from illnesses and stayed at home in the company of adults, visiting his father's laboratory, where he learned the use of scientific instruments. At the age of ten he attended Notre Dame school in Copenhagen, where he received a good education in humanities and mathematics. In 1656, he entered the University of Copenhagen, where he studied medicine and natural sciences. Rasmus and Thomas Bartholin, Ole Borch, and Simon Paulli were his professors. During his youth, Steno went through difficult times; his father died when he was six, the Thirty Years War ended when he was ten, the plague in 1656 killed many in Copenhagen, including in Steno's family. The Danish–Swedish wars in 1657–60 took place while he was at the University. In 1660 he went to Amsterdam where he made his first anatomical discovery, a duct in the salivary glands, still known as Steno's duct. From there he went on to Leyden, where he pursued his anatomical studies under the guidance of famous anatomists. He had to return to Copenhagen in 1664 for family reasons, but, disappointed not to have been appointed Professor of Anatomy at the University of Copenhagen, he set off for Paris, where he spent one year, under the patronage of Melchisedec Thevenot (1620–1692), in whose house he stayed, continuing his studies of the anatomy of the brain. After a long stay in France, he went to Pisa, Rome, and Florence where he became attached to the court of the Grand Duke Ferdinand II, who appointed him anatomist of the Santa Maria Novella Hospital (1665). It is in Florence that Steno spent the happiest and most fruitful years of his life (he called it 'his second home'); it was there that he made his geological and mineralogical studies. It was also there that he converted to Catholicism (1667). He was invited back to Denmark in 1667 but did not go immediately. In 1668 he wrote his book, *De solido intra solidum*, made a long trip in Europe during which he visited mining sites in Southern Germany, Austria, Bohemia and Hungary, collecting specimens, came back to Florence for a while, and finally only arrived in Copenhagen in July 1672, where he had been appointed Professor of Anatomy. The latter years of Steno's life were austere and miserable. Being a Catholic in a deeply Protestant country, he became involved in religious controversies and chose to come back to Florence in 1674. In 1675 he was ordained a priest. In 1677 he was appointed Bishop of Hannover, and in 1679 vicar apostolic for Northern Europe. He was beatified in 1988.

MAIN PUBLICATIONS

1659 *Chaos*. Unpublished diary.

1661 *Observationes anatomicae*.

1664 *De musculis et glandulis*.

1667 *Elementorum Myologiae Specimen seu Muscoli Descriptio Geometrica*.

1669 *Discours sur l'anatomie du cerveau*.

1669 *De solido intra solidum naturaliter contento dissertationis prodromus*.

1673 *Historia Musculorum Aquilae*.

but he never had the time nor the opportunity to develop it further. In its original concise form, however, it is a self-contained, scientifically sound, and rigorous account of his observations.[69]

The turning point of Steno's research career may have occurred when he dissected, in 1666, the head of a big shark which had been caught in the harbour of Leghorn (Livorno). The shape of its teeth reminded him of a well-known 'tongue-shaped' stone, called *glossopetra*, used in medicine. His attention had been called to it by his master Thomas Bartholin, who had brought back a similar specimen from Malta. At that time, fossils were considered as tricks of nature and, in the particular case of the *glossopetra*, its curative powers were attributed to some legend. It was Steno's exceptional merit to have guessed on anatomical and geological bases that this feature was in fact a fossilized shark tooth and that the ground where it was found had once been covered by a sea. It was the presence of the fossil in the encasing rock that gave him the idea for the title of the book: 'a solid body enclosed ... within a solid'. Like Hooke, Steno understood that fossils were petrified living organisms. This led him to be interested in geology and to travel all over Tuscany, observing the landscape and its geological features. With a revolutionary intuition, he stated the principles of stratigraphy: the superposition of the deposited layers, with the oldest on the bottom and the youngest on the top, their initial horizontality, and their lateral continuity. By his reasoning, he was able to show that fossils and crystals must have solidified before the host rock that contains them was formed. He also noted, on the other hand, that veins (mineral-filled cracks) and many crystals must have been formed after the surrounding rock was a solid. Changes in the position of strata account for the formation of mountains under the action of earthquakes, volcanic activity or erosion.

Steno gave a lot of thought to the growth mechanism of solids, working down from the geological features to crystals. The following excerpts taken from Winter's (1916) translation show how his reasoning developed:

> 'A natural body is an aggregate of imperceptible particles subject to the action of forces. ... While the solid body is being produced, its particles are in motion from place to place (while in a fluid they are in perpetual motion) ... A [solid] body grows while new particles secreted from an external fluid are being added to its particles ... The additions are only to certain places in the case of *angular bodies*' (the term used by Steno to designate crystals).

Steno then comes to the specific case of quartz crystals (*crystallus montium*, mountain crystal). He first states that everything he has read in other writers on the formation of rock-crystal 'is not to the point, for neither irradiations nor a shape of the particles resembling the shape of the whole[70] nor the perfection of the hexagonal form[71] and the assembling of the parts about a common centre ... agree with fact'. After which, he describes the most frequent habit of quartz crystals: hexagonal prisms terminated by hexagonal 'pyramids' (in fact interpenetrating rhombohedra) and goes on to state the rules by which the crystal grows:

I. 'A crystal grows while new crystalline matter is being added to the external planes of the crystal already formed' (not like a plant, 'which draws its nourishment from the side on which it is attached to the matrix').

---

[69]For detailed discussions of the work, see, for instance, Burke (1971), Schneer (1971), Shafranovskii (1971), Ellenberger (1988), and Pedersen (1991).

[70]In opposition to the view which was to be taken a century later by T. Bergman and R.-J. Haüy.

[71]This may be understood as a refutation of Kepler's analysis.

II. 'The new crystalline matter is not added to all the planes but for the most part to the planes of the apex [the planes of the 'pyramids'], with the results that the intermediate planes [the planes of the prism] are larger in some crystals, smaller in others, and wholly wanting in others. The intermediate planes are almost always striated'.

III. 'The crystalline matter is not added to the terminal planes [the planes of the prisms] at the same time, nor in the same amount'.

IV. 'An entire plane is not always always covered by crystalline matter, but exposed places are left sometimes towards the angles, sometimes towards the sides, and sometimes in the centre of the plane'(the crystal faces may be stepped).

Steno had therefore clearly established that the growth rate is not the same for all the faces and may also vary with time. As a result, a crystal of quartz may take the different shapes represented on Figure 11.23, which reproduces Steno's original diagram. The key sentence in Steno's figure caption is:

> 'Figures 5 and 6 belong to the class of those which I could present in countless numbers to prove that in the plane of the axis both the number and the length of the sides [the lateral expansion of the faces] are changed in various ways *without changing the angles* [72] (*non mutatis angulis*).

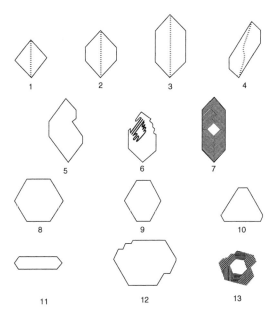

Fig. 11.23   Steno's sketches of the possible shapes of quartz crystals. 1 to 7: the hexagonal axis lies in the plane of the figure; 8 to 13: the plane of the figure is the basal plane. 5, 6, and 9 to 13 show that the lateral extension of the faces may vary without the angles between them varying, 7 and 13 show the variation of the external shape of the crystal during its growth. An enlargement of 13 is shown in Fig. 11.24. After Steno (1669).

[72]The emphasis is mine.

Fig. 11.24   Growth horizons in a quartz plane normal to the *c*-axis. *Left*: Enlargement of Steno's sketch 13. After Steno (1669). *Right*: X-ray topograph of a quartz crystal showing growth bands and dislocations. Photo by the author.

It is striking that sketches 7 and 13 representing successive growth horizons look amazingly like modern-time X-ray topographs showing growth bands (Fig. 11.24).[73]

Steno has thus understood that crystals grow layer by layer and has clearly shown that the lateral extension of crystal faces only depends on the conditions of growth, while the interfacial angles do not. This is the only place where Steno clearly states the constancy of interfacial angles. He presents it as a fact of observation, without proof, and not as a universal law, and he refrains from relating it to any atomistic hypothesis about the inner structure of the crystal. We have seen that he did the same in *Chaos* for the shape of snowflakes. Pedersen (1991) thinks that Steno wanted 'to avoid committing himself to any specific hypothesis which might prove untenable in the light of new experience'. This is different from Hooke who, as we have seen, had already implicitly observed the constancy of interfacial angles, noting the extension of crystal faces depended on the number of spheres added on each plane during the growth of the crystal (see Section 11.5.2). Schneer (1960, 1971) finds that some of Steno's sketches in Fig. 11.23 look very much like some of Hooke's in Fig. 11.17, and discusses whether Steno might have seen Hooke's *Micrographia* during his stay in Paris in 1665 and have been influenced by it, but he does not have a conclusive answer.

Steno, who was very familiar with the works of Kepler, Descartes, and Bartholin, does, however, use the concept of 'imperceptible particles' (*particulae insensibiles*) as the constituent of matter to make a clear distinction between a fluid and a solid:

> 'A solid differs from a fluid in that in a fluid the imperceptible particles are in constant motion ... while in a solid, although the imperceptible particles may sometimes be in motion, they hardly ever withdraw from one another so long as that solid remains a solid and intact'.

Descartes had already made a similar distinction. Steno uses the same concept to explain the mechanism of crystal growth; his view is that the crystal is the result of an aggregation of 'imperceptible particles' moving in a 'permeating fluid' and oriented under the operation of a force, 'whereby the new crystalline matter is spread forth over the plane', in the same way as iron filings are moved and oriented by the action of a magnet. For Steno, the growth of quartz is analogous to that of crystals of salts, such as nitre, in water: 'the fluid, in which the

---

[73]For X-ray topography see, for instance, Authier (2003).

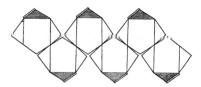

Fig. 11.25 Steno's template of a hematite crystal. After Steno (1669).

crystal is formed, bears the same relation to the crystal that ordinary water bears to salt'. And he concludes that 'the efficient cause of the crystal is not extreme cold', as was the belief of the Ancients.

Steno was a very careful observer. He described liquid inclusions and noted that quartz crystals may present zones with different hues, and ascribed it to a different composition of the external fluid in which the crystal grew.

Along with quartz, Steno studied the habits of numerous crystals of hematite from the island of Elba ('angular bodies of iron': *angulata ferri corpora*), showing that the shape of the crystals can be deduced by suitable truncations and drawing templates to reconstruct the complete shape of some crystals (Fig. 11.25)[74]. The constancy of interfacial angles is here implicit. He also described crystals of diamond and pyrite cubes.

11.8 Christiaan Huygens and the structure of calcite, 1678

Huygens' life and his wave theory of light have been presented in Section 3.4. His views concerning the structure of calcite are detailed at the end of Chapter V of the *Traité de la Lumière* (Section 3.4.2). He first mentions a few crystals with a regular shape: quartz with its hexagonal prisms, diamond and its square points (octahedra), rock-salt from the sea in little cubes, crystallized sugar, snowflakes as six-pointed little stars or hexagonal platelets. He adds, as if it were an accepted fact, that this 'regularity comes from the arrangement of the small, invisible and identical particles which compose them'. Fig. 11.26 shows his representation of the structure of calcite. It is a close-packed stacking (left) of oblate ellipsoids of revolution (middle), so as to form a cube compressed along one of its diagonals, the cleavage rhombohedron (right).[75] The ratio of the two axes, EF and GH, is equal to $1/\sqrt{8}$ (1/2.828) and has been calculated so that the obtuse angles ADB, BDC and CDA each be equal to the 101° 52' measured by Huygens. The *pyramid* (as it was called by Huygens) $ABCD$ on the left is simply a corner of the cleavage rhombohedron on the right. By simple considerations related to the number of first neighbours, Huygens showed that cleavage should be easy along the sides of the rhombohedron, but difficult along other directions, such as the plane of the stacking, ABC, (111) in modern notation, or the diagonal plane $GHKL$, (011). The fact that it is indeed the case was for Huygens a confirmation of his model of the structure. In a manuscript reproduced in *Oeuvres complètes de Christiaan Huygens* (*Société Hollandaise des Sciences*, The Hague: Martinus Nijhoff, Tome XIX (1937), p. 545), Huygens showed that he has understood how the successive layers are put over the interstices of the preceding one, the fourth layer coming above the first one (see the sketch in the left-hand side of Fig. 3.6), and that he has constructed models: 'these things are visualized much better

---

[74]Such templates were later used in a systematic way by Linnaeus (1735) and Romé de l'Isle (1772).

[75]Note the similarity of Huygens' stacking with Kepler's (Fig. 11.12) and with Hooke's sketches (Fig. 11.17).

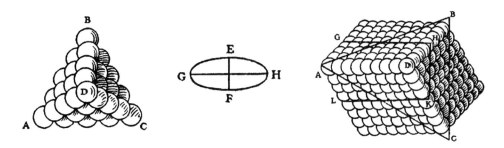

Fig. 11.26   Huygens' structure of calcite. After Huygens (1690).

by constructing the layers with real spheroids than by imagining them on the figure'. He also made a comparison with rock-salt, guessing that its structure is similar to that of calcite, but with a stacking of spherical particles.

In his conclusion, Huygens avoided committing himself as to 'the way so many equal particles are formed and how they are set in such a beautiful order; whether they are first formed and then assembled or order themselves thus as they are being produced, which seems to me more probable. To develop such hidden truths, a much greater knowledge of nature than that we have would be required'.

11.9   D. Guglielmini and *Riflessioni filosofiche dedotte dalle figure de' sali*, 1688

Domenico Guglielmini was an Italian physician, mathematician, and physicist, best known for his studies of hydraulics and the physics of rivers. He was, however, a man of wide interests. He made acute observations on salts, and was among the first to have described the constancy of their shapes and to imagine them to be built up by identical elementary particles. In a first publication (Guglielmini 1688) he had observed that the earliest forms of the crystals of common salt, vitriol (copper sulphate), nitre, and alum were always a cube, a rhombohedral parallelepiped (see footnote 1, Section 2.1), a hexagonal prism, and an octahedron, respectively. In his main publication on crystals (Guglielmini 1705), which refers to Democritus and Descartes, he went further, stating that these earliest forms were always the same, inalterable, and indivisible, limited by faces of which the inclinations are constant. They are too small to be perceptible by our senses, and the crystals we see are in fact aggregates of them, with the same form. Their shapes are characteristic of the corresponding substances, and constitute the 'essential difference' between the salts (*differentia essentialis*). He noted that the hexagonal prism could be considered as the coalescence of six triangular prisms and the octahedron that of two triangular pyramids (tetrahedra). Triangular prisms could indeed be observed, although rarely, in the early stages of the growth of nitre. He concluded that the primitive forms of these four crystals are, respectively, the cube, the rhombohedron, the triangular prism, and a half octahedron; all the crystalline forms taken by these substances would be modifications of these basic forms. In practice, the crystals one observes do not always have exactly the shape of the earliest form. For instance, sodium chloride often presents a rectangular shape instead of a square one, because of a difference in the accumulation of the small elemental cubes in

Fig. 11.27   Domenico Guglielmini by F. Rosaspina, Wellcome Library, with permission.

Domenico Guglielmini: born 27 September 1655 in Bologna, Italy; died 27 July 1710 in Padua, Italy. He studied at the same time mathematics with G. Montanari (1633–1687) and medicine with M. Malpighi (1628–1694) at the University of Bologna, where he obtained his Doctorate in Medicine in 1678. In 1686 he was appointed Professor of Mathematics at the University of Bologna and general manager of the water system of the Bologna state. He made important theoretical and experimental investigations in hydraulics, which incited Bologna University to create for him a Chair of Hydrometry in 1694. His reputation led the Republic of Venice to call him to Padua in 1698, as Professor of Mathematics at the University.

He continued his studies in medicine there, and was appointed Professor of Medicine in 1702. He was elected an associated member of the French Academy of Sciences in 1686 and of the Royal Society in 1687.

MAIN PUBLICATIONS

1688 *Riflessioni filosofiche dedotte dalle figure de' sali.*

1690 *Aquarum fluentium mensura nova methodo inquisita.*

1697 *Della natura de' fiumi trattato fisico-matematico.*

1701 *De sanguinis natura & constitutione exercitatio physico-medica.*

1705 *De salibus dissertatio epistolaris.*

different directions. What matters is the inclination of faces (*inclinatio planorum*). Details of Guglielmini's life and works are given in his obituary by B. de Fontenelle.[76]

11.10   T. O. Bergman and 'On the various crystalline forms of Iceland spar', 1773

11.10.1   One of the founding fathers of quantitative chemical analysis

Torbern Olof Bergman (Fig. 11.28) was a Swedish chemist and mineralogist, to whom are due great advances in the field of chemical analysis.[77] He was an excellent experimenter; he devised systematic methods of examining compounds by the wet way and by means of the blow-pipe, and first made it possible to analyse minerals that are insoluble in acids by fusing them in alkali. He and his co-workers discovered and described many new substances, including metals such as nickel or bismuth, organic acids: uric, tartaric, oxalic, citric . . . He studied carbonic acid, which was called *fixed air* at the time, and showed that it had the properties of an acid; for that reason,

[76]*Histoire de l'Académie Royale de Sciences, année 1710*, pp. 152–166.

[77]For biographies of T. O. Bergman, see, for instance, N. Condorcet, Eloge de M. Bergman, *Histoire de l'Académie Royale des Sciences*, 1784; J. A. Schufle (1985). *Torbern Bergman: a man before his time.* Lawrence, Kan.: Coronado Press.

he called it *acido aereo*, acid of the air; in 1770 he developed an apparatus to produce carbonated water. He is also one of those who, in the 1770s, showed that air (one of Aristotle's elements!) is in fact composed of three elastic fluids, vitiated air (nitrogen), fixed air, and pure air (oxygen), an air necessary for fire and animal life.[78] In his dissertation on Elective Attractions (1775), he developed the notion of chemical affinity and chemical reaction, illustrated by a table of affinities and a series of diagrams.

Bergman is very important for our story because he was the first to derive the external form of a crystal, namely calcite, from an elementary cleavage rhombohedron. He was very interested in mineralogy and he extended Linnaeus's[79] classification of minerals (Linnaeus, 1735), which had been based on morphology, by sorting them according to their chemical composition: salts, earths, bitumen or phlogistic materials[80] (*i.e.* burning materials such as sulphur or diamond) and metals (*Sciagraphy of the mineral kingdom*, 1782).

11.10.2 Derivation of the crystalline forms of calcite from the cleavage rhombohedron, 1773

Bergman's study of the crystalline forms of calcite was first published in 1773: *Variae crystallorum formae a spatho orthae*, 'On the various crystalline forms derived from spar' (Bergman 1773), where 'spar' is to be understood as the cleavage rhombohedron of calcite[81] It was reproduced in 1780 in considerably extended form, as part of his *Physical and Chemical Essays*, under title *De formis crystallorum, praesertim e spatho orthis*, 'On crystalline forms, mainly derived from spar' (Bergman 1780).

Bergman started the 1773 paper by noting that there is a very large variety of crystalline forms, but that he is convinced that they can be derived from a small number of simpler forms that should be called *primitive*. Indeed, forms, which are very unlike each other, can be derived from 'a spar structure which is the cleavage rhombohedron of calcite, of facial angles 101.5 and 78.5 degrees. Let us consider how various crystals can be generated by a suitable accumulation of identical rhombohedra and let *ACDEGFBO* be a spar nucleus crossed by axis *HI* at *D* and *O'*[82] (Fig. 11.29). Bergman first showed how to generate the prism. 'By putting along this axis, above and below, successive identical rhombs, such as *MLPN, MNQR, MLTR*, one generates an hexahedral prism, a very usual form found in basaltic crystals, and in some calcareous crystals' (Fig. 11.29, *Middle*).

---

[78]See, for instance, Bergman's introduction to his co-worker and friend, C. W. Scheele's Treatise on Air and Fire (*Chemische Abhandlung von der Luft und dem Feuer*, 1777, Uppsala and Leipzig).

[79]Carl Linnaeus (Carl von Linné), born 23 May 1707 in Sweden, died 10 January 1770 in Sweden, was a Swedish botanist, physician, and zoologist. He received his higher education in the Universities of Lund and Uppsala, and after many travels in Europe, took a Doctorate in Medicine in The Netherlands in 1735. He was back in Sweden in 1738, and in 1741 was appointed Professor of Medicine at Uppsala University. He is well-known for having laid the foundations for the modern scheme of binomial nomenclature. His new system of classification was published under the title *Systema Naturae*, reedited many times, which included three 'kingdoms': plant, animal and mineral. The mineral kingdom was itself divided into three parts: *Petrae et Lapides simplices* (rocks), *Minerae et Lapides compositi* (minerals and ores), *Fossilia et Lapides agregati* (fossils and aggregates).

[80]From phlogiston, a hypothetical fluid thought to be part of combustible bodies, before the role of oxygen in combustion was recognized.

[81]A French translation was proposed by de M. Morveau (1792). *Journal de Physique* **XL**, 258–270.

[82]The quotes from the 1773 and 1780 papers by Bergman are translated from the Latin originals by the author.

Fig. 11.28   Torbern Olof Bergman. Artist unknown. Source: Wikimedia commons.

Torbern Olof Bergman: born 20 March 1735 in Katrineberg, Sweden, the son of Barthold Bergman, a district tax collector; died 8 July 1784 in Medevi, Sweden. After his secondary school he entered Uppsala University in 1752. Following his parents' wish, he read law and theology, but also, secretly, according to his own inclination, mathematics and physics. Due to overwork, he became ill and had to go back home after one year. There, he continued studying but also collected insects. Back in Uppsala he was allowed to resume his science studies. He had found some rare specimens of insects that were not in Linnaeus's classification and he dared to show them to the famous man who commended him highly and with whom he studied entomology. He graduated in 1758 and in 1761 was appointed Associate Professor of Mathematics at Uppsala University, where he taught both physics and mathematics. He published a number of papers on various physical phenomena, such as the rainbow, the *aurora borealis*, and lightning. In 1764 he was elected a member of the Swedish Academy of Sciences, and in 1765 a Fellow of the Royal Society of London. In 1766 he published a paper on the pyroelectric properties of tourmaline, showing that it is the variation of temperature, and not the temperature by itself which produces the appearance of electric charges. In 1766 also, he published an important geological work, *Physical description of the Earth*, introducing a new classification of rock strata. In 1767, the famous Swedish Professor of Chemistry and Mineralogy, J. G. Wallerius (1709–1785), author of a *Treatise on Mineralogy* (1747), resigned, and his position became open. Bergman was a candidate but, not being known as a chemist, it is only thanks to the strong support of the future King of Sweden, Gustav III, Chancellor of the University, that he was appointed. He founded an important school of chemistry, among which members one may mention J. Afzelius (1753–1837) whose pupil was J. J. Berzelius (1779–1848), J. Gadolin (1760–1852), a Finnish chemist who discovered yttrium, C. W. Scheele (1742–1786), one of the discoverers of oxygen and a close friend of Bergman's, and J. G. Gahn (1745–1818) who isolated manganese and whose observation of calcite's cleavage inspired Bergman. A new mineral, torbernite, a beautiful emerald green hydrated copper uranium phosphate, was dedicated to Torbern Bergman by the German mineralogist A. G. Werner (1749–1817) in 1793.

MAIN PUBLICATIONS

1766 *Physick Beskrifning Ofver Jordklotet.* (Physical Description of the Earth.)	1778 *De Analysi Aquarum.*
	1780 *Opuscula Physica et Chemica.*
1775 *Disquisitio de Attractionibus Electivis.*	1782 *Sciagraphia regni mineralis.*

Bergman then generated a rhomb-dodecahedron, 'which is the form proper to garnets', by 'stopping the accumulation of successive planes when $PNFE$ is a rhomb' (Fig. 11.29, *Middle*). The form thus obtained by Bergman (Fig. 11.30 (1)) was in fact nothing more than another hexagonal prism terminated by rhombohedral faces. Bergman was misled because its facial angles, equal to 101.5°, do not differ too much from the facial angles of a rhomb-dodecahedron, which are equal to 109.5°. The form of hyacinth (a yellow variety of zircon, also cubic!) is obtained in a similar way. In the next step, Bergman derived the 'dog tooth' scalenohedron of calcite (Bergman uses the expression *dens suilli*, pig's tooth, in French *dent de cochon*, in

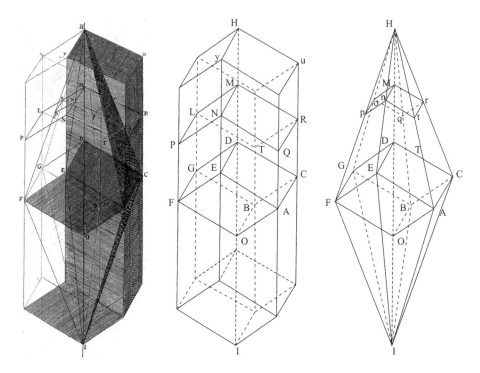

Fig. 11.29   Bergman's derivation of crystalline forms of calcite from the cleavage rhombohedron. *Left*: Bergman's original figure (after Bergman 1773). *Middle*: derivation of the prism and *Right*: derivation of the scalenohedron (redrawn for clarity from Bergman's figure). *ACDEGFBO*: primitive spar nucleus.

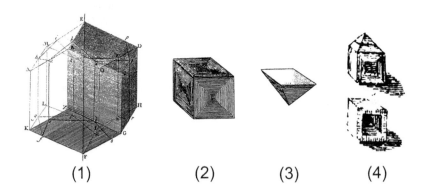

(1)   (2)   (3)   (4)

Fig. 11.30   (1) Bergman's derivation of a garnet rhomb-dodecahedron from the cleavage rhombohedron. After Bergman (1773). (2) to (4): Derivation of the rock-salt structure from hollow pyramids. (2) Crystal of common salt and (3) pyramidal salt. After Bergman (1780). (4) Rock-salt. After Cappeller (1723).

German *Schweinszähne*) 'adding regularly decreasing planes such as *Mopn, Mntr, Moqr*, according to a certain law' (Fig. 11.29, *Right*). Bergman used the term *planum*, plane, but he obviously meant 'layer'. He added that the form is the more elongated, the slower the decrease of these layers is. Bergman then described various cases where the added layers are different from those of the spar nucleus. For instance, by truncating the rhombs, he obtained a pentagonal dodecahedron of which he had observed irregular forms in calcareous crystals and of which complete forms are common in pyrite-type crystals.

In the last part of the 1773 paper, Bergman wrote that 'if someone thinks that this theory is purely geometric and speculative, he should examine the calcareous crystals whose loose constitution progressively and cautiously broken apart reveals to curious eyes its internal structure'. Here, Bergman added a note: 'The central nucleus of pyramidal calcareous crystals [calcite scalenohedra] has first been observed by our very dear disciple, J. G. Gahn'. According to Groth (1926), the testimony of a traveller, Hausmann, who visited Scandinavia in 1806–1807, when Gahn was still alive, Bergman's views are in fact to a large part to be attributed to Gahn.[83] Bergman goes on: 'Once it is known, one understands easily and fully the phenomena which couldn't have been understood otherwise, to which it is enough to add only one. If one strikes carefully the edges *AH, BH* and *FH* of the scalenohedron (Fig. 11.29, *Right*), the crystal breaks up into small rhombohedra, while it would be very difficult or impossible for the edges *CH, EH* and *GH*. The reason is obvious. In the first case, the blow is parallel to the accumulation of planes, in the second case, it is against the intersection of two different stacks of planes. It is the same for the lower pyramid, with *AI* playing the role of *AH*. As to harder crystals with the same shapes, they cannot be split. Nevertheless, basaltic crystals have clearly a structure based on the spar nucleus and garnets are probably also composed of lamellae'. Bergman added that the same probably holds also for artificial salts.

11.10.3 Derivation of the rock-salt structure from hollow pyramids, 1780

Bergman started the extended 1780 version in *Opuscula Physica et Chemica* by stressing that it would be impossible to make a systematic classification of the large variety of crystalline forms without taking into account the small number of 'primitive' forms they are derived from. After reproducing the contents of the first paper, adding some more examples, such as tourmaline and staurolite, he made a generalization: 'since so many various forms arise from spar particles accumulated one way or another, one can be certain that the different external shapes presented by all crystals come from a small variety of mechanical elements (*elementi mechanici*). One may rightly ask whether the smaller molecules (*moleculae*) of the integrant parts, and so to speak the first threads, had by nature a certain angular form or whether they acquired it during crystallization'. A first indication is given by the striations often exhibited by larger crystals, and which reveal their inner structure. A further proof is provided by the small calcite rhombohedra which a careful eye can observe at the surface of calcareous water when it evaporates.

---

[83]Johann Gottlieb Gahn, born 17 August 1745, died 8 December 1818, was at first a miner, before studying mineralogy with T. O. Bergman in Uppsala. In 1770 he went to Falun, where he worked on copper smelting and set up several factories. He isolated manganese in 1784 and, together with another student of Bergman's, C. W. Scheele, discovered the presence of phosphoric acid in bones. The mineral gahnite, $ZnAl_2O_4$, a spinel, is named after him. In 1784 he was elected a member of the Royal Swedish Academy of Sciences.

Another example is given by the detailed observation of the crystallization of common salt. It often exhibits regularly decreasing squares, which are the traces of its internal structure (Fig. 11.30 (2)). Each cube is in fact composed of four quadrangular pyramids (Fig. 11.30 (3)) joined along their external surfaces and filled with smaller and smaller pyramids encased in one another, with their apices coinciding in the centre of the cube. Such crystals are now called 'hopper' crystals; separate pyramids can be obtained by evaporation. Cubes terminated by a pyramid can also be observed, as mentioned by Cappeller (1723), quoted by Bergman (Fig. 11.30 (4)). The same type of structure is also found in sylvite (KCl), silver chloride, sodium nitrate (which is in fact trigonal, isomorphous with calcite), galena (PbS) from the copper mines in Falun, Sweden, and Rochelle salt. Going into further detail, Bergman says that upon close examination each pyramid of salt appears to be composed of four triangles, each of these triangles of lines parallel to the base, themselves being series of small cubes, but wonders whether the smaller parts that are not to be seen have the same internal structure. He then considers water; 'when it freezes, threads are formed which arrange themselves so as to make 60° angles between one another, which explains the particular shape of snow . . . If the atoms which can produce crystalline forms have a tendency to produce threads making a constant angle with one another, triangles and pyramids composed of the triangles will be formed. According to the angle between the threads, a big variety of forms can be obtained'.

In the last part of the chapter, Bergman made a long development on the action of heat on crystallization in the different modes of crystal growth, from the melt, from the vapour phase or from the solution; it is when the molecules have become free to move that they are attracted and become agglomerated to form crystals. He found that the process of formation of threads and hollow pyramids occurs in the same way in the three modes, for instance as well for galena as for rock-salt. 'According to all probability', he added, 'the pyramid structure is common to all crystals and it is not contradictory to what has been said previously regarding crystals derived from a spar nucleus since the spar elements easily produce suitable pyramids when they are arranged properly'. He found, however, some difficulty is generating hexagonal prisms from the pyramids. Finally, he raised the question of whether the crystallization could be due to the presence of foreign salts, a point of discussion among mineralogists at the time, but concludes negatively.

11.10.4   Bergman's contributions and shortcomings

Bergman, or rather Gahn, did observe the cleavage of calcite in rhombohedra and measured the angle between their edges, but so did Bartholin and Huygens. Bergman conceived, and this was new, that many crystalline forms could be derived from this rhombohedron. He imagined that, by adding identical layers on the sides of the rhombohedron, one could generate the association of an hexagonal prism and a rhombohedron, and that by adding layers decreasing according to a certain law, one could generate a scalenohedron. In his extended 1780 work he emphasized hollow pyramids, triangles, and threads as the mechanical elements of the inner structure of crystals. He clearly stated that, at the root of the process of crystal growth, there are identical building units, 'molecules', 'atoms' or 'threads', and, in doing so, he certainly made a big advance in the conception of the inner structure of crystals. But he did not think explicitly of the building layers as constituted of two-dimensional arrays of identical rhombohedra. He was confused into thinking that the structure of other minerals than calcite could be derived from the spar rhombohedron and was in fact far from the concept of space lattices.

11.11 J.-B. L. Romé de l'Isle and the law of the constancy of interfacial angles, 1783

J.-B. L. Romé de l'Isle (Fig. 11.31) was a French mineralogist who established the general character of the law of the constancy of dihedral angles between crystal faces, which earlier had only been described in specific cases, in particular by N. Steno (see Section 11.7). The importance of Romé de l'Isle's work was stressed by Haüy (1795) who wrote: 'To the exact descriptions he gave of the crystalline forms, he added the measure of their angles, and,

Fig. 11.31   Jean-Baptiste Louis Romé de l'Isle. Courtesy IMPMC, Université P. et M. Curie, Paris.

Jean-Baptiste Louis Romé de l'Isle: born 26 August 1736 in Gray, Haute-Saône, France, the son of an officer; died 7 March 1790 in Paris, France. After schooling in the Collège Sainte Barbe in Paris, he took part, as a naval officer, in the Indian wars and was taken prisoner by the British in 1761. Back in France, in 1764, he studied mineralogy with the chemist and mineralogist Balthazar Georges Sage (1740–1824), who was to be the Director of the newly founded Paris School of Mines. At first, he earned his living by writing catalogues for the auction of mineral collections, but at the end of the 1760s he met A. J. d'Ennery, a Secretary of the King, a collector of old medals and coins, who invited him to live in his house where he stayed for more than twenty years, until d'Ennery's death in 1787. He had his own collection of minerals, well-stocked with more than 5000 pieces, which was auctioned after his death and bought by the Agence des Mines. Haüy, who was in charge of the collections of the Agency, made use of it when he wrote his *Traité de Minéralogie*. Romé de l'Isle was elected a member of the Science Academies of Stockholm and Mainz in 1775 and of Berlin in 1780. He failed to be admitted to the French Academy of Sciences, which made him very bitter, all the more so that Haüy got elected. The injustice was underlined by J. Cl. de la Métherie in his 1790 obituary of Romé de l'Isle.

MAIN PUBLICATIONS

1767 *Catalogue systématique et raisonné des curiosités de la nature et de l'art qui composent le cabinet de M. Davila.*

1772 *Essai de Cristallographie ou description des figures géométriques propres à différents corps du règne minéral connus vulgairement sous le nom de cristaux.*

1773 *Description méthodique d'une collection de minéraux du cabinet de M. D. R. D. L...*

1779 *Action du feu central bannie de la surface du globe, et le soleil rétabli dans ses droits*

contre les assertions de MM. Buffon, Bailly, de Mairan....

1783 *Cristallographie, ou description des formes propres à tous les corps du règne minéral.*

1784 *Des caractères extérieurs des minéraux, ou Réponse à cette question: existe-t'il des substances du règne minéral des caractères que l'on puisse regarder comme spécifiques?*

1789 *Métrologie ou tables pour servir à l'intelligence des poids et mesures des anciens et principalement à déterminer la valeur des monnaies grecques et romaines.*

which was essential, showed that these angles were constant for each variety. In one word, his crystallography was the fruit of an immense work, almost entirely new and most precious for its usefulness'.

The context of the time was not very favourable for such studies, for had not the famous French naturalist G.-L. L. Buffon (1717–1788) written in his *Natural History of Minerals*, Volume I (1783): 'One has pretended that rhombohedra constitute a specific character of calcareous spar, without paying attention to the fact that some vitreous or metallic substances also crystallize in rhombohedra and that if calcareous spar does crystallize often in rhombohedra, it also takes different other forms; and our *crystallographers*,[84] when trying to borrow from geometricians the way to transform a rhombohedron into an octahedron, a pyramid or a lens, have done nothing more than substitute ideal combinations to the real facts of Nature. This crystallization in rhombohedra, like all others, will never have a specific character. Not only there isn't any crystallization form which is not common to different substances, but, conversely, there are few substances which do not present different crystallization forms, as shown by the prodigious variety of forms of calcareous spar itself'.

J.-B. L. Romé de l'Isle had started collecting minerals during his travels as a naval officer. Back in Paris after the Indian wars, he was introduced into mineralogy by the apothecary, chemist and mineralogist, Balthazar Georges Sage (1740–1824), who became his friend. It was very fashionable at the time in Paris to have a mineral collection. The owner of an important private collection, P. Davila, wanted to sell his. At Sage's suggestion, he asked Romé de l'Isle to draw up the inventory. Romé made a very thorough job of it, the inventory running up to three thick volumes. This was his first work in mineralogy, published in 1767. It gave him the opportunity to study crystalline forms in detail and led to his *Essai de Cristallographie ou description des figures géométriques propres à différents corps du règne minéral connus vulgairement sous le nom de cristaux* (Romé de l'Isle 1772, Fig. 11.32 (*Left*)). In the preface, he noted that 'of the curious phenomena of the mineral kingdom those which struck him most were the regular and constant forms taken by some bodies designated by the name of crystals'. It was encouraged by the works of Linnaeus, he added, that he undertook the study of the angular forms of crystals and of their transformations. Their polyhedral shape was only known of the Ancients for quartz, diamond, and a few others, and Romé widely extended this observation. Minerals are sorted by him in four classes, salts, stony, pyritic and metallic. For each mineral, the most frequent forms observed are described, with a reference to Linnaeus's classification in *Systema naturae*. In general, with a few exceptions (calcite, garnet, gypsum, quartz), there are no values of facial or interfacial angles and those given are old ones. Steno's ideas relative to the growth of quartz layer by layer (see Section 11.7.2) are quoted at length and Romé de l'Isle felt they could be applied to all crystals. The book was a success, acclaimed by Linnaeus himself,[85] and brought international fame to Romé de l'Isle.

The second crystallographic treatise, *Cristallographie, ou description des formes propres à tous les corps du règne minéral* (Fig. 11.32 (*Right*) and 11.33), dedicated to the Prussian Royal Academy of Sciences, contains a description of a much larger number of crystal forms (more than 500) than his first one (110) and Linnaeus's (about 40). It is in this work that Romé de

---

[84]The emphasis is Buffon's.

[85]See Linnaeus's congratulatory letter to Romé de l'Isle in May 1773, quoted in Romé de l'Isle (1783).

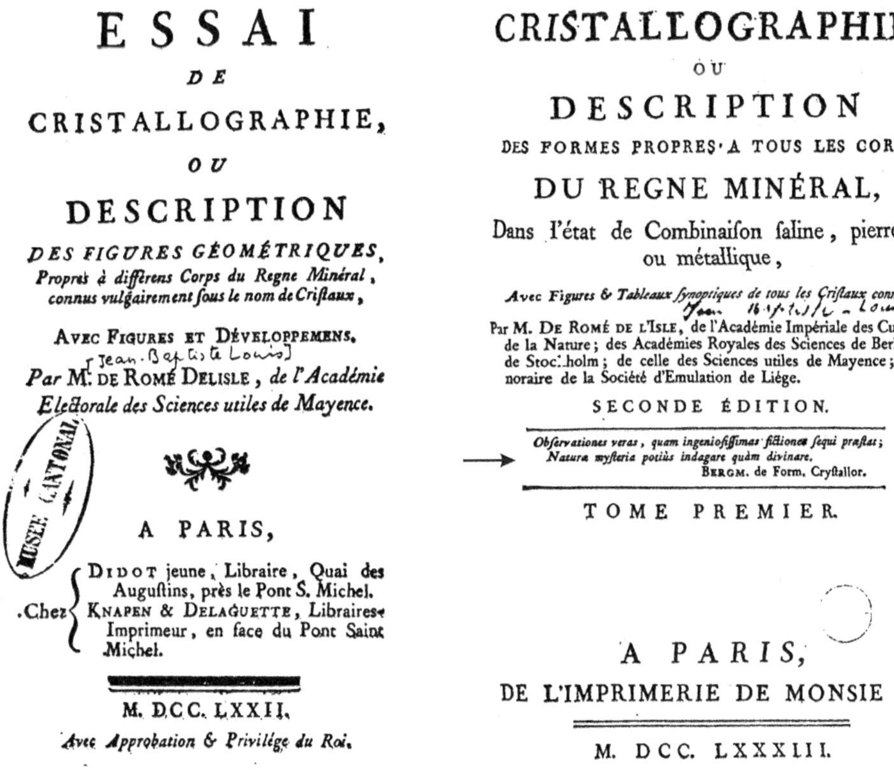

ESSAI

DE

CRISTALLOGRAPHIE,

OU

DESCRIPTION

DES FIGURES GÉOMÉTRIQUES,

Propres à différens Corps du Regne Minéral,
connus vulgairement sous le nom de Cristaux,

AVEC FIGURES ET DÉVELOPPEMENS.

*Par M. DE ROMÉ DELISLE , de l'Académie
Electorale des Sciences utiles de Mayence.*

A PARIS,

Chez { DIDOT jeune , Libraire , Quai des
Augustins, près le Pont S. Michel.
KNAPEN & DELAGUETTE, Libraires
Imprimeur, en face du Pont Saint
Michel.

M. DCC. LXXII.

Avec Approbation & Privilége du Roi.

CRISTALLOGRAPHIE,

OU

DESCRIPTION

DES FORMES PROPRES·A TOUS LES CORPS

DU REGNE MINÉRAL,

Dans l'état de Combinaison saline , pierreuse
ou métallique ,

Avec Figures & Tableaux synoptiques de tous les Cristaux connus.

Par M. DE ROMÉ DE L'ISLE, de l'Académie Impériale des Curieux
de la Nature ; des Académies Royales des Sciences de Berlin &
de Stockholm ; de celle des Sciences utiles de Mayence ; Ho-
noraire de la Société d'Emulation de Liége.

SECONDE ÉDITION.

Observationes veras , quam ingeniosissimas fictiones sequi præstat ;
Natura mysteria potiùs indagare quàm divinare.
BERGM. de Form. Cryftallor.

TOME PREMIER.

A PARIS,

DE L'IMPRIMERIE DE MONSIE

M. DCC. LXXXIII.

Fig. 11.32   First pages of Romé de l'Isle's treaties of crystallography. *Left*: 1772. *Right*: 1783. Note the quotation
from T. Bergman (arrow). After Romé de l'Isle (1783).

l'Isle (1783) states the constancy of interfacial angles: *Nothing is easier than to show, with the
help of the goniometer, which we owe to M. Carangeot*[86] (Fig. 11.33), 'the constancy of the
[interfacial] angles and of the crystalline form of every species'. More precisely, 'the faces of a
crystal may vary in their shape and in their extension, but their respective inclination is constant
and invariable in each species'. Romé de l'Isle's aim was to put some order in the confusion
created by the large variety of forms exhibited by most crystals. He notes that 'at the imitation
of the famous Bergman, some physicists among us busy themselves at present demonstrating
by geometrical figures and calculations the mechanism by which are constructed some crystals
that are easy to divide with a cutting tool'. He had Haüy specifically in mind, whom he calls

[86]Arnould Carangeot (1742–1806), a student and co-worker of Romé's, actually preceded his master in observing
the constancy of angles. In his paper published in March 1783 (Carangeot 1783), he told how, when preparing models of
crystals in clay for Romé de l'Isle, he used a piece of cardboard to measure interfacial angles on a quartz crystal with a
complicated habit and observed their constancy. He communicated the news to Romé de l'Isle, who was very interested
and encouraged him to repeat the observations. In order to be able to make accurate measurements, he then had the idea
of the goniometer, which he constructed.

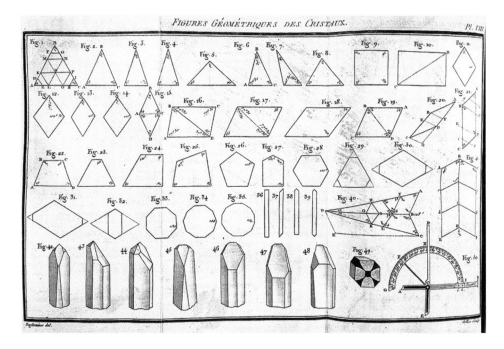

Fig. 11.33   Table showing the result of Romé de l'Isle's measurements of crystal angles. Figure 40 (lower right) is a sketch of the gypsum dovetail twin, and Figure 50 (also lower right) a drawing of Carangeot's contact goniometer. After Romé de l'Isle (1783).

a *cristalloclaste*, who 'mutilates the few crystals that can be divided mechanically, looking for an alleged nucleus which, even if it existed, could not be explained by geometry alone'. Romé de l'Isle also criticized Haüy's interpretation of the garnet forms. For him, on the contrary, 'one should start by investigating all the various forms of a given species'. This he did by carefully measuring the interfacial angles with the Carangeot contact goniometer and identifying the individual forms associated in a given specimen of a crystal. He defined six primitive forms and showed that each of the individual forms can be derived from a primitive form by suitable truncations.[87] Romé de l'Isle was, however, himself criticized for the arbitrary choice of the primitive forms, but also by mineralogists such as Bergman and Kirwan, who described him as a mere 'catalogue maker'!

Romé de l'Isle was also the first mineralogist to give a rational description of twins, preliminary in 1772, and more detailed in 1783. He described, for instance, the dovetail twin of gypsum (Fig. 11.33) and the feldspars twins. He introduced the term *macle* to designate a crystal in which 'one half is produced by the inversion in the opposite sense of the other half of the same crystal'—a property which he demonstrated by the concordance of angles.

[87]It seems it is Démeste (1779) who was the first to use the expression truncation of angles and edges. Werner (1774) had used the terms *Abstumpfung, Zuschärfung, Zuspitzung*.

For details of Romé de l'Isle's life, the reader may consult, for instance, his obituary by the French naturalist and mineralogist, J. Cl. de la Mètherie (1743–1817) in *Observations sur la physique, sur l'histoire naturelle et sur les arts*, **36**, April 1790, pp. 315–323, Groth (1926) and his notice by L. Touret in *Travaux du Comité Français d'Histoire de la Géologie* (1997), **11**.[88]

---

[88] <http://annales.org/archives/cofrhigeo/rome-de-lisle.html>.

12

THE BIRTH AND RISE OF THE SPACE-LATTICE CONCEPT

A crystal—considered as indefinitely extended—consists of interpenetrating regular point systems, each of which is formed from similar atoms.

P. von Groth (1904)

12.1 R.-J. Haüy and the *Theory of the structure of crystals*, 1784

René-Just Haüy (Fig. 12.1), a French mineralogist and crystallographer, is one of the major contributors to the development of the concept of space lattice. He was also a physicist who studied the double refraction of crystals and their electrical, in particular pyroelectric, and magnetic properties. His interest in physics and in botany started at a young age. In 1780, Haüy was introduced to mineralogy by the lectures of L. J.-M. Daubenton[1] at the Jardin du Roi in Paris. There is a well-known anecdote, told by Haüy himself (Haüy 1784, 1793, 1795, 1801, 1822), and repeated with a slightly embellished variation by G. Cuvier (1769–1832) in his 1823 Eulogy of Haüy,[2] that it was his observation of the oblique fracture of a fragment accidentally broken off a larger prismatic calcite crystal which was at the origin of his investigations, mirroring Gahn's story (see Section 11.10.2). Whether true or not, the anecdote symbolizes the fact that Haüy's theory was at the start based on the observation of the various orientations of the cleavage planes of crystals.

It is noteworthy that, roughly at the same time as Bergman and Haüy, other mineralogists considered that the structure of the various forms of calcite could be related to the cleavage rhombohedron. One is the German mineralogist Christian Friedrich Gotthard Westfeld (1746–1823). In the sixth of his mineralogical essays (Westfeld 1767), he wrote: 'Most calcareous spars break under the hammer into rhombohedron-shaped pieces ...All crystals of spar are composed of rhombohedra-shaped pieces or, rather, it is really Nature which constituted them that way; the main origin of this construction is therefore not to be cared about. One then asks only why rhombohedron-shaped crystals build up other specific forms and it seems to me that, put that way, the question becomes much more complicated'. Another one is William Pryce, MD,

---

[1]Louis Jean-Marie Daubenton, born 29 May 1716, died 31 December 1799, was a French naturalist and physician. He obtained his MD in 1741 in Reims, and in 1742 was asked by G.-L. L. Buffon (1717–1788) to assist him by providing anatomical descriptions for his *Natural History*. In 1744, Daubenton became a member of the French Academy of Sciences as an adjunct botanist, and Buffon appointed him keeper and demonstrator of the King's cabinet in the Jardin du Roi, which became the Museum d'Histoire Naturelle at the French Revolution, and of which he was the first Director. In 1755, he was elected a Fellow of the Royal Society and, in 1778, a Professor at the Collège de France. Daubenton was one of the pioneers of comparative anatomy and introduced it in the study of fossils. He was also a mineralogist, Professor of Mineralogy at the Museum d'Histoire Naturelle, and author of a classification of minerals.

[2]Georges Cuvier, *Eloge historique de M. Haüy*, in *Mémoires de l'Académie royale des sciences de l'Institut de France.* (1829), **VIII**, 123–175.

Fig. 12.1   René-Just Haüy holding a calcite crystal. After Société Française de Minéralogie, *René Just Haüy*, Masson, Paris, 1945.

René-Just Haüy: born 28 February 1743 in Saint-Just-en-Chaussée, Oise, France, the son of a poor weaver, Just Haüy; died 01 June 1822 in Paris, France; his brother, Valentin (1745–1822), devoted his life to the blind. As a boy, René-Just loved music, and his frequent attendance at the services of the local church drew the attention of the prior of an abbey of the Premonstrants. Thanks to his recommendation, Haüy obtained a scholarship at the Collège de Navarre in Paris around 1755. After completing his studies, he was appointed, in 1764, as regent (master) in the College and in 1770 in the Cardinal Lemoine College where he taught until his retirement in 1784. In 1770, he was ordained priest. He was elected at the Académie Royale des Sciences in 1783 as *adjoint* in the class of botany and, in 1788, as *associé* in the class of Natural History and Mineralogy. In 1791 he became a member of the commission in charge of elaborating the metric system. During the French Revolution, all priests were required to take an oath, which Haüy refused to do, with the result that he was briefly imprisoned in August 1792, freed thanks to the help of his friend, the French naturalist Geoffroy Saint-Hilaire (1772–1844). He was appointed Professor of Physics at the Ecole Normale Supérieure in 1795, when the Ecole was created. His lectures served as the basis of his treatise on physics for high schools (1803), commissioned by Bonaparte; the book was an immediate success and was re-edited several times. In 1795 he was appointed Professor of Mineralogy at the Ecole des Mines and curator of its mineralogy collection, and, in 1802, at the Museum of Natural History. His fame extended beyond the frontiers and attracted many students from every part of Europe. In 1809, he became the first Professor of Mineralogy at Paris University, where he created the still existing Laboratory of Mineralogy. After the Restoration he lost his pension and finished his life in poor conditions.

MAIN PUBLICATIONS

1784 *Essai d'une théorie sur la structure des crystaux.*

1787 *Exposition raisonnée de la théorie de l'électricité et du magnétisme, d'après les principes d'Æpinus.*

1792 *Exposition abrégée de la théorie de la structure des cristaux.*

1793 *Théorie de la structure des cristaux.*

1801 *Traité de minéralogie.*

1803 *Traité élémentaire de physique.*

1809 *Tableau comparatif des résultats de la cristallographie, et de l'analyse chimique relativement á la classification des minéraux.*

1817 *Traité des caractères physiques des pierres précieuses.*

1822 *Traité de minéralogie.* Second edition.

1822 *Traité de cristallographie.*

a Cornishman who wrote in his treatise on minerals, mines and mining (Pryce 1778): 'The formation of Spar is yet a subject of enquiry. Its atoms are all Spar; each particle, into which we can without violence divide it, is the same in all respects as the whole: and as the Fossil world admits of no generation by egg or seed, it seems most probable, that all the variety of forms, in which we behold this Protean Mineral, are owing to no cause but the arrangement of rhombs,

Fig. 12.2   Structure of a calcite lamella, exhibiting half rhombs along the edge. After Haüy (1782*b*).

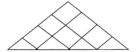

into as many forms as they are capable of producing'. Neither Westfeld nor Pryce went any further and they are not referred to by either Bergman or Haüy.

For biographical details on Haüy, the reader may consult the books by Metzger (1918), Groth (1926), Burke (1966), Burckhardt (1988), and the special issues dedicated to him by the *American Mineralogist* (1918, Volume 3), the *Société Française de Minéralogie (René Just Haüy*, 1945. Paris: Masson), the *Revue d'histoire des sciences (René-Just Haüy (1743–1822), physicien*, 1997, Tome 50 n°3); see also Wiederkehr (1977), Scholz (1989), Maitte (2001) and the historical introduction to the recently published lectures of Haüy at the Ecole Normale Supérieure (Guyon Editor 2006).

12.1.1   Haüy's preliminary theory, 1781

Haüy's first works were two memoirs submitted to the Académie Royale des Sciences on 10 January 1781 (*Sur la formation des spaths calcaires et des grenats*) and 22 August 1781 (*Sur la cristallisation des spaths calcaires*). The first manuscript and the reports presented on them by Academicians Daubenton and Bezout on 21 February 1781 and 22 December 1781, respectively, are in the archives of the *Académie des Sciences* in Paris.[3] Extracts of the manuscripts were published in 1782 (Haüy 1782*a*, 1782*b*). In the two papers, Haüy develops a lamellar theory of the structure of crystals, which is in line with Bergman's, but is much more rigorous and elaborate, and constitutes a real step forward. The first paper starts with the remark that 'the same substance may present a large variety of forms that at first sight do not seem to be reducible to a given form', while 'very different crystals may present the same form, for instance the cube or the octahedron'. A good indication of their structure may be obtained from the 'sections that can be made in them by means of a cutting tool' (cleavage), 'which only happens along specific directions' and 'shows the polish of nature', or, if they are too hard to be cleaved, from the various 'striations and lineaments' that can be observed at their surface. The results of these observations is that 'every crystal of a given kind, whatever its shape, contains, as a nucleus, a form that can be considered as primitive; namely, the form obtained after cleaving off successive lamellae from the crystal is always the same, a cube, a rhombohedron or an octahedron'. These lamellae are identical to those which constitute the primitive form, with the difference that, if the primitive form is a rhombohedron, for instance, the matter surrounding the nucleus may be bounded by half rhombs or isosceles triangles instead of full rhombs (Fig. 12.2). Haüy's preliminary theory is therefore indeed a lamellar theory and distinct from his future molecular theory where the side faces are stepped. It is also shown that different crystal species crystallizing with the same shape may have different primitive forms. For instance, both common salt and fluorspar (fluorite) crystallize in cubes but their respective primitive forms, obtained by cleavage, are the cube and the octahedron. Haüy then proceeds to

---

[3]The reports are available on the web site <http://gallica.bnf.fr/>.

analyse the structure of garnet. Like Bergman, he thinks that the most common form, the rhomb-dodecahedron, can be conceived as composed of four identical rhombohedra, but, contrary to Bergman, he calculates the facial angles to have the correct value, 109° 29′.[4] Two other forms, the tetragonotrioctahedron and a combination of the two, are analysed in the same way.

According to the Academicians' report, the first part of the second manuscript is devoted to fluorspar for which the primitive form is an octahedron. In particular, Haüy shows that Démeste (1779) had imagined wrong truncations to derive the octahedron from the cube. In the second part he derives the most common habits of calcite such as the rhomboidal lenticular spar, the hexagonal prism, the scalenohedra, and associations of these forms. Haüy used for that the indications provided by the cleavage, the striations, and the various features exhibited by the natural faces of the crystal, criticizing in passing Bergman's derivations, except for the dog's-tooth scalenohedron (Fig. 11.29). He also calculated the facial angle of the cleavage rhombohedron (101° 32′) from the geometrical construction used to derive the hexagonal prism, in good agreement with the values given by La Hire (1710),[5] 101° 30′, and by Newton (1704), 101° 52′, who in fact used the value measured by Huygens (see Section 3.4.2). The report concludes that Haüy has shown in a convincing way that the primitive form is the same for every crystal of the same sort and that all the matter surrounding the nucleus is composed of identical particles, which are also identical to the particles which compose the nucleus. 'Much skill and shrewdness', they add, 'was required to unravel the complexities of the structure of calcareous crystals'. (Abbé Haüy) 'is the first one to have explained in a satisfactory way the forms of calcareous spar, with the exception of the dog's tooth which had already been explained by Bergman'. The success of Haüy's preliminary theory was acclaimed by all the scientists of his time and led to his election to the Académie Royale des Sciences in 1783.

Naturally enough, given the impact of Haüy's work on the development of crystallography, historians of science have discussed Bergman's influence on him (see, for instance, Hooykaas 1955 and Burke 1966). From the first manuscript, it is clear that Haüy was aware of Bergman's work when he wrote it. According to Haüy (1795, 1801), it is some time after he started attending Daubenton's mineralogy lectures that he happened to make the chance observation of an oblique cleavage plane on a prismatic crystal of calcite. This led him to start cleaving calcite crystals, and the examination of the cleavage rhombohedra started a train of thoughts in his mind. Haüy discussed them with Daubenton, who drew his attention to Bergman's articles and, along with mathematician and Academician P.-S. de Laplace (1749–1827), encouraged him to pursue his studies and to publish his ideas. Haüy did criticize Bergman's failed attempt at a general theory, but paid tribute to his pioneering work: 'he considered the various forms of a given substance as produced by the superposition of decreasing planes around a primitive form. He applied this idea to a few simple forms and tested it on a particular variety of calcareous spar. But he stopped there and did not try to determine by calculus the laws of the structure. This was a simple sketch drawn as if in passing, from the most beautiful viewpoint of mineralogy, and where one recognizes

---

[4]A similar wrong interpretation of the garnet form was also given by the Belgian physician and mineralogist, J. Démeste (1748–1783), who worked with Romé de l'Isle (Démeste 1779).

[5]Philippe de La Hire (1640–1718) was a French astronomer, physicist, and mathematician, who made accurate interfacial angle measurements and studied the dovetail twin of gypsum found in Paris quarries.

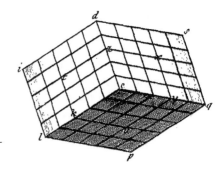

Fig. 12.3   Structure of the nucleus as an aggregation of 'constituting molecules'. After Haüy (1784).

the hand who worked so successfully to improve the picture of chemistry' (Haüy 1795, 1801).

The development of Haüy's own general theory is based on the systematic study of crystalline forms and on Romé de l'Isle's law of the constancy of interfacial angles. Carangeot's contact goniometer was to be his favourite tool, and he is often portrayed with it (Fig. 12.1).

12.1.2 Haüy's theory of crystal structure

The law of decrements, 1784.   With his 1784 essay, *Essai d'une théorie sur la structure des cristaux, appliquée à plusieurs genres de substances cristallisées*, Haüy breaks away radically from his own and Bergman's lamellar theory and introduces his molecular theory: in the case of calcite, for example, 'the nucleus, or primitive form, has as constituting molécules small rhombohedra, rather than lamellae' (Fig. 12.3). The constituting molecules ('molécules constituantes') are those which were floating in the fluid where the crystal was dissolved, which were mutually attracted and became agglomerated during crystallization. The shape of these constituting molecules can be found with the help of the mechanical division of the crystal, or the striations on the surface. The cleavage takes place between two molecules and can be done at any place in the crystal; if the mechanical division is pushed to its ultimate limit, therefore, it leads necessarily to identical and uniform molecules. Their arrangement, which constitutes the 'structure' of the crystal (Fig. 12.3), spreads regularly throughout the crystal. This is a crucial point, since the notion of three-dimensional periodicity of crystals is here clearly introduced for the first time.

The secondary forms, which are the external forms actually observed, are obtained by the deposition of layers over layers of constituting molecules. A single layer or a group of two or more layers is shifted with respect to the layers beneath it by one, two, three or, rarely, four rows. In theory, the number of possibilities is very large, but in practice the law describing how the secondary form is derived from the primitive form is always simple, it is the *law of decrements*, or *law of simple rational truncations* ('Haüy's law'). The French mineralogist and crystallographer, G. Friedel[6] (1904, 1907) pointed out that the term 'simple' in the wording of the law is essential. Without it, the law would be a simple mathematical statement, impossible to verify, and not a physical law.

[6]See footnote 3, Section 7.2.

Haüy observes that the resultant faces of the secondary forms will be stepped, and look like a staircase, but that these steps are so small that they are invisible to the naked eye. Due to the irregularities of the growth, they may in fact sometimes be larger and give rise to the striations one observes at the surface of the crystal. There are no longer half-rhombs along the edges of the layers, as represented in Fig. 12.2. The decrements can take place along an edge, around a summit or both (mixed decrements), as shown below.

The habit of a particular crystal is the result of the association of various secondary forms which can be interpreted by means of appropriate laws of decrement. After having described the different forms of calcite, the primitive forms of various minerals are determined: the rhombic prism for barite, the octahedron for fluorspar (fluorite), a prism with a rectangular base for gypsum, the rhomb-dodecahedron for garnet, which Haüy recalls is the shape of the honeycomb cell (the cleavage along {110} is poor, but was observed by Haüy), a rhombic prism for topaz. There is a particular difficulty in the case of fluorspar. A common habit is the cube, and Haüy notes that the smallest crystals, at the start of growth, are little cubes. But the primitive form, obtained by cleavage, is an octahedron. During growth, the layers which agglomerate are parallel to the surface of the cube, but they are not smooth, and are covered with tiny spikes corresponding to the summits of octahedra or tetrahedra. The case of fluorite is further developed below.

Decrements along an edge. Fig. 12.4, *Left*, shows the principle of the derivation of a form by decrements along an edge, with the cube as the primitive form. Successive layers are added, in this example, on two faces of the cube, of which the area decreases by one or more layers at a time. Haüy (1792) insists that the process of the derivation of the secondary forms from the primitive form is a purely theoretical and geometrical one and that it does not pretend to represent the way a crystal grows. Nevertheless, this scheme proved to be a source of inspiration for the Stranski–Kossel surface nucleation model of crystal growth (Fig. 12.4, *Right*). Two examples are shown in Fig. 12.5. In the one at left, the rhomb-decahedron (Miller indices {110}, to use the modern notation) of hyacinth (zircon) is obtained by a shift of one row of little cubes. In the second one, at right, the pentagon-dodecahedron, or pyritohedron, of pyrite is obtained by a shift of a single layer by two rows along two of the four edges of the cubic nucleus and a shift

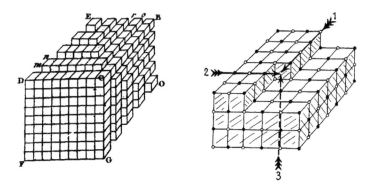

Fig. 12.4   *Left*: Principle of the derivation of a form by decrements along the edges. After Haüy (1792). *Right*: Kossel crystal. After Kossel (1927).

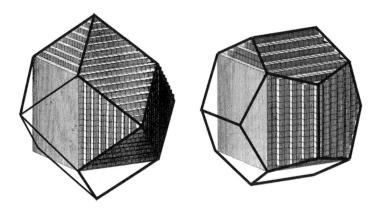

Fig. 12.5   Decrements along the edges. *Left*: rhomb-dodecahedron. *Right*: pentagon-dodecahedron. The contours of
the forms have been added for clarity. After Haüy (1801).

of a double layer by one row along the two other edges. As the modern crystallographer knows, this form is characteristic of the centrosymmetric hemihedry of the cubic system (class $m\bar{3}$) and its Miller indices are 1/2 {210}. Haüy showed that this pentagon-dodecahedron is slightly different from the regular pentagon-dodecahedron of geometry with which the other authors had confused it, and that it is not possible to find a law of decrement describing the regular pentagon-dodecahedron.

As another example, Fig. 12.6 shows the formation of the faces of the calcite dog's tooth, which Haüy called *métastatique* (it is a scalenohedron of Miller indices {1$\bar{2}$0} with rhombohedral axes and {21.1} with hexagonal axes), and the orientation of the nucleus, or primitive form, with respect to it. The layers which are added successively decrease by two rows at a time. It is interesting to note the improvement with respect to Bergman's model (Fig. 11.29).

Decrements around a summit.   The law of decrements around a summit can be illustrated by the derivation of the octahedron from a cubic nucleus in common salt. It is achieved by adding successive lamellae decreasing around the corners on the faces of the cube (Fig. 12.7, see also Fig. 12.26). The faces of the octahedron thus obtained are not smooth but bristly with the spikes formed by the corners of the small cubes. Haüy thus imagined three types of crystal faces: flat for the primitive forms, the most frequent ones, stepped for the faces resulting from a decrement along an edge and spiked for the faces resulting from a decrement around a summit (or a mixed decrement). This seemed unthinkable to Haüy's contemporaries (see, for instance, Bernahrdi 1809; Glocker 1831). It is, however, the first attempt at relating crystal structure and morphology. The next step is due to Bravais, who said that the most important faces or cleavage planes are the planes with highest density (see Section 12.11.3). It is noteworthy that in the Periodic Bond Chain (PBC) theory of crystal growth (Hartman and Perdok 1955), precisely three types of faces are considered; namely, *F* faces (for flat), parallel to at least two periodic bond chains, *S* faces (for stepped), parallel to one periodic bond chain, and *K* faces (for kinked), which are not parallel to any and look just like the face represented in Fig. 12.7, top right (Fig. 12.8).

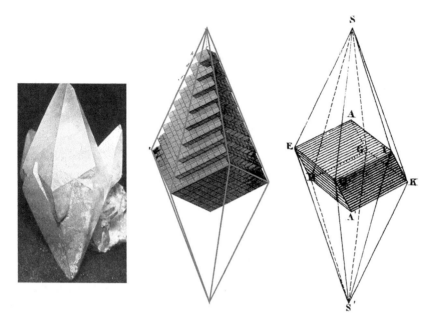

Fig. 12.6 Haüy's derivation of the dog's tooth (*chaux carbonatée métastatique*). *Left*: calcite crystal (photo by the author). *Middle*: derivation of the scalenohedron, and *right*: the primitive form inscribed in the scalenohedron. After Haüy (1822).

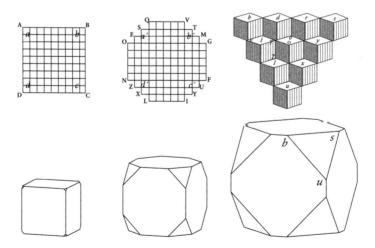

Fig. 12.7 Decrements around the summits. *Bottom*: Passage of the cube to the octahedron. *Top*: *left*, top layer of the cubic nucleus and *middle*, layer showing the decrements around the corners, after Haüy (1792), *right*: structure of facet *bsu* of the octahedron, after Haüy (1822).

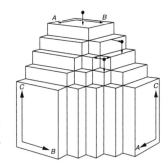

Fig. 12.8   Hypothetical crystal with three P.B.C. vectors, $A \parallel [100]$, $B \parallel$ [010], $C \parallel [001]$, showing F-faces, (100), (010), (001), S-faces, (110), (101), (011) and K-face (111). After Hartman and Perdok (1955).

Reduction of the various primitive forms to a parallelepiped, 1792, 1793.   When developing the law of decrements in his 1784 book, which implied layers of small parallelepipeds, Haüy had come across a difficulty in the case of crystals having other solids as primitive form, for instance the octahedron for fluorspar (fluorite). In 1792, instead of 'constituting molecules', he used the term 'integrant molecules' (*molécules intégrantes*)[7] to designate the smallest corpuscles which would be obtained if mechanical division was pushed to its ultimate limit, had we sufficiently sharp tools. The concept of integrant molecule as the smallest building unit of minerals struck the imagination of Haüy's contemporaries. For instance, Dolomieu[8] (1801) considered the integrant molecule to be a complete individual and defined a mineral species by the constitution of its integrant molecule. In a similar way, the French naturalist, J.-B. Lamarck (1744–1829), an early proponent of the concept of evolution, stated in his quest for the notions of species and 'individual' in living beings that 'the individual of any mineral resides in its integrant molecule'.[9]

Having, since 1784, investigated many more crystals, Haüy listed in his 1792 tract various possible primitive forms: a), the cube, the rhombohedron and all the parallelepipeds, b) the regular tetrahedron, the octahedron, the hexagonal prism, the rhomb-dodecahedron, the dodecahedron with triangular faces. For the law of decrements to be applicable, the nucleus must have the shape of a parallelepiped and Haüy stated that it is always possible to extract a parallelepiped as nucleus from any crystal. Either it is the shape obtained by mechanical division and also the shape of the integrant molecule or it can be deduced from it. In the latter case, two situations may arise. The first one is illustrated by apatite. Its primitive form is a hexagonal prism. By cleavage, it can be divided into six triangular prisms. By joining two of them, one obtains a prism with a rhomb as base, which can be used as a suitable parallelepiped. This operation does

---

[7]This term was usual at time; it was used, for instance, by Romé de l'Isle (1772).

[8]Déodat Gratet de Dolomieu, born 23 June 1750, died 28 November 1801, was a French mineralogist and geologist, and a knight of the order of St John of Jerusalem. After a career as an officer until 1775, he devoted himself to earth sciences, in particular volcanology and mineralogy. Dolomieu studied in particular the calcareous mineral which was called dolomite, after him. In 1794 he was appointed Professor at the School of Mines, and in 1795 he was elected a member of the Academy of Sciences. In 1801, following L. J.-M. Daubenton's death, he was appointed Professor at the Museum of Natural History, but died soon afterwards and was succeeded by Haüy.

[9]*Recherches sur l'organisation des corps vivans et particulièrement sur son origine*, 1802, Paris: Maillard.

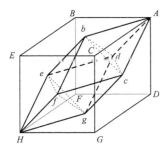

Fig. 12.9   Structure of fluorite: *ABCDEFGH*, growth cube; *AbcdefgH*, elemental parallelepiped or subtractive molecule; *bcdefg*, octahedron (primitive form); *Abcd* or *Hefg*, tetrahedra. (Redrawn for clarity from Fig. 38 of Haüy, 1792.)

not leave any void in the structure. The second situation is that where the integrant molecules cannot be combined to form a parallelepiped without leaving voids.

As an example of the latter situation, Haüy took the case of fluorspar of which the primitive form is an octahedron. When the crystal starts growing, it takes the shape of a small cube, *ABCDEFGH* (Fig. 12.9). By cleavage along the faces of this cube, it is possible to obtain an acute rhombohedron *AbcdefgH* whose summits, *A* and *H*, are the two opposite summits of the cube, and its other summits the centres of the faces of the cube (Fig. 12.9). It is quite remarkable that Haüy has thus found, without knowing, the relation between the unit cell and the cubic multiple cell of a face-centred cubic lattice, which is to be credited to his geometric intuition. The cubic faces can then easily be obtained by superposing layers of small rhombohedra with a very simple law of decrements. This rhombohedron constitutes the elemental parallelepiped, what we now call the unit cell. By cleaving further, it can be divided into an octahedron, *bcdefg*, and two tetrahedra, *Abcd* and *Hefg*. The structure of fluorite can therefore be described either by a stacking of octahedra surrounded by eight tetrahedra or by tetrahedra surrounded by four octahedra. Haüy was here up against a difficulty since, for him, there can only be one sort of integrant molecule, which must be uniform throughout the crystal. He could not really solve it satisfactorily except by saying that the stacking of any of the two possible integrant molecules leaves voids that are so small as to be indistinguishable. At the time, the constitution of fluorite was not known, and Haüy always made it a point not to make any hypothesis or conjecture as to the nature of the integrant molecules, but one cannot help but be struck by the fact that the analysis of the situation led Haüy to consider as repeating unit a rhombohedron containing an octahedron and two tetrahedra as possible integrant molecules, thus hitting without realizing it the structure of CaF_2, more than a hundred years before Bragg (1913*b*, see Section 8.3.1). However, the same difficulty arises for any structure having an octahedron as primitive form so that it is in fact just a happy coincidence. In the *Traité de Minéralogie* (1801), Haüy concluded that the primitive form of fluorspar is the octahedron and the integrant molecule a tetrahedron.

Molécules soustractives (subtractive molecules), 1801.   The structure to which the law of decrements is applied is a regular assembly of adjacent elemental parallelepipeds (Figs. 12.3 and 12.4). As has just been said, either they have the same shape as the integrant molecule or not. In the latter case, the integrant molecule is a subdivision of the elemental parallelepiped. This happens in fluorspar described above, quartz or garnet, which both have a tetrahedron as primitive form, irregular for quartz (Haüy 1786), regular for garnet. Haüy has therefore found necessary to introduce a special name for these parallelepipeds: *molécules soustractives*, or 'subtractive molecules' (Haüy 1801). Those who objected to Haüy's theory of integrant molecules (Bernhardi 1809; Whewell, 1837) and thought that Haüy's resort to the concept of subtractive molecules was a confirmation of their criticism did not realize, as Seeber (1824) did, that it opened the way for the notion of space lattice.

12.1.3   Treatise on Mineralogy (Traité de Minéralogie), 1801

Haüy's major work is his *Traité de Minéralogie* (1801). He started by developing his theory of crystal structure and the law of decrements. The decrements governing each face of a form were denoted by a system of letters and numbers, which he had introduced earlier (Haüy 1796), and which indicated the edge or the summit where the decrement takes place, the width and the height of the steps. By defining the various criteria, physical, chemical, mineralogical, and geometrical, serving for a classification of minerals, he endeavoured to establish the notion of mineral species: 'the mineral species is a collection of bodies the integrant molecules of which have similar forms, and are composed of the same principles, united in the same proportions'. Haüy could thus show that beryl and emerald belong to the same species and that heavy spar (barite, barium sulfate) and strontia (celestite, strontium sulfate) are different species. A systematic description of all known minerals is given, sorted in four classes: acidiferous minerals (such as the carbonates), earths (such as quartz), non-metallic burning minerals (such as sulphur, diamond, or bitumen), and metallic substances. For each mineral, the physical (specific weight, hardness, eventually optical[10]), chemical, and geometrical properties (primitive form and integrant molecule) are given, and all the known forms are described, with their locations, the appropriate law of decrements, and reference to their description, in other treatises on mineralogy, by Wallerius, Werner, Romé de l'Isle, Daubenton, Kirwan, and others.

Haüy used the measure of the dihedral angles to determine the law of decrements. In some cases, conversely, he guessed the law of decrement of a form first, by analogy with other secondary forms, and then calculated the dihedral angles. From the dihedral angles, the shape only of the primitive form can be obtained, and not its dimensions. The quantity that came into all his calculations was the ratio between the two diagonals of a face (see Fig. 12.19). For Haüy, this ratio had to be simple, and the values of the dihedral angles that he gave were always adjusted accordingly. As a consequence, they were not always in perfect agreement with those measured accurately by other mineralogists, in particular after the introduction of Wollaston's reflection goniometer (see Section 12.2). Unfortunately, this led to much criticism and loss of credibility in Haüy's theory (see, for instance, Mohs 1823).

A second, updated, edition of the *Treaty of Mineralogy* was published in 1822.

12.1.4   The law of symmetry, 1815

The notion of symmetry was introduced early on by Haüy, for instance in his first lesson at the Ecole Normale Supérieure (Haüy 1795): 'The way in which nature produces crystals is always that of the greatest symmetry, in that opposite and corresponding parts are always similar in number, arrangement and shape'. In answering a question by a student, Haüy specified that

---

[10]Haüy found twenty minerals presenting double refraction.

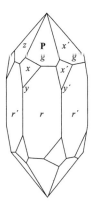

Fig. 12.10   Habit of quartz. *P*, primitive form, {10.1} and *z*, {01.1}, rhombohedra; *x*, {15.1} and *x'*, {51.1}, trigonal trapezohedra (plagihedral faces). After Haüy (1801).

this property does not apply to crystals which become polar under the influence of heat (now called 'pyroelectric'). The law of decrements and his notation of them was modified accordingly (Haüy 1815) by taking the symmetry of the crystal into account: 'a given type of decrement repeats itself on all the parts of the nucleus that are so similar that they can be substituted one for the other, when changing the position of this nucleus with respect to the eye. I call these parts *identical* ... For instance, if the parallelepiped is a cube, any decrement along an edge of the cube will repeat itself along all the other edges of the cube'. In his treatise on crystallography, Haüy (1822) defines thus an axis of symmetry: 'I define axis in general, a straight line drawn from the centre of the crystal and of which the direction is such that the various parts of the crystal are arranged symmetrically with respect to it'.

Haüy used this law of symmetry to determine more precisely the shape of the primitive form of some minerals; for instance, it is by application of the law of symmetry that he showed that the primitive form of hematite is a rhombohedron and not a cube, while both Steno and Romé de l'Isle had described the habit of hematite with truncations of the cube (Haüy 1822). But, Haüy noted, the law of symmetry suffers exceptions. One example is that of the facets denoted *x* and *x'* (of Miller indices {15.1} and {51.1}, respectively) that he observed in a variety of quartz which he called 'plagihedral' (Fig. 12.10) and which did not follow the symmetry of the hexagonal prism; in some crystals, the facets were inclined to the right, in others to the left. The structural and optical implications of this observation will be discussed in Section 12.10. Another example is that of tourmaline, of which the primitive form is the rhombohedron and for which some of the expected facets are missing. Haüy knew about the pyroelectric effect and the polarity induced in tourmaline by a change of temperature; he thought therefore that the absence of half the number of facets might have something to do with a different action of some electric forces on the two extremities of the axis of the crystal during the growth (Haüy 1822).

12.1.5   *Treatise on Crystallography (Traité de Cristallographie), 1822*

An improved version of Haüy's theory of the structure of crystals was published a few months before his death, the Treatise on Crystallography (1822). Haüy developed there his law of decrements taking into account the law of symmetry and simplified his classification in terms

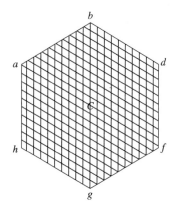

Fig. 12.11   Structure of a crystal having a triangular prism as integrant
molecule. After Haüy (1822).

of the shape of the primitive form. He distinguished: 1) the primitive forms having the shape
of a parallelepiped: the rhombohedron, the cube, the oblique rhombic prism, the right rhombic
prism, the oblique rectangular prism; 2) the primitive forms different from a parallelepiped: the
hexagonal prism of which the integrant molecule is a prism with an equilateral triangle as base
(Fig. 12.11), the rhombic dodecahedron, the octahedron and the tetrahedron.

This new treatise was also the occasion for Haüy to clarify his definition of a mineral species
and of the integrant molecule. A mineral is the result of 'the aggregation of molecules bound
together by affinity'. He distinguished two stages in this operation and takes as an example
the formation of calcareous spar: (a) 'in the first one, the molecules of carbonic gas and lime
combine according to a certain proportion to form tiny regular solids, invariable in shape,
namely the integrant molecules of calcareous spar, carbonic gas being itself composed of
molecules of carbon and oxygen',[11] (b) 'in the second one, the integrant molecules align face
to face in rows and layers in such a way that the external shape of the aggregate is that of
a geometrical body'. It follows that a mineralogical species is defined by two characters, its
geometrical type and its *chemical type*. The ultimate limit of the mechanical division of the
crystal is also that of the physical division of the substance. The geometrical type is determined
by the relative quantities of the constituents of the integrant molecule and the actions they exert
on one another. The ascription of the chemical type has been made easy by the progresses of
chemical analysis since the beginnings of Haüy's work in the 1780s and by the establishment
of the law of definite proportions at the end of the eighteenth century by the French chemist
Joseph Louis Proust (1754–1826) and that of the law of multiple proportions in 1803 by the
English chemist John Dalton (1766–1844).[12]

It is important to note that the type of crystal thus defined by Haüy is what we would
call today a molecular crystal, and this is why he was not prepared to accept the concept of

---

[11]A view already present in his lessons at the Ecole Normale (Haüy 1795). In a similar way, he had noted there that
pyrite is composed of iron and sulphur, crystallizing respectively in the forms of a cube and a kind of octahedron. Cubic
and octahedral molecules must therefore combine in a certain way to form the cubes of pyrite.

[12]Dalton's atomic theory is developed in *A New System of Chemical Philosophy*, 1808.

isomorphism introduced in 1819 by E. Mitscherlich[13] in a conference at the Prussian Academy of Science in Berlin. His observation that phosphate and arsenate of potassium crystallize with the same primitive form and the same angles prompted him to further study many other salts. The work was first published in a restricted form (Mitscherlich 1819, 1820), and it is to this first paper that Haüy reacted, through a letter written by one of his students to the Editors of the *Annales de Chimie et de Physique.*[14] An extension of Mitscherlich's work devoted to arsenates and phosphates was published in Swedish in 1821 and was later translated into German and French (Mitscherlich 1821). It is in that paper that he introduced the term *isomorphism* and stated the general law for the relation of crystallography to chemical composition: 'An equal number of atoms, if they are bound in the same way, produce similar crystal forms, and the crystal form depends not on the nature of the atoms but on the number and method of combination'. Mitscherlich (1824) also observed the variation of crystal angles due to thermal expansion, which had important implications for the physicists.

Another important new concept was developed by Mitscherlich: that of dimorphism, based on his own observations of different crystalline forms of sodium phosphate (Mitscherlich 1821) and of sulphur (Mitscherlich 1823), as well as on Berzelius'[15] observation of yellow and white (marcasite) pyrites mentioned in the latter paper. As noted by L. Pasteur (1848*b*) in his work on dimorphism, it is an error to think that this concept brought a blow to Haüy's ideas. Haüy conceded that substances with the same chemical composition but different molecular arrangements could have different crystalline forms, and was the first to establish the existence of two different crystalline forms for two minerals, calcite and aragonite, which the chemists had shown to have the same chemical composition, calcium carbonate (Haüy 1808, 1822).

12.1.6 Haüy's legacy

Haüy's work spread rapidly and widely. The mineralogists and crystallographers who followed him in the first half of the nineteenth century were deeply influenced by him, and his work was always their basic reference. His works were translated into German by C. S. Weiss[16] (see Section 12.3) and by J. F. C. Hessel[17] (see Section 12.9). It is striking that Christian Weiss, whose philosophical position regarding the nature of crystals was radically different

---

[13]Eilhard Mitscherlich (1794–1863) was a German chemist and mineralogist who developed the concepts of isomorphism and polymorphism. After studies in Heidelberg and Göttingen, he went to Berlin where he worked on phosphates and arsenates. His observations on their crystallization led him to the notion of isomorphism. He continued his work with J. J. Berzelius in Sweden, and in 1821 returned to Berlin, where he was appointed Professor in 1822.

[14]*Réflexions sur le Mémoire de M. Mitscherlich concernant l'identité de la forme cristalline dans plusieurs substances différentes, Ann. Chim. Phys.* Paris, **14**, 303–308.

[15]Jöns Jacob Berzelius was a Swedish chemist, born 20 August 1779 near Linköping, in Sweden, died 7 August 1848 in Stockholm, Sweden. He studied medicine at Uppsala University and obtained his MD in 1802. In 1807 he was appointed Professor of Chemistry and Pharmacy at the Karolinska Institute, and in 1808 was elected a member of the Royal Swedish Academy of Sciences. Berzelius is known in particular for his discovery of the law of constant proportions and of several elements (Si, Se, Th), and for the determination of atomic weights.

[16]*Lehrbuch der Mineralogie*, 4 Volumes by R.-J. Haüy, translated by C. J. B. Karsten and C. S. Weiss, with an introduction by D. L. G. Karsten (1804–1810). Paris and Leipzig: C. H. Reclam.

[17]*Haüy's Ebenmassgesetz der Krystall Bildung* (The law of symmetry), 1819, translated with annotations by J. F. C. Hessel. Frankfurt am Main: in der Hermannschen Buchhandlung.

from Haüy's, chose to present his own views in the form of an appendix to his translation of the Treatise on Mineralogy.[18] It is to be noted, however, that in his dissertation (Weiss 1809), his derivations of crystalline forms are entirely based on Haüy's description of the same forms.

Haüy was generally admired and respected: 'his method of derivation of the crystallization systems was one of the most important enrichments of the physical sciences' (Bernhardi 1808), 'his genius raised mineralogy to the rank of sciences' (Berzelius in the dedication of his 1819 book), *Nouveau système de minéralogie*), 'having found the law of decrements is enough to raise Haüy to the status of founder of crystallography' (Frankenheim 1835), 'Haüy is to be considered without contest as the founder of the science of crystallography' (Hessel 1830), and his system of classification of minerals was 'the basis of the best succeeding works in mineralogy' (Whewell 1837), but he was also subject to many criticisms. They are of different orders. Some are more technical, such as the already mentioned inaccuracies in his measurements of angles. Others are more fundamental, in general related to the concept of integrant molecule. Haüy took it to be not only an element of the geometrical and physical structure of the crystal, but also the molecule itself of the substance, and this led to many misunderstandings. This hypothesis was, however, unnecessary for the development of the theory of decrements, which is independent of the nature of the integrant particles and of their relation with the actual molecules of the crystal. Haüy (1822) himself wrote: 'if the corpuscles obtained by pushing the mechanical division to its ultimate limit do not represent exactly those upon which acts the affinity during crystal growth, they are their equivalent and the results that we get are in such agreement with observation that we could assume not to be mistaken by taking the molecules of the theory for the molecules of nature'. Haüy was also criticized, but that was after Weiss and Mohs had introduced the notion of crystal systems, for not having enough taken symmetry into account in his classification of crystalline forms and because his list of primitive forms seemed somewhat arbitrary (Sohncke 1879) and included too many cubic ones (Whewell 1837), although some found there a prefiguration of the crystal systems (Kupffer 1831, Frankenheim 1835).

An objection of quite a different nature was made by Weiss (1816–17) who rejected atomism and wrote that he developed his own theory 'while seeking a dynamical ground for the existence of the primitive form, instead of the untenable' (*verwerflich*) 'atomistic way of thinking'. This view was generally held in Germany at the time. The German mineralogist Ernst Friedrich Glocker (1793–1858), Professor of Mineralogy in Breslau, who found that Haüy's molecular theory was 'quite unphilosophical', wrote in his *Handbuch der Mineralogie* (Glocker 1831) that the atomistic view 'betrays an unworthy' (*unwürdige*) 'representation of the way nature operates . . . How are the small particles linked to one another and what brings them together'? In a similar way, the German physician, mineralogist and botanist, Johann Jakob Bernhardi (1774–1850), Professor of Medicine at the University of Erfurt, noted: 'the fact that so many special forces of attraction must combine to form a crystal is unreasonable; which are these attraction forces that bring elementary molecules of so many different shapes together'? (Bernhardi 1809). This view was not, however, general among German crystallographers and

---

[18]*Dynamische Ansicht der Krystallisation* (Dynamical view of the crystallization), by C. S. Weiss, in *Lehrbuch der Mineralogie*, (1804), Vol. I, pp. 365–389.

Haüy's main legacy is certainly the postulate that crystals can be considered as a triply periodic assembly of small identical parallelepipeds, integrant or subtractive molecules, joined together, their relative shapes being determined using mechanical division and their natural striations, or deduced from the law of decrements applied to the secondary forms (Figs. 12.3, 12.5, 12.6). The concept of subtractive molecules was criticized by some of his contemporaries and also by Sohncke (1879). It had been introduced by Haüy because of the difficulties raised by the presence of the voids which appear in a stacking of integrant particles that are not parallelepipeds. One sees now what a remarkable intuition that was. Haüy's geometrical construction of crystals implies a space lattice and it is in Haüy's work that this notion first appears; for that reason, he does deserve to be considered as the founding father of modern crystallography. The concept of space lattice was developed in different forms independently by Seeber in 1824 (see Section 12.7), Frankenheim in 1835 (see Section 12.8), and Delafosse in 1840 (see Section 12.10), starting from Haüy's atomistic vision of crystals. It was to be further developed by A. Bravais (see Section 12.11) in its modern form. Seeber's and Frankenheim's works remained, however, unnoticed in Germany for several decades.

Through the law of decrements, Haüy created a notation using a system of letters and fractions to designate a crystalline form, which contained all the information required to characterize the orientation of its faces.[19] C. S. Weiss (1814–15) (see Section 12.3) replaced it by a set of indices referring to a set of crystal axes. Weiss' system was at its turn modified and improved by Neumann (1823), W. Whewell (1825) and W. H. Miller (1939) (see Section 12.6).

The law of simple rational intercepts (*loi des indices rationnels simples*), or law of rationality (*Gesetz der Rationalität der Indices*), is the fundamental law of crystallography and the basic property of lattice planes. It was not stated as such by Haüy, but is directly implied by the law of decrements. It was stated in various forms by Haüy's followers: G. Delafosse (1825), M. L. Frankenheim (1826, 1832, 1835), A. T. Kupffer[20] (1826, 1831), W. H. Miller (1839), Sohncke (1882), and E. S. Fedorov (1894). Bravais's 'law' (see Section 12.11.3) is, in fact, also equivalent to it.

12.2   W. H. Wollaston and *On the elementary particles of certain crystals*, 1812

12.2.1   *The inventor of the reflecting goniometer*

William Hyde Wollaston (Fig. 12.13) was an English chemist, metallurgist, physicist and crystallographer, discoverer of rhodium and palladium. He is an important actor of our story because of his close-packing suggestion for the inner structure of crystals. Wollaston also invented a number of optical devices, of which several are very useful in crystallography:

---

[19]This notation was improved by the French mineralogist Armand Lévy (1795–1841), and was still used by some mineralogists in the middle of the twentieth century, under the name Haüy–Lévy notation.

[20]Adolf Kupffer (1798–1865) was a Russian mineralogist and crystallographer from Latvia. After his thesis in Göttingen (1821) he visited Haüy in Paris and Weiss in Berlin. In 1823 he won a prize of the Prussian Academy of Science on the accurate measurement of the dihedral angles of crystals. He was appointed Professor of Physics and Chemistry in Kazan in 1824 and Director of the Mineralogy Cabinet in St Petersburg in 1826.

Fig. 12.12   Wollaston's   goniometer.   After   Wollaston
(1809).

The refractometer (Wollaston 1802*a*) is an instrument for determining refractive indices with the
aid of total reflection. Wollaston used it to repeat and confirm Huygens' measurements of the
double refraction of calcite (Wollaston 1802*b*).

The reflecting goniometer (Wollaston 1809) is used to measure interfacial angles between crystal
faces belonging to the same zone. It is mounted on a rotatable horizontal axis *cc* (Fig. 12.12),
and the crystal is set at *n* with the zone axis parallel to that horizontal zone axis. A ray of light is
reflected into the eye successively by the two faces and the angle is measured on the graduated
circle, *ab*. The Wollaston goniometer is very accurate and a considerable improvement over the
Carangeot goniometer. Wollaston (1812) used it, for instance, to carefully measure the interfacial
angles of the cleavage rhombohedra of three carbonates, carbonate of lime (calcite, $CaCO_3$),
bitter-spar (magnesite, $MgCO_3$), and iron spar (siderite, $FeCO_3$), and to show that they are slightly
different. His measurements, more accurate than Haüy's, confirmed the latter's view that different
chemical compounds could have similar forms.

The Wollaston prism is a polarizing beam splitter consisting of two orthogonal calcite prisms,
cemented together on their base to form two right triangle prisms with perpendicular optic
axes.

12.2.2   On the elementary particles of certain crystals: close packing

Wollaston's reflections on the inner structure of crystals stemmed from his difficulty in imagining what could be the primitive form associated with a cleavage octahedron or tetrahedron
(Wollaston 1813). If one takes fluorspar (fluorite, CaF_2), for instance, it is easy to obtain
a rhombohedron by cleavage and one can fill space with it, but, by further cleaving that
rhombohedron, one obtains an octahedron, which can be cleaved into two tetrahedra, which
can themselves be cleaved into octahedra, and so on. But neither of these two latter forms

Fig. 12.13   William Hyde Wollaston by John Jackson, Source: *Platinum Metals Review* (2003).

William Hyde Wollaston: born 6 August 1766 in East Dereham, Norfolk, England, the son of Reverend Francis Wollaston, Fellow of the Royal Society; died 22 December 1828 in Chislehurst, England. After schooling at Charterhouse, he entered Caius College, Cambridge, in 1782. He graduated in medicine in 1788 and obtained his Doctorate of Medicine in 1793; he was elected a Fellow of the Royal Society the same year. He started medical practice at Huntingdon, Cambridgeshire, in 1789, continued at Bury St Edmunds, Suffolk, finally settling in London in 1797, relinquishing medical practice for experimental science in 1800. His interests were very varied. In 1797 he described the main components of urinary calculi; in 1801 he showed that static and voltaic electricity are identical; in 1802 he developed the refractometer; in 1802–4, he discovered rhodium and palladium; in 1807 he patented the camera lucida, which makes use of a half-silvered mirror tilted at 45° and helps microscopists and artists to make accurate drawings of what they see—in fact, it had already been described by Kepler in *Dioptrice* (1611). In 1809 Wollaston described a vibratory action of muscular activity and invented the reflective goniometer; in 1813 he gave his views on the inner structure of crystals. In his last paper, published after his death, in 1829, he described his method for obtaining malleable platinum.

MAIN PUBLICATIONS

1802   *On the Oblique Refraction of Iceland Crystal.*

1804   *On a New Metal, Found in Crude Platina.*

1805   *On the Discovery of Palladium; with Observations on Other Substances Found with Platina.*

1808   *On Super-Acid and Sub-Acid Salts.*

1809   *The description of a reflective goniometer.*

1813   *On the Elementary Particles of Certain Crystals.*

1829   *On a Method of Rendering Platina Malleable.*

forms can fill space without leaving voids.[21] It did not seem to Wollaston that the elementary particles could be either tetrahedra with octahedral cavities or, on the contrary, octahedra with tetrahedral cavities, which would be 'an unstable arrangement, ill adapted to form the basis of any permanent crystal'. The solution, which he found three years before his 1813 Bakerian lecture, was that 'all difficulty is removed by supposing the elementary particles to be perfect spheres, which by mutual attraction have assumed that arrangement which brings them as near to each other as possible'. Wollaston found out later that Hooke had already presented the same idea in his *Micrographia* (see Section 11.5.2). Figure 12.14 illustrates how Wollaston built tetrahedra (*5, 7, 8*), octahedra (*6*), and rhombohedra (*9, 10*) by close packing of hard spheres. It is easy to show how these forms can be deduced from one another by cleavage; for instance, the octahedron may be obtained by removing a tetrahedral group from each extremity of the rhombohedron (*10, 11*).

[21]As we have seen (see Section 12.1.2), Haüy (1784, 1792, 1793) had made the same observations and come to the same difficulty with the voids, but he considered they were so small as to be undetectable.

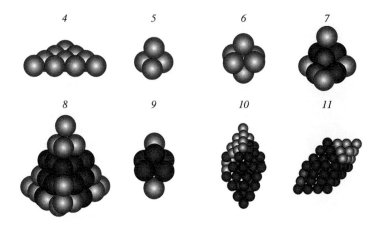

4   *5*   *6*   *7*

8   *9*   *10*   *11*

Fig. 12.14   Close-packing of hard spheres. After Wollaston (1813).

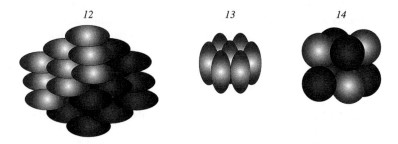

12   *13*   *14*

Fig. 12.15   Close-packed structures. After Wollaston (1813).

In the last part of the paper, Wollaston considered the type of structures one could obtain by packing other objects than spheres, for instance spheroids. He noted that regular rhombohedra could be derived from oblate spheroids (Fig. 12.15, *12*) as had been done by Huygens, and he recalculated the required ratio between the axes of the ellipsoid in order to get the calcite rhombohedron; he obtained 1 to 2.87, to be compared with the value of 1 to 2.83 obtained by Huygens (see Section 11.8).[22] Hexagonal prisms could be derived in similar fashion from oblong spheroids (Fig. 12.15, *13*).

Wollaston then addressed the question of cubic crystals. He discarded the simple cubic packing as not being 'the form which simple spheres are naturally disposed to assume' but misses the face-centred cubic packing. He went round the problem by introducing the idea of an intermixed binary compound, and considered a stacking of 'spherical particles all of the same size, but of two different kinds, represented by black and white balls', as shown in Fig. 12.15, *14*; the configuration of both the black and the white balls is a regular tetrahedron,

[22]Huygens had in fact given a ratio of 1 to $\sqrt{8}$, but in the edition of Huygens' book that Wollaston had read, a ratio of 1 to 8 had been given, due to a misprint.

which Wollaston considered as stable. It is interesting to note he had thus obtained a portion of the structure of a face-centred cubic binary compound such as sodium chloride, well before Barlow (see Section 12.13)!

The close packing of balls was in fact for Wollaston an image of the structure rather than a real description. 'Though the existence of ultimate physical atoms absolutely indivisible may require demonstration', he explained, 'their existence is by no means necessary to any hypothesis here advanced, which requires merely mathematical points endued with powers of attraction and repulsion equally on all sides, so that their extent is virtually spherical, for from the union of such particles the same solids will result as from the combination of spheres impenetrably hard'. Wollaston admitted very honestly that it was difficult to understand why the structure of fluorspar could be explained with one kind of sphere only, real or virtual, while it contained at least two elements and he concluded 'that any attempts to trace a general correspondence between the crystallographical and supposed chemical elements of bodies must, in the present state of these sciences, be premature'.

Wollaston's crystal theory was further developed by the English chemist John Frederic Daniell (1790–1840), the inventor of an electric battery called the Daniell cell, who studied at length the figures formed by crystals after a slight dissolution in their solvents (Daniell 1817).

12.3 C. S. Weiss and the *Kristallisations-Systeme*, 1815

12.3.1 The founder of the dynamistic school of crystallography

Christian Samuel Weiss (Figs. 12.16, 12.17) was a German mineralogist and crystallographer, considered to be the founder of the dynamistic tradition. He did not contribute to the development of the concept of crystal lattice as such, but he played a major role in the development of crystallography with the introduction of the notion of crystal systems and of that of hemihedry. Like Haüy, he had a profound influence on his followers. His approach to the nature of crystalline matter was fundamentally different from that of Haüy; he considered it as resulting from the action of a distribution of attractive and repulsive forces in the volume of the crystal, and this led him to a very fruitful analysis of its symmetries. This was done from a purely geometric viewpoint, without any speculation as to the inner structure of matter, which was considered as continuous. Weiss was strongly influenced by the German philosophers of his

Fig. 12.16   Christian Samuel Weiss as a young man. © Universitätsbibliothek Leipzig, Portraitstisch C. S. Weiss.

Fig. 12.17   Christian Samuel Weiss. © UB der Humboldt Universität zu Berlin, Porträtsammlung (Chr. S. Weiss).

Christian Samuel Weiss: born 26 February 1780 in Leipzig, Germany, the son of the archdeacon of Saint-Nicholas church; died 1 October 1856 in Eger, Bohemia. After schooling at the Evangelischen Gnadenschule in Hirschberg, Silesia, he started medicine studies at the University of Leipzig in 1796. Following his bachelor's degree in 1798, he shifted to natural sciences, first in Leipzig, then in Berlin in the laboratory of the chemist M. H. von Klaproth (1743–1817) where he met two former pupils of A. G. Werner's, L. von Buch (1774–1853) and D. L. G. Karsten (1768–1810), Director of the Academy of Mines. He obtained his Magistergrad (PhD) in 1800 and his habilitation on 23 September 1801. On Karsten's advice he undertook to translate Haüy's treatise on Mineralogy and spent a few years in Freiberg where he became one of Werner's favourite students. In the years 1806–8 he travelled through Austria, Switzerland, and France where he visited the Auvergne volcanoes and met Haüy in Paris. He was appointed ordinary Professor of Physics at the University of Leipzig in August 1808 and returned to Leipzig in March 1809 where he had to present a Dissertation. In 1810, he was called to Berlin to become Professor of Mineralogy in the University of Berlin (Friedrich-Wilhelms-Universität, founded in 1810 by Wilhelm v. Humboldt, renamed Humboldt-Universität Berlin in 1849), in replacement of D. L. G. Karsten, previously Director of the Berlin Bergakademie (Mining Academy), whose untimely death had prevented the appointment. The post also included the directorship of the university's Mineralogy Museum. Among his many students, A. T. Kupffer, M. L. Frankenheim, F. E. Neumann, F. A. Quenstedt, and G. Rose are to be specially mentioned. Weiss was highly rated as a professor and was elected five times Dean of the Faculty of Philosophy of the Friedrich-Wilhelms University, and twice Rector. He became a member of the Munich Academy of Sciences in 1803, of the Prussian Academy of Sciences in 1815, and of the Deutsche Akademie der Naturforscher Leopoldina in 1818.

MAIN PUBLICATIONS

1801 *De notionibus rigidi et fluidi accurate definiendis.*

1809 *De indagando formarum crystallinarum charactere geometrice principali.*

1809 *De charactere geometrico principali formarum crystallinarum octaedricarum pyramidibus rectis basi rectangula oblonga.*

1819 *Krystallographische Fundamentalbestimmung des Feldspathes.*

1819 *Über eine verbesserte Methode für die Bezeichnung der verschiedenen Flächen eines Krystallisations systemes.*

1820 *Betrachtung der Dimensionsverhältnisse in den Hauptkörpern des sphäroëdrischen Systems.*

1822 *Grundzüge der Theorie der Sechsundsechskantner und Dreiunddreikantner, entwickelt aus den Dimensionszeichen für ihre Flächen.*

1834 *Über das Gypssystem.*

1836 *Über rechts und linksgewundene Bergkrystalle.*

time: Immanuel Kant (1724–1804)[23] and Friedrich Wilhelm Joseph Schelling (1775–1854),[24] whom he had the opportunity of meeting while he was studying medicine in Leipzig in the late 1790s.

In the German romantic philosophy of nature (*romantische Naturphilosophie*), the general trend was towards a dynamical view of matter, in opposition to the atomistic and mechanical viewpoint. A definition of the crystalline state according to this conception is given by Bernhardi (1809): 'the main difference between fluids and solids does not come so much from their different degree of cohesion than from the different ability of their constituents to move apart under repulsion forces without losing their cohesion. In a fluid, the attraction forces are identical in every direction, but not so in solids. If the directions along which the particles are pulled are specific, the body is crystallized in the broader sense of the word. We speak of a crystal in the stricter sense if we observe external flat faces assembled with specific angles between them. The directions of these faces and of the cleavage planes are those along which the particles are pulled'.

Weiss's interests soon shifted from medicine towards physics, chemistry, and mineralogy. His 1801 habilitation dissertation was concerned with the, at the time controversial, notions of solid, crystalline, and fluid states. It is there that the roots of his ideas relative to the dynamistic theory of crystallization and to the classification according to main and secondary axes are to be found (Höbler 2006). In 1801–2, he went to Berlin where he continued his studies at the Academy of Mines with the chemist M. H. von Klaproth (1743–1817), who discovered uranium in pitchblende in 1789, the geologist L. von Buch (1774–1853) and the mineralogist D. L. G. Karsten (1768–1810). The latter two, who were former pupils of A. G. Werner[25], acquainted him with the more practical aspects of mineralogy, and Karsten commissioned him and his own nephew, C. J. B. Karsten, to translate Haüy's *Treaty of Mineralogy*, which took them several years, Weiss being the main contributor.[26] Doing so, Weiss became aware of the weaknesses of Haüy's theory. Meanwhile, at von Buch's and Karsten's suggestion, he spent the years 1802–3 with A. G. Werner in Freiberg. In 1807 and 1808, Weiss visited Paris, where he met many French men of science; he became friendly with the French mineralogist and mining engineer A. J. F. Brochant de Villiers (1772–1840) and had many discussions with Haüy, who was at first very fond of the young scientist; however, their relations deteriorated when it became clear that their conceptions of the nature of crystallization differed sharply.

The development of Weiss's concept of the properties of the crystalline state can be followed in his successive writings: appendices to his translation of Haüy's Treaty (1804–10), his 1809

---

[23]See, for instance, *Metaphysische Anfangsgründe der Naturwissenschaft* (Metaphysical Foundations of Natural Science), 1786.

[24]See, for instance *Erster Entwurf eines Systems der Naturphilosophie* (First Outline of a System of the Philosophy of Nature), 1799.

[25]Abraham Gottlob Werner, born 25 September 1749 in Wehrau, now Osiecznica, Lower Silesia, died 30 June 1817 in Dresden, Germany, was a German geologist, well known for his theory of stratification and for his classification of minerals (Werner 1774). After receiving his education in Freiberg and Leipzig, he became inspector and instructor at the Freiberg Bergakademie (Mining Academy) in 1775, and later Professor. Among his students, A. von Humboldt, L. von Buch, F. Mohs, R. Jameson, and C. S. Weiss are to be specially mentioned.

[26]See footnote 16.

dissertation when he became Professor of Physics at the Leipzig University, and his 1814–15, 1816–17, 1818–19 papers in the *Transactions of the Berlin Royal Academy of Sciences*.[27]

For further details on the life and works of Christian Weiss, the reader may consult Metzger (1918), Groth (1926), Fischer (1962), Burke (1966), Paufler (1981), Hoppe (1982), Scholz (1989), and Höbler (2006).

12.3.2 Weiss's dynamistic theory of crystallization, 1804

In his appendix to the translation of the first volume of Haüy's Treaty of Mineralogy,[28] Weiss endeavoured to explain the solidification of fluids from a dynamistic viewpoint, rather than from the atomistic one. He considered matter to be continuous but polar and constituted of positive and negative quantities. It is submitted to conflicting and orientation-dependent (anisotropic) chemical forces, both attractive and repulsive, which are responsible for the cohesion of matter and the external shape of crystals: 'There are in nature not only chemical attractions that strive to induce heterogeneous matter to interpenetrate and to combine, but also opposite repulsive forces that tend to break up homogeneous bodies into heterogeneous parts. The crystallization of fluids is a phenomenon of repulsion by which the dissociation and the separation of the constituents has not been completed and where instead these repulsive forces have been inhibited without reaching their goal. The forces of chemical attraction and repulsion are in all bodies in eternal conflict and are in steady equilibrium in perfect fluids. As soon as the chemical tendency to separate prevails, matter strives to break up into separate poles. But the repulsion force becomes progressively weaker and the opposite unification force grows with equal intensity until the repulsion is reduced and stopped'. The stage thus reached characterizes crystallization. These forces act in specific directions and account for the existence of the natural faces of the crystal and of the cleavage planes. In relation to the angle between the cleavage planes, Weiss speaks of an *Abstossungswinkel* (literally, repulsion angle). These 'repulsion regions' can be two- or three-dimensional. In the first case, there are only individual sets of cleavage planes, such as in barite (barium sulfate),[29] in the second one, there are several sets of equivalent cleavage planes, for instance three in galena (lead sulphide)[30] and four in fluorite (calcium fluoride).[31] Besides these main cleavage directions, which represent *primitive directions of crystallization*, there are also secondary ones, along natural crystal faces or imperfect cleavage planes. These secondary directions are linked to the primitive ones and are related to them by specific relations. The directions of the normals to the crystal faces also play an important role.

Like Haüy, Weiss was criticized by his contemporaries. Bernhardi (1809), who did not see whence come the forces that bring together Haüy's integrant molecules (see Section 12.1.6), wrote that there is no proof of the chemical repulsion forces invoked by Weiss and that the process by which cleavage planes make different angles in different substances is not known.

---

[27]Königlich-Preussische Akdemie der Wissenschaften (founded 1700 in Berlin), now named Berlin-Brandenburgische Akademie der Wissenschaften.

[28]See footnote 18.

[29]Barite is orthogonal and its three sets of cleavage planes along {100} are independent.

[30]The cubic planes, {100} in modern notation.   [31]The octahedral planes, {111}.

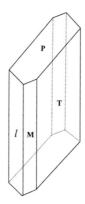

Fig. 12.18   Cleavage planes of feldspar. After Haüy (1801).

Weiss, however, all his life held to his ideas about the crystallization process (see, for instance, Weiss 1832).

12.3.3   The zone law, 1804–6

The concept of zones is introduced in an appendix to Volume II (1804) of the translation of Haüy's treatise.[32] It follows a discussion of the properties of feldspar, which Weiss finds one of the most important existing minerals. Haüy had found two good cleavage directions, along faces P, {001}, and M, {010}, and an imperfect one, along T, {110} (Fig. 12.18). Weiss showed that one could easily be mistaken and that l, {1$\bar{1}$0}, is in fact also a good cleavage plane. It should therefore be taken into account when determining the shape of the primitive form. Generalizing, Weiss thinks that the nature of the crystallization is not the result of one primary direction of repulsion, but always of two interrelated ones, namely of two directions of perfect cleavage. The other, secondary, crystallization directions follow naturally from them by truncation at the summits and edges. A zone is a set of crystallographic planes having one direction in common, and three planes having a common direction are necessary to define a zone.

Weiss came back to the properties of zones in annotations to Volume III (1806) of the translation (Groth 1925, Fischer 1962). It is here that the law of zones is stated. Weiss introduced 'crystal elements' to describe the crystal forms and, through their angle dependence, the anisotropy of the crystal. In each crystal it is possible to find three main forces from which the others can be derived by suitable combinations. They are expressed by three mutually orthogonal principal axes along which a suitable metric has been defined. Crystal faces are characterized by their intersections with the three axes. They are grouped in zones but, to Weiss, a zone is not simply a geometric assembly of faces parallel to a given direction; more importantly and physically, it is represented by a plane normal to the zone axis and in which lie the directions perpendicular to the cleavage planes. All possible crystal faces can be deduced, even without making any angle measurement, from the already known faces by considering the planes parallel to two zone axes. This is the zone law (*Zonenverbandgesetz*).

---

[32]*Ueber die Krystallisation des Feldspathes*, by C. S. Weiss, in *Lehrbuch der Mineralogie* (1804), Vol. II, pp. 711–723.

In modern notation, the so-called Weiss law, $hu + kv + lw = 0$, expresses the fact that the direction $[u, v, w]$ of the zone axis is parallel to the crystal face (h, k, l). Weiss further states that only crystal faces can exist whose intersections with the axes are multiples of the intersections of the primitive faces with these axes. This is the law of rationality (*Rational-itätsgesetz*). These two laws are of course implied by Haüy's law of decrements, but were restated by Weiss from a different viewpoint. The concept of zones is further developed in later papers, in particular in Weiss (1820–21) where the full habit of feldspar is described in detail. The zone law was applied, among others, by J. F. L. Hausmann (*Untersuchungen über die Formen der leblosen Natur*, Göttingen 1821, see Groth 1926) and by F. E. Neumann; see Section 12.6.1.

12.3.4   Determination of the main geometric character of the crystalline shapes, 1809

When Weiss was appointed Professor of Physics at the University of Leipzig, he presented on 8 March 1809 a dissertation, *De indagando formarum crystallinarum charactere geometrice principali*, and, on 11 March, for his inaugural lecture, a second part devoted to the 'dipyramids with an elongated rectangular basis' (*De charactere geometrico principali formarum crystalli-narum octaedricarum pyramidibus rectis basi rectangula oblonga*)[33].

The dissertation was the occasion for Weiss to reassert his views on the dynamistic inter-pretation of the crystallization process and to analyse the crystalline forms of a large number of minerals, based on Haüy's descriptions. In the course of the memoir, he paid tribute to the contributions of Romé de l'Isle and of Haüy, who 'opened a new era for the science of Crystallography'. Weiss starts by the *a priori* statement that the crystalline forms are the necessary result of the geometric distribution of the generating forces which determine the formation of the crystals, adding that his experience has convinced him that it is indeed so. His aim was to find the simplest way to characterize a crystalline species. For that, he introduced what he calls the *main geometric character* of each primitive form. It is obtained by means of the ratio of certain *crystal elements* associated with the directions of the main forces acting within the crystal, namely its axes: 'an axis is a straight line governing all the faces and around which they are disposed in equivalent positions'. In general, he chose the sine s and the cosine c of the angle of a face of the primitive form with the axis. For instance, in the case of calcite, it is the angle made by a face of the primitive rhombohedron with the crystal axis (see Fig. 12.19, *Left*), while Haüy used the ratio of the two diagonals, g and p, of that face, and Weiss calculated the relations between s and c on one side and g and p on the other. If the primitive form was a prism, Weiss chose another primitive form related to it, for instance a 'dodecahedron' or an 'octahedron' (in modern terms, a hexagonal or a tetragonal dipyramid) that have faces inclined with respect to the axis.

Weiss analysed in that way a large number of minerals, following each time Haüy's descrip-tions, using the angles measured or determined by Haüy, and relating his crystal elements to Haüy's. He set aside the case of the crystals having a regular polyhedron as primitive form, such as the cube, the octahedron, the tetrahedron, or the rhomb-dodecahedron, since the angles between faces are the same whatever the crystal species. Weiss then considered the crystals

---

[33]Translated into French by A. J. F. Brochant de Villiers, *Journal des Mines* (1811), **29**, 349–390 and 401–437, with a subject index and a table of the minerals described by Weiss added by the translator.

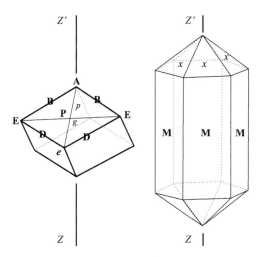

Fig. 12.19   Weiss's crystal elements: the sine and the cosine of the angle between the crystal axis, zz', and faces P in calcite (*Left*), x in apatite (*Right*). After Haüy (1801). Haüy's parameters, the diagonals g and p, and the axis zz' have been added to Haüy's original drawings.

which have as primitive form a hexagonal prism, such as emerald or apatite (Fig. 12.19, *Right*), a rhombohedron, such as calcite, corundum, quartz, tourmaline, dioptase, or an octahedron with a square base (a tetragonal dipyramid), such as zircon or anatase; in other words, crystals having one main axis. In the second memoir presented at his inaugural lecture, Weiss discussed crystals that have more than one axis, such as barite, strontianite, anglesite, or olivine, which are in fact orthorhombic, or euclase, which is monoclinic.

12.3.5   The crystal systems, 1815

In the years following his 1809 dissertation and his appointment in Berlin, Weiss deepened his analysis of crystal forms and concluded that he could classify all the crystallization systems[34] in a few divisions (*Abteilungen*), that are now called crystal systems. These he presented on 12 December 1815 in a talk to the Berlin Royal Academy of Sciences (Weiss 1814–15) and their list is given in Table 12.1. Weiss distinguished two main divisions. In the first one the systems are related to three orthogonal axes and in the second one to four axes, three lying in a plane at 60° of one another and the fourth one orthogonal to that plane; this latter division comprises the hexagonal and trigonal systems, anticipating the present notion of 'hexagonal crystal family'and the later introduction of the four Miller–Bravais indices for the hexagonal system (see Section 12.11.1). The terms *viergliedriges, zweigliedriges* ... have been translated by Whewell (1837) literally by 'four-membered', 'two-membered', and so on, but what they really mean is fourfold, twofold, and so on.

[34]The term *crystallization system*, already used by Haüy, denotes all the forms presented by a given crystal species.

Table 12.1   Weiss crystal systems

Weiss crystal systems	English translation	Modern systems
Gleichgliedriges oder sphäroedrisches (kugelartiges)	Isotropic or spheroidal	Cubic
Viergliedriges	Four-fold	Tetragonal
Zwei-und-zwei-gliedriges	Two- and two-fold	Orthorhombic
Ein-und-zwei-gliedriges	One- and two-fold	Monoclinic
Zwei-und-ein-gliedriges	Two- and one-fold	
Ein-und-ein-gliedriges	One- and one-fold	Triclinic
Sechsgliedriges	Six-fold	Hexagonal
Drei-und-drei-gliedriges	three- and three-fold	Trigonal

Gleichgliedriges oder sphäroedrisches System: Cubic System. The first system considered by Weiss is designated by several equivalent names: *reguläres, sphäroedrisches, tessulares, kuge-lartiges*. It is characterized by opposition to the other, irregular, systems, in that all directions are equivalent, which is also true of all the directions passing through the centre of a sphere, hence the names *sphäroedrisches* or *kugelartiges* (spheroidal); the term *tessulares*, tessular, evokes the cube but is too restrictive in Weiss's mind because it does not carry the idea of the equivalence of all directions, which is expressed by the name *gleichgliedriges* (isotropic, isometric) coined by him. All the crystal elements are referred to three main axes at right angles along which the unit length is the same (he calls them *dimensions*). The main forms are the octahedron, its inverse, the cube, and a *dynamic, not mechanic*, intermediary between them, the granatohedron, namely the form characteristic of garnets (rhomb-dodecahedron,[35] see in Fig. 12.20 the picture of a type of garnet, grossular). All the other forms are derived from them, some of which are observed in nature, such as the leucitohedron (tetragon-trioctahedron), typically exhibited by leucite (potassium and aluminium silicate, see Fig. 12.20). Weiss described all the possible forms obtained by truncation of the primary forms, the most general one being the *Pyramiden-Granatoeder* (hexaoctahedron), with forty-eight faces, a cube truncated by the octahedron and the rhomb-dodecahedron. He showed that this number can be explained easily. Since the three axes and both directions of each axis are equivalent, one obtains the planes equivalent to a given one by considering all the permutations of the intersections of that plane with the positive and negative sides of the three axes, namely, $3! \times 2^3 = 48$.

In the spheroidal system, besides the complete forms, there are also *incomplete* forms having half the number of faces. For instance, the octahedron reduces in this way to the tetrahedron (for instance in zinc-blende or sphalerite, tetrahedrite, boracite) and the tetrahexahedron to the pentagon-dodecahedron (in pyrite and cobaltite). All the possible reductions of the various forms by halving their number of faces were then reviewed by Weiss. The spheroidal system can thus be subdivided into two subdivisions: the complete, or *homo-spheroidal*, spheroidal system and the *hemi-spheroidal* system.

---

[35]It is amusing to note that Weiss called *superfluous* the term rhomb-dodecahedron, now officially in use.

Grossular   Leucite   Cassiterite

Staurolite   Sphene   Axinite

Fig. 12.20   Minerals exhibiting characteristic forms of Weiss's crystal systems.

Viergliedriges System: Tetragonal System. In the *Viergliedriges* or four-fold system (tetragonal), two of the three main directions are equivalent, with the same unit length, and the third one is independent. The main form is the octahedron (tetragonal dipyramid), present for instance in zircon, cassiterite (tin oxide, see Fig. 12.20) or mellite (honeystone, an aluminium benzene hexacarboxylate hydrate). The most general form has sixteen faces only. Weiss did not find examples of a mineral exhibiting a hemihedral form, with the possible exception of staurolite (a complex silicate of aluminium and iron, which is fact monoclinic, pseudo-orthorhombic, and presents a very frequent and characteristic cross-shaped twin, see Fig. 12.20), for which Haüy had given the irregular tetrahedron as primitive form.

Zwei-und-zwei-gliedriges System: Orthogonal System. In the *zwei-und-zwei-gliedriges* or two- and two-fold (orthorhombic system), all three main directions are independent, with different unit lengths. The general form is the rhomb-octahedron (rhombic dipyramid) and its inverse the rhombic prism. As examples, Weiss mentioned topaz, olivine, barite, and anglesite.

Zwei-und-ein-gliedriges and ein-und-zwei-gliedriges System; Monoclinic System. The *zwei-und-ein-gliedriges* or the *ein-und-zwei-gliedriges*, two- and one-fold or one- and two-fold (monoclinic) system is related to the two- and two-fold (orthorhombic system) in the same way as the hemi-spheroidal to the spheroidal system, namely by a halving of the number of faces. It is characterized by a prism terminated by oblique faces which are equivalent, but opposite (Fig. 12.18). Typical examples given by Weiss were titanite, or sphene (calcium titanium silicate, see Fig. 12.20), hornblende, augite and feldspar. The two- and one-fold or one- and two-fold system is still referred to three axes at right angles by Weiss.

Ein-und-ein-gliedriges System: Triclinic System.   Some crystals, such as axinite (calcium alu-minium boro-silicate, see Fig. 12.20) and copper sulphate, have so little symmetry that the only kind of form they exhibit is an irregular parallelepiped. They are grouped in the 'one- and one-fold' (triclinic) system. Weiss still kept, however, as for the monoclinic system, a system of three axes at right angles.

Sechsgliedriges System: Hexagonal System.   The *sechsgliedriges* or six-fold (hexagonal) sys-tem is referred to a set of four axes, three equivalent and coplanar ones at 60° of one another and the fourth one normal to that plane. The primary form is a hexagonal dipyramid, as in apatite (faces *x*, Fig. 12.19). As other examples of minerals in this system, Weiss mentioned beryl, graphite, molybdenite, and mica (in fact, monoclinic).

Drei-und-Drei-gliedriges System: Trigonal System.   The *Drei-und-Drei-gliedriges* or three-and three-fold (trigonal) system, like the hexagonal one, is referred to a set of four axes—three equivalent and coplanar ones at 60° of another and the fourth one normal to that plane. It is is a hemihedry of the hexagonal system. Its primary form is the rhombohedron, which has half the number of faces of the hexagonal dipyramid. As examples, Weiss mentioned calcite, magnesite, siderite, corundum, and tourmaline.

Weiss had a difficulty with quartz. In his list of minerals related to each crystal systems, he put it in the hexagonal system, but he had always been preoccupied by the two rhombohedra which constitute the terminations of quartz. These terminations had been considered to be dipyramids (Steno, Section 11.7.2; Cappeller 1723; Romé de l'Isle 1783), but shown by Haüy to be two rhombohedra rotated by 60° around the axis (Fig. 12.10). Each of these rhombohedra results from the hexagonal dipyramid by removing half of its faces. Haüy had chosen one of them as the primitive form but, to Weiss, they were equivalent and none of them could be deemed as dominant; both should be taken into account for the derivation of the habit of quartz. Weiss had already alluded to this difficulty in his 1804 appendix to the translation of Volume I of Haüy's Treaty and had further developed it in his dissertation (Weiss 1809).

12.3.6   *The Weiss numbers (weisssche Zahlen) – 1817*

Bernhardi (1808) was the first to show the inadequacies of Haüy's method for denoting crystal shapes by means of the law of decrements and proposed some improvements. In his memoir presented to the Berlin Royal Academy of Sciences on 20 February 1817, Weiss made similar criticisms and proposed a new system to describe the orientation of a crystal face (Weiss 1816–17). A face is denoted simply by its three intercepts, ma, nb, pc, with the three axes where a, b, c, are unit lengths along these axes: $\boxed{ma{:}nb{:}pc}$, in modern notation, $\{\frac{1}{m}\frac{1}{n}\frac{1}{p}\}$.

For instance, an octahedron, a tetragonal dipyramid and a rhombic dipyramid, {111}, are represented respectively by $\boxed{a{:}a{:}a}$, $\boxed{a{:}a{:}c}$, and $\boxed{a{:}b{:}c}$, a rhombic prism, {110}, by $\boxed{a{:}b{:}\infty c}$, and plane (100) by $\boxed{a{:}\infty b{:}\infty c}$.

For the hexagonal and trigonal systems with four axes, the following notations are used:

$$\boxed{\frac{c}{a{:}a{:}\infty a}} \; , \; \boxed{\frac{c}{2a{:}2a{:}\infty a}}$$ for the hexagonal dipyramids $\{1\bar{1}.1\}$ and $\{1\bar{1}.2\}$.

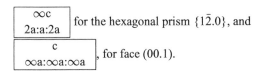

$\dfrac{\infty c}{2a:a:2a}$ for the hexagonal prism $\{1\bar{2}.0\}$, and

$\dfrac{c}{\infty a:\infty a:\infty a}$, for face (00.1).

12.4   D. Brewster and *On the laws of polarization and double refraction in regularly crystallized bodies*, 1818

At the same time as Weiss and Mohs (see Section 12.5) were elaborating their classifications of minerals in crystal systems, the Scottish physicist Sir David Brewster (Fig. 12.22), well-known for his study of polarization and the development of scientific instruments, was busy observing the optical properties of many minerals. He found 165 which produced double refraction, while Haüy had found only twenty. Noting at the same time the shape of the primitive form of each of them, as given by Haüy, he came to the very important conclusion that 'there is a connection by no means equivocal between the primitive form of a crystal and the number of its axes of extraordinary refraction' (Brewster 1818). Brewster distinguished three types of crystals, depending on the number of optical axes:

1. *Crystals with one apparent axis of polarization*: they present a uniaxial optical axis figure (Fig. 12.21, *Right*).
2. *Crystals with two or more axes of polarization*: they present a biaxial optical axis figure (Fig. 12.21, *Left*).
3. *Crystals that have the cube, the regular octahedron, and the rhomboidal dodecahedron for their primitive form*. These crystals *seem to be entirely destitute of the polarizing structure*, in other words are optically isotropic.

Brewster also observed that the colour of many crystals, both uniaxial and biaxial, illuminated in polarized light, depended on the direction of the light with respect to the optical axes (Brewster 1819). According to him, the arrangement of the 'colouring particles' which constitute the crystal, instead of being uniform, is related to the 'ordinary and extraordinary forces which they exert upon light'.

When Brewster wrote his 1818 paper (1 June 1817), he was not aware of Weiss's publications. In the spring of 1818, F. Mohs visited his friend R. Jameson (see Section 12.5) in Edinburgh and revealed to him his method of crystal systems, of which an account 'by one of

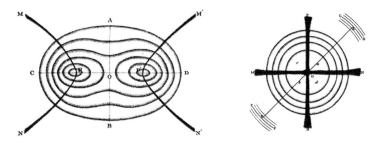

Fig. 12.21   Optical axis figures. *Left*: Nitre (biaxial). *Right*: Calcite (uniaxial). After Brewster (1818).

Table 12.2   Comparison of Haüy's primitive forms, Mohs's crystal systems and Brewster's optical systems. After Brewster (1821).

Haüy's primitive forms	Examples of minerals	Mohs's crystal systems	Brewster's Optical systems
Rhombohedron	calcite, magnesite	rhomboidal	*One axis of double refraction*
	tourmaline, corundum		
Hexagonal prism	apatite, beryl		
Right prism with square base	idocrase	pyramidal	
Dipyramid with square base	zircon, mellite		
Right rectangular prism	peridot, anhydrite	prismatic	*Two axes of double refraction*
Right rhombic prism	barite, topaz		
Rectangular dipyramid	aragonite, nitre		
Oblique quadrangular prism	feldspar		
Oblique prism	axinite		
Cube	rock-salt, boracite	tessular	*Three axes (no double refraction)*
Octahedron	diamond, fluorite		
Rhomb-dodecahedron	garnet, zinc-blende		

his pupils' was published for the first time in Volume III of the *Edinburgh Philosophical Journal* (1820), prior to its publication in German in 1821. Brewster was struck by the remarkable agreement between his classification according to the optical properties of crystals and Mohs' crystal systems, as he explained in his talk to the Wernerian Natural History Society on 5 August 1820: 'The new and beautiful system of crystallography, proposed by Professor Mohs of Freiberg, harmonises in a very singular manner with the optical arrangement of minerals' (Brewster 1821). Table 12.2 shows the correspondence established by Brewster between the optical and crystal systems, and Haüy's primitive forms.

In a very interesting paper written some time later, Brewster (1830) tried to correlate the property of double refraction and the arrangement of the molecules in crystals, noting that this property is not inherent in the molecules themselves. For instance, melted quartz is not birefringent. 'From the phenomena of crystallization and cleavage', he writes, 'it is obvious that the molecules of crystals have several axes of attraction, or lines along which they are most powerfully attracted, and in the directions of which they cohere with different degrees of force'. Furthermore, 'supposing the molecules to be spherical or spheroidal, we infer that their axes are three in number and at right angles to each other, and are related in position to the geometrical axis of the primitive form'. Brewster's main point was that the phenomena of double refraction are related to these axes. If the three axes are equivalent (cubic crystals), there is no double refraction. If two are equal and the third different, the crystal is optically uniaxial; Brewster referred there to Huygens' explanation of the calcite double refraction by the stacking of oblate spheroids. If the three axes are different, the crystals are optically biaxial.

Fig. 12.22   Sir David Brewster in his younger years. Engraved by W. Holl from a painting by Sir H. Raeburn. Source: Images from the History of Medicine (N.I.H. USA).

Sir David Brewster: born 11 December 1781 in Jedburgh, Scotland, the son of the rector of the local grammar school; died 10 February 1868 in Melrose, Scotland. A child prodigy, he constructed a telescope when only ten years old. He went to the University of Edinburgh at the age of twelve to follow theological studies. After finishing his studies he turned towards the natural sciences, and in 1799 began studies on the diffraction of light. He established the laws of light polarization by reflection ('the Brewster angle') and refraction, independently of the works of F. Arago, A. Fresnel, and E. L. Malus. He discovered the laws of metallic reflection and the double refraction induced by heat and pressure. He showed that some crystals have two axes of double refraction, and established the connection between the polarizing structure of crystals and their primitive forms. He also invented the kaleidoscope. He was elected a Fellow of the Royal Society of London in 1815. From 1819 to 1824 he was one of the editors of the *Edinburgh Philosophical Journal*, which he established jointly with Robert Jameson (see Section 12.5) and, from 1824 to 1832, editor of the *Edinburgh Journal of Science*. He was awarded the Rumford Medal by the Royal Society in 1818.

MAIN PUBLICATIONS

1819 *Treatise on the Kaleidoscope.*
1824 *Notes and Introduction to Carlyle's translation of Legendre's Elements of Geometry.*
1831 *Treatise on Optics.*

1832 *Letters on Natural Magic, addressed to Sir Walter Scott.*
1841 *The Martyrs of Science, or the Lives of Galileo, Tycho Brahe, and Kepler.*
1854 *More Worlds than One.*

12.5   F. Mohs and the crystal systems, 1821

Friedrich Mohs (Fig. 12.23) was a German mineralogist, 'one of the last representatives of the old natural-historical school of Mineralogy' (Groth, 1926). He established a new system for the classification of minerals (Mohs 1812, 1820, 1822) founded on the 'external characters of minerals', crystallographic and physical, and totally independent of any aid from chemistry.[36] Among the physical characters should be noted the hardness of minerals for which he designed the scale still in use today (Mohs 1812, 1820) and for which he is best known. The first hardness scale is due to the Swedish mineralogist Quist (1768).

Mohs' other important contribution is the introduction of a classification of the crystal forms in crystal systems, quite independently of Weiss, although the latter accused him of plagiarism in a letter to the *Edinburgh Philosophical Journal* (Weiss 1823). Mohs replied rather sharply in the same Journal that he had only become acquainted with Weiss's crystal systems in 1822 (Mohs 1823). According to him, it was during the years 1812–14, in Graz, that he introduced the notion of crystal systems, which he taught his students there. During a stay in the spring

---

[36]This was criticized at the time; see, for instance, the anonymous report in the *Monthly Review or Literary Journal*, London, 1825, **107**, pp. 141–147, following the publication of the English translation of the *Treatise of Mineralogy*.

Fig. 12.23   Friedrich Mohs, from a lithograph by Joseph Kriehuber, 1832. Source: Wikimedia commons.

Friedrich Mohs: born 29 January 1773 in Gernrode, in the Harz mountains, Germany; died 29 September 1839 in Agordo, Tyrol, now Italy. After studies of mathematics, physics, and chemistry in the University of Halle, in 1798 he joined the Mining Academy (Bergakademie) in Freiberg, where he was A. G. Werner's student. From 1801 to 1802 he was a foreman in the Neudorf galena mines of his home district, Harz. He made at that time a trip to England and Ireland at the invitation of two former English students of Werner's, G. Mitchell and R. Jameson, with the purpose of establishing an Institute analogous to Freiberg's Mining Academy; but, due to the deaths of G. Mitchell and of the English mineralogist, R. Kirwan (1733–1812), the project never materialized. He was then called to Vienna to take care of and sort the minerals collection of a banker, J. F. van der Null. In the period 1802–11 he travelled in Austria to perform geological mining studies, meeting again with Werner in Carlsbad (Bohemia). During this time he worked to establish an entirely new mineralogical classification system, which led to his 1812 book in which he defined minerals by specific 'characters' and devised his hardness scale. In 1811 he was appointed curator of the mineralogy collection of the newly established Johanneum University in Graz, Austria, and in 1812 Professor of Mineralogy in that university. In the spring of 1818 he visited England and Scotland where he met Sir David Brewster and R. Jameson. The same year, following A. G. Werner's death in 1817, Mohs was appointed to succeed him in Freiberg. In 1826 he returned to Vienna to be curator of the main mineralogical collection and to be Professor of Mineralogy at the university. He resigned in 1835 to become Bergrath, imperial counsellor in charge of mining affairs. He was elected a member of the Deutsche Akademie der Naturforscher Leopoldina in 1822.

MAIN PUBLICATIONS

1804 *Über die oryktognostiche Klassification, nebst Versuch eines auf blosse äussere Kennzeichen gegründenten Mineralsystem.*

1812 *Versuch einer Elementarmethode zur naturhistorischen Erkennung und Bestimmung der Fossilien.*

1820 *Die Charaktere der Klassen, Ordnungen, Geschlechter, und Arten der naturhis-* *torischen Mineral-Systeme.* Second edition, 1821.

1822 *Grundriss der Mineralogie. Erster Teil. Terminologie, Systematik, Nomenklatur, Charakteristik.*

1824 *Grundriss der Mineralogie. Zweiter Teil. Physiographie.*

1832 *Leichtfässliche Anfangsgründe der Naturgeschichte des Mineralreichs.*

Bergakademie in Freiberg, the Scottish mineralogist Robert Jameson (1774–1854), Professor of Natural History at the University of Edinburgh, and founder in 1808 of the Wernerian Natural History Society of Edinburgh, named in honour of A. G. Werner. Jameson had subsequently introduced Mohs's system in the third edition of his treatise on mineralogy (Jameson 1820), giving, for each mineral, the crystal system to which it belonged, according to Mohs. This prompted Mohs to publish rather hastily, both in German and in English, his classification of minerals with, for each of them, the system to which it belongs, its physical characteristics, including the hardness and main cleavage, and the names used by Werner, Jameson, and Haüy, but without defining what he meant by crystal system (Mohs 1820). He gave a summarized explanation of his crystal systems in the second edition, published in 1821, and a detailed description in Volume I of his fundamental work, *Grundriss der Mineralogie*.[37] A brief account was also given 'by one of his pupils' in the *Edinburgh Philosophical Journal* (1820, **III**, 154–176, 317–342; 1821, **IV**, 55–67).

A crystal was defined by Mohs as 'homogeneous matter contained with continuity in a regularly limited space' (the crystal form), produced by the action of a 'crystallization force'. His position regarding crystallography as a science was radically different from Haüy's; for him, 'its object is the study of this regularly limited space, and not the matter itself which fills out that space'; it is therefore 'a purely geometrical science'. His description of forms was based on their symmetry, although he did not identify the symmetry elements as such. His definition of symmetry was similar to Haüy's, but he did not refer to him: faces are homologous (*gleichnamig*) if they are similar and occupy similar positions. Mohs stated from the outset that the foundation of crystallography was the constancy of angles, but without any justification nor reference to any previous author. His system of mineralogy was 'independent of any measurement' and was based purely on geometrical derivations; however, its account in the *Edinburgh Philosophical Journal* by one of his pupils showed that it relied entirely, besides his own observations, on Haüy's descriptions.

Mohs distinguished the simple forms, the faces of which are repeated by the symmetry of the crystal and the combinations of simple forms. There are two types of simple forms, those having one main axis, such as the rhombohedron, dipyramids, tetragonal and hexagonal scalenohedra, prisms, and those having several main axes, namely the cubic forms, including the hexaoctahedron with forty-eight faces. The axis was defined as 'a straight line passing through the middle points of parallel sections and normal to the planes of these sections'. All the simple forms could be derived from four fundamental forms according to several possible elaborate geometrical processes,[38] but not following the method of truncations: the rhombohedron (*Rhomboeder*), the tetragonal dipyramid (*gleichschenkliche vierseitige Pyramide*), the rhombic dipyramid (*ungleichschenkliche vierseitige Pyramide*) and the cube (*Hexaeder*). All the simple forms derived from one of these fundamental forms, and their combinations constituted a *crystal system*. There are therefore four crystal systems:

1. **Rhombohedral system**. Mohs distinguished two types of combinations, the *rhombohedral combinations*, strictly speaking, which correspond to the trigonal system, and the *dirhombohedral combinations* where the rhombohedron or the scalenohedron appear in two different orientations, at 60° to one another, which correspond to the hexagonal system. In both cases, combinations could be found where one or several forms present half the number of faces, such as the trigonal trapezohedron, which presented half the number of faces of the scalenohedron. These combinations were called *hemi-rhombohedral combinations* or *hemi-dirhombohedral combinations*. They did not represent, however, in Mohs's mind, subsystems. For instance, he said that quartz presented both *hemi-rhombohedral combinations* and *hemi-dirhombohedral combinations*.

---

[37]English translation: *Treatise on Mineralogy, or the Natural History of the Mineral Kingdom*, F. Mohs, translated with considerable additions by W. Haidinger. 3 Vol. (1825). Edinburgh: Archibald Constable. Wilhelm Ritter von Haidinger (1795–1871) was an Austrian mineralogist, who started to study with Mohs in 1812 in the Johanneum University in Graz, and followed him in Freiberg after Werner's death.

[38]A simplified method is given by Whewell (1825).

2. **Pyramidal system**. It corresponded to the tetragonal system, and zircon was a typical example. There are, according to Mohs, two types of combinations where some forms present half the number of faces:

- *Hemi-pyramidal combinations with inclined faces* with forms such as the tetragonal tetrahedron, for instance in chalcopyrite.
- *Hemi-pyramidal combinations with parallel faces* with forms such as the tetragonal trapezohedron.

3. **Prismatic system**. Mohs distinguished three types of combinations:

- *Prismatic combinations*, which correspond to the orthorhombic system.
- *Hemi-prismatic combinations* where one or several forms present half the number of faces (*hemi-prismatic combinations with parallel faces*) or if their faces are referred to an axis *inclined* with respect to the normal to the base (*hemi-prismatic combinations with inclined faces*). They correspond to the monoclinic system.
- *Tetarto-prismatic combinations* where one or several forms present the fourth of the number of faces and all axes are inclined. They correspond to the triclinic system. Mohs gave as examples axinite and chalcanthite (copper sulfate pentahydrate).

4. **Tessular system**. It is the cubic system. Mohs distinguished the *tessular combinations*, which correspond to the holohedry and to Weiss's *homo-spheroidal system*, and the *semi-tessular combinations*, which are of two types, *semi-tessular combinations with inclined faces* with forms such as the tetrahedron, and *semi-tessular combinations with parallel faces* with forms such as the pentagon-dodecahedron.

Table 12.3 compares Weiss's and Mohs' crystal systems. It is to be noted that Mohs distinguished four systems only, including the monoclinic and triclinic systems within the prismatic system, while Weiss had identified our seven systems. Mohs was the first to find that monoclinic and triclinic crystals should be referred to inclined axes, which Weiss did not, but Weiss rightly noted that hexagonal and trigonal crystals should be referred to a system of four axes, three at 60° to one another and the fourth one normal to them.

The fact that monoclinic and triclinic crystals should be referred to inclined axes was confirmed by Mohs' student Carl Friedrich Naumann[39] (1824) in a paper in which he discussed how to calculate dihedral angles when the axes make angles α, β, and γ and the parameters a, b, c along these axes are different, taking as examples sphene, epidote and mirabilite (Glauber salt, sodium sulfate), which are all monoclinic. In the books he wrote a few years later (Naumann 1826, 1830),[40] he made a synthesis of Weiss's and Mohs' systems, considering four possible configurations for the axes:

- Orthogonal (*Orthoedrisch*): three right angles.
- Monoclinic (*Monoklinoedrisch*): two right angles and one oblique one.
- Diclinic (*Diklinoedrisch*): one right angle and two oblique ones.
- Triclinic (*Triklinoedrisch*): three oblique ones.

---

[39]Carl Friedrich Naumann (1797–1873) was a German mineralogist, the son of a composer. He was Mohs's student at the Bergakademie in Freiberg. After his thesis in Jena, he became Professor of Mineralogy in Leipzig in 1824 and succeeded Mohs in Freiberg in 1826. In 1846 he returned to Leipzig as Professor of Mineralogy at the university. He was elected a member of the Deutsche Akademie der Naturforscher Leopoldina in 1863.

[40]His 1830 book, *Lehrbuch der reinen und angewandte Kristallographie* is dedicated to Professors Weiss and Mohs, *the coryphaei of the German crystallographers!*

Table 12.3   Comparison of the Weiss and Mohs crystal systems.

Fundamental form	Mohs crystal systems	Combinations	Weiss crystal systems
Rhombohedron	*rhombohedral*	*a*1. rhombohedral	I. *three- and three-fold*
		*a*2. hemi-rhombohedral	
		*b*1. dirhombohedral	II. *six-fold*
		*b*2. hemi-dirhombohedral	
Tetragonal dipyramid	*pyramidal*	*a*. pyramidal	III. *four-fold*
		*b*1. hemi-pyramidal with inclined faces	
		*b*2. hemi-pyramidal with parallel faces	
Rhombic dipyramid	*prismatic*	*a*. prismatic	IV. *two- and two-fold*
		*b*1. hemi-prismatic with parallel faces	V. *two- and one-fold*
		*b*2. hemi-prismatic with inclined faces	
		c. tetarto-prismatic	VI. *one- and one-fold*
Cube	*tessular*	*a*. tessular	VII. *spheroidal*
		*b*1. semi-tessular with inclined faces	hemi-spheroidal
		*b*2. semi-tessular with parallel faces	

Combining Weiss's and Mohs' systems, Naumann defined seven crystal systems:

1. *Tesseral*
2. *Tetragonal*
3. *Hexagonal*. Naumann considered the rhombohedral combinations (the trigonal system) to constitute a hemihedry of the hexagonal system.
4. *Rhombic*
5. *Monoclinic*
6. *Diclinic*
7. *Triclinic*

The diclinic system did not survive, and a discussion of the literature related to the diclinic system is given in Pertlik (2006). Naumann was the first to use the terms *holohedry, hemihedry,* and *tetartohedry*. In his later book on theoretical crystallography (Naumann 1856), he used black and white to represent the complementary hemihedral forms (Fig. 12.24).

12.6   F. E. Neumann, W. Whewell, W. H. Miller, and crystal indices, 1823–1839

12.6.1   F. E. Neumann and the Beiträge zur Krystallonomie, 1823

Franz Neumann (Fig. 12.25), the most outstanding of Weiss's pupils, and a well-known theoretical physicist and crystallographer, introduced in his first book, *Beiträge zur Krystallonomie*, a very important modification of Weiss's notations of crystal faces by using the reciprocals of the intercepts with the crystal axes (Neumann 1823). A face making the intercepts $1/ma$, $1/nb$,

Fig. 12.24   Complementary hemihedral forms, group $m\bar{3}$. After Naumann (1856).

Fig. 12.25   Franz Ernest Neumann in 1886, by Carl Steffeck. Source: Nationalgalerie Berlin.

Franz Ernest Neumann: born 11 September 1798 near Joachimsthal, Brandenburg, Germany, the son of a farmer and a divorced countess; died 23 May 1895 in Königsberg, Prussia. He attended school at the Berlin Werder Gymnasium. A very patriotic young man, he volunteered in the Prussian army at the age of 16. After the Napoleonic Wars during which he was seriously wounded, he returned to the Berlin Gymnasium. In 1817 and 1818, following his father's wish, he started theological studies in Berlin and Jena. He returned to Berlin University in 1819 to study mineralogy and crystallography with C. S. Weiss. Encouraged by the latter, he published his projection theory in 1823 and submitted his dissertation in 1825, which he defended on 16 March 1826. He was immediately called to Königsberg, where he was successively Privatdozent (1826), extraordinary (1828) and ordinary Professor of Mineralogy and Physics (1829). Neumann made important contributions in the fields of specific heat, elasticity, electromagnetism, crystal optics, and the theory of light; he is one of the pioneers of the use of the concept of symmetry in physics; for instance, he used symmetry to reduce elastic constants, showed that the optical and thermal axes coincide with the crystal axes (Neumann 1833) and stated the well-known principle of symmetry, usually called Neumann principle, during his course at the University of Königsberg in 1873–74 (Neumann 1885). He had many students, notably W. Voigt (1850–1918), B. Minnigerode (1837–1896) and L. Sohncke (see Section 12.12), and was the adviser of the German physicist G. Kirchhoff (1824–1887). At Königsberg, he had as colleagues the German mathematician C. G. Jacobi (1804–1851), with whom he founded his famous mathematics and physics seminar, and the German mathematician and astronomer F. W. Bessel (1784–1846). He did not publish much, and his works were mostly edited by his son and his students.

MAIN PUBLICATIONS

1823 *Beiträge zur Krystallonomie.*

1826 *De lege zonarum, principio evolutionis systematum crystallinorum.* Dissertation.

1878 *Beiträge zur Theorie der Kugelfunctionen.*

1881 *Vorlesungen über die Theorie des Magnetismus.* Edited by C. G. Neumann.

1885 *Vorlesungen über theoretische Optik.* Edited by E. Dorn.

1885 *Vorlesungen über die Theorie der Elastizität der festen Körper und des Lichtäthers.* Edited by O. E. Meyer.

1887 *Vorlesungen über die Theorie des Potentials.* Edited by C. G. Neumann.

1894 *Vorlesungen über die Kapillarität.* Edited by A. Wangerin.

$1/pc$ with the three axes is thus noted $\boxed{\frac{1}{m}\text{a}:\frac{1}{n}\text{b}:\frac{1}{p}\text{c}}$, in modern notation, (mnp). With remarkable intuition, Neumann found that this notation made the application of the zone law much easier, namely to find the indices of a plane parallel to two zone axes. Neumann considered two sets of planes, $\boxed{\frac{1}{m'}\text{a}:\frac{1}{n'}\text{b}:\frac{1}{p'}\text{c}}$, $\boxed{\frac{1}{m''}\text{a}:\frac{1}{n''}\text{b}:\frac{1}{p''}\text{c}}$ and $\boxed{\frac{1}{m_{\prime}}\text{a}:\frac{1}{n_{\prime}}\text{b}:\frac{1}{p_{\prime}}\text{c}}$, $\boxed{\frac{1}{m_{\prime\prime}}\text{a}:\frac{1}{n_{\prime\prime}}\text{b}:\frac{1}{p_{\prime\prime}}\text{c}}$. They define two zone axes. He further let:

$$M = n'p'' - n''p'; \qquad M' = n_{\prime}p_{\prime\prime} - n_{\prime\prime}p_{\prime};$$

$$N = p'm'' - p''m'; \qquad N' = p_{\prime}m_{\prime\prime} - p_{\prime\prime}m_{\prime};$$

$$P = m'n'' - m''n'; \qquad P' = m_{\prime}n_{\prime\prime} - m_{\prime\prime}n_{\prime}.$$

The modern crystallographer recognizes $[M, N, P]$ and $[M', N', P']$ to be vectors parallel to the two zone axes, given respectively by the vector products of the reciprocal lattice vectors $[m', n', p']$ and $[m'', n'', p'']$ on the one hand, and $[m_{\prime}, n_{\prime}, p_{\prime}]$ and $[m_{\prime\prime}, n_{\prime\prime}, p_{\prime\prime}]$ on the other. Neumann correctly gave the indices of the faces parallel to these two zones axes:

$$m = NP' - N'P;$$

$$n = PM' - P'M;$$

$$p = MN' - M'N,$$

which the modern crystallographer recognizes as the coordinates of the reciprocal lattice vector given by the vector product of the direct lattice vectors $[M, N, P,]$ and $[M', N', P',]$. Bravais obtained similar results by means of his polar lattice (see Sections 12.11.1 and 12.11.3).

Neumann noted in his book that all the normals to the planes of a given zone lie in the same plane. Following up the same idea, Neumann showed that the most fruitful way of getting a view of the whole habit of a crystal was to consider the normals to these faces, drawn from a common point, and their intersections with a sphere. He was thus the first to use polar axes and the stereographic projection to represent crystal faces. A crystallographic projection using polar axes, inspired by Neumann's, was also developed by another of Weiss's students, F. A. Quenstedt (1809–89). Neumann's complete study of zones was the topic of his dissertation, submitted in 1825: *De lege zonarum, principio evolutionis systematum crystallinorum*. Neumann's early crystallographic studies were republished by one of his sons, the mathematician Carl Neumann, in 1916. For further reading about Neumann's life and works, the reader may consult Wangerin (1896, 1907), Groth (1926), and Fritsch *et al.* (2005);[41] for his contribution to the concept of symmetry, see, for instance, Katzir (2004).

12.6.2 W. Whewell and the indices of crystal faces, 1825

William Whewell (Fig. 12.27) was an English scientist, Anglican priest, and philosopher, sometimes described as the first modern historian of science (Kragh 1987), who made important

---

[41]*Franz Ernst Neumann (1798–1895). Zum 200. Geburtstag des Mathematikers, Physikers und Kristallographen*, Editors R. Fritsch, E. Neumann-Redlin-von Neumann and T. J. Schenk. Kaliningrad and Munich: Terra Baltica and Ludwig-Maximilians-Universität. Bilingual German–Russian.

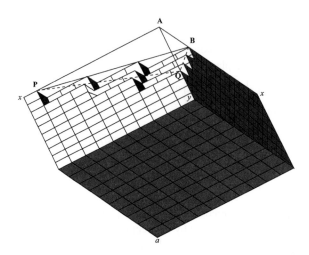

Fig. 12.26   Truncation at an angle of a rhombohedron. After Whewell (1825).

contributions to crystallography. He found Haüy's notation of truncations *very inelegant and imperfect*, and proposed, independently of Neumann, a simpler system exempt of the inconveniences of Haüy's (Whewell 1825). Each crystal plane was represented by a set of indices describing any type of truncation through its intercepts with three axes. According to Haüy's law of simple rational truncations, these intercepts were small integers, and the law became the law of simple rational intercepts or law of rationality. Consider, for example, the truncation PQR at the summit A of a rhombohedron (Fig. 12.26). The equation of that plane is:

$$x/9 + y/6 + z/3 = 1 \qquad (12.1)$$

The equation can be generalized to read:

$$x/h + y/k + z/l = m, \qquad (12.2)$$

When calculating the cosine of the dihedral angle between two planes defined by h, k, l and h', k', l', respectively, Whewell found that it was always expressed in terms of the reciprocals, $1/h, 1/k$ and $1/l$ and $1/h', 1/k'$ and $1/l'$, respectively. He therefore defined the indices of the plane as these reciprocals:

$$p = 1/h; \quad q = 1/k; \quad r = 1/l, \qquad (12.3)$$

and the equation of the plane can be written

$$px + qy + rz = m. \qquad (12.4)$$

The plane will be denoted $(p; q; r)$. If one plane exists, the corresponding planes according to the law of symmetry exist also. For instance, in a trigonal crystal, all the planes whose indices are any of the six permutations of p, q, r will exist. Whewell denotes them by (p, q, r), with a comma instead of a semicolon. A plane truncating an edge of the primitive form will be denoted

Fig. 12.27   William Whewell. Source: *The Life and Selections from the Correspondence of William Whewell* (1881).

William Whewell: born 24 May 1794 in Lancaster, Lancashire, England, the eldest son of a carpenter; died 6 March 1866 in Cambridge, England. After schooling at Heversham grammar school, he entered Trinity College, Cambridge, in 1812. He graduated Second Wrangler in the Mathematical Tripos in 1816, was elected to a fellowship the next year, in 1818 became tutor of his college, and in 1820 was elected a Fellow of the Royal Society. He was Professor of Mineralogy from 1828 to 1832 at the University of Cambridge, and from 1841 to 1866 was Master of Trinity College. His main interests were crystallography and mineralogy, mechanics, physics, geology, astronomy, and economics, as well as the history of science.

MAIN PUBLICATIONS

1819 *An Elementary Treatise on Mechanics.*

1823 *A Treatise on Dynamics.*

1828 *Essay on Mineralogical Classification and Nomenclature.*

1833 *Astronomy and general physics considered with reference to Natural Theology.*

1837 *History of the Inductive Sciences, from the Earliest to the Present Times.*

1840 *The Philosophy of the Inductive Sciences, founded upon their history.*

1852 *Lectures on the history of Moral Philosophy.*

$(p; q; 0)$. Whewell then gave the correspondence between the values of p, q, r and Haüy's notations for the various types of truncations. He applied his system to a number of examples drawn from Haüy's treatise on mineralogy, and made calculations of dihedral angles. Among them, he took the case of the calcite dog's-tooth scalenohedron which has, as indices, $(2, \bar{1}, 0)$. Whewell also introduced the inverse problem, namely finding the indices of a secondary plane from the angles it makes with planes.

12.6.3   The Miller indices, 1839

Whewell's former student, the English mineralogist W. H. Miller (Fig. 12.29), who succeeded him as Professor of Mineralogy at Cambridge, introduced the notation of crystal planes which is still in use today. Following Whewell (1825) and Neumann (1823), the indices of a crystal plane are defined as three integers inversely proportional to the intercepts of the plane with the three axes of the crystal (Miller 1839). They are called h, k, l, and if a, b, c, are the parameters of the crystal, they are given by, referring to Fig. 12.26:

$$h = m\,\frac{a}{AP}; \ k = m\,\frac{b}{AQ}; \ l = m\,\frac{c}{AR}. \tag{12.5}$$

Miller's indices h, k, l are identical to Whewell's p, q, r and Neumann's m, n, p.

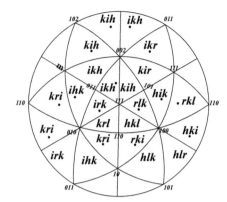

Fig. 12.28   Stereographic projection of the octahedral sys-
     tem. After Miller (1839).

Miller made a systematic use of the stereographic projection inspired by the method of
projection introduced independently by F. E. Neumann (1823), and J. G. Grassmann[42] (1829).
Calculations of dihedral angles using spherical trigonometry are greatly simplified that way.
Figure 12.28 gives an example of Miller's stereographic projections of the poles of faces (*hkl*)
in the octahedral (cubic) system, using [111] as the axis of the projection.

Neither Whewell nor Neumann nor Miller mentioned Weiss as the first author to have denoted
crystal faces by numbers, and Weiss complained bitterly about this in the latter part of his life,
after the Miller notation had become universal (Groth 1826). He himself used his own notations
all his life (see, for instance, Weiss 1855).

12.7   L. A. Seeber and the notion of space lattice, 1824

Ludwig August Seeber was a German mathematician and physicist, a former student of C. F.
Gauss, born 14 November 1793 in Karlsruhe, Germany, died 9 December 1855 also in Karlsruhe. From 1819 to 1822 he was a teacher at the school of cadets in Karlsruhe, and from
1822 to 1824 Professor of Physics at the University of Freiburg im Bresgau. He then came
back to Karlsruhe to become Professor of Physics at the Polytechnicum and the Lyceum, where
he remained until his retirement in 1840. Seeber did not publish much; his main publications
are his crystallographic paper on an *Attempt to explain the inner structure of solid bodies*
(Seeber 1824), which did not receive much attention at the time and was only rediscovered
more than fifty years later by Sohncke (1879), and his mathematical book on the *Properties of
positive ternary quadratic forms* (Seeber 1831), dedicated to his 'revered friend and teacher',
the mathematician C. F. Gauss. This book has been quoted often, in particular because of
Gauss's review of it (Gauss 1831), but by mathematicians; its importance for crystallography
was recognized only much later. He showed that the distance between two equivalent points
of a lattice can be expressed as a ternary quadratic form, as noted by Bravais (1850), and

---

[42]Julius Gunther Grassmann (1779–1852) was a German mathematician, Professor of Mathematics and Physics at the
University of Stettin. Not knowing Weiss's and Neumann's works (Hessel 1830), he described the symmetry relations
of the crystal systems by means of an algebraic system of linear combinations which he called the geometric science of
combinations (*geometrische Combinationslehre*).

Fig. 12.29   William Hallowes Miller. Source: Wikimedia commons.

William Hallowes Miller: born 6 April 1801 near Llandovery, in Wales, UK, the son of Captain Miller, who had served in the American Civil War; died 20 May 1880 in Cambridge, England. After receiving his earlier education at private schools, he entered St John's College, Cambridge, and graduated in 1826 Fifth Wrangler in the Mathematical Tripos. In 1829 he was elected a fellow of his college, and from 1830 to 1844 was a College tutor. He had been a student of Whewell's and, when the latter resigned in 1832, he succeeded him as Professor of Mineralogy. His main scientific activities were devoted to crystallography and mineralogy, but he was also involved in the experiments and investigations connected with the restoration of the standards of measurement and weight to replace those which were destroyed by the burning of the Houses of Parliament in 1834. In 1838 he was elected to the Royal Society, and in 1856 became its Foreign Secretary.

MAIN PUBLICATIONS

1831 *The Elements of Hydrostatics and Hydrodynamics.*

1833 *An Elementary Treatise on the Differential Calculus.*

1839 *A Treatise on Crystallography.*

1852 *An Elementary Introduction to Mineralogy.* With W. Phillips.

1863 *A Tract on Crystallography.*

this contribution to the concept of space lattice was a major step forward between Haüy's representation and Delafosse's and Bravais' more developed theories.

12.7.1   Attempt at an explanation of the inner structure of solid bodies, 1824

Seeber's approach was that of a physicist, in contrast to that of Haüy, which is geometric and crystallographic, and to that of Weiss, which is philosophical. The first part (pp. 239–248) starts with the observation that solid bodies undergo modifications under the action of an electric or a magnetic field, of temperature, or of a stress, and that these changes are necessarily associated with displacements of the individual particles which constitute them. 'In order to establish an exact theory of these observations, a more precise knowledge of the nature and of the inner constitution of the solid bodies than that we have at our disposal at present is required'. In fact, Seeber explained, 'the two hypotheses known under the names of atomistic and dynamistic limit themselves to a discussion of the type of space filling and are not adequate to account for the physical observations'. His conception of the nature of crystals borrowed from Haüy the atomistic description and from Weiss the notion of attractive and repulsive forces: 'solid bodies are constituted of minute particles, which is confirmed by the cleavage properties, but the most probable hypothesis is that their cohesion results from the reciprocal attraction these particles exert on one another'.

The particles which constitute a solid body are not in direct contact with one another. For Seeber, it was impossible for the individual particles to be in direct contact with one another:

'in order to explain the variations in density presented by different solid bodies, one would have to admit that there are as many types of individual particles as there are of bodies with different density'. All the properties of the solid bodies can best be explained and in the simplest way by the assumption that 'the individual particles lie at a certain distance from one another, under the action on the one hand of a repulsive force tending to move them apart, and, on the other hand, of an attractive force which tends to bring them together'. The property of cleavage can be explained with this model in the same way as with Haüy's.

The stable equilibrium of a solid body requires a regular arrangement of the particles.   For a solid body 'to be in a state of stable equilibrium, the resultant of the forces acting on each particle must be equal to zero, and this requires that the particles should be arranged according to a certain law. In the inorganic bodies, one has clear signs of a regular arrangement of the particles. They are either crystals, namely bodies limited by planar faces, or aggregates of small crystals', which have 'the property of being cleaved easily along certain directions, specific of their species'. This implies a regular arrangement of the particles, as already shown by Haüy. Seeber added that the stable equilibrium depended not only on the arrangement of the atoms, but also on the law governing the reciprocal action between them. He then explained Haüy's theory of integrant molecules according to which the mutual orientation of the crystal faces is governed by the law of decrements, applied to a system of parallelepipeds, the subtractive molecules (see Section 12.1.2).

Haüy's polyhedral molecules can be substituted by spherical atoms: parallelepipedic arrangement.   In order to reconcile Haüy's hypothesis that explains satisfactorily crystal faces and cleavage planes, and the physical observations which require that the individual particles should not be in direct contact, Seeber proposed to 'substitute spherical atoms to Haüy's polyhedral molecules, located at their centre, and leaving between them enough space as required by the physical properties' (Fig. 12.30). In this way neither the type nor the properties of the arrangement of molecules is altered. The cleavage was explained in the same way, as well as the existence of planar crystal faces, and Haüy's explanations were made even simpler. The arrangement obtained by substituting spheres to Haüy's polyhedral molecules was called the *parallelepipedic arrangement*, a 'lattice' in the words of Delafosse and Bravais. Seeber's parallelepipedic arrangement is shifted with respect to Haüy's by half the diagonal of the unit parallelepiped (Fig. 12.31), with the atoms lying at the summits of the parallelepipeds.

The hypothesis of central forces.   'The spherical shape is the simplest and the most probable one, and, more importantly, it makes it easier to find the law governing the conditions of stable equilibrium. The attractive and repulsive forces exerted by two atoms on one another pass through their centres' (in other words, central forces, although the expression was not used by Seeber) 'and cannot induce rotations of the atoms. The resultant force is the difference between the attractive and repulsive forces and is either positive or negative. For any portion of the solid body to be in a state of stable equilibrium, the resultant of the forces exerted on one another by any two spherical atoms must be equal to zero, and it must change from a repulsive one to an attractive one when the distance between the two atoms increases. This can be obtained with the parallelepipedic arrangement'.

Expression of the distance between two lattice points as a ternary quadratic form.   In the second part (pp. 349–371) of his 1824 paper, Seeber calculates the distance s between the centres

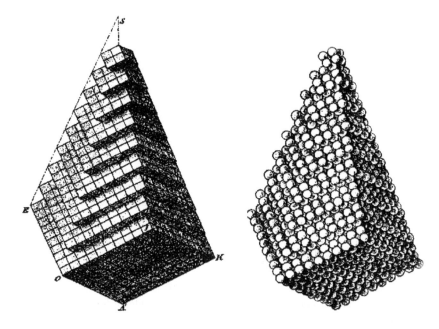

Fig. 12.30   Calcite scalenohedron. *Left*: Haüy's arrangement of parallelepipeds. *Right*: Seeber's corresponding arrangement of spherical atoms. After Seeber (1824).

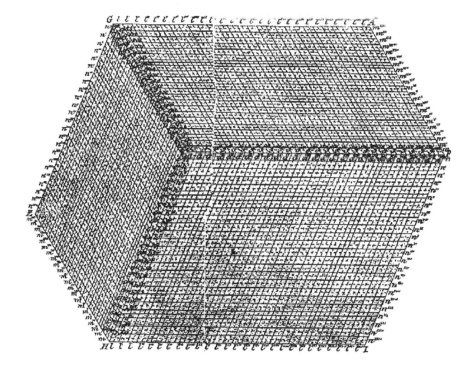

Fig. 12.31   Haüy's (full lines) and Seeber's (broken lines) arrangements of parallelepipeds. After Seeber (1824).

of two spherical atoms, M and M', referred to a system of coordinates parallel to the edges of the parallelepipedic arrangement. Its square is given by

$$f = s^2 = (MM')^2 = a^2 x^2 + b^2 y^2 + c^2 z^2 + 2bc \, \cos\alpha \, yz + ca \, \cos\beta \, zx + ab \, \cos\gamma \, xy$$

$$(12.6)$$

where a, b, c and α, β, γ are the parameters and the angles of the unit parallelepiped and x, y, z are the numerical coordinates of M' with respect to M. It is a positive ternary quadratic form.

At equilibrium, the reciprocal force exerted by the two atoms on one another must be equal to zero when these coordinates are respectively equal to three integers, u, v, w. It should furthermore be positive when $s < s_o$ where s_o is the value of s when M and M' are at summits of the system of parallelepipeds, and negative when $s > s_o$, and decrease with increasing distance. This can be achieved, for instance, if the force is expressed by:

$$F(x, y, z) = -\exp\left(-gs^2\right) sin\left(\frac{2\pi s^2}{v^2}\right)$$

$$(12.7)$$

and s^2/v^2 is an integer and v an arbitrary parameter. This is obvious for a cubic crystal ($a = b = c$, $\alpha = \beta = \gamma = 90°$), since $s^2/v^2 = x^2 + y^2 + z^2$ is an integer for $x = u$, $y = v$, $z = w$ and if one takes $v = a$. For the general case, Seeber discussed the inequalities which must be satisfied by the six terms a^2/v^2, b^2/v^2, c^2/v^2, $(bc/v^2) \cos\alpha$, $(ca/v^2) \cos\beta$ and $(ab/v^2) \cos\gamma$ for $F(x, y, z)$ to take zero values when M and M' lie at summits of the parallelepipedic arrangement. He then took as an example a number of crystals for which the values of the angles α, β, and γ were taken from Haüy's measurements.

12.7.2   C. F. Gauss's review of Seeber's book Properties of positive ternary quadratic forms, 1831

Quadratic forms.   The topic of quadratic forms constitutes a very interesting chapter of the theory of numbers. It is also very important for the theory of lattices. Given a binary positive quadratic form $f \equiv Ax^2 + 2Bxy + Cy^2$, in which $B^2 - C^2 \geq 0$, an important problem is to find under which conditions f can be transformed into an equivalent form $f \equiv A'x'^2 + 2B'x'y' + C'y'^2$ by a change of variables $x = \alpha x' + \beta y'$; $y = \gamma x' + \delta y'$ where α, β, γ and δ are integers and $\alpha\delta - \beta\gamma = 1$. The problem was solved by J.-L. Lagrange (1772)[43] and in 1801 by C.-F. Gauss[44] in Section V of his *Disquisitiones Arithmeticae*. In the same work, Gauss took up the case of the ternary quadratic forms to which he came back thirty years later, at the

---

[43]Joseph-Louis Lagrange, born 25 January 1736 in Turin, Italy, died 10 April 1813 in Paris, was an Italo-French mathematician. He was appointed Professor of Mathematics at the School of Artillery in Turin in 1755 and founded the Academy of Sciences of Turin in 1758. In 1766 he was called to Berlin by the King of Prussia Frederick II to become Director of the class of mathematics of the Berlin Academy of Sciences. He left Berlin for Paris in 1787, when he was elected to the French Academy of Sciences.

[44]Carl Friedrich Gauss, born 30 April 1777 in Braunschweig, Germany, died 23 February 1855, Göttingen, Germany, was one of the most important mathematicians of the nineteenth century. A child prodigy, he studied at the Universities of Braunschweig and Göttingen and was awarded his doctorate in 1799 by Academia Julia in Helmstedt. Soon afterwards he wrote his most important work, *Disquisitiones Arithmeticae*, published in 1801. He became Professor of Mathematics in Göttingen in 1807.

occasion of his review of Seeber's book. By a suitable change of variables, quadratic forms can also be transformed in a linear combination of quadratic terms:

$$f \equiv D_1 X_1^2 + \frac{D_2}{D_1} X_2^2 + \frac{D_3}{D_2} X_3^2 \qquad (12.8)$$

The coefficients D_1, D_2, D_3 are called the determinants of the form. As seen above, Seeber showed that quadratic forms can be interpreted, under certain conditions, as the square of the distance between two summits of the parallelepipedic arrangement. Expressed in terms of the coefficients of 12.6, the determinants are given by, using modern notations:

$$D_1 = a^2;$$

$$D_2 = \begin{vmatrix} a^2 & ab \cos \gamma \\ ab \cos \gamma & b^2 \end{vmatrix} = a^2 b^2 \sin^2 \gamma \qquad (12.9)$$

$$D_3 = \begin{vmatrix} a^2 & ab \cos \gamma & ac \cos \beta \\ ab \cos \gamma & b^2 & bc \cos \alpha \\ ac \cos \beta & bc \cos \alpha & c^2 \end{vmatrix} \qquad (12.10)$$

$$= (a^2 b^2 c^2)(1 - \cos^2 \alpha - \cos^2 \beta - \cos^2 \gamma + 2 \cos \alpha \cos \beta \cos \gamma), \qquad (12.11)$$

where the determinant D_2 can be interpreted as the square of the area of the elemental parallelogram (in modern notation, $|\mathbf{a} \wedge \mathbf{b}|$ where $\mathbf{a}$ and $\mathbf{b}$ are the two unit vectors), and the determinant D_3 as the square of the volume of the elemental parallelepiped (in modern notation, $(\mathbf{a}, \mathbf{b}, \mathbf{c})$). Equivalent forms are simply forms based on different sets of basic vectors defining different unit cells having the same volume.

Seeber's theorem. Seeber (1831) derived a certain number of properties and inequalities related to the ternary quadratic forms, in particular that the smallest value that can be taken by the determinant D_3 is half the product of its diagonal terms:

$$D_3 \geq a^2 b^2 c^2 / 2 \qquad (12.12)$$

Some of the auxiliary conditions derived by Seeber were expressed in terms of the *adjoint* lattice, which is in fact the reciprocal lattice, and this is the first time the reciprocal lattice was introduced in crystallography (Engel 1986). Gauss (1801, *Disquisitiones Arithmeticae*) had defined the adjoint form, which can be interpreted as minus the square of the distance between two reciprocal lattice points, and the corresponding *adjoint* determinant, which can be interpreted in a similar way as the square of the volume of the unit cell of the reciprocal lattice.

Gauss's geometrical interpretation of Seeber's theorem: The face-centred cubic lattice is the densest lattice. Seeber had demonstrated his theorem by induction after testing its result on over 600 practical cases. Gauss (1831) found Seeber's result beautiful but the demonstration far too lengthy, and gave a much shorter and elegant proof. He also gave, at the end of his paper, a geometrical interpretation: 'the volume of the elemental parallelepiped $\times \sqrt{2}$ is never smaller than that of a cube having as edge the shortest edge of that parallelepiped'. The lower limit is reached, although Gauss did not state it in his paper, when $a = b = c$ and $\alpha = \beta = \gamma$; that is if the parallelepiped is a rhombohedron, and therefore, from 12.11, if

$$1 - 3\cos^2\alpha + 2\cos^3\alpha = 1/2. \qquad (12.13)$$

This is obtained for $\alpha = 60°$; in other words, for the unit cell of a face-centred cubic lattice. This shows that it is the densest type of lattice, and this result is called the 'Kepler conjecture' (see Section 11.4.8).

Other proofs of Seeber's theorem have been given by, among others, Hermite (1850), Dirichlet (1850) in relation with the Dirichlet domain, and Lebesgue (1856); for a modern view see, for instance, Rousseau (1992). Selling (1877) discussed the relations between the coefficients of a ternary quadratic form for each of Bravais' lattice modes.

12.8   M. L. Frankenheim, crystal classes, 1826, and crystal lattices, 1835

Moritz Frankenheim was a German physicist and crystallographer, born 29 June, 1801 in Braunschweig, Germany; died 14 January 1869 in Dresden, Germany. He obtained his PhD in 1823 at Berlin University, where he had been one of Weiss's students. In 1827, he was appointed Assistant Professor of Physics, Geography, and Mathematics at the University of Breslau, and in 1850 full Professor. He was elected a member of the Deutsche Akademie der Naturforscher Leopoldina in 1841. Frankenheim was the first to derive the thirty-two crystal classes and to introduce the crystal lattices. His work was, however, forgotten for many decades but was quoted by Bravais (1850, 1851), in relation to the lattice modes of the monoclinic system, Delafosse (1867), Sohncke (1879), and Fedorov (1892). It is briefly described in Burckhardt (1984, 1988) and Scholz (1989).

12.8.1   The thirty-two crystal classes, 1826

The law of rationality.   In the first part of his 1826 paper (pp. 498–515), Frankenheim started by expressing the law of rationality. Consider the equations of two different crystal faces in an orthogonal system of coordinates:

$$ax + by + cz = f^2,$$
$$a'x + b'y + c'z = f^2$$

The law of rationality states that the ratios:

$$\alpha = \frac{a'}{a}; \quad \beta = \frac{b'}{b}; \quad \gamma = \frac{c'}{c}, \qquad (12.14)$$

are rational.

Order of the axes of rotation.   Frankenheim deduced from the law of rationality that the number of equivalent faces around an axis cannot be equal to 5, 7, 8 *etc.*, in modern terms, that the order of an axis of rotation can only take the values 1, 2, 3, 4, 6. He did this by calculating the cosine of the projections of the normals to two equivalent faces on the plane normal to the axis of rotation and proving that it can only take the values 0, ∓ 0.5, ∓ 1. He was the first to have shown this basic property of crystals. It is interesting to note that the demonstration did not refer to the existence of a crystal lattice, which is of course implied by the law of rationality.

Notation of crystal faces. Frankenheim denoted a face a', b', c' by $\boxed{a'+b'+c'}$ or, using 12.14, by $\boxed{\alpha a + \beta b + \gamma c}$. In order to express that α, β, and γ can take positive or negative values, this is written: $\boxed{\mp \alpha a \mp \beta b \mp \gamma c}$. If the eight corresponding faces are equivalent, they are denoted by $\boxed{\alpha a : \beta b : \gamma c}$. A prism parallel to axis Oz is denoted by $\boxed{\alpha a : \beta b : 0c}$ and the terminal faces respectively perpendicular to the axes $0x$, $0y$, $0z$, will be denoted $\boxed{a : 0b : 0c}$, $\boxed{0a : b : 0c}$ and $\boxed{0a : 0b : c}$. For instance, in the tetragonal system, a face is represented by $\boxed{\alpha a : \beta a : \gamma c}$ where α and β can be exchanged. One can thus easily calculate the number of each equivalent faces of each form. In the general case, the ditetragonal dipyramid, it is $2 \times 8 = 16$.

The thirty-two crystal classes. In the next step, Frankenheim listed all the crystal forms that have the full symmetry of each of the four crystal systems that he considered: *reguläre System* (cubic), characterized by three equal parameters along the three axes, *viergliedrige System* (tetragonal), characterized by two equal parameters along two of the axes, *sechsgliedrige System* (comprising the hexagonal and trigonal systems) referred to a system of four axes, with equal parameters in along the three axes at 60° degrees in the plane normal to the fourth one, and *zweigliedrige System* (comprising the orthorhombic, monoclinic and triclinic systems). In the last part of the paper (pp. 542–565), Frankenheim looked for all the possible forms compatible with the characteristics of each system.

For instance, in the cubic system, equivalent faces have the same orientation with respect to the three axes, so that the three faces $\boxed{\alpha a : \beta a : \gamma a}$ obtained by exchanging α, β, and γ are equivalent. Furthermore, to each face corresponds a face similarly inclined with respect to the main axes: $\boxed{\alpha a : \beta a : \gamma c}$ and $\boxed{-\alpha a : -\beta a : \gamma c}$ are equivalent. Combining the two conditions, one finds a form with twelve faces (the pentagon-tritetrahedron). It has the minimum symmetry compatible with the cubic system. Frankenheim then looked for intermediary symmetries and found three, corresponding respectively to the symmetries of the pentagon-dodecahedron, the tetrahedron, and the pentagon-trioctahedron or gyroid. He then grouped all the possible forms compatible with the cubic systems in five divisions, which he called *Ordnungen*, the crystal classes:

1. Full symmetry, the general form has 48 faces (in modern notation $m\bar{3}m$).
2. *Kantenhälfte* with half the number of faces ($m\bar{3}$).
3. *Diagonalhälfte* with half the number of faces ($\bar{4}3m$).
4. *abwechselnde Hälfte* with half the number of faces (432).
5. *Viertheil* with the fourth of the number of faces (23).

Frankenheim did the same thing for the other systems, finding 7 classes for the tetragonal system (one with the full symmetry, four with half the number of faces and two with one fourth), twelve in the hexagonal system (one with the full symmetry, five with half the number of faces, five with one fourth, one with one eighth) and eight in the *zweigliedrige System* (one with the full symmetry, three with half the number of faces, three with one fourth, one with one eighth).

12.8.2   The crystal lattices, 1835

In his book on the cohesion of matter (Frankenheim 1835), Frankenheim introduced the notion of lattice, independently of L. A. Seeber, although he did not use the word. He started by giving a nicely balanced account of the developments of crystallography in the first three decades of the nineteenth century, presenting R.-J. Haüy as the founder of that science and crediting C. S. Weiss for having systematized it through the classification in crystal systems. Contrary to Weiss and other German crystallographers, he tended to agree with Haüy in thinking that crystals are composed of molecules. According to him, 'the atomistic theory requires that the molecules should not touch other but be separated by empty spaces. These spaces change in various ways under the action of external forces. In order to explain how these forces vary with direction, two elements have to be taken into account: the definite shape of the molecule, and the variation of their mutual distances with direction'. Frankenheim then put forward an hypothesis which represented very well the aspect of crystals and was in agreement with the other disciplines of physics:

1) 'The solid bodies are constituted by small particles separated by spaces. For crystallography, neither the shape of the particles nor the respective sizes of these molecules and the spaces need be known. For that reason, and taking an atomistic viewpoint, I shall consider these small particles as mathematical points which are the points of action of the forces'.

2) 'In the crystallized bodies, the small particles lie symmetrically next to each other. That is to say, if one draws a line through two of them inside the crystal, it will meet other particles at regular intervals'. This is a lattice row!

From this basic concept, Frankenheim deduced, without any proof, that the molecules could be arranged according to one of fifteen crystal families, which all belonged to the six crystal systems. These corresponded to the Bravais lattices of which the number is in fact fourteen. He was, however, not fully convinced at the time by the first statement, namely the atomistic hypothesis. He showed himself more convinced of the atomistic hypothesis a few years later (Frankenheim 1842), and fully convinced fourteen years later in a paper entitled *Ueber die Anordnung der Molecule im Kristall* (Frankenheim 1856), in which he stated that despite the philosophical standpoint, evidence showed that the 'elementary parts of bodies are constituted by molecules separated from each other'. He considered that they formed rows (*Molecular-Linie*) equally spaced in planes, and he discussed the symmetry of the arrangement of these rows and planes in the different crystal systems. This work came after Bravais' 1848–50 works (see Section 12.11) and seems to have been influenced by it.

The orientations of the faces which provide the easiest way of reconstructing a given crystal and the directions of the main cleavage planes determine what Frankenheim (1842) called the fundamental forms (*Grundformen*) of the crystal, from which all others could be derived. For each crystal system, he found which are these fundamental forms, and grouped all the crystals belonging to that system in corresponding orders called *Ordnungen*.[45] Frankenheim found fifteen of them, which coincide with the fifteen types of arrangement of the molecules,

[45]In his 1842 paper, Frankenheim denotes 'crystal system' by *Krystall Classe*, 'crystal class' by *Familie* and 'Bravais lattice' by *Ordnung*.

Table 12.4   Comparison of Frankenheim's orders, or fundamental forms, and Bravais's lattice modes (written in modern notation).

Frankenheim's crystal systems	number of classes	Bravais's crystal systems	Frankenheim orders (*Grundformen*)	Bravais lattice modes
Tesseral	5	*terquaternaire*	cube	P
(cubic)			octahedron	F
			rhomb-dodecahedron	I
Tetragonal	7	*quaternaire*	quadratic prism	P
			quadratic octahedron	I
Hexagonal	12	*sénaire*	hexagonal prism	HP
		ternaire	rhombohedron	RP
Isoclinic	3	*terbinaire*	right rectangular prism	P
(orthorhombic)			right rhombic prism	A, B or C
			rectangular octahedron	I
			rhombic octahedron	F
Monoclinic	3	*binaire*	right rhombic prism	P
			rhombic octahedron	I
			oblique rhombic prism	A or C
Triclinic	2	*asymétrique*	asymmetric parallelepiped	P

not described by him. Their correspondence with the fourteen lattice modes given by Bravais (1850), listed with their modern notation, is given in Table 12.4.[46]

For each of the fifteen orders, Frankenheim gave examples of minerals chosen according to their habit, but of course, they are usually wrong. For instance, the lattice mode P was wrongly attributed to rock-salt, sylvite, and galena, but quartz was rightly given as trigonal.

12.9   J. F. C. Hessel and the thirty-two crystal classes, 1830

Johann Hessel (Fig. 12.32) was a German mineralogist who determined the thirty-two crystal classes (Hessel 1830), independently of Moritz Frankenheim, but with quite a different approach, by combining symmetry elements. As the translator of Haüy's 1815 'law of symmetry', he had become interested in problems of symmetry when quite young. He seems to have been the first to introduce inverse symmetry elements, but, like Frankenheim's, his work was forgotten for many decades. It was rediscovered by Sohncke (1891*a*), and was republished with annotations in 1897 by his former student, Edmond Hesse.[47] It is briefly commented upon by Burckhardt (1988) and Scholz (1989).

---

[46]Bravais (1850) showed that Frankenheim's fundamental forms 'rhombic octahedron' and 'oblique rhombic prism' of the monoclinic system (I and A or C, respectively) are in fact identical.

[47]J. F. C. Hessel, *Krystallometry*, edited by Edmond Hesse, 2 volumes (1897). Leipzig: Verlag von Wilhelm Engelmann.

Fig. 12.32   Johann Friedrich Christian Hessel. Source: Wikimedia commons.

Johann Friedrich Christian Hessel: born 27 April 1796 in Nürnberg, Germany, the son of a merchant; died 3 June 1872 in Marburg, Germany. After schooling at the industrial school in Nürnberg, he studied science and medicine in Erlangen (1813) and in Würzburg (1814), where he received his MD in 1817. From there he went to Munich to continue his studies. He had the opportunity of meeting the mineralogist C. C. von Leonhard, who recognized the originality of his mind and his dedication. When the latter became Professor of Mineralogy in Heidelberg in 1818, he persuaded him to become his assistant. Hessel then devoted himself to the study of mineralogy and crystallography, and, at Leonhard's request, translated several of Haüy's seminal publications. He obtained his PhD on 24 January 1821, and was appointed extraordinary Professor of Mineralogy at the University of Marburg in the autumn of the same year, and ordinary Professor in 1825.

MAIN PUBLICATIONS

1819 *Haüy's Ebenmassgesetz der Krystallbildung*, translated with annotations by J. F. C. Hessel.

1830 *Krystallometrie oder Krystallonomie und Krystallographie.*

1853 *Die Anzahl der Parallelstellungen und jene der Coinzidenzstellungen eines jeden denkbaren Raumgebildes ....*

1862 *Ueber gewisse merkwürdige statische und mechanische Eigenschaften der Raumgebilde.*

1871 *Uebersicht der gleicheckigen Polyeder und Hinweisung auf die Beziehungen dieser Körper den gleichflächigen Polyedrn.*

Hessel's book introduced an original nomenclature but contained many lengthy developments. It included, at the end, an interesting account of the development of crystallography since Romé de l'Isle. Hessel started by general considerations relative to the symmetry of a figure:

1) *Mirror images*: if two objects, A and B, are superposable, B is said to be the image of A, which Hessel denoted by $A \cong B$; if B is superposable on the mirror image of A, it is denoted by $A| = |B$; if A and B are both superposable and identical to their mirror images, they are symmetrical ($A| \cong |B$).

2) *Axes of rotation*: if the object is left unchanged by a rotation of $2\pi/p$ around a line, this line is said to be an axis of rotation of order p (*p-gliedrige Achse*).

In order to describe the symmetry of a crystal, Hessel associated to each face its oriented normal, drawn from a central point, which he called 'ray' (*Strahl*). The ensemble of these rays was called a *Strahlensystem*, and Hessel's purpose was to find all its possible symmetries. He proceeded in two stages:

1. *Combinations with one main axis (Hauptachse)*. The main axis may combine several of the following properties:

   • Each object or oriented ray repeated by the axis of order p is associated with its mirror image. In that case, the axis is called *zweifach p-gliedrig*. If not, it is simply *einfach p-gliedrig*. This

Table 12.5 Hessel's twenty-seven crystal classes with one main axis (two appear twice, m and $\bar{6}$). The more recent notations, Schoenflies and Hermann-Mauguin, have been added for clarity.

N°	Extremities of the axis	mirror perpendicular to the axis (gleichstellig)	rotoinversion (gerenstellig)	mirror parallel to the axis	Schoenflies symbols	Hermann-Mauguin symbols
I	zwei-endig	Yes	No	Yes (zweifach)	D_{ph}	mmm, $3/m\,m\,2 = \bar{6}m\,2$, $4/m\,m, 6/m\,m$
II	zwei-endig	Yes	No	No (einfach)	C_{ph} C_{ph}	$m, 2/m, 3/m = \bar{6}$, $4/m, 6/m$
III	zwei-endig	No	Yes	Yes (zweifach)	D_{pd}	$\bar{3}m, \bar{4}2m$
IV	zwei-endig	No	Yes	No (einfach)	C_i, S_4	$\bar{1}, \bar{3}, \bar{4}, \bar{6} = 3/m$
V	zwei-endig	No	No	No (einfach)	D_p	$222, 32, 422, 622$
VI	ein-endig	No	No	Yes (zweifach)	C_{pv}	$m, 2m, 3m, 4mm$, $6mm$
VII	ein-endig	No	No	No (einfach)	C_p	$1, 2, 3, 4, 6$

indicates the presence or absence of a mirror plane parallel to the axis. Hessel never used the expression 'mirror plane'.

- The two extremities of the main axis are identical (zwei-endig) or not (ein-endig). This implies that to any normal to a crystal face above the horizontal plane perpendicular to the axis corresponds, or not, an equivalent one in the half-space below that plane.
- To each normal corresponds, or not, its mirror image with respect to the horizontal plane (gleichstellig or ungleichstellig).
- To each normal corresponds its transform by a 'rotoinversion' (using the modern term); the axis is in this case called by Hessel gerenstellig.

Table 12.5 shows all the possible combinations of these properties. They result in twenty-seven crystal classes (Krystallgestalten or Krystallabteilungen), which Hessel distributed among the three crystal systems: twelve in the hexagonal system (including hexagonal proper and trigonal), called Ein- und dreimaassiglge Systeme, seven in the tetragonal system, called Ein- und zweimaassigige Systeme and eight in the rhombic system (including orthorhombic, monoclinic and triclinic), called Ein- und einmaassigige Systeme.

2. *Combinations with more than one main axis*. Hessel obtained in a similar way the five crystal classes of the cubic system, called dreigliedrig vieraxige Systeme.

12.10   G. Delafosse, the notion of unit cell and the structural interpretation of hemihedry, 1840

Gabriel Delafosse (Fig. 12.33) was a French mineralogist and crystallographer, the last of Haüy's pupils, with whom he collaborated on some chapters of the latter's last treatises, and was Louis Pasteur's teacher. He played a very important role in the development of the concept of crystal lattice, and gave a structural interpretation of hemihedry, emphasizing the relations between symmetry, molecular structure, morphology, and physical properties of crystals. In that

Fig. 12.33   Gabriel Delafosse. Courtesy IMPMC, Université P. et M. Curie, Paris.

Gabriel Delafosse: born 24 April 1796 in Saint-Quentin, France; died 13 October 1878 in Paris, France. He was received at the Ecole Normale Supérieure in Paris in 1813 and was Haüy's last pupil. He obtained his PhD in 1840, was appointed Professor of Mineralogy at Paris University in 1841, where he succeeded to François Sulpice Beudant (1787–1850), a French mineralogist who had been another of Haüy's pupils, and at the Museum d'Histoire Naturelle in 1857. He was elected a member of the French Academy of Sciences in 1857.

Main Publications

1833 *Précis élémentaire d'Histoire Naturelle.*
1840 *De la structure des cristaux, considérés comme base de la distinction et de la classification des systèmes cristallins.*
1843 *Recherches sur la cristallisation.*

1852 *Mémoire sur une relation importante qui se manifeste, en certains cas, entre la composition atomique et la forme cristalline.*
1858 *Nouveau cours de Minéralogie.*
1867 *Rapport sur les progrès de la Minéralogie.*

sense he was Haüy's continuator, but he was also strongly influenced by the German school of Weiss and Mohs, despite his indirect response to the rejection of Haüy's atomistic theory by the German school (see Section 12.1.6): its 'idealistic philosophy, a kind of metaphysics of nature, led [German physicists] to prefer, for the interpretation of natural phenomena, this type of vague and obscure explanations, which they call dynamical, to the clear-cut and precise views provided by the atomistic system' (Delafosse 1843). But he did recognize, in fact, the virtue of the views expressed by the German crystallographers, whom he quotes often, and, like Seeber (see Section 12.7), endeavoured to combine the atomistic and dynamical approaches. Delafosse's views on the phenomena of crystallization, seen both from the geometrical and physical points of view were the topic of his thesis, defended in 1840, and were published first in abridged form (Delafosse 1840) and later more completely (Delafosse 1843).

12.10.1   The notion of space lattice and of unit cell

Romé de l'Isle and the German school had concentrated their study on the geometrical aspects of crystals and neglected cleavage and other physical properties. Haüy's ambition had been to attain the molecular structure of crystals, but in practice he limited himself to a determination of the elemental particle of the crystal, and to a classification of minerals based on their geometrical form. Delafosse's work stems from the reflection, quite independently of Seeber, that the notions of integrant molecule and physical molecule, which had been confused by Haüy, should be distinguished. The integrant molecule is an element of the geometrical and physical structure of the crystal, the shape of which can be deduced from the cleavage properties, and which plays the major role in the application of the law of decrements (or of that of the law of rationality,

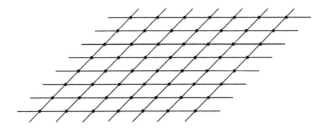

Fig. 12.34   Delafosse's crystal lattice. The physical molecules are symbolically represented by points. Redrawn after Delafosse (1843).

which is equivalent to it). But these properties cannot give any indication as to the nature of its physical content, the physical molecule. They only tell us 'which regular configuration results from the way they are arranged'. Delafosse was the first to use the terms of lattice (*réseau*) and unit cell (*maille*) – Delafosse 1836: during crystallization, 'the molecules are spaced out in a uniform and symmetric way, along parallel rows, with their centres of gravity at the intersections of three series of parallel planes, constituting a continuous lattice of parallelepipedic cells' (Fig. 12.34). 'The molecules are not fixed in an absolute way to these points but are in a state of more or less stable equilibrium'. From this view of the crystallization, Delafosse deduced two important statements about the structure of crystals: 1) the orientation of the molecular axes is constant, which implies symmetry of translation, 2) the external symmetry reflects the inner symmetry, namely the symmetry of the molecules and of their arrangement. In other words, the law of symmetry applies in the first place to the inside of the crystal, and, in the second place, to the outside.

The shape of the physical molecules may be completely different from that of the integrant molecule, but is related to it by their symmetry. As an example, Delafosse took the case of boracite, magnesium borate chloride, $Mg_3B_7O_{13}Cl$, which crystallizes in the form of little cubes, often with tetrahedral facets. The modern mineralogist knows, which the nineteenth century mineralogists did not, that it is in fact orthorhombic, with point group *mm2*, and that it has a high-temperature cubic phase, above 265°C, of point group $\bar{4}3m$. It is strongly pyroelectric and piezoelectric. Its phase transition was observed by H. Haga and F. M. Jaeger (1914) by means of X-ray powder diffraction.

For Haüy, who had described the pyroelectric properties of boracite, the primitive form and the integrant molecule were both a cube. Weiss had put it in the hemi-spheroidal system (see Section 12.3.5). Delafosse also considered its integrant molecule to be a cube but, in order to take its habit and its physical properties into account, assumed the shape of the physical molecule to be a regular tetrahedron, and its structure to be a uniform assembly of aligned and similarly oriented tetrahedra (Figs. 2.4 and 12.35). This structure is as compatible with the properties of cleavage as the assembly of integrant molecules, which is simply a geometrical framework. The same remark would apply to the assembly of oblate spheroids considered by Huygens to explain the cleavage and optical properties of calcite. Delafosse, of course, had no idea of the actual distribution of the atoms in the molecule, the shape of which could be more

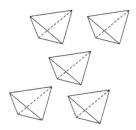

Fig. 12.35   Structure of boracite in plane (011), according to Delafosse. Redrawn
after Delafosse (1843).

complicated than a tetrahedron, but the main thing was that its symmetry should be that of a
tetrahedron.

In a later work, Delafosse attempted to relate the structure of the molecule to the external shape of the crystal. Mitscherlich (1819, 1822) had shown that the analogy in chemical
composition of two substances leads in general to an analogy of their crystalline forms. The
French physicist, A.-M. Ampère[48] (1814) had published a theory of the chemical combinations of substances, based on Haüy's description of primitive forms, in which he assumed
the molecules (he used the term *particule*) had the shape of polyhedra, with atoms (he used
the term *molecule*) at their summits. These polyhedra were constructed by combinations of
Haüy's five primitive forms: tetrahedron, octahedron, parallelepiped, hexagonal prism, rhomb-
dodecahedron. Ampère's hypothesis was, however, hardly noticed by chemists at the time.
Delafosse (1852) tried to extend it by considering that the summits of the polyhedra could
be occupied by more than one atom and that the centre of the polyhedra could also be occupied
by atoms. For instance, he suggested that in hydrated salts the molecule was constituted by a
central nucleus surrounded by water molecules.

12.10.2   Structural interpretation of hemihedry

Haüy had observed that the law of symmetry suffers exceptions, for instance in quartz or
tourmaline (see Section 12.1.4). Weiss (see Section 12.3.5) had also observed in some minerals
incomplete forms having half the number of faces (sphalerite or zinc-blende, tetrahedrite,
boracite, pyrite, cobaltite), speaking of a hemi-spheroidal system. So had Mohs (see Section 12.5), who described hemi-rhombohedral, hemi-pyramidal or hemi-prismatic combinations. His student Naumann was the first to use the terms holohedry, hemihedry and tetardohedry
(see Section 12.5). Another of his students, G. Rose,[49] studied in detail the electric polarization
in crystals of tourmaline of many different origins (Rose 1836). These authors did not give any
physical explanation for these observations. Haüy had thought that in the case of tourmaline it
had something to do with the action of electric forces during growth.

For Delafosse, the reason was that the law of symmetry must be completed by saying that 'the
identity of two parts [of the crystal] should include two conditions, one geometrical (identity of

---

[48]André-Marie Ampère, born 20 January 1775 in Lyon, France, died 10 June 1836 in Marseille, France, Professor at
Ecole Polytechnique, is one of the main discoverers of electromagnetism.

[49]Gustav Rose, born 18 March 1798, died 15 July 1873, was a German mineralogist who studied in Berlin with
C. S. Weiss and with J. Berzelius in Stockholm. He succeeded Weiss as Professor of Mineralogy at Berlin University.

 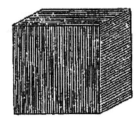 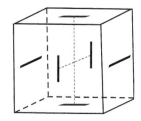

Fig. 12.36   *Left*: Crystals of pyrite triglyphe. Photo by the author. *Middle*: Pyrite triglyphe after Delafosse (1843). *Right*: Structure of pyrite redrawn after Delafosse (1843).

forms) and the second one physical (identity of physical properties). The origin of hemihedry as well as that of the physical properties of some crystals is to be found in the fact that a change in the symmetry of the inner structure entails a change in the external symmetry of the crystal: *the physical properties of crystals must be related to their molecular structure.* According to Fig. 12.35, the rows of molecules of boracite are polar, with the extremities of the tetrahedra all pointing in the same direction, and the electrical polarity of the crystal acquired when heated must be related to this structural characteristic. Delafosse adds, in a prefiguration of Curie's principle, expressed fifty years later by P. Curie[50] (1894), but not further exploited by Delafosse: 1) 'the electrical polarity would not appear if the rows of molecules were symmetric with respect to a plane perpendicular to them, and 2) if this seems to be a *sine qua non* condition, it is not a sufficient one for the phenomenon to happen'. Boracite is a case of *polar* hemihedry in which the two extremities of the axes have different properties; so is tourmaline, which Delafosse discussed also.

The second type of hemihedry analysed by Delafosse is *lateral* hemihedry in which the dissymmetry appears *around* the axis. Pyrite and quartz are two examples. In the case of pyrite (point group $m\bar{3}$), he suggested that the distribution of atoms in the molecule should be something as shown in Fig. 12.36, *Right*, the atoms being elongated along the directions of the striations of pyrite triglyphe (Fig. 12.36, *Left* and *Middle*), which correspond to alternate facets of the pentagon-dodecahedron {210}. The structure of pyrite was determined by W. L. Bragg (1914*a*, see Section 8.3.1), and refined by P. P. Ewald (1914*b*, see Section 8.3.2).

The case of quartz.   Haüy was the first to observe the exception of symmetry presented by the plagihedral facets x and x' of quartz (see Section 12.1.4 and Fig. 12.10). Left- and right-spiralled quartz crystals were also discussed in detail by Weiss (1836). The property of optical rotation of quartz was first observed by F. Arago[51] (1811). His experiments were repeated in a more systematic way, with various plates of different thicknesses, by J.-B. Biot,[52] a strong supporter of Newton's emission theory (see Section 3), and the inventor of the polarimeter, who

---

[50]See footnote 78 in Section 12.14.   [51]See footnote 41 in Section 3.4.2.

[52]Jean-Baptiste Biot, born 21 April 1774 in Paris, France, died 3 February 1862 in Paris, France, was a French physicist and astronomer, a former student of Ecole des Ponts et Chaussées and Ecole Polytechnique, he was appointed Professor at Ecole Centrale in 1797 and at the Collège de France in 1801. He was elected at the French Academy of Sciences in 1803. The mineral biotite was dedicated to him by the German mineralogist J. F. L. Hausmann, in honour of his studies on the optical properties of micas.

showed that some crystals of quartz turned the plane of polarization towards the right and others towards the left (Biot 1812, 1818); Biot had concluded that it was due to a property 'inherent to the molecules themselves, independent of their regular arrangement in the crystalline body', as in the case of liquids in which he had discovered optical rotation. It was J. F. L. Herschel[53] who established the relation between the optical rotation of quartz and the plagihedral faces (Herschel 1822), and it is interesting to follow his reasoning. He felt that there appeared 'to exist an intimate connection between crystallographical and optical properties of bodies'. His idea was that deviation from the symmetry of the full crystallographic form 'might arise from a preference given to more rapid laws of decrement on some of the angles and edges than on those adjacent to them' and that this would result in 'faces unsymmetrically situated with respect to that axis'. This was the case of Haüy's plagihedral faces, and Herschel, as he told the story, happened to have in his possession fine plates of a rock-crystal of which he had 'fortunately preserved the summit, on which were two small but very distinct and brilliant faces of the plagihedral kind, leaning to the left'. He found that they turned the plane of polarization also to the left. Looking at thirteen other crystals exhibiting plagihedral faces, six leaning to the left and seven leaning to the right, he was happy to find that the sense of rotation of the polarization was each time in agreement with that of the facets, and he could consider it was safe to conclude to the generality of this result. His view as to the origin of the phenomenon was unclear, but he felt that the plagihedral 'faces are produced by the same cause which determines the displacement of the plane of polarization of a ray traversing the crystal parallel to its axis' and conjectured that 'there might possibly exist in every molecule a direction or axis in which the force of rotation is a maximum'.

A. Fresnel (1827) went further and gave an interpretation of optical rotation (see Section 3.6). He considered linearly polarized waves to be combinations of right-hand and left-hand circularly polarized waves, and suggested that, in optically active media, the rotation of the plane of polarization was due to a difference in the indices of refraction of the two waves. To account for the optical rotation, Fresnel assumed 'a helicoidal arrangement of the molecules, which would present inverse properties, depending on whether these helices would be dextrorsum or sinistrorsum'.

Delafosse, noticing that fused quartz did not have the property of optical rotation, thought that the origin of the phenomenon did not lie in the molecules only, but also in their crystalline arrangement. It could not be helicoidal since the molecules should all have the same orientation. He suggested that the distribution of atoms in the rhombohedral molecule be such that the edges B and C have different properties (Fig. 12.37, Left). The rows of molecules along the axis would thus have a rotary arrangement (and thus the proper point group, which we know to be 32). Delafosse's vision of crystals, like Haüy's, was that of what we call now a 'molecular crystal'. The modern crystallographer knows that its structure is in reality constituted by a tri-dimensional array of SiO_4 tetrahedra linked by their summits, forming helicoidal chains parallel to the trigonal axis, either all left-handed or all right-handed. The structure (Fig. 12.37, Right)

---

[53]Sir John Frederick William Herschel, born 7 March 1792, died 11 May 1871, was an English mathematician and astronomer. He graduated Senior Wrangler at Cambridge University in 1813. He conducted his research privately and never held an academic position. In 1820 he was one of the founding members of the Astronomical Society of London (later the Royal Astronomical Society).

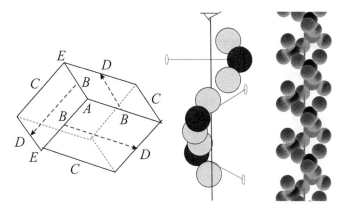

Fig. 12.37   Structure of quartz. *Left*: according to Delafosse. Redrawn after Delafosse (1843). *Middle*: Determined by X-ray diffraction. Redrawn after Bragg and Bragg (1915). *Right*: Modern view.

was studied by X-ray diffraction by Bragg the father as early as 1914 (Bragg 1914*d*), and was determined by Bragg and Gibbs (1925).

Delafosse's work was prolonged by his former student, L. Pasteur,[54] who made a complete chemical, physical, and crystallographic study of thirteen compounds of tartrate salts and acids, establishing the chirality of the molecules and its correspondence with the morphology of the crystals and their optical activity (Pasteur 1848*a*, *b*, *c*). Details can be found, for instance, in Flack (2009).

12.11   A. Bravais and *Systèmes formés par des points distribués régulièrement sur un plan ou dans l'espace*, 1848

Auguste Bravais (Fig. 12.38) was a French physicist and astronomer with a strong mathematical background, who developed the theory of crystal lattices and derived the fourteen lattice modes. According to his presentation of the first of his three major memoirs (Bravais 1848), he first became interested in systems of regularly distributed points when studying the arrangement of leaves on the stems of plants. He came back to that topic a few years later when the study of the parhelia and solar halos led him to become interested in the morphology of ice crystals and in crystal lattices. A naval officer, he had taken part in a rescue expedition of the ship *La Recherche* to the Spitzberg looking for another ship, *La Lilloise*. He had been wounded in the leg and forced to hibernate in Lapponia. This had been the occasion for him to observe sky phenomena.

[54]Louis Pasteur, born 27 December 1822 in Dole, France, died 28 September 1895 in Marnes-la Coquette, France, was a French chemist and one of the pioneers of microbiology. A former student of Ecole Normale Supérieure, he received his doctorate in 1847. He was Professor at the Universities of Strasbourg, Lille and Paris, and Director of the Ecole Normale Supérieure. In crystallography, his work bore on molecular chirality and dimorphism, and this led him to associate dissymmetry and life. He studied fermentation and developed the method of pasteurization. He is also well-known for his breakthroughs in the causes and prevention of diseases, such as the puerperal fever, which were decisive in the acceptation of the germ theory of diseases; he created the first vaccine for rabies and anthrax. He was elected at the French Academy of Sciences in 1862.

Fig. 12.38   Auguste Bravais. Courtesy IMPMC, Université P. et M. Curie, Paris.

Auguste Bravais: born 23 August 1811 in Anonay, France, the son of a medical doctor; died 30 March 1863 in Le Chesnay, France. After schooling at Collège Stanislas in Paris, he was admitted at Ecole Polytechnique in 1829. He began his professional life as a Naval officer but, while serving, did original work in mathematics and was able to defend his doctorate in 1837. He then took part in a naval expedition in the northern seas and Lapponia. In 1841 he was appointed Professor of Mathematics applied to Astronomy at the University of Lyon, and in 1845 Professor of Physics at Ecole Polytechnique. He was elected to the French Academy of Sciences in 1854.

MAIN PUBLICATIONS

1839 *Essai sur la disposition générale des feuilles rectisériées.*

1843 *Mémoire sur le mouvement propre du soleil dans l'espace.*

1847 *Mémoire sur les halos et les phénomènes optiques qui les accompagnent.*

1849 *Mémoire sur les polyèdres de forme symétrique.*

1850 *Mémoire sur les systèmes formés par des points distribués régulièrement sur un plan ou dans l'espace.*

1851 *Etudes cristallographiques.*

1854 *Sur l'influence qu'exerce la rotation de la terre sur le mouvement du pendule conique.*

Bravais' knowledge of the crystallography of the time was quite extensive. He was familiar with the works of Haüy, Beudant, and Delafosse as well as with those of the German school (Weiss, Neumann, Naumann, Seeber, Rose, and Frankenheim) and the Cambridge school (Whewell and Miller). The importance of Bravais's memoir on a 'system formed by regularly distributed points' was immediately recognized by his contemporaries, but, being a mathematical construction consisting in a series of theorems and lemmas, it was considered disheartening and difficult to read, in particular by mineralogists. It was found 'exemplary' (*musterhaft*) by Sohncke (1879) and leaving nothing to desire from the viewpoint of the strength of the argument, but abstract and difficult to take in (Sohncke 1867). It is his former student at Ecole Polytechnique, F.-E. Mallard,[55] whose teaching made it more accessible and who popularized it for the French public (Mallard 1879). Details of Bravais' life can be found in his eulogy read at the French Academy of Sciences by its Secretary, E. de Beaumont.[56]

[55]François-Ernest Mallard (1833–1894) was a French mineralogist and crystallographer. A former student of Ecole Polytechnique, he was appointed Professor of Geology and Mineralogy at the Saint-Etienne Ecole des Mines in 1859 and, in 1872, Professor of Mineralogy at the Paris Ecole des Mines. Besides his purely mineralogical studies, he continued Bravais' work and developed the theory of twins and twinning. He was elected at the French Academy of Sciences in 1890.

[56]'Eloge historique d'Auguste Bravais'. (1866). *Mémoires de l'Académie des Sciences*. **35**, pp. XXIII–XCIX.

12.11.1 Systems formed by regularly distributed points in the plane or in space and the fourteen lattice modes, 1848

The theory of lattices. A. Bravais 'was the first to treat the symmetric distribution of points in the plane and in space from a purely geometric viewpoint, without any speculation about the structure of matter' (Burckhardt 1988). In his memoir, read to the French Academy of Sciences on 11 December 1848 (Bravais 1850), he started by defining lattice rows, lattice planes, and three-dimensional assemblies of 'summits' (lattice nodes) and establishing the invariability of the lattice by a lattice translation, which can be expressed in the following way:

Bravais postulate: *Given any point of the system, there is in the medium a discrete, unlimited in the three directions of space, number of points around which the arrangement of atoms is identical, with the same orientation.*

Bravais then defined the notions of unit cell, conjugate rows, and multiple cells, calculated the volume of the unit cell and the interplanar distance, and gave the relation between Miller indices in a change of coordinates.

The fourteen lattice modes. Axes of symmetry are introduced in the second part of Bravais's memoir, with the proof that only axes of order 2, 3, 4, and 6 are possible, and the now classical properties of symmetry of lattices are derived (every axis of symmetry is a row, every lattice plane normal to an axis and passing through a node is a plane of symmetry, *etc.*). In a second step, Bravais made a classification of all the possible symmetries of lattices, giving all the symmetry elements for each of the seven 'classes' (systems), as shown in Table 12.6.

Bravais then showed that by centring the unit cell or lateral faces, one generates new lattice modes without changing the symmetry. The list of the fourteen lattice modes thus obtained is given in Table 12.6. Bravais preferred the term 'modes' to the term 'orders' used by Frankenheim, because *it better expresses the geometrical fact to which they correspond*. The comparison with Frankenheim's orders is given in Table 12.4.

Table 12.6   Bravais' seven lattice symmetries and fourteen lattice modes. A^q: main axis; L^q: other axes; C: centre of symmetry; Π: main plane of symmetry; P: other planes of symmetry. After Bravais (1850).

Bravais classes	Systems	Symmetry elements	Lattice modes
Terquaternaire	Cubic	$3L^4, 4L^3, C, 3P^1, 6P^2$	hexahedral, octahedral, dodecahedral
Sénaire	Hexagonal	$A^6, 3L^2, 3L'^2, C, \Pi, 3P^2, 3P'^2$	1 mode
Quaternaire	Tetragonal	$A^4, 2L^2, 2L'^2, C, \Pi, 2P^2, 2P'^2$	hexahedral, octahedral
Ternaire	Trigonal	$A^3, 3L^2, C, \Pi, 3P^1$	1 mode
Terbinaire	Orthorhombic	$A^2, L^2, L'^2, C, \Pi, P^2, P'^2$	rectangular hexahedral, rhombic hexahedral rectangular octahedral, rhombic octahedral
Binaire	Monoclinic	A^2, C, Π	hexahedral, octahedral
Asymétrique	Triclinic	$0L, C, 0\Pi$	1 mode

The polar lattice. The last part of Bravais' memoir is devoted to the 'polar assemblies', in other words the reciprocal lattice or, rather, a lattice homothetic to the reciprocal lattice. Surprisingly, this aspect of Bravais' work is seldom quoted. Laue (1960) is a rare exception.[57] Bravais' definition of the polar lattice is similar to that of the reciprocal lattice, except for the lengths of the edges of the elemental parallelepiped. They are equal to the areas of the elemental parallelograms divided by what Bravais calls, after Poisson (1831), the *average interval between summits*, namely the cubic root of the volume of the elemental parallelepiped. Using modern notations, one may compare the lengths, a_B^* and a_R^*, respectively, of the Bravais polar lattice vector and of the usual reciprocal lattice vector:

$$a_B^* = \frac{|\mathbf{b} \wedge \mathbf{c}|}{(\mathbf{a}, \mathbf{b}, \mathbf{c})^{1/3}}; \quad a_R^* = \frac{|\mathbf{b} \wedge \mathbf{c}|}{(\mathbf{a}, \mathbf{b}, \mathbf{c})} \tag{12.15}$$

Bravais then derived the well-known properties of the reciprocal lattice vector, such as the polar lattice of the polar lattice of the polar lattice is the initial lattice, the indices of a row of the initial lattice are equal to the indices of a lattice plane of the polar lattice, the polar lattice of a body-centred lattice is a face-centred lattice, *etc.* Given his definition of the metric of the polar lattice (12.15), he found correctly that the volume of the elemental parallelepiped of the polar lattice is equal to that of the elemental parallelepiped of the initial lattice (and not to its reciprocal, as in the usual reciprocal lattice). He also found that the square of the distance between two polar lattice points is given by the adjoint ternary quadratic form, as defined in Gauss's *Disquisitiones Arithmeticae* (see Section 12.7.2), and that the square of the volume of the polar elemental parallelepiped is equal to the determinant of that form. Finally, he derived the polar lattices for each of the seven classes. In particular, he noted that the rhombohedron which constitutes the elemental parallelepiped of the polar lattice of a trigonal lattice is what Weiss (1829) called *Invertierungs-Rhomboeder* and Haüy, *rhomboèdre inverse*. Applications of the polar lattice for crystallographic calculations are given in Bravais' third memoir, *Etudes cristallographiques* (see Section 12.11.3).

12.11.2   Symmetric polyhedra, 1849

Symmetric polyhedra were first discussed by A.-M. Legendre[58] (1794, Book VI), who defined the symmetry with respect to a plane. In his second memoir (Bravais, 1849), Bravais extended Legendre's study by looking systematically for all possible combinations of symmetry elements, axes of even or odd order, centre of symmetry, and mirror planes, and arrived at Table 12.7. He did not include rotoinversions among the symmetry elements, and group $\bar{4}$ is absent, although he had described the corresponding symmetry, but dismissed it as he did not think it could be observed in nature. He therefore obtained only thirty-one possible crystallographic groups, but he also described polyhedra with five-fold axes.

[57]See also: G. Rigault, *Il reticulo polare di Bravais e il reticulo reciproco. Acc. Sc. Torino: Atti. Sc. Fis.* (1999), **133**, 1–8. The author is grateful to F. Abbona for bringing this publication to his attention.

[58]Adrien-Marie Legendre, born 18 September 1752 in Paris, France, died 9 January 1833 in Auteuil, near Paris, France, was a French mathematician. After schooling at Collège Mazarin, he became Professor of Mathematics at Ecole Militaire. He is well known for his contributions to analysis and number theory, and gave his name to the Legendre

Table 12.7 Possible symmetries of polyhedra, after Bravais (1849).

Polyhedra			Symmetry elements	Hermann-Mauguin symbols
	Asymmetric		$0L,0C,0P$	1
Symmetric	Main axes			
	0		$0L,C,0P$	$\bar{1}$
			$0L,0C,P$	m
	1	order $2q$	$A^{2q},0L^2,0C,0P$	2, 4, 6
			$A^{2q},0L^2,0C,\Pi$	$2/m, 4/m, 6/m$
			$A^{2q},qL^2,qL'^2,0C,0P$	222, 422, 622
			$A^{2q},0L^2,0C,qP,qP'$	$2mm, 4mm, 6mm$
			$A^{2q},qL^2,qL'^2,0C,\Pi,qP^2,qP'^2$	$2/m\,m, 4/m\,m, 6/m\,m$
			$A^{2q},2qL^2,0C,2qP^2$	$\bar{4}2m$
		order $2q+1$	$A^{2q+1},0L^2,0C,0P$	3
			$A^{2q+1},0L^2,C,0P$	$\bar{3}$
			$A^{2q+1},0L^2,0C,\Pi$	$3/m = \bar{6}$
			$A^{2q+1},(2q+1)L^2,0C,0P$	32
			$A^{2q+1},0L^2,0C,(2q+1)P$	$3m$
			$A^{2q+1},0L^2,C,(2q+1)P$	$\bar{3}m$
			$A^{2q+1},(2q+1)L^2,0C,\Pi,(2q+1)P$	$\bar{6}2m$
	> 1	cubic	$4L^3,3L^2,0C,0P$	23
			$4L^3,3L^2,C,3P$	$m\bar{3}$
			$4L^3,3L^2,0C,6P$	$\bar{4}3m$
			$3L^4,4L^3,6L^2,0C,0P$	432
			$3L^4,4L^3,6L^2,C,3P^4,6P^2$	$m\bar{3}m$
		icosahedral	$6L^5,10L^3,15L^2,0C,0P$	235
			$6L^5,10L^3,15L^2,0C,15P^2$	$m\bar{3}\bar{5}$

We have seen that both Frankenheim (see Section 12.8.1) and Hessel (Table 12.5) had found the thirty-two classes. They were also found later, quite independently, by A. Gadolin[59] (1871). The hemihedries and tetartohedries described by Naumann (1856) constituted his starting base, which he completed, using the stereographic projection to represent the symmetry elements of each of the thirty-two groups. Group $\bar{4}$ had already been included by Naumann, who had named it *sphenoidische Tetartoëdrie*. Gadolin called it *tétartoèdrie sphénoïdale*, and its present name is 'tetragonal disphenoidal'.

polynomials and the Legendre transformation. He was elected at the French Academy of Sciences in 1783 and at the Royal Society in 1787.

[59]Axel Gadolin, born 24 June 1828, died 27 December 1892, was a Finnish artillery general in the Russian army, and a mineralogist. He was appointed Professor at the Artillery Academy in St Petersburg in 1867. On 8 February 1869 he obtained his PhD in mineralogy at the University of St Petersburg, and was appointed Professor at the St Petersburg Technological Institute in 1873.

12.11.3   Crystallographic studies, 1849

Bravais' first two memoirs had been purely geometrical constructions; in his third one, *Etudes cristallographiques* (Bravais 1851), he established the relations between crystals and assemblies of regularly distributed points: 'It is now admitted by all physicists that solid bodies are aggregations of molecules of the same type, held apart from one another by attractive and repulsive forces, the resultant of which is null'. By 'of the same type', it was understood that the molecular polyhedra all have the same chemical composition and the same geometrical arrangement around their centre of gravity.

Crystals considered as assemblies of points.   In the first part of the memoir, read to the French Academy of Sciences on 26 February 1849, Bravais only considered the centres of gravity of each molecule: 'the crystal, thus reduced, is but a system of mathematical points equally distributed on parallel rows', it is identical to the lattice described in the first memoir; for instance, the crystal faces coincide with lattice planes. The seven classes correspond to the six systems of the mineralogists, the trigonal and the hexagonal systems, which had been separated by Weiss, having been grouped by Mohs and his followers. The 'law of symmetry' is deduced by mineralogists *empirically* from many measurements on many samples and, as noted by Delafosse, without taking physical properties into account. In fact, it simply results from the application of the symmetry elements of the lattice to the assemblies of molecules which constitute minerals. Bravais then calculated the number of faces of all possible complete forms for each of the seven systems, extending the Miller indices to four in the case of hexagonal lattices (the 'Miller–Bravais indices').

Bravais next showed how the polar lattice can be applied to the method of zones and to crystallographic calculations: to find the zone axis of two lattice planes, (g, h, k) and (g', h', k'), the indices of the lattice plane parallel to two rows, mnp and $m'n'p'$, crystallographic angles between faces, rows, faces and rows.

Bravais 'law'.   Bravais then made the important assumption that 'the lattice planes easiest to separate by cleavage are, in general, those with the highest interplanar distance', namely with the highest density: their *tangential cohesion* is highest, and their *cohesion normal to the plane* smallest. He defined the density of nodes in a lattice plane, which is the inverse of the area of the elemental parallelogram of that plane, and calculated the square of this area in the different modes of each systems, in terms of the Miller indices of the plane, listed by increasing values. The primitive forms may thus be predicted in a rational way, rather than by Haüy's arbitrary way, or, for that matter, by that of the mineralogists who criticized him. Bravais' assumption, called the Bravais 'law', went nearly unnoticed for more than fifty years, until G. Friedel (1907) substantiated it by experimental observation on many examples. Bravais expressed it in the following way: *The directions of planes which are revealed in a homogeneous crystal by discontinuous vector properties are among the lattice planes with the smallest unit cell area* (Bravais–Friedel' law). It was used systematically by Fedorov (1912) in establishing his tables of the crystallographic characteristics of more than 10 000 substances. The order of occurrence of the crystal forms present sometimes lacks with respect to the list determined according to their reticular density. This is due to the fact that Bravais' law applies to the lattice and not to the space group. Friedel (1907) showed that in the case of tourmaline, the first face lacking is the 21st; Fedorov (1912) gave also several examples, in particular that of quartz.

Bravais-Friedel's law was generalized by J. D. H. Donnay and D. Harker (1937) by taking into account the space group of the crystal. As we have seen in Section 12.1.2, the relation between crystal morphology and crystal structure has been handled from a different angle by Hartman (Hartman and Perdock, 1955; Hartman, 1973), by consideration of the directions of Periodic Bond Chains.

Crystals considered as assemblies of polyatomic molecules. In the second part of the memoir, read to the French Academy of Sciences on 6 August 1849, Bravais related the symmetry of the crystal, considered as an assembly of molecules,[60] to that of its constituting molecules and formalized the reticular interpretation of hemihedry given by Delafosse: 'the external signs [of the crystal] which may reveal the structure of the molecule result from the action of attractive and repulsive forces exerted by the molecules along different directions. Bravais quoted here Poisson (1831): 'in solid bodies, the cause that holds together the molecules on their directions of alignment can only come from that part of their action which depends on their form and their relative positions', and concluded that the symmetry of the crystalline structure is the result of the influence of the direction of the molecular forces: 'it is the molecular polarity which produces the phenomenon of hemihedry, as was first indicated by Delafosse'. This polarity is the result of their polyhedral shape. These polyhedra may have elements of symmetry, centres, axes, and mirrors.

Two important situations may occur: 'either the molecular polyhedron has the same elements of symmetry as the lattice, or not'. The first one corresponds to *holohedral* crystals, the second one to *merohedral* crystals (*hemihedral* or *tetartohedral*). It is Bravais who coined the word *merohedry*. The molecular polyhedra may have any of the symmetries described in the second of Bravais' memoirs (Bravais, 1849) and were classified by him among the seven crystal systems, with the remark that trigonal crystals may have either a hexagonal or a trigonal lattice. The number of faces of the various crystalline forms are then given for each class.

Finally, Bravais gave examples of minerals belonging to hemihedral classes, mostly taken from the literature of the time: pyrite ($m\bar{3}$), tetrahedrite ($\bar{4}3m$), and quartz (hexagonal lattice, class 32), for which he imagined a molecular structure rather like that suggested by Delafosse (Fig. 12.37), scheelite ($4/m$), chalcopyrite ($\bar{4}2m$), scapolite, or wernerite, a calcium sodium aluminum silicate carbonate chloride (422, but in reality $4/m$), dioptase ($3m$, but in reality $\bar{3}$), tourmaline ($3m$), manganite (222, but in reality $2/m$), topaz ($2mm$), epsomite, sodium ammonium tartrate tetrahydrate (Pasteur 1848*a*) and Epsom salt, magnesium sulphate (222), tartaric acid (Pasteur, 1848*c*) and saccharose (2), axinite (1).

Twinning. The last part of the memoir, read to the French Academy of Sciences on 8 June 1850, was devoted to twins and twinning (*macles et hémitropie*). As seen in Section 11.11, Romé de l'Isle was the first to study them rationally. Several examples were also given by Haüy (1822), who used the term *hémitrope* for the association of two individuals which could be deduced from one another by a rotation of 180°, such as in calcite, amphibole, or pyroxene, and 'crystal groupings' for other types of twins, as in quartz, staurolite, rutile, and aragonite. Twins were also described by Weiss and Mohs, but it is to Bravais that one owes the first reticular interpretation of twinning. He distinguished three cases:

---

[60]Designated by Schoenflies (1891) under the term *Molekelgitter*.

Fig. 12.39   *Left*: Quartz Dauphiné law twin. *Right*: Rutile elbow twin. Photo A. Jeanne-Michaud. Collections de Minéraux de' UPMC-Sorbonne Universités, with permission.

- *Twinning by molecular hemitropy*, when there are two possible orientations of the molecules with respect to the lattice. This is what has been called 'Twinning by merohedry' by G. Friedel (1926). One of the examples given by Bravais, quoting Rose (1844), is that of the twinning of quartz according to the *Dauphiné* law (Fig. 12.39, *Left*).
- *Twinning by molecular inversion*, when the molecules have the same orientation, but are inverse of one another; it is also twinning by merohedry. Here again Bravais gave the examples of quartz (this twin is now called the *Brazil* law twin) and of Pasteur's tartrate crystals.
- *Twinning by reticular hemitropy*, when the whole crystalline assembly, lattice and molecular polyhedra, occupies two positions deduced from one another by a rotation of 180° and there is a coincidence plane between the two lattices. This corresponds to 'Twinning by reticular merohedry' and 'Twinning by reticular pseudo-merohedry'. Examples given by Bravais are those of the twins of calcite, rutile (Fig. 12.39, *Right*), aragonite, and the feldspars.

Bravais' successful analysis constituted a proof of the soundness of the space-lattice concept. It is still valid today. It was expanded and developed by Mallard (1879) and Friedel (1926). A modern treatment of twinning is given in Hahn and Klapper (2003); for a general overview and a historical introduction, see, for instance, Boulliard (2010) and Hardouin-Duparc (2011).

12.12   L. Sohncke and *Entwicklung einer Theorie der Kristallstruktur*, 1879

Leonhard Sohncke (Fig. 12.40) was a German mathematician and physicist who showed that molecules in the crystalline arrangement did not necessarily all have the same orientation and derived the sixty-five proper groups of movements. Bravais's contribution had been to formalize the theory of lattices, Sohncke's, which paved the way for the modern vision of crystalline structures, was to take into account the positions of the molecules within the elemental parallelepiped.

Fig. 12.40   Leonhard Sohncke at the end of his life. Photo Deutsches Museum, Munich.

Leonhard Sohncke: born 22 February 1842 in Halle, Germany, the second son of a Professor of Mathematics at the University of Halle; died 1 November 1897 in Munich, Germany. After schooling at the Gymnasium in Halle, he studied mathematics and physics at the University of Halle, where he obtained his PhD in mathematics in 1866, and then at the University of Königsberg, where he was a student of Franz Ernst Neumann (see Section 12.6.1). He defended his habilitation in 1869 on the *Cohesion of rock-salt*. In 1872, he became Professor of Physics at the Karlsruhe Technische Hochschule (Technical University), on G. Kirchoff's recommendation. He was then appointed Professor of Physics at the University of Jena in 1883 and in Munich in 1886, where he was Director of the Physics Department of the Munich Technische Hochschule till his death. In 1887 he was elected a member of the Bavarian Academy of Sciences.

MAIN PUBLICATIONS

1867 *De aequatione differentiali seriei hypergeo-metrica.*

1869 *Über die Cohäsion des Steinsalzes in krystallographisch verschiedenen Richtungen.*

1875 *Über Stürme and Sturmwarnungen.*

1876 *Die unbegrenzten regelmässigen Punktsysteme als Grundlage einer Theorie der Krystallstruktur.*

1876 *Universalmodell der Raumgitter.*

1879 *Entwickelung einer Theorie der Krystallstruktur.*

1892 *Gemeinverständliche Vorträge aus dem Gebiete der Physik.*

1895 *Gewitterstudien auf Grund von Ballonfahrten.*

Sohncke's father had been a Professor of Mathematics and Leonhard's upbringing was mathematical. His PhD thesis in his home town of Halle bore on hypergeometric series. It was in Königsberg, where he was a student of Franz Neumann's, that he was introduced to crystallography. Sohncke (1879) knew the works of the main actors of the development of the concept of space lattice: Hooke, Guglielmini, Huygens, Bergmann, Haüy, Wollaston, Brewster, Seeber, Dana, Frankenheim, Delafosse, and Bravais. He openly put his work in the footsteps of Delafosse, and Bravais: 'the most probable representation of the inner constitution of crystalline bodies that has prevailed till now was put forward by Delafosse and thoroughly followed up by Bravais' (Sohncke 1876). His first work (Sohncke 1867) was based on Bravais' analysis, but, as has been noted above (see Section 12.11), he found it rigorous but too mathematical, too long, and, for the first two memoirs, unrelated to crystallography. Sohncke felt that the results should be expressed in a more compact and more comprehensible way. This he did, starting with the statement known as the Sohncke postulate (Sohncke 1867):

Sohncke postulate: *The distribution of points in a crystalline assembly of points (which will be taken as unlimited in space) is around one point the same as around another one.*

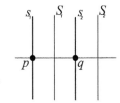

Fig. 12.41   The periodicity of the mirror planes (s_1, S_1, s_2, S_2) is half that of the lattice
points (p, q). Redrawn after Sohncke (1867).

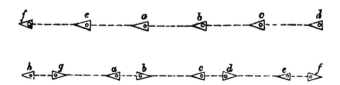

Fig. 12.42   First graphical representation of one-dimensional space-groups. After Wiener (1863).

It is nearly the same as Bravais' (see Section 12.11), with the notable difference, which will be very important in Sohncke's future work, that the last phrase, *with the same orientation*, present in Bravais' formulation, is absent in Sohncke's. His derivation of the fourteen lattice modes was not based on the axes of symmetry, as Bravais had done, but on the planes of symmetry. Sohncke is the first to have shown that if there are mirror planes at points p, q, *etc.* of a lattice row, there are also mirrors half-way in between (Fig. 12.41). Like Bravais, Sohncke considered seven, and not six, crystal systems, and his proof that five-fold symmetry is impossible for crystal lattices was slightly different.

After his habilitation work on the variation of the tensile strength of rock-salt crystals with the crystallographic direction (Sohncke 1869), Sohncke moved to Karlsruhe where he resumed his work on the 'regular systems' of points (*regelmässigen Punktsysteme*), which he defines thus (Sohncke 1874):

A point system of unlimited extension is called regular when the bundles formed by all the lines drawn from each point to all others are either all congruent or partially congruent and partially symmetric.

The same idea had already been expressed a few years before by another Professor of Physics at the Karlsruhe Technical University, C. Wiener[61] (1863), who described various possible regular arrangements (*regelmässige Anordnungen*) of atoms. Fig. 12.42 shows, as for example, two one-dimensional space-groups. Sohncke (1879) wrote that [Wiener] 'made the first step forward since Bravais's theory'.

In his 1874 article, Sohncke concentrated on all the possible regular systems of points in the plane and found fourteen, out of the seventeen plane groups, which were determined by E. S. Fedorov (1891*b*); see Section 12.14. It seems that it is in this paper that the word *Raumgitter*

---

[61]Christian Wiener, born 7 December 1826 in Darmstadt, Germany, died 31 July 1896 in Karlsruhe, Germany, was a German mathematician and physicist. He was educated in Darmstadt, Giessen, and Karlsruhe. After teaching at the Technische Hochschule in Darmstadt, he became Professor of Descriptive Geometry at the Karlsruhe Technische Hochschule. He is best known for his work on Brownian motion.

(space lattice) first appears. At that time, Sohncke had not yet heard of the work on groups of motions by C. Jordan.[62] Jordan had become interested in the notion of groups, introduced by the short-lived French mathematician E. Galois[63] when the works of the latter were published in 1846. Jordan was the first to develop a systematic approach to the theory of finite groups. In particular, starting from the fact that any displacement of a solid body results from a helicoidal motion, and spurred by Bravais' *Etudes cristallographiques*, he discussed the problem of 'a group of motion such that two movements of the group result in a movement which is also part of the group' (Jordan 1867). In a reference to Bravais, he called *merohedral* all subgroups of the *main* groups. In his main paper (Jordan 1868), he detailed the 174 groups of motion he obtained by combining rotations, translations, and helicoidal motions resulting from the association of a rotation and a translation parallel to its axis ('screw axes'), without any reference to a possible crystallographic application.

It was only a little later that Sohncke became acquainted with Jordan's work and found (Sohncke 1876) that there was a treatment by Jordan, related to his concerns about the regular point systems, that discussed in a much more general manner the geometry of motions. 'With the help of Jordan's work', he wrote, 'I was in a position to present an overall view of all the possible unlimited regular point systems'. He found that there were some groups missing, corrected inaccuracies, and applied Jordan's analysis to the crystal systems, discarding the more than 100 groups that were not relevant for crystallography. He could thus describe all possible mapping motions (*Deckbewegungen*) of the arrangements of atoms (Sohncke 1879), and enumerate 66 proper groups of motion (the chiral groups), reduced to 65 by Schoenflies (1887), without taking into account the rotation-reflection and rotation-inversion axes. These were considered a few years later independently by E. S. Fedorov (1891*b*) and A. Schoenflies (1886)—see Section 12.14—and by W. Barlow (1894)—see Section 12.13—who derived the 230 space groups.

Sohncke's theory of regular point systems was, however, not completely general and missed some of the thirty-two point groups. This was corrected a few years later (Sohncke 1888), when, goaded in particular by L. Wulff's discussion of the regular point systems (Wulff 1887), he generalized his systems of single spherical atoms by stating that 'a crystal of infinite extension consists in a limited number (1, 2, 3, … n) of interpenetrating regular systems of points, all characterized by the same set of translations, which, in general, are not congruent'. Fully aware of Wollaston's (see Section 12.2) and Barlow's close-packing (see Section 12.13) models, he suggested that the building blocks of the crystal structure could be made of individual atoms (Fig. 12.43, *Left*). He applied this notion to the interpretation of the structure of optically active crystals, such as quartz, as consisting of parallel helicoidal assemblies of atoms. For instance, in Fig. 12.43 *Right*, the positions of a group of two atoms in the successive layers of the helicoidal

---

[62]Camille Jordan, born 5 January 1838 in Lyon, France, died 22 January 1922 in Paris, France, was a French mathematician. A former student of Ecole Polytechnique, he was Professor successively at Ecole Polytechnique and Collège de France in Paris. He introduced the mathematicians Sophus Lie (1842–1899), from Norway, and Felix Klein, from Germany, to the theory of groups during their stay in Paris in 1870. Lie was to study continuous groups and Klein discrete groups.

[63]Evariste Galois, born 25 October 1811 in Bourg-la-Reine, France, killed in a duel 31 May 1832 in Paris, France, student of Ecole Normale Supérieure, then called Ecole préparatoire.

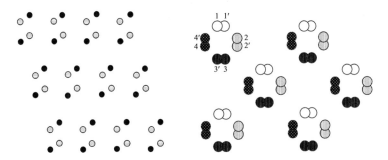

Fig. 12.43   *Left*: model atomic structure of a crystal of point group 2; the building blocks are constituted by a set of individual atoms. After Sohncke (1888). *Right*: model atomic structure of an optically active crystal of space group $I4_1$. After Sohncke (1891*b*).

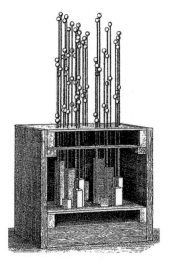

Vollständige Sammlungen der
Sohncke'schen Krystallstruktur-Modelle
liefert für Mark 180
      Mechaniker **Martin**, Polytechnikum Karlsruhe.

Fig. 12.44   *Left*: Sohncke's model to demonstrate screw axes. Source: Wikicommons. *Right*: Advertisement in *Annalen der Physik*, 1882, **252**.

arrangement are designated by 1 and 1′, 2 and 2′, 3 and 3′, 4 and 4′, respectively, in the particular case of a tetragonal crystal.

In 1876, Sohncke had models constructed from cigar boxes to illustrate screw axes (Fig. 12.44, *Left*), which were commercially available (Fig. 12.44, *Right*). He took them with him to Munich where some were still around in 1912.

12.13   W. Barlow and the *Probable nature of the internal symmetry of crystals*, 1883

William Barlow was an English amateur crystallographer who had a rich scientific production. He developed a theory of close packing of separate atoms, which did away with the molecular theories of the time; he derived, after Fedorov and Schoenflies, the 230 groups of movements,

Fig. 12.45   William Barlow. After Pope (1935).

William Barlow: born 8 August 1845 in Islington, in London, England, the son of a builder and building surveyor; died 28 February 1934 in Great Stanmore, Middlesex, England. He received his first education at home from a private tutor. Later, he attended the City and Guilds Technical College in South Kensington, where he met the English mineralogist H. Miers (1858–1942) and the English chemist William Pope (1870–1339), with whom he published several papers. When his father died in 1875, he and his brother, George, inherited a considerable sum of money and he could devote himself to his researches in crystallography, without having to work for a living. He did not have an academic career, but worked from home. He was, however, always in close contact with the scientists of his time, and became a member of several scientific societies: the British Association for the Advancement of Science (1885), the Geological Society (1887), and the Mineralogical Society (1894), of which he was President from 1915 to 1918. He was elected a Fellow of the Royal Society in 1908.

MAIN PUBLICATIONS

1883 *Probable nature of the internal symmetry of crystals.*

1885 *New theories of matter and force.*

1894 *Über die geometrischen Eigenschaften homogener starrer strukturen und ihre Anwendung auf die Krystalle.*

1898 *A Mechanical Cause of Homogeneity of Structure and Symmetry Geometrically investigated; with special Application to Crystals and to Chemical Combination.*

1901 *Crystal symmetry. The actual basis of the 32 classes.*

1906 *A development of the atomic theory which correlates chemical and crystalline structure and leads to a demonstration of the nature of valency.* With W. Pope.

1910 *The relation between the crystal structure and the chemical composition, constitution, and configuration of organic substances.* With W. Pope.

1914 *On the Interpretation of the Indications of Atomic Structure Presented by Crystals when Interposed in the Path of X-rays.*

1923 *Partitioning of space into enantiomorphous polyhedra.*

and, together with his friend William Pope,[64] related the crystallographic structure of a substance to the valencies of its chemical constituents. Barlow came from a wealthy family and was educated privately. His father was a land surveyor and, at a young age, he was interested in the geometrical problems associated with the distribution and arrangement of houses in a given piece of land. It was only after his father's death in 1875, when he had become independent financially, that he started doing research. His lack of rigid disciplinary training, noted Pope in his obituary,[65] was 'in some aspects a hindrance, but in others an advantage; it left a beautiful

---

[64]William H. Pope was an English chemist, born 31 March 1870 in London, England, died 17 October 1939 in Cambridge, England. After studying at the City and Guilds College, South Kensington, he worked at first in industry, as chief of the Chemistry Department of Goldsmiths' Company's Institute at New Cross (1897–1901). In 1901 he was appointed Professor of Chemistry in Manchester, and in 1908 in Cambridge, where he remained till the end of his life. His contributions were mainly in the field of stereochemistry. He gave judicious advice to Lawrence Bragg at the very beginning of the studies of crystal structures with X-rays (see Section 7.7). He was elected a Fellow of the Royal Society in 1902.

[65]Pope, W. J. (1935). William Barlow 1845–1934. *Obituary Notices of Fellows, Roy. Soc.* (1935–1936), **1**, 367–370.

intellect unhampered by authority'. He wrote his first papers on close packing without being yet fully acquainted with chemistry and crystallography, and received serious training in these two disciplines only around 1888, at the City and Guilds Technical College, in chemistry by the chemist H. E. Armstrong (1848–1937), and in crystallography by Pope and the latter's former teacher, the mineralogist H. Miers.[66] Afterwards, he and his family spent some time in Germany, where he had the opportunity of meeting P. Groth,[67] who was then very interested in the molecular constitution of crystals (Groth, 1888). It was on his return to England that Barlow started his study of theoretical crystallography, and it is not surprizing that it is in *Zeitschrift für Kristallographie* that he published his derivation of the 230 groups. The interested reader may find more details about Barlow's life and researches in his biographies by W. T. Holser (1970)[68] and P. Tandy (2004).

Barlow's first work (1883, 1886) is based on the belief that 'in the atom groupings [molecules] which modern chemistry reveals to us the several atoms occupy distinct portions of space and do not lose their individuality'. His model of close-packed atoms, he felt, might explain the symmetrical forms of crystals and the relations between internal symmetry and cleavage. Starting from the assumption that 'crystals are built up of minute masses of different elements symmetrically disposed', Barlow looked for all the possible *very symmetrical* arrangements of points or spheres of equal size[69] and found five possible arrangements of close packed spheres (a sixth one is described in Barlow 1886):

1. The spheres occupy the summits and centres of an array of cubes regularly stacked together; each atom has eight first neighbours. If the two types of position are occupied by the same kind of atoms, the arrangement is what is called today body-centred cubic. In Fig. 12.46 (a), the atoms which occupy the centres are of a different nature from that of those occupying the summits; we know now that this is the case of caesium chloride for instance.
2. One kind of atom occupies the summits and the centres of the faces of the array of cubes, and the other kind the middle of the sides (Fig. 12.46 (b)); each atom has six first neighbours. This type of structure is face-centred cubic. This is the structure proposed by Barlow (1897) for sodium chloride and confirmed by W. L. Bragg (1913).
3. The third type of symmetry corresponds to the face-centred cubic close packing of one kind of atom, as we now know is the case for gold, for instance.
4. The fourth kind of symmetry corresponds to the hexagonal close-packed stacking.
5. The fifth kind of symmetry corresponds to the stacking of planes of close-packed spheres of two kinds according to the scheme of Fig. 12.46 (c), and is appropriate for compounds of composition AB_2, such as ice (H_2O), which crystallizes in six-sided prisms. Barlow suggested that the structure of quartz, SiO_2, was the result of a helicoidal stacking, left- or right-handed, of such planes (Figs. 12.46 (d) and (e)).

---

[66]Sir Henry A. Miers, born 25 May 1858 in Rio de Janeiro, Brazil, died 10 December 1942 in England, was an English mineralogist and crystallographer. He was educated at Eton College and Trinity College, Oxford. He joined the British Museum in 1882, and was elected a Fellow of the Royal Society in 1896. He was later successively Professor of Mineralogy and Crystallography at Oxford University (1894), Principal of London University (1908) and Vice-Chancellor and Professor of Crystallography at the Victoria University of Manchester (1915–26).

[67]See Section 6.1.1.

[68]*Dictionary of Scientific Biography*, Charles Scribner's Sons, New York, 460–463.

[69]The Swedish chemist J. J. Berzelius (1779–1848) had assumed that atoms were all spherical and were all of the same size (Berzelius 1813).

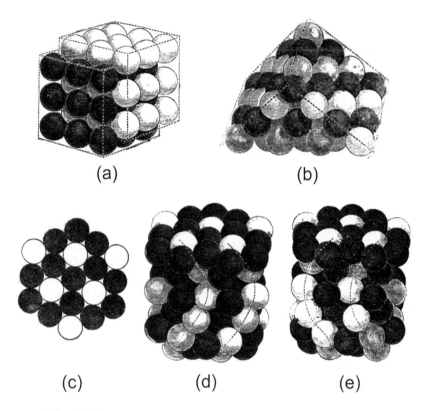

Fig. 12.46 Barlow's arrangements of closed-packed atoms. After Barlow (1883).

Barlow (1886) noted that binary compounds AB are usually cubic, and considered that their most probable form of crystallization is one of his first two types of symmetry. He was the first to imagine a crystal that was not molecular, and he was also the first to consider hexagonal close packing. He found that his model was well suited to account for isomorphism and dimorphism, as well as for twinning. His description of the structure of alkali halides was on the right track, but his models of ice and quartz were wrong, although he was right in trying to interpret the optical polarization of quartz with a helicoidal arrangement; worse, he proposed for calcite and aragonite that their composition should not be, as usually accepted, $CaCO_3$, but CaC_2O_3 (*sic*), and that their structure should be constituted by alternate layers of oxygen and of calcium and carbon arranged according to Fig. 12.46 (c). He was, however, to suggest the correct structure later (Barlow and Pope 1908).

Sohncke was prompt to react: Barlow's letters to *Nature* were dated 20 and 27 December 1883, and Sohncke's criticisms were published in the same journal on 21 February of the following year (Sohncke 1884). His objections were three-fold. 1) *Geometrical*: four of Barlow's 'very symmetric' arrangements correspond to four of Sohncke's 66 regular systems of points,[70]

---

[70]At that time, the number of 66 systems had not yet been reduced to 65 by Schoenflies.

but not the fifth one; it is expected that any of Sohncke's systems is suitable for substances to crystallize in, and Barlow's choice seems quite arbitrary. Sohncke had also already suggested that the group of movement of quartz contained a three-fold helicoidal axis. 2) *Chemical*: if a compound such as NaCl crystallized, for instance, in Barlow's first arrangement, then each Na atom would be surrounded by eight Cl atoms, and, in the same way, each Cl atom by eight Na atoms. The atoms of Cl and Na therefore would appear to be *octovalent*, and not univalent. 'By this example', added Sohncke, 'we see that from Mr Barlow's point of view both the notion of chemical valency and of chemical molecule completely lose their present import for the crystallized state'. This kind of objection was made for a long time, even after the determination of the structures of alkali halides by W. L. Bragg, and as late as the mid 1920s (Armstrong 1927 – see Section 1.2). 3) *Physical*. Sohncke refuted that Barlow's model could explain the isomorphism of, for instance, $CaCO_3$ and $FeCO_3$.

Barlow (1884) briefly replied one week later by saying that at the first steps of crystallization stable equilibrium requires that the atoms be *very evenly distributed throughout the space allotted to them*, which is the case in the systems he had proposed. Concerning the valency in the solid state, he answers that there is no clear knowledge of the nature of the union between the various sorts of atoms, and that his view is supported by the phenomenon of electrolysis.

This work had really been a beginner's work, with many imperfections, but Barlow's intuition about the close packing of individual atoms proved to be very fruitful. His next study, ten years later, after his visit to Germany, is of quite a different standard. Starting from Sohncke's work, he derived at his turn the 230 types of homogenous structures (Barlow 1894), as is briefly told in Section 12.14. In a long paper presented at the Dublin Royal Society (Barlow 1897), he investigated geometrically the mechanical cause of the homogeneity of structures and of their symmetry, and described many cases of close packings of spheres of two or more different sizes, as well in holohedral crystals as in hemihedral and tetartohedral ones. As an example, Fig. 12.47 shows the structure of rock-salt as predicted by Barlow, and as determined by Bragg with X-ray diffraction (Bragg 1913*b*).

As a further step, Barlow and his friend Pope elaborated on Barlow's early theory of close packing in a series of papers (Barlow and Pope 1906, 1907, 1908, 1910) devoted to the correlation between chemical constitution and crystalline form. In order to establish it on a firmer basis, they represented the space occupied by atoms by a sphere of influence, and emitted two fundamental hypotheses.

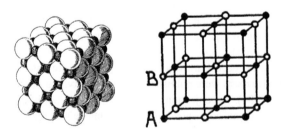

Fig. 12.47   Structure of NaCl. *Left*: After Barlow (1897). *Right*: After W. L. Bragg (1913*b*).

1. *A crystalline structure is a close-packed, homogeneous assemblage of the spheres of influence of the component atoms.* (Barlow and Pope 1906). It is in equilibrium under the action of two opposite centred forces, of the nature of kinetic repulsion and gravity attraction, exerted on the atomic centres, as suggested in R. Boscovich's atomic theory (see Section 2.3). In their second paper, Barlow and Pope (1907) qualified their statement by saying that the spheres of influence representing the atoms are in fact not strictly spherical in form, 'but are polyhedra enclosed by interfaces drawn at right angles to lines joining adjacent centres of atomic influence' (these polyhedra are Dirichlet domains). To picture these polyhedra, Barlow and Pope imagined that rubber balls of appropriate sizes and suitable arrangement were squeezed together so that the interstices between them became practically eliminated. This is reminiscent of Kepler's explanation of the rhomb-dodecahedral shape of pomegranate seeds (see Section 11.4.3), although they do not seem to have been aware of it. In practice, however, they found it convenient to keep the term 'spheres of influence'.

2. The atoms, supposed spherical, occupy volumes in the crystal structure proportional to their valency (the 'valence–volume theory'). If a sphere of volume m is removed, it can be replaced by several spheres of total volume m, and this corresponds to replacing an atom of valency m by several atoms of total valency m.

In this way they derived structures of complicated compounds from simpler ones. For instance, in their first paper Barlow and Pope (1906) related the organization of the methane and ethane molecules to the structure of carbon. There are two ways in which the assembly of equal spheres of carbon can be arranged. The first one is that of diamond, which Barlow (1897) had proposed to be cubic close-packed, and in which they assume the atoms to be tetrahedrally arranged—a structure confirmed by the Braggs (Bragg and Bragg 1913*b* and *c*). The other one is that of graphite, which they proposed is built up by the stacking together of groups of six atoms linked as in benzene or naphthalene, forming an assemblage of rhombohedral symmetry. If now one half of the carbon atoms is homogeneously replaced, each by four hydrogen spheres, the resultant assemblage can be partitioned into units of composition CH_4, in agreement with the tetrahedral character attributed to the molecule of methane independently by J. H. van't Hoff[71] and J. A. Le Bel[72] in 1874. If in this methane assembly one hydrogen sphere is removed from each unit, and if close-packing is restored by shrinking the assemblage together, the subsequent partitioning resolves itself into units of composition C_2H_6 attributed to ethane by van't Hoff and Le Bel.

In a similar way, in their second paper, among many other examples, Barlow and Pope (1907) deduced the structure of ammonium chloride, NH_4Cl, from that of potassium chloride, KCl, by replacing the atoms of potassium the atom groupings NH_4.

The third paper (Barlow and Pope 1908) was devoted to the isomorphism and polymorphism of $NaNO_3$ and $CaCO_3$. The structures of sodium nitrate, calcite and dolomite, $(Ca,Mg)CO_3$, were there correctly described (Fig. 12.48), as confirmed a few years later by Bragg (1914). The fourth paper (Barlow and Pope 1910) concerned organic crystals.

Barlow and Pope's assumption of the proportionality between valency volumes and valencies was criticized by T. W. Richards[73] (1913), who had previously developed a theory of the

---

[71]See footnote 16 in Chapter 7.   [72]See footnote 17 in Chapter 7.

[73]Theodore William Richards, born 31 January 1868 in Germantown, PA, USA, died 2 April 1928 at Cambridge, MA, USA, was an American chemist. He studied at Harvard University, where he received his PhD in 1888, and

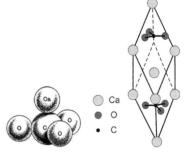

Fig. 12.48   Structure of calcite. *Left*: After Barlow (1908). *Right*:
As determined by X-ray diffraction.

compressibility of atoms and found no justification to it. An exchange of letters to the *American Journal of Chemistry* followed, trying to clear up various misunderstandings between them, but the two parties could not come to terms. Barlow and Pope (1914) terminated the argument by the sound remark that W. H. and W. L. Bragg were at that time developing a method for the practical determination of the arrangement of atoms in crystal structures, and that further discussion should therefore be postponed until its results were known. X-ray measurements have shown that Barlow and Pope's postulate of the proportionality of the volume of an atom to the valency was very far from reality (Tutton 1917; Goldschmidt 1929).

12.14   The 230 space groups: E. S. Fedorov (1891), A. Schoenflies (1891), and W. Barlow (1894)

The saga of the development of the space lattice concept before 1912 was crowned by the derivation, independently, by E. S. Fedorov and A. Schoenflies in 1891, and by W. Barlow in 1894 of the 230 space groups. Henry Miers, at that time Professor of Mineralogy and Crystallography at Oxford University, suggested to a young mathematical Tutor at Magdalen College, Harold Hilton, to write a comparison of their notations (Hilton 1902). Hilton (1903) also published a synthesis in English of their treatments, based, however, primarily on Schoenflies' for the derivation and on Fedorov's for the drawings. The story of the discovery is told by Barlow and Miers (1901), I. I. Shafranovskii and N. V. Belov in Ewald (1962a), Burckhardt (1967, 1988), Senechal (1990), Galiulin (2003), and Zimmermann (2004). A biography of A. Schoenflies and his family is given by his grandson, Thomas Kaemmel (2006), who also details the chronology of the exchanges between Fedorov and Schoenflies. The correspondence between them was published in Russian by Bokij and Shafranovskii[74] and in German by Burckhardt (1971).

Schoenflies and Fedorov became involved in the problem for different reasons—Schoenflies (Fig. 12.49), under the influence of F. Klein[75] in Göttingen, was interested in the group

where he became Professor in 1901. He received the Nobel Prize for Chemistry in 1914 'in recognition of his accurate determinations of the atomic weight of a large number of chemical elements'.

[74]*Correspondence of E. S. Fedorov with A. S. Schoenflies*, published by G. B. Bokij and I. I. Shafranovskii, *Nauchnoe Nasladenie (Scientific Heritage) Izdaniya Akademii Nauk SSSR*, (1951), Vol. **2**, pp. 314–343.

[75]See footnote 5 in Chapter 6.

Fig. 12.49   Arthur M. Schoenflies in the early 1880s. Photo L. Haase & Co. Courtesy T. Kaemmel.

Arthur Moritz Schoenflies: born 17 April 1853 in Landsberg an der Warthe, Brandenburg, now Gorzòw, Poland, the youngest son of a manufacturer of cigars; died 27 May 1928 in Frankfurt am Main, Germany. After schooling at the Gymnasium in Landsberg from 1862 to 1870, he became a student of mathematics at the Friedrich-Wilhelms University of Berlin, now Humboldt University, with the mathematicians E. E. Kummer (1810–1893) and K. T. W. Weierstrass (1815–1897) as professors, obtaining his PhD in mathematics on 2 March 1877. At the same time, he also attended lectures by the mathematician S. H. Aronhold (1819–1884) at the Berlin Industrial Institute (Berliner Gewerbeinstitut), which later became the Technical University of Berlin, and was awarded a degree entitling him to teach. He taught first at the Friedrich-Wilhelms-Gymnasium in Berlin, and from 1880 at the Gymnasium in Colmar, Alsace, then Germany. During that time, however, he continued doing research and was awarded his habilitation as Privatdozent at the University of Göttingen in 1884. Felix Klein, who had been appointed Professor at that University in 1886, sought to establish Göttingen as a leading mathematics centre and ensured that Schoenflies was appointed Extraordinary Professor of Applied Mathematics there in 1892. This was not without difficulty, Schoenflies being Jewish. It was under Klein's influence that Schoenflies did his work on the crystallographic groups of motion. In the mid-1890s, he changed topics and started working on topology and set theory. He left Göttingen in 1899 to become Professor of Mathematics at the Albertus-University in Königsberg, from where he moved to Frankfurt am Main in 1911 as Professor at the Academy for Social and Commercial Science. He was one of the founders of the University of Frankfurt am Main, establishing a Faculty of Science of which he was the first Dean, and inviting the Nobel Prize winner M. von Laue to occupy the chair of Theoretical Physics. In 1920 he became the rector of the Univerity. He was elected a member of the Deutsche Akademie der Naturforscher Leopoldina in 1896 and a fellow of the Bayerische Akademie der Wissenschaften in 1918.

MAIN PUBLICATIONS

1877 *Synthetisch-geometrische Untersuchungen über Flächen zweiten Grades und eine aus ihnen abgeleitete Regelfläche.*

1886 *Geometrie der Bewegung in synthetischer Darstellung.*

1891 *Kristallsysteme und Kristallstruktur.*

1895 *Einführung in die mathematische Behandlung der Naturwissenschaften. Kurzgefasstes Lehrbuch der Differential- und Integralrechnung.* With W. Nernst.

1900, 1908 *Die Entwicklung der Lehre von den Punktmannigfaltigkeiten I, II.*

1908 *Einführung in die Hauptgesetze der zeichnerischen Darstellungsmethoden.*

1913 *Entwicklung der Mengenlehre und ihrer Anwendungen.* With H. Hahn.

1923 *Theorie der Kristallstruktur.*

1925 *Einführung in die analytische Geometrie der Ebene und des Raumes.*

theoretical aspect, and Fedorov (Fig. 12.50), in the wake of his early work on parallelohedra, in the regularity of real crystal structures. In his first letter, dated 14 December 1889,[76] Schoenflies sent Fedorov his 1888 papers in *Nachrichten Kgl. Gesell. Wiss. Göttingen* and told him he was happy to see the convergence of their ideas about the geometrical systems of points. *Die Priorität gebe ich Ihnen gern zu*, 'I gladly recognize your priority', he added. For his part, Fedorov wrote, 'It is with pleasure that I see a repetition of the important underlying features of my theory of crystal structures' (quoted by I. I. Shafranovskii and N. V. Belov in Ewald, 1962). From then on they exchanged views, but reached the final result separately; Fedorov was the first, which Schoenflies acknowledged in his book, *Krystallsysteme und Krystallstruktur*, submitted in August 1891, by mentioning the 'publication' by Fedorov of the complete list of the space groups (*Raumgruppen*). In fact, Fedorov's memoir only appeared in December 1891.

Fedorov became interested in geometry at a young age, and started writing his first book, *The elements of the study of figures*, before entering the Mining Institute. It was completed in 1883, and finally published in 1885, thanks to help of A. V. Gadolin[77] and despite the opposition of some leading scientists, such as the mathematician and Professor at St Petersburg University, P. Chebyshev (1821–1894), who sneered that 'contemporary science is not interested in these questions' (quoted by Senechal 1981) and the mineralogist and Professor at the Mining Institute of St Petersburg, P. V. Eremeev (1830–1899) who found it too speculative (E. Scholz, in Burckhart 1988). The book is a treatise on fundamental geometry, making use of the concept of regularity. Fedorov introduced there a new kind of symmetry elements, *composite symmetry elements*, which are combinations of axes of symmetry and planes of symmetry, and are now called 'rotoinversion' elements. They were also introduced independently by P. Curie[78] (1884), who called them elements of 'alternate symmetry' (*symétrie alterne*). Curie made use of them to derive the thirty-two classes of symmetry. These were also derived a little later by B. Minnigrerode[79] (1887), by means of the theory of groups and substitutions. Fedorov first investigated the symmetry of finite rigid bodies in general and then, using a condition equivalent to the law of rational indices, determined the thirty-two classes of symmetry or 'symmetry arts'. The symmetry elements of the second kind were also suggested to Schoenflies by F. Klein in 1887 (see below). Schoenflies wrote to Fedorov on 4 July 1890,[80] after receiving Fedorov's

[76]Burckhardt, 1971, p. 92.

[77]Gadolin described in 1871 the thirty-two point groups, independently of Frankenheim and Hessel; see Section 12.11.1.

[78]Pierre Curie, born 15 May 1859 in Paris, France, the son of a physician who educated him, died 19 April 1906 in Paris, France. After his Baccalauréat (1875), he studied at the Paris Faculty of Science, obtaining his BSc in 1877. He was at first laboratory assistant and, together with his brother Jacques, discovered piezoelectricity in 1880 in the Mineralogy Laboratory of the Faculty of Sciences (Sorbonne), then headed by the mineralogist and chemist Charles Friedel (1822–1899). He was appointed chief assistant in charge of the physics laboratory at the newly founded Ecole Supérieure de Physique et Chimie Industrielle de Paris in 1882, and in 1895 obtained his Doctorate on magnetism and on the symmetry principle that bears his name. The same year he became Professor of Physics. With his wife, Maria Skłodowska (1867–1934), they discovered polonium and radium and shared the 1903 Nobel Prize for Physics with Henri Becquerel (1852–1908). In 1904 he was appointed Professor of Physics at the Paris Faculty of Sciences, and in 1905 he was elected to the French Academy of Sciences. He died at the age of 47. hit and run over by a horse-drawn cart.

[79]Bernhard Minnigerode (1837–1896), a pupil of F. E. Neumann.   [80]Burkhardt 1971, p. 93.

Fig. 12.50   Evgraf S. Fedorov at the age of 45. Courtesy L. Aslanov.

Evgraf Stepanovich Fedorov: born 10 (22) February 1853 in Orenburg, on the Ural river, Russia, the son of a sapper major-general; died 21 May 1919 in Petrograd (formerly St Petersburg), Soviet Union. He lost his father at a young age and was educated at a military engineering school, from which he graduated in 1872. After serving two years as a lieutenant in a sapper unit, he attended briefly the military Medical and Surgery Academy, and then followed the course of chemistry at the Technological Institute of St Petersburg, where he studied D. I. Mendeleev's famous *Foundations of chemistry*. Fascinated by crystallography, he entered the Mining Institute (Institut Gorny), also in St Petersburg, in 1880, where he learnt geology and mineralogy, and from which he graduated in 1883 with honours. He was, however, not permitted to remain there, maybe because of the hostility of one of the professors, P. V. Eremeev, and went through difficult times financially, earning his living with small translation jobs. In 1885, he joined the staff of the Geological Committee and took part in geological expeditions to the Northern Urals during the summers, under rather strenuous conditions. In 1894 he was invited to Turjinsk in the Urals where he looked successfully for new ore deposits and founded the Mineral Museum. In 1895 he was appointed Professor of Geology at the Moscow St Peter's Agricultural Institute (now Timiriazev Academy). At the same time, he was travelling every week from Moscow to St Petersburg to give lectures at his old school, the Mining Institute. In 1905 he became its first elected director, for a period of three years, and moved to St Petersburg permanently. He founded there the *Transactions of the Mining Institute* (*Zapiski Gornogo Instituta*), and was visited by many scientists from abroad. The scientific council wanted to have his position of director renewed, but that was not accepted by the government, for fear of the development of revolutionary feelings among the students. In 1896, he was elected, on Groth's and Sohncke's recommendation, a corresponding member of the Bavarian Academy of Sciences. In 1901 he became an adjunct of the Imperial Academy of Sciences in St Petersburg and, shortly before his death in 1919, a member of the Russian Academy of Sciences.

MAIN PUBLICATIONS

1885 *The elements of the study of figures.* (in Russian)

1890 *Report on the progress of theoretical crystallography during the last decade.* (in Russian)

1891 *The symmetry of regular systems of figures.* (in Russian)

1893 *The theodolite method in mineralogy and petrology.* (in Russian)

1895 *Theorie der Realstruktur.*

1914 *First experimental demonstration of the asymmorphous real systems.*

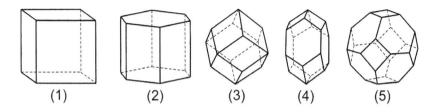

Fig. 12.51   Fedorov parallelohedra. (1): parallelepiped, (2): hexagonal prism, (3): rhomb-dodecahedron, (4): hexarhombic (Fedorov) dodecahedron, (5): cubooctahedron.

1885 publication: 'concerning the point systems, we have the following historic development: Curie (1884), Fedorov (1885), Klein (1887)', to which Fedorov replied on 30 September 1890[81] that his paper had in fact been submitted in 1883 and officially presented on 7 January 1884 to the Imperial Mineralogical Society in St Petersburg, so that he had the priority! In another letter, dated 21 October 1890,[82] Fedorov agrees with Schoenflies that 'there is nothing new in Minnigerode's 1887 paper'.

In the last part of his memoir, Fedorov derived the properties of the convex polygons and polyhedra that fill the plane and space in parallel orientation. In the plane, there are two types of such polygons: the rectangle and the hexagon. In space, Fedorov showed that there are five topological types of such polyhedra, called 'parallelohedra': the parallelepiped, the hexagonal prism, the rhomb-dodecahedron, the hexarhombic dodecahedron, which was discovered by Fedorov, and the truncated octahedron or cubooctahedron[83] (see Fig. 12.51). The publication earned him to be appointed a staff member of the Geological Committee in St Petersburg; he could continue his scientific work, in spite of his tiring field trips to the Northern Urals where he drew up geological maps of the region. It is also during that time that he developed a four-axis universal stage for the identification of feldspars. It was based on the principle of the theodolite, and allowed angles and optical properties of minerals to be measured accurately (Fedorov 1893). Fedorov's work on parallelohedra was pursued later by a study of their subdivision into congruent convex regions, 'stereohedra' (Fedorov 1899).

Meanwhile, Schoenflies (1883, 1886) had studied the geometrical properties of rigid-body motion in the plane and in space. From there, he moved on to the study of the groups of motion. He took up Sohncke's 1879 derivation of the proper groups and derived them again using the tools of group theory (Schoenflies 1887). He sorted them into cyclic groups ($\mathfrak{C}$), the Klein four-group ($\mathfrak{V}$), now denoted D_2^n, dihedral ($\mathfrak{D}$), tetrahedral ($\mathfrak{T}$) and octahedral groups ($\mathfrak{O}$), using notations which are nearly the same as those in use today.[84] *En passant*, he showed that two of the 66 groups found by Sohncke were in fact identical. For each group, he gave the correspondence with Jordan's and Sohncke's groups.

[81]Burkhardt, p. 94.   [82]Burkhardt, p. 97.

[83]Also described by Lord Kelvin, 1894*b*.

[84]*International Tables of Crystallography*. Vol. A. *Space-Group Symmetry*. (2005). Editor Theo Hahn. Dordrecht: Springer and IUCr.

This led Schoenflies to become interested in the partitioning of space with congruent domains (Schoenflies, 1888a) and to construct models of the corresponding polyhedra, which were analogous to Fedorov's. They were even sold commercially (*Modelle zur Darstellung von regulären Gebietstheilungen des Raumes*), for a price of 140 marks for the whole series (Kaemmel 2006). In a very important step, he next related the partitioning of space and the regular point systems, showing that to each partitioning of space into elemental crystalline building blocks (*Kristallbaustein*) corresponds a point system (Schoenflies 1888b).

Mirror planes (*Spiegelungen*) are introduced in these 1888 articles, but it is F. Klein (Schoenflies 1889) who brought to Schoenflies' attention that the groups of motion should be extended to groups with improper operations, *räumliche Operationen der zweiter Art*: rotations of angle ω, followed by a symmetry with respect to a mirror plane normal to the axis of rotation, $\mathfrak{S}(\omega)$ ('rotoinversions'), inversions, $\mathfrak{I} = \mathfrak{S}(\pi)$, and glide planes, $\mathfrak{S}(t)$. Schoenflies combined these symmetry elements with the 65 proper groups, following their classification in the 1887 paper, and obtained 162 improper groups, which makes up a total of 227 groups of motion (Schoenflies 1889). The paper was submitted in November 1888 and published in the second 1889 issue of *Mathematische Annalen*.

During this time, E. S. Fedorov was looking for new ore deposits in Turjinsk in the Urals, but was nevertheless continuing his scientific work on the tessellation of space (Fedorov 1887, 1889) and on the algebraic description of the symmetry of crystals. He read with interest Schoenflies's publications, in particular the paper dealing with the partitioning of space (Schoenflies 1888a), which, he noted, had obviously been written without knowledge of his own 1885 paper, probably because it was in Russian (Fedorov 1892). He then set out to complete the derivation of the groups of motion.

The exact chronology of the final stages of the derivation of 230 groups by E. S. Fedorov and Schoenflies is difficult to ascertain because there is some confusion in the sources. On 2 December 1889[85] Fedorov communicated to the St Petersburg Mineralogical Society a first version of his paper, *The symmetry of regular systems of figures*, in which 228 regular systems of figures were described. Between December 1889 and November 1890, Schoenflies and Fedorov exchanged several letters, comparing their views. In October 1890 Fedorov sent to Schoenflies a preprint of the communication he submitted on 22 November to the *Neues Jahrbuch für Mineralogie*, which is a short summary of his work with the number of groups of motion in each of the thirty-two point groups, adding up to 230 (Fedorov 1891a). It contains, however, some errors: one group too many in point group, *mm2*, and one missing in point group $\bar{4}3m$. On 25 November he sent to the St Petersburg Mineralogical Society a new version of his manuscript, with corrections at the end. It is not absolutely clear when he made the last correction including the missing group, 103a (T_d^6 in Schoenflies notation, $I\bar{4}3d$ in Hermann–Mauguin notation).[86] The same day (25 November), he wrote to Schoenflies[87] that 'though [the latter] said in his last two letters his happiness with the complete agreement of their results, it is not so'. Fedorov

---

[85]All dates are given in the Gregorian calendar.

[86]See the discussion in T. Kaemmel (2012). Arthur Schoenflies, promoter of mineralogy, one of the explorers of the 230 space groups and the later priority dispute. *European Mineralogical Conference*. **1**, EMC2012-323.

[87]Burckhardt 1971, p. 101.

corrected some errors of Schoenflies's and announced that he had found 230 systems, while Schoenflies had found only 227! On 19 March 1891, Fedorov sent belatedly to Schoenflies the final version of his manuscript, with group 103*a* and the corresponding algebraic formulæ.

Schoenflies completed his book, *Krystallsysteme und Krystallstruktur*, with his own derivation of the 230 groups, in the first half of 1891, writing the preface in August 1891. The book was published in the autumn. He sent a copy to Fedorov who, in a letter sent to Groth on 18 October 1891, acknowledged receipt of 'Schoenflies' new extensive book'.[88] Together with the letter, Fedorov sent Groth the manuscript in German of a paper submitted to *Zeitschrift für Kristallographie und Mineralogie* with a table of the 230 groups. He sent back the proofs on 22 December 1891, and the paper was published in 1892. The final version of Fedorov's work in Russian was published in December 1891 in the *Transactions of the Minereralogical Society of Saint Petersburg* (Fedorov 1891*b*). Immediately afterwards, Fedorov (1891*c*) also gave a derivation of the seventeen plane groups.

In Fedorov's (1892) paper in *Zeitschrift für Kristallographie und Mineralogie*, his results were compared with those of Schoenflies. It included a table giving the list of the 230 groups with their corresponding Schoenflies notation, and it expressed the complete identity of views between Schoenflies and himself with respect to the regular partitioning of space[89]. Schoenflies replied the same year, giving his own position (Schoenflies 1892). Fedorov came back on his results in later publications (Fedorov 1895, 1896).

The sequence of the space groups in Schoenflies' book (Schoenflies 1891) is different from that in his 1888 paper, and is given according to the crystal systems. Niggli (1919) considered Schoenflies' sequence to be unsatisfactory and suggested a more appropriate sequence.

Fedorov had found the 230 regular systems by looking for all the possible types of basic design which a crystal can present and a classification of the regular systems of figures according to their symmetry. These were defined as follows (Fedorov 1891): 'Under regular system of figures, I understand an assembly, unlimited in all directions, of finite figures which, if two of them are made congruent by the symmetry operations, the whole system is also made congruent' (quoted by E. Scholz in Burckhardt, 1988). The symmetry operations are rotations, screw axes, mirror planes, and glide planes. The resulting 230 systems (*Raumgruppen* = space groups) are sorted in three categories (Fedorov 1892):

1. *Symmorphic* systems in which, apart from the lattice translations, all generating symmetry operations leave one common point fixed; they are obtained by the combination of the Bravais lattices with symmetry elements with no translational components, e.g. Pm, $P2$, Cm, $P2/m$, $P422$, $P4mm$, $P32$, $P23$. They are in one-to-one correspondence with the arithmetic crystal classes. Symmorphic systems have the same symmetry as the unit cell. There are 73 symmorphic systems.
2. *Hemisymmorphic* systems in which two analogous symmorphic systems combine to form a double non symmorphic group: the equivalent points may be divided into two categories, 1) those which

---

[88]The letter is published in Otmar Faltheiner (1973). *Briefwechsel E. S. v. Fedorow St. Petersburg: P. H. v. Groth, München. Ein historisches Dokument zum Ende der klassischen Kristallographie.* Dissertation der Technischen Universität München. The author is grateful to T. Kaemmel, Schoenflies's grandson, for communicating the letter to him.

[89]It is interesting to note that Lord Kelvin (Sir William Thomson, 1824–1907), in England, was also interested in homogeneous partitioning of space, quite independently, and from a different viewpoint (Lord Kelvin 1887, 1894*a* and *b*).

are deduced from a given point by translations, rotations and screw rotations, and 2) those which are deduced from that point by operations of the second kind (inversions, glide reflections and rotoinversions). In other words, the hemisymmorphic systems only have the elements of simple rotation and symmetry in common with their unit cell. For instance, group 20h in Fedorov notation ($Pnnn$, Hermann–Mauguin notation, $\mathfrak{B}_h^2$ in original Schoenflies notation, D_{2h}^2 in modern Schoenflies notation) is hemimorphic, while group 18a in Fedorov notation ($Pmmm$, Hermann–Mauguin notation, $\mathfrak{B}_h^1$ in original Schoenflies notation, D_{2h}^1 in modern Schoenflies notation) is symmorphic. There are 54 hemisymmorphic systems.

3. *Asymmorphic* systems, all other groups. Fig. 12.52 represents amorphous group 31a in Fedorov notation, $P4_122$ in Hermann-Mauguin notation, D_4^3 in Schoenflies notation. There are 103 asymmorphic systems.

Fedorov had also derived the regular systems of points algebraically, and a special feature of his description was the analytical description of each analytical system: for each group a series of algebraic expressions was given, which enabled the coordinates of all the equivalent points to be determined.

Although Fedorov's work was also published in German, Schoenflies' version was better accepted by the contemporaries and eventually formed the basis of the International Tables of Crystallography. All commentators agree that Fedorov's was more refined but also more complicated for groups of higher order, and this is possibly the reason why Schoenflies's version was preferred.

Barlow (1894) knew the works of Fedorov and Schoenflies (he quoted them, pp. 38 and 41, and he had certainly been informed by Groth during his stay in Germany), but his derivation followed a different path. In order to find all possible regular systems of points, he looked for all possible regular systems of asymmetric patterns. Starting from Sohncke's 65 groups, he noted that some are enantiomorphous, namely not identical with their mirror-image. He then combined each enantiomorphous system with its image by a rotoinversion. There are usually

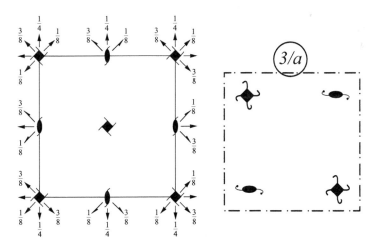

Fig. 12.52 Space group $P4_122$ (Hermann–Mauguin), D_4^3 (Schoenflies), 31a (Fedorov). *Left*: After International Tables of Crystallography, Vol. A. *Right*: After Fedorov (1895).

several methods of doing this, and he found 165 new systems, bringing the total to 230. As an aid to work out these additional systems, he used his talents as cabinet-maker, acquired during his youth, to fabricate models of space groups, with gloves to represent the objects repeated by the symmetry elements. These, which he bought in vast quantities, served to figure the repeated patterns of each group. There are still four such models, out of more than 200, in the Natural History Museum, London (Tandy 2004).

REFERENCES

Abbona, F. (2000). *Percezione e pensiero nel mondo dei cristalli.* pp. 97–168. In *Epistemología de las Ciencias. El conocimiento. Aproximación al orden ontológico.* CIAFIC Ediciones, Buenos Aires.

Abbona, F. (2002). Steensen, Niels (1638–1686). *Documentazione Interdisciplinare di Scienza e Fede. G. Tanzella-Nitti e A. Strumia, eds., Urbaniana University Press e Città Nuova.* **2**, 2009–2100.

Abbona, F. (2004). Niccolò Stenone, un modello di ricercatore. *Emmeciquadro.* **21**, 64–86.

Abrahams, S., Robertson, J. M., and White, J. G. (1949). The crystal and molecular structure of naphthalene. II. Structure investigation by the triple Fourier series method. *Acta Crystallogr.* **2**, 239–244.

Allen, F. H. (2002). The Cambridge Structural Database: a quarter of a million crystal structures and rising. *Acta Crystallogr.* **58B**, 380–388.

Allen, F. H. and Motherwell, W. D. S. (2002). Applications of the Cambridge Structural Database in organic chemistry and crystal chemistry. *Acta Crystallogr.* **58B**, 407–422.

Allen, H. S. (1902). A preliminary note on the relation between primary and secondary Röntgen radiation. *Phil. Mag.* Series 6. **3**, 126–128.

Allison, S. K. (1932). The reflecting and resolving power of calcite for X-rays. *Phys. Rev.* **41**, 1–20.

Aminoff, G. (1922). Über Lauephotogramme und Struktur von Zinkit. *Z. Kristallogr.* **57**, 495–505.

Aminoff, G. (1923). Untersuchungen über die Kristallstrukturen von Wurtzit und Rotnickelkies. *Z. Kristallogr.* **58**, 203–219.

Ampère, A.-M. (1814). Lettre de M. Ampère à M. le Comte Berthollet, sur la détermination des proportions dans lesquelles les corps se combinent, d'après le nombre et la disposition respective des molécules dont leurs particules intégrantes sont composées. *Ann. Chimie.* **90**, 43–86.

Andrade Da C., E. N. (1950). Wilkins Lecture: Robert Hooke. *Proc. Roy. Soc. B.* **137**, 153–187.

Andrews, M. R. (1921). X-ray analyses of three series of alloys. *Phys. Rev.* **17**, 261–261.

Anterroches d', C. (1984). High resolution T.E.M. study of the Si/SiO$_2$ interface. *J. Microscop. Spectrosc. Electron.* **9**, 147–162.

Arago, F. (1811). Mémoire sur une modification remarquable qu'éprouvent les rayons lumineux dans leur passage à travers certains corps diaphanes et sur quelques autres nouveaux phénomènes d'optique. *Mémoires de la classe des sciences mathématiques et physiques de l'Institut de France* **12**, 93–134.

Arago, F. (1816). Sur le phénomène remarquable qui s'observe dans la diffraction de la lumière. *Ann. Chim. Phys. Paris* **1**, 199–202.

Arago, F. and Fresnel, A. (1819). Sur l'action que les rayons de lumière polarisée exercent les uns sur les autres. *Ann. Chim. Phys. Paris* **10**, 288–305.

Armstrong, H. E. (1927). Poor common salt! *Nature,* **120**, 478–478.

Aroyo, M. I., Müller, U., and Wondratschek, H. (2004). Historical introduction. In *International Tables of Crystallography.* Vol. **A1**, pp. 1–5. Editors H. Wondratschek and U. Müller. Published by the International Union of Crystallography. Kluwer Academic Publishers, Dordrecht.

Astbury, W. T. (1923*a*). The crystalline structure and properties of tartaric acid. *Proc. Roy. Soc. A* **102**, 506–528.

Astbury, W. T. (1923*b*). The crystalline structure of anhydrous racemic acid. *Proc. Roy. Soc. A* **104**, 219–235.

Astbury, W. T. and Street, A. (1932). X-Ray studies of the structure of hair, wool, and related fibres. I. General. *Phil. Trans. Roy. Soc. A* **230**, 75–101.

Astbury, W. T. and Woods, H. J. (1934). X-Ray studies of the structure of hair, wool, and related fibres. II. The molecular structure and elastic properties of hair keratin. *Phil. Trans. Roy. Soc. A* **232**, 333–394.

Astbury, W. T. and Yardley, K. (1924). Tabulated data for the examination of the 230 space-groups by homogeneous X-rays. *Phil. Trans. Roy. Soc. A* **224**, 221–257.

Authier, A. (1960). Mise en évidence expérimentale de la double réfraction des rayons X. *C. R. Acad. Sci. Paris* **251**, 2003–2005.

Authier, A. (1961). Etude de la transmission anomale des rayons X dans des cristaux de silicium. *Bull. Soc. Fr. Miner. Crist.* **84**, 51–89.

Authier, A. (2003). *Dynamical theory of X-ray diffraction*. Oxford University Press, Oxford.

Authier, A. and Klapper, H. (2007). Gerhard Borrmann (1908–2006) and Gerhard Hildebrandt (1922–2005): their life stories and their contribution to the revival of dynamical theory in the 1950s. *Phys. Sta. Sol. (a)* **204**, 2515–2527.

Authier, A. (2009). 60 years of IUCr journals. *Acta Crystallogr.* **65**, 167–182.

Bäcklin, E. (1935). Absolute Wellenlängenbestimmung der Al Kalpha 1,2 Linie nach der Plangittermethode. *Zeit Physik*, **93**, 450–463.

Bain, E. C. (1921*a*). Studies of crystal structure with X-rays. *Chem. Met. Eng.* **25**, 729.

Bain, E. C. (1921*b*). What the X-ray tells us about the structure of solid solutions. *Chem. Met. Eng.* **25**, 657–664.

Bain, E. C. (1922). X-ray data on martensite formed spontaneously from austenite. *Chem. Met. Eng.* **26**, 543–545.

Bain, E. C. (1923*a*). The nature of solid solutions. *Chem. Met. Eng.* **28**, 21–24.

Bain, E. C. (1923*b*). Cored crystals and metallic compounds. *Chem. Met. Eng.* **28**, 65–69.

Bain, E. C. (1924). The nature of martensite. With discussion. *Trans. AIME*. **70**, 25–46.

Bain, E. C. and Jeffries, Z. (1921). Mixed orientation developed in crystals of ductile metals by plastic deformation. *Chem. Met. Eng.* **25**, 775–777.

Barkla, C. G. (1903). Secondary radiation from gases subject to X-rays. *Phil. Mag.* Series 6. **5**, 685–698.

Barkla, C. G. (1904). Energy of secondary Röntgen radiation. *Phil. Mag.* Series 6. **7**, 543–560.

Barkla, C. G. (1905*a*). Polarized Röntgen radiation. *Phil. Trans. Roy. Soc. A* **204**, 467–479.

Barkla, C. G. (1905*b*). Polarized Röntgen radiation. *Proc. Roy. Soc. A* **74**, 474–475.

Barkla, C. G. (1905*c*). Secondary Röntgen radiation. *Nature*. **71**, 440–440.

Barkla, C. G. (1906*a*). Polarisation in secondary Röntgen radiation. *Proc. Roy. Soc. A* **77**, 247–255.

Barkla, C. G. (1906*b*). Secondary Röntgen radiation. *Phil. Mag.* Series 6. **11**, 812–928.

Barkla, C. G. (1908). Note on X-rays and scattered X-rays. *Phil. Mag.* Series 6. **15**, 288–296.

Barkla, C. G. (1909). Phenomena of X-ray transmission. *Proc. Camb. Phil. Soc.* **15**, 257–268.

Barkla, C. G. (1911). The spectra of the fluorescent Röntgen radiations. *Phil. Mag.* Series 6. **22**, 739–760.

Barkla, C. G. (1916). Note on experiments to detect refraction of X-rays. *Phil. Mag.* Series 6. **31**, 257–260.

Barkla, C. G. and Ayres, T. (1911). The distribution of secondary X-rays and the electromagnetic pulse theory. *Phil. Mag.* Series 6. **21**, 270–278.

Barkla, C. G. and Martyn, G. H. (1912). Reflection of Röntgen radiation. *Nature* **90**, 435–435.

Barkla, C. G. and Martyn, G. H. (1913). An X-ray fringe system.*Nature* **90**, 647–647.

Barkla, C. G. and Sadler, C. A. (1907). Secondary X-rays and the atomic weight of nickel. *Phil. Mag.* Series 6. **16**, 550–584.

Barkla, C. G. and Sadler, C. A. (1908). Homogeneous secondary Röntgen radiations. *Phil. Mag.* Series 6. **1**, 408–422.

Barkla, C. G. and Sadler, C. A. (1909). The absorption of Röntgen rays. *Phil. Mag.* Series 6. **17**, 257–260.

Barlow, W. (1883). Probable nature of the internal symmetry of crystals. *Nature*, **29**, 186–188 and 205–207.

Barlow, W. (1884). Probable nature of the internal symmetry of crystals. *Nature*, **29**, 404.

Barlow, W. (1886). A theory of the connection between the crystal form and the atom composition of chemical compounds. *Chemical News* **53**, 3–6, 16–19.

Barlow, W. (1894). Über die geometrischen Eigenschaften homogener starrer strukturen und ihre Anwendung auf die Krystalle. *Z. Kristallogr.* **23**, 1–93.

Barlow, W. (1897). A mechanical cause of homogeneity of structures and symmetry geometrically investigated; with special application to crystals and to chemical combination. *Sci. Proceedings of the Royal Dublin Society* **8**, 527–690.

Barlow, W. and Miers, H. A. (1901). The structure of crystals. *Report of the Seventy-first Meeting of the British Association for the Advancement of Science*. London: John Murray. pp. 237–396.

Barlow, W. and Pope, W. J. (1906). A development of the atomic theory which correlates chemical and crystalline structure and leads to a demonstration of the nature of valency. *Journ. Chem. Soc.* **89**, 1675–1744.

Barlow, W. and Pope, W. J. (1907). The relation between the crystalline form and the chemical constitution of simple inorganic substances. *Journ. Chem. Soc.* **89**, 1150–1214.

Barlow, W. and Pope, W. J. (1908). On polymorphism with especial reference to sodium nitrate and calcium carbonate. *Journ. Chem. Soc.* **93**, 1528–1560.

Barlow, W. and Pope, W. J. (1910). The relation between the crystal structure and the chemical composition, constitution, and configuration of organic substances. *Journ. Chem. Soc.* **97**, 2308–2388.

Barlow, W. and Pope, W. J. (1914). The chemical significance of crystalline form. *J. Am. Chem. Soc.* **36**, 1675–1686, 1694–1695.

Bartholin, R. (1661). *De figura nivis dissertatio*. In *De nivis usu medico*. T. Bartholin. Hafniae, Copenhagen.

Bartholin, R. (1669). *Experimenta crystalli islandici disdiaclastici quibus mira & insolita refractio detegitur*. Hafni, Copenhagen.

Bassler, E. (1909). Polarisation der X-Strahlen, nachgewiesen mittels Sekundärstrahlung. *Ann. Phys.* **333**, 808–884.

Battelli, A. and Garbasso, A. (1897). Sur quelques faits se rapportant aux rayons de Röntgen. *C. R. Acad. Sci. Paris* **122**, 603.

Bearden, J. A. (1927). Measurements and Interpretation of the Intensity of X-Rays Reflected from Sodium Chloride and Aluminum. *Phys. Rev.* **29**, 20–33.

Bearden, J. A. (1931). The grating constant of calcite crystals. *Phys. Rev.* **38**, 2089–2098.

Bearden, J. A. (1932). Status of X-ray wavelengths. *Phys. Rev.* **41**, 399.

Bearden, J. A. (1967). X-ray wavelengths. *Rev. Mod. Phys.* **39**, 78–124.

Becker, K. (1927). Der röntgenographische Nachweis von Kornwachstum und Vergütung in Wolframdrähten mittels des Debye–Scherrer–Verfahrens. *Z. Phys.* **42**, 226–245.

Becker, K., Herzog, R. O., Jancke, W., and Polanyi, M. (1921). Über Methoden zur Ordnung von Kristallelementen. *Z. Phys.* **5**, 61–62.

Becquerel, H. (1896). Sur les radiations émises par phosphorescence. *C. R. Acad. Sci. Paris* **122**, 420–421.

Becquerel, H. (1903). Recherches sur une Propriété Nouvelle de la Matière. Activité radiante spontanée ou radioactivité de la matière. *Mém. Acad. Sci. Paris* **46**, 1–360.

Beevers, C. A. and Lipson, H. (1934). A rapid method for the summation of a two-dimensional Fourier synthesis. *Phil. Mag.* Series 7. **17**, 855–859.

Bergengren, J. (1920). Über die Röntgenabsorption des Phosphors. *Z. Phys.* **3**, 247–249.

Bergman, T. O. (1773). Variae crystallorum formae a spatho orthae. *Nova Acta Regiae Societatis Scientarum Upsaliensis*. **I**, 150–155.

Bergman, T. O. (1780). *De formis crystallorum, praesertim e spatho orthis*, in *Opuscula Physica et Chemica*. Vol. **II**, 1–25. Johan Edman, Uppsala.

Bernal, J. D. (1924). The structure of graphite. *Proc. Roy. Soc. A* **106**, 749–773.

Bernal, J. D. (1926). On the interpretation of X-ray, single crystal, rotation photographs. *Proc. Roy. Soc. A* **106**, 117–160.

Bernal, J. D. (1928). The complex structure of the copper-tin intermetalic compouds. *Nature*. **122**, 54.

Bernal, J. D. (1929). The problem of the metallic state. *Trans. Far. Soc.* **25**, 367–379.

Bernhardi, J. J. (1808). Darstellung einer neueun Methode, Kristalle zu beschreiben. *J. Chem. Phys. Mineral. Berlin* **5**, 157–198, 492–564, 625–654.

Bernhardi, J. J. (1809). Gedanken über die Krystallogenie und Anordnung der Mineralien, nebst einigen Beilagen über die Krystallisation verschiedener Substanzen. *J. Chem. Phys. Mineral. Berlin* **8**, 360–423.

Berzelius, J. J. (1813). Essay on the Cause of Chemical Proportions, and on Some Circumstances Relating to Them: Together with a Short and Easy Method of Expressing Them. *Ann. of Philosophy* **2**, 443–454.

Bethe, H. (1928). Theorie der Beugung von Elektronen an Kristallen. *Ann. Phys.* **87**, 55–129.

Bijvoet, J. M. (1949). Phase determination in direct Fourier synthesis of crystal structures. *Proc. Roy. Neth. Acad. (KNAW)* **52**, 313–314.

Bijvoet, J. M. (1954). Structure of optically active compounds in the solid state. *Nature* **173**, 888–891.

Bijvoet, J. M., Bernal, J. D., and Patterson, A. L. (1952). Forty years of X-ray diffraction. *Nature* **169**, 949–950.

Bijvoet, J. M., Burgers, W. G., and Häggs, G., Editors (1969). *Early papers on X-ray Diffraction of Crystals*. Published for the I.U.Cr. by A. Oosthoek's Uitgeversmaatschappij N. V., Utrecht.

Bijvoet, J. M., Peerdeman, A. F., and van Bommel, A. J. (1951). Determination of the absolute configuration of optically active compounds by means of X-rays. *Nature* **168**, 271–272.

Bijl, J. and Kolkmeijer, N. H. (1919). Investigation by means of X-rays of the crystal-structure of white and grey tin. *Proc. Roy. Neth. Acad. (KNAW)* **21**, 405–408 and 494–500.

Biot, J.-B. (1812). Sur un nouveau genre d'oscillations que les molécules de la lumière éprouvent en traversant certains cristaux. *Mémoires de la classe des sciences mathématiques et physiques de l'Institut de France* **13**, 1–371.

Biot, J.-B. (1818). Mémoire sur les lois générales de la double réfraction et de la polarisation dans les corps régulièrement cristallisés. *Mémoires de l'Académie Royale des Sciences* **13**, 1–371.

Blake, F. C. and Duane, W. (1917). The value of *h* as determined by means of X-rays. *Phys. Rev.* **10**, 624–637.

Bloch, F. (1928). Über die Quantenmechanik der Elektronen in Kristallgittern. *Z. Phys.* **52**, 555–600.

Blythswood, Lord (1897). On the reflection of Röntgen light from polished speculum-metal mirrors. *Proc. Roy. Soc.* **39**, 330–332.

Boece de Boot, A. (1609). *Gemmarum et lapidum Historia.* Claudius Marnius, Hanau. French translation: (1644). *Le parfaict Joaillier ou Histoire des Pierreries.* Antoine Huguetan, Lyon.

Bohlin, H. (1920). Eine neue Vorrichtung für röntgenkristallogr. Untersuchungen von Pulvern. *Ann. Phys.* **366**, 421–439.

Böhm, J. (1926). Das weissenbergsche Röntgengoniometer. *Z. Phys.* **39**, 557–561.

Bohr, N. (1913). On the constitution of atoms and molecules. *Phil. Mag.* Series 6. **26**, 1–25, 476–502, 857–875.

Bohr, N. (1915). On the quantum theory of radiation and the structure of the atom. *Phil. Mag.* Series 6. **30**, 394–415.

Bohr, N. (1961). The Rutherford Memorial Lecture 1958: Reminiscences of the founder of nuclear science and of some developments based on his work. *Proc. Phys. Soc.* **78**, 1083–1115.

Bonse, U. and Hart, M. (1965). Principles and design of Laue-case X-ray interferometers. *Z. Phys.* **188**, 154–164.

Born, M. (1915). *Dynamik der Kristallgitter.* B. G. Teubner, Leipzig.

Born, M. (1928). Sommerfeld als Begründer einer Schule. *Naturwiss.* **16**, 1035–1036.

Born, M. and Kàrmàn, T. (1912). Über Schwingunen in Raumgitter. *Phys. Zeit.* **13**, 297–309.

Borrmann, G. (1935). Röntgenlichtquelle im Einkristall. *Naturwiss.* **23**, 591–592.

Borrmann, G. (1936). Über die Interferenzen aus Gitterquellen bei Anregung durch Röntgenstrahlen. *Ann. Phys.* **419**, 669–693.

Borrmann, G. (1938). Über die Röntgeninterferenzen des selbstleuchtenden Eisens. *Z. Kristallogr.* **100**, 229–238.

Borrmann, G. (1941). Über Extinktionsdiagramme der Röntgenstrahlen von Quarz. *Phys. Zeit.* **42**, 157–162.

Borrmann, G. (1950). Die Absorption von Röntgenstrahlen in Fall der Interferenz. *Z. Phys.* **127**, 297–323.

Borrmann, G. (1954). Der kleinste Absorption Koeffizient interfierender Röntgenstrahlung. *Z. Kristallogr.* **106**, 109–121.

Borrmann, G. (1955). Vierfachbrechung der Röntgenstrahlen durch das ideale Kistallgitter. *Naturwiss.* **42**, 67–68.

Borrmann, G. (1959a). Max von Laue und das Fritz Haber Institut – 80 Geburtstag. *Phy. Bl.* **15**, 453–456.

Borrmann, G. (1959b). Röntgenwellenfelder. *Beitr. Phys. Chem. 20 Jahrhunderts.* Braunschweig: Vieweg & Sohn. pp. 262–282.

Borrmann, G., Hartwig, W., and Irmler, H. (1958). Schatten von Versetzungslinien im Röntgen-diagramm. *Z. Naturforsch.* **13a**, 423–425.

Borrmann, G., Hildebrandt, G. and Wagner, H. (1955). Röntgenstrahlfächer in Kalkspat. *Z. Phys.* **142**, 406–414.

Boscovich, R. (1758). *Theoria Philosophiae Naturalis.* Typographia Remondiniana, Venice.

Boulliard, J.-C. (2010). *Le cristal et ses doubles.* CNRS Editions, Paris.

Boyer, C. B. (1991). *A history of mathematics.* Second Edition, revised by U. C. Merzbach. Wiley and sons, New York.

Boyle, R. (1661). *The Sceptical Chemyst, or Physical Doubts and Paradoxes.* J. Crooke, London.

Boyle, R. (1664). *Experiments and considerations touching colours.* Henry Herringman, London.

Boyle, R. (1666). *The origine of forms and qualities.* Oxford.

Bradley, A. J. (1925). The allotropy of manganese. *Phil. Mag.* Series 6. **50**, 1018–1030.

Bradley, A. J. and Thewlis, J. (1926). The structure of γ-brass. *Proc. Roy. Soc. A.* **112**, 678–692.

Bradley, A. J. and Thewlis, J. (1927). The structure of α-manganese. *Proc. Roy. Soc. A.* **115**, 456–471.

Braga, D. (2003). Crystal engineering, where from, where to. *Chem. Comm.* 2751–2754.

Bragg, W. H. (1907). Properties and nature of various electrical radiations. *Phil. Mag.* Series 6. **14**, 429–449.

Bragg, W. H. (1910). Consequences of the corpuscular hypothesis of the γ- and X-rays and the range of β-rays. *Phil. Mag.* Series 6. **20**, 385–416.

Bragg, W. H. (1912a). X-rays and crystals. *Nature* **90**, 219–219.

Bragg, W. H. (1912b). X-rays and crystals. *Nature* **90**, 360–361.

Bragg, W. H. (1913a). X-rays and crystals. *Nature* **90**, 572.

Bragg, W. H. (1913b). The reflection of X-rays by crystals. II. *Proc. Roy. Soc. A* **89**, 246–248.

Bragg, W. H. (1913c). The reflection of X-rays by crystals. *Nature* **91**, 477.

Bragg, W. H. (1913d). Crystals and X-rays. *Report of the Eighty-third Meeting of the British Association for the Advancement of Science.* London: John Murray. pp. 386–387.

Bragg, W. H. (1913e). La réflexion des rayons X et le spectromètre à rayons X. In *Rapport du Conseil de Physique de l'Institut International Solvay,* Bruxelles, 27–31 Octobre 1913. (1921). Gauthier-Villars, Paris. pp. 113–120.

Bragg, W. H. (1914a). The influence of the constituents of the crystal on the form of the spectrum in the X-ray spectrometer. *Proc. Roy. Soc. A* **89**, 430–438.

Bragg, W. H. (1914b). The X-ray spectrometer. *Nature* **94**, 199–200.

Bragg, W. H. (1914c). The intensity of reflection of X-rays by crystals. *Phil. Mag.* Series 6. **27**, 881–899.

Bragg, W. H. (1914d). The X-ray spectra of crystals given by sulphur and quartz. *Proc. Roy. Soc. A* **89**, 575–580.

Bragg, W. H. (1915a). The structure of magnetite and spinels. *Nature* **95**, 561.

Bragg, W. H. (1915b). The structure of the spinel group of crystals. *Phil. Mag.* Series 6. **30**, 305–315.

Bragg, W. H. (1915c). X-rays and crystal structure (Bakerian lecture). *Phil. Trans. Roy. Soc.* **215**, 253–274.

Bragg, W. H. (1915d). The relation between certain X-ray wave-lengths and their absorption coefficients. *Phil. Mag.* Series 6. **29**, 407–412.

Bragg, W. H. (1921a). Application of the Ionisation Spectrometer to the Determination of the Structure of Minute Crystals. *Proc. Phys. Soc.* **33**, 222–224.

Bragg, W. H. (1921b). The intensity of X-ray reflection by diamond. *Proc. Phys. Soc.* **33**, 304–311.

Bragg, W. H. (1921c). The structure of organic crystals. *Proc. Phys. Soc.* **34**, 33–50.

Bragg, W. H. (1922a). The significance of crystal structure. *J. Chem. Soc.* **121**, 2766–2787.

Bragg, W. H. (1922b). The crystal structure of ice. *Proc. Phys. Soc.* **34**, 98–103.

Bragg, W. H. (1929). Organic compounds. *Trans. Faraday Soc.* **25**, 346–347.

Bragg, W. H. and Bragg, W. L. (1913a). The reflection of X-rays by crystals. *Proc. Roy. Soc. A* **88**, 428–438.

Bragg, W. H. and Bragg, W. L. (1913b). The structure of diamond. *Nature* **91**, 557.

Bragg, W. H. and Bragg, W. L. (1913c). The structure of diamond. *Proc. Roy. Soc. A* **89**, 277–291.

Bragg, W. H. and Bragg, W. L. (1915). *X-rays and crystal structure*. G. Bell and Sons, Ltd, London.

Bragg, W. H. and Bragg, W. L. (1937). The discovery of X-ray diffraction. In *Laue Diagrams, 25 Years of research on X-ray diffraction following Prof. Max von Laue discovery*, pp. 9–10. *Current Science*. Indian Academy of Sciences, Bengalore.

Bragg, W. H. and Gibbs, R. E. (1925). The Structure of α- and β-Quartz. *Proc. Roy. Soc. A* **109**, 405–427.

Bragg, W. H. and Madsen, J. P. V. (1908a). An experimental investigation of the nature of the γ-rays. *Phil. Mag.* Series 6. **14**, 663–675.

Bragg, W. H. and Madsen, J. P. V. (1908b). An experimental investigation of the nature of the γ-rays. – No 2. *Phil. Mag.* Series 6. **16**, 918–939.

Bragg, W. L. (1912). The specular reflection of X-rays. *Nature* **90**, 410.

Bragg, W. L. (1913a). The Diffraction of Short Electromagnetic Waves by a Crystal. *Proc. Cambridge Phil. Soc.* **17**, 43–57.

Bragg, W. L. (1913b). The Structure of Some Crystals as indicated by their Diffraction of X-rays. *Proc. Roy. Soc. A* **89**, 248–277.

Bragg, W. L. (1914a). The Analysis of Crystals by the X-ray Spectrometer. *Proc. Roy. Soc. A* **89**, 468–489.

Bragg, W. L. (1914b). The crystalline structure of copper. *Phil. Mag.* Series 6. **28**, 355–360.

Bragg, W. L. (1914c). Eine Bemerkung über die Interferenzfiguren hemiedrischer Kristalle. *Phys. Z.* **15**, 77–79.

Bragg, W. L. (1920a). The crystalline structure of zinc oxide. *Phil. Mag.* Series 6. **39**, 647–651.

Bragg, W. L. (1920b). The arrangement of atoms in crystals. *Phil. Mag.* Series 6. **40**, 169–189.

Bragg, W. L. (1924). The Structure of Aragonite. *Proc. Roy. Soc. A.* **105**, 16–39.

Bragg, W. L. (1925a). The Crystalline Structure of Inorganic Salts. *Nature* **116**, 249–251.

Bragg, W. L. (1925b). The interpretation of intensity measurements in X-ray analysis of crystal structure. *Phil. Mag.* Series 6. **50**, 306–310.

Bragg, W. L. (1926). Interatomic distances in crystals. *Phil. Mag.* Series 7. **2**, 258–266.

Bragg, W. L. (1929a). The determination of parameters in crystal structures by means of Fourier series. *Proc. Roy. Soc. A* **123**, 537–559.

Bragg, W. L. (1929b). An optical method of representing the results of X-ray analysis. *Z. Kristallogr.* **70**, 475–492.

Bragg, W. L. (1929c). Atomic arrangement in the silicates. *Trans. Faraday Soc.* **25**, 291–314.

Bragg, W. L. (1930). The structure of silicates. *Z. Kristallogr.* **74**, 237–304.

Bragg, W. L. (1937). *Atomic structure of minerals*. Cornell University Press, Ithaca, N.Y.

Bragg, W. L. (1943). *The history of X-ray analysis*. Longmans Green and Company, London.

Bragg, W. L. (1948). Recent advances in the study of the crystalline state. *Science* **108**, 455–463.

Bragg, W. L. (1949). Acceptance of the Roebling Medal of the Mineralogical Society of America. *Am. Miner.* **34**, 238–241.

Bragg, W. L. (1952). Forty years of X-ray diffraction. A note by Sir Lawrence Bragg. *Nature* **169**, 950–951.

Bragg, W. L. (1959). The diffraction of Röntgen rays by crystals. *Beitr. Phys. Chem. 20 Jahrhunderts*. Braunschweig: Vieweg & Sohn. pp. 147–151.

Bragg, W. L. (1961). The Rutherford Memorial Lecture, 1960. The development of X-ray analysis. *Proc. Roy. Soc. A* **262**, 146–158.

Bragg, W. L. (1967). *The start of X-ray analysis*. Longmans Green and Co, London.

Bragg, W. L. (1968a). Professor P.P. Ewald. *Acta Crystallogr.* **A24**, 4–5.

Bragg, W. L. (1968b). X-ray Crystallography. *Scientific American*, **219**, 58–70.

Bragg, W. L. (1970). Early days. *Acta Crystallogr.* **A26**, 171–172.

Bragg, W. L. (1975). *The development of X-ray analysis*. Bell & Sons, London.

Bragg, W. L. and Brown, G. B. (1926a). The crystalline structure of chrysoberyl. *Proc. Roy. Soc. A* **110**, 34–63.

Bragg, W. L. and Brown, G. B. (1926b). Die Struktur des Olivins. *Z. Kristallogr.* **63**, 538–556.

Bragg, W. L. and Caroe, G. M. (1962). Sir William Bragg, F. R. S. *Notes and Records. Roy. Soc.* **17**, 169–182.

Bragg, W. L. and Nye, F. F. (1947). A dynamic model of crystal structures. *Proc. Roy. Soc. A* **190**, 474–481.

Bragg, W. L. and West, J. (1926). The structure of beryl, $Be_3Al_2Si_6O_{18}$. *Proc. Roy. Soc. A* **111**, 691–714.

Bragg, W. L. and West, J. (1927). The structure of certain silicates. *Proc. Roy. Soc. A* **114**, 450–473.

Bragg, W. L. and West, J. (1929). A technique for the X-ray examination of crystal structures with many parameters. *Z. Kristallogr.* **69**, 118–148.

Bragg, W. L. and West, J. (1930). A note on the representation of crystal structure by Fourier series. *Phil. Mag.* Series 7. **10**, 823–841.

Bragg, W. L. and Williams, E. J. (1934). The effect of thermal agitation on atomic arrangement in alloys. *Proc. Roy. Soc. A* **145**, 699–730.

Bragg, W. L., Darwin, C. G., and James, R. W. (1926). The intensity of reflexion of X-rays by crystals. *Phil. Mag.* Series 7. **1**, 897–922.

Bragg, W. L., James, R. W., and Bosanquet, C. H. (1921a). The intensity of reflexion of X-rays by rock-salt. *Phil. Mag.* Series 6. **41**, 309–337.

Bragg, W. L., James, R. W., and Bosanquet, C. H. (1921b). The intensity of reflexion of X-rays by rock-salt. II. *Phil. Mag.* Series 6. **42**, 1–17.

Bragg, W. L., James, R. W., and Bosanquet, C. H. (1922a). Über die Streuung der Röntgenstrahlen durch die Atome eines Kristalles. *Zeit. Phys.* **8**, 77–84.

Bragg, W. L., James, R. W., and Bosanquet, C. H. (1922b). The distribution of electrons around the nucleus in the sodium and chlorine atoms. *Phil. Mag.* Series 6. **44**, 433–449.

Bravais, A. (1848). Sur les propriétés géométriques des assemblages de points distribués régulièrement dans l'espace. *C. R. Acad. Sci. Paris* **27**, 601–604.

Bravais, A. (1849). Mémoire sur les polyèdres de forme symétrique. *J. Math. Pures Appl.* **14**, 137–180.

Bravais, A. (1850). Mémoires sur les systèmes de points distribués régulièrement sur un plan ou dans l'espace. *Journ. de l'Ecole Polytechnique*. **19**, 1–128.

Bravais, A. (1851). Etudes Cristallographiques. *Journ. de l'Ecole Polytechnique*. **20**, 101–276.

Brentano, J. C. M. (1919). Sur un dispositif pour l'analyse spectrographique des substances à l'état de particules désordonnées par les rayons Röntgen. *Arch. Sci. Phys. et Nat. Genève* **1**, 550–552.

Brentano, J. C. M. (1924). Focussing method of crystal powder analysis by X-rays. *Proc. Phys. Soc. A* **37**, 184–193.

Brewster, D. (1818). On the laws of polarisation and double refraction in regularly crystallized bodies. *Phil. Trans. Roy. Soc.* **108**, 199–273.

Brewster, D. (1819). On the laws which regulate the absorption of polarised light by doubly refracting crystals. *Phil. Trans. Roy. Soc.* **109**, 11–28.

Brewster, D. (1821). On the connection between the primitive forms of crystals and the number of their axes of double refraction. *Memoirs of the Wernerian Natural History Society* **3**, 50–74, 337–350.

Brewster, D. (1830). On the Production of Regular Double Refraction in the Molecules of Bodies by Simple Pressure; with Observations on the Origin of the Doubly Refracting Structure. *Phil. Trans. Roy. Soc.* **120**, 87–95.

Brewster, D. (1855). *Memoirs of the Life, Writings, and Discoveries of Sir Isaac Newton*. Edmonton and Douglas, Edinburgh.

Brill, R. (1928). Teilchengrössenbestimmungen mit Hilfe von Röntgenstrahlen. *Z. Kristallogr.* **68**, 398–405.

Brill, R. (1964). Technische Anwendung der Röntgenanalyse: Organische Chemie. *Z. Kristallogr.* **120**, 118–124.

Brill, R., Grimm, H. G., Herman, C., and Peters, C. (1939). Anwendung der röntgenographischen Fourieranalyse auf Fragen der chemischen Bindung. *Ann. Phys.* **426**, 393–445.

Brillouin, M. (1913). Sur la structure des cristaux et l'anisotropie des molécules. Dimorphisme du carbonate de calcium. In *Rapport du Conseil de Physique de l'Institut International Solvay*, Bruxelles, 27–31 Octobre 1913. (1921). Gauthier-Villars, Paris. pp. 185–217.

Brockway, L. O. and Robertson, J. M. (1939). The crystal structure of hexamethylbenzene and the length of the methyl group bond to aromatic carbon atoms. *J. Chem. Soc.* 1324–1332.

Broglie, M. de (1913*a*). Sur les images multiples que présentent les rayons de Röntgen après avoir traversé des cristaux. *C. R. Acad. Sci. Paris* **156**, 1011–1012.

Broglie, M. de (1913*b*). Reflection of X-rays and X-ray fringes. *Nature* **91**, 161–162.

Broglie, M. de (1913*c*). Sur un nouveau procédé permettant d'obtenir la photographie des spectres de raies des rayons Röntgen. *C. R. Acad. Sci. Paris* **157**, 924–926.

Broglie, M. de (1913*d*). The reflection of X-rays. *Nature* **92**, 423.

Broglie, M. de (1913*e*). Enregistrement photographique continu des spectres des rayons de Röntgen; spectre du tungsten. Influence de l'agitation thermique. *C. R. Acad. Sci. Paris* **157**, 1413–1416.

Broglie, M. de (1914*a*). Sur la spectroscopie des rayons Röntgen. *C. R. Acad. Sci. Paris* **158**, 177–180.

Broglie, M. de (1914*b*). Sur l'obtention des spectres des rayons de Röötgen par simple passage des rayons incidents au travers de feuilles minces. *C. R. Acad. Sci. Paris* **158**, 333–334.

Broglie, M. de (1914*c*). Sur la spectroscopie des rayons secondaires émis hors des tubes à rayons de Röntgen, et des spectres d'absorption. *C. R. Acad. Sci. Paris* **158**, 1493–1495.

Broglie, M. de (1914*d*). Sur l'analyse spectrale directe des rayons secondaires des rayons de Röntgen. *C. R. Acad. Sci. Paris* **158**, 1785–1788.

Broglie, M. de (1914*e*). La spectrographie des rayons de Röntgen. *J. Phys. Theor. Appl.* **4**, 101–116.

Broglie, M. de (1916*a*). La spectrographie des phénomènes d'absorption des rayons X. *J. Phys. Theor. Appl.* **6**, 161–168.

Broglie, M. de (1916*b*). Sur un sytème de bandes d'absorption *L* des spectres de rayons X des éléments, et sur l'importance des phénomènes d'absorption sélective en radiographie. *C. R. Acad. Sci. Paris* **163**, 352–355.

Broglie, M. de and Friedel, G. (1923). La diffraction des rayons X par les corps smectiques. *C. R. Acad. Sci. Paris* **176**, 738–740.

Broglie, M. de and Lindemann, F. A. (1913). Sur les phénomènes optiques présentés par les rayons de Röntgen rencontrant des milieux cristallins. *C. R. Acad. Sci. Paris* **156**, 1461–1463.

Broglie, M. de and Lindemann, F. A. (1914). Observation fluoroscopique par vision directe des spectres des rayons Röntgen. *C. R. Acad. Sci. Paris* **158**, 180–181.

Buchwald, J. Z. (1980). Experimental Investigations of Double Refraction from Huygens to Malus. *Arch. Hist. Exact Sci.* **21**, 311–373.

Buckley, H. E. (1934). On the mosaic structure in crystals. *Z. Kristallogr.* **89**, 221–241.

Buerger, M. J. (1934). The lineage structure of crystals. *Z. Kristallogr.* **89**, 195–220.

Buerger, M. J. (1942). *X-Ray Crystallography*. John Wiley, New York.

Buerger, M. J. (1964). The development of methods and instrumentation for crystal-structure analysis. *Z. Kristallogr.* **120**, 3–18.

Burckhardt, J. J. (1967/1968). Zur Geschichte der Entdeckung der 230 Raumgruppen. *Arch. Hist. Exact Sci.* **4**, 235–246.

Burckhardt, J. J. (1971). Der Briefwechsel von E. S. von Fedorow und A. Schoenflies, 1889–1908. *Arch. Hist. Exact Sci.* **7**, 91–141.

Burckhardt, J. J. (1984). Die Entdeckung der 32 Kristallklassen durch M. L. Frankenheim im Jahre 1826. *N. Jb. Miner. Mh.* **11**, 481–482.

Burckhardt, J. J. (1988). *Die Symmetrie der Kristalle: Von René-Just Haüy zur kristallogrpahischen Schule in Zürich*. Birkhäuser-Verlag, Basel.

Burdick, C. L. and Ellis, J. H. (1917). The crystal structure of chalcopyrite determined by X-rays. *J. Amer. Chem. Soc.* **39**, 2518–2525.

Burdick, C. L. and Owen, E. A. (1918). The atomic structure of carborundum determined by X-rays. *J. Amer. Chem. Soc.* **40**, 1749–1759.

Burke, J. G. (1966). *Origins of the science of crystals*. University of California Press, Berkeley.

Burke, J. G. (1971). *The work and influence of Nicolaus Steno in Crystallography*. pp. 163–174. In *Dissertations on Steno as Geologist*, edited by G. Scherz, Acta Historica Scientiarum Naturalium Et Medicinalium, Edidit Bibliotheca Hauniensis, Vol. **23**. Odense University Press, Copenhagen.

Cappeller, M. A. (1723). *Prodromus crystallographiae de crystallis improprie sic dictis commentarium.* H. R. Wyssing, Lucerne.

Carangeot, A. (1783). Goniomètre, ou mesure-angle. *Observations sur la physique, sur l'histoire naturelle et sur les arts.* **XXII**, 193–197.

Caspar, M. (1993). *Kepler,* translated and edited by C. Doris Hellman, with a new introduction by Owen Gingerich. Dover, New York.

Cauchois, Y. (1964). La spectroscopie X. *Z. Kristallogr.* **120**, 182–211.

Chapman, A. (2005). *England's Leonardo, Robert Hooke, and the Seventeenth-Century Scientific Revolution.* Institute of Physics Publishing Ltd.

Chapman, A. (2009). A new perceived reality: Thomas Harriot's Moon maps. *Astronomy and Geophysics,* **50**, 127–133.

Clark, G. L. and Duane, W. (1923). The reflection by a crystal of X-Rays characteristic of chemical elements in it. *Proc. Nat. Acad. Sci. (Washington)* **9**, 126–130.

Collins, E. H. (1924). Effect of temperature on the regular reflection of X-rays from aluminium foil. *Phys. Rev.* **24**, 152–157.

Compton, A. H. (1915). The distribution of the electrons in atoms. *Nature* **95**, 343–344.

Compton, A. H. (1917a). The intensity of X-ray reflection, and the distribution of the electrons in atoms. *Phys. Rev.* **9**, 29–57.

Compton, A. H. (1917b). The reflection coefficient of monochromatic X-rays from rock-salt and calcite. *Phys. Rev.* **10**, 95–96.

Compton, A. H. (1921). The degradation of γ-ray energy. *Phil. Mag.* **41**, 479–769.

Compton, A. H. (1923a). The total reflexion of X-rays. *Phil. Mag.* Series 6. **45**, 1121–1131.

Compton, A. H. (1923b). A quantum theory of the scattering of X-rays by light elements. *Phys. Rev.* **21**, 483–502.

Compton, A. H. (1926). *X-rays and electrons.* D. Van Nostrand Company, New York.

Compton, A. H. (1931). A precision X-ray spectrometer and the wavelength of $MoK\alpha_1$. *Rev. Sci. Instrum.* **2**, 365–376.

Compton, A. H. and Doan, R. L. (1925). X-Ray spectra from a ruled reflection grating. *Proc. Nat. Acad. Sci. USA* **11**, 598–601.

Coppens, P. (1992). Electron density from X-ray diffraction. *Ann. Rev. Phys. Chem.* **43**, 663–692.

Coster, D. (1920). On the rings of connecting-electrons in Bragg's model of the diamond crystal. *Proc. Roy. Neth. Acad. (KNAW)* **22**, 536–541.

Coster, D., Knol, K. S. and Prins, J. A. (1930). Unterschiede in der Intensität der Röntgenstrahlen-reflexion an den beiden 111-Flächen der Zinkblende. *Z. Phys.* **63**, 345–369.

Cox, E. G. (1928). The crystalline structure of benzene. *Nature* **122**, 401–401.

Crookes, W. (1879). Radiant matter. *Nature* **25**, 419–423.

Crowther, J. A. (1912). On the distribution of the scattered Röntgen radiation. *Proc. Roy. Soc. A.* **86**, 478–494.

Cruickshank, D. W. J. (1999). Aspects of the history of the International Union of Crystallography. *Acta Crystallogr.* **54**, 687–696.

Cruickshank, D. W. J., Juretschke, H. J., and Kato, N., Editors (1992). *P.P. Ewald and his dynamical theory of X-ray diffraction.* Oxford University Press, Oxford.

Curie, P. (1884). Sur la symétrie. *Bull. Soc. Fr. Miner.* **7**, 418–457.

Curie, P. (1894). Sur la symétrie dans les phénomènes physiques, symétrie d'un champ électrique et d'un champ magnétique. *J. Phys. Theor. Appl.* **3**, 393–415.

Curie, P. and Curie, J. (1880). Développement, par pression, de l'électricité polaire dans les cristaux hémièdres à faces inclinées. *C. R. Acad. Sci. Paris* **91**, 294–295.

Curie, P. and Sagnac, G. (1900). Electrisation négative des rayons secondaires produits au moyen des rayons Röntgen. *C. R. Acad. Sci. Paris* **130**, 1013–1016.

Cuvelier, P. (1977). Les *Experimenta crystalli Islandici disdiaclastici* d'Erasme Bartholin. *Rev. Hist. Sci.* **30**, 193–224.

Dana, J. D. (1836). On the formation of compound or twin crystals. *Am. J. Science and Arts* **30**, 275–300.

Daniell, J. F. (1817). On some phenomena attending the process of solution, and their application to the laws of crystallization. *The Journal of the Science and the Arts* **I**, 24–49.

Darwin, C. G. (1914a). The theory of X-ray reflection. *Phil. Mag.* Series 6. **27**, 315–333.

Darwin, C. G. (1914b). The theory of X-ray reflection. Part II. *Phil. Mag.* Series 6. **27**, 675–690.

Darwin, C. G. (1922). The reflection of X-rays from imperfect crystals. *Phil. Mag.* Series 6. **46**, 800–829.

Davis, B. (1917). Wavelength energy distribution in the continuous X-ray spectrum. *Phys. Rev.* **9**, 64–77.

Davis, B. and Purks, H. (1929). Unusual reflecting power of a pair of calcite crystals. *Phys. Rev.* **34**, 181–184.

Davis, B. and Slack, C. M. (1925). Refraction of X-Rays by a prism of copper. *Phys. Rev.* **25**, 881–882.

Davis, B. and Slack, C. M. (1926). Measurement of the refraction of X-rays in a prism by means of the double X-Ray spectrometer. *Phys. Rev.* **27**, 18–22.

Davis, B. and Stempel, W. M. (1921). An experimental study of the reflection of X-rays from calcite. *Phys. Rev.* **17**, 608–623.

Davis, B. and Terrill, H. M. (1922). The refraction of X-rays in calcite. *Proc. Nat. Acad. Sci. USA* **8**, 357–361.

Davisson, C. J. and Germer, L. H. (1927*a*). The scattering of electrons by a single crystal of nickel. *Nature* **119**, 558–560.

Davisson, C. J. and Germer, L. H. (1927*b*). Diffraction of electrons by a crystal of nickel. *Phys. Rev.* **30**, 705–740.

Debye, P. (1912). Zur Theorie der spezifischen Wärmen. *Ann. Phys.* **344**, 789–839.

Debye, P. (1913*a*). Über den Einfluss der Wärmebewegung auf die Interferenzerscheinungen bei Röntgenstrahlen. *Verh. Dtsch. Phys. Gesell.* **15**, 678–689.

Debye, P. (1913*b*). Über die Intensitätsvereilung in den mit Röntgenstrahlen erzeugen Interferenzbilder. *Verh. Dtsch. Phys. Gesell.* **15**, 738–752.

Debye, P. (1913*c*). Spektrale Zerlegung der Röntgenstrahlung mittels reflexion und Wärmebewegung. *Verh. Dtsch. Phys. Gesell.* **15**, 857–875.

Debye, P. (1913*d*). Interferenz von Röntgenstrahlen und Wärmebewegung. *Ann. Phys.* **348**, 49–95.

Debye, P. (1915). Zerstreuung von Röntgenstrahlen. *Ann. Phys.* **351**, 809–823.

Debye, P. (1923). Zerstreuung von Röntgenstrahlen und Quantumtheorie. *Phys. Zeit.* **24**, 161–166.

Debye, P. and Scherrer, P. (1916*a*). Interferenzen an regellos orientierten Teilchen im Röntgenlicht. *Nachrichten Kgl. Gesell. Wiss. Göttingen*, I. 1–15; II. 16–26.

Debye, P. and Scherrer, P. (1916*b*). Interferenzen an regellos orientierten Teilchen im Röntgenlicht. I. *Phys. Zeit.* **17**, 277–283.

Debye, P. and Scherrer, P. (1917). Interferenzen an regellos orientierten Teilchen im Röntgenlicht. III. *Phys. Zeit.* **18**, 291–301.

Debye, P. and Scherrer, P. (1918). Atombau. *Phys. Zeit.* **19**, 474–483.

Dehlinger, U. (1927). Über die Verbreiterung der Debye-linien bei Kaltbearbeiteten Metallen. *Z. Kristallogr.* **65**, 615–631.

Delafosse, G. (1825). Observations sur la méthode générale du Rév. Whewell pour calculer les angles des cristaux. *Annales des Sciences Naturelles* **6**, 121–126.

Delafosse, G (1836). *Précis Elémentaire d'Histoire Naturelle.* 3rd edition, Librairie classique de L. Hachette, Paris.

Delafosse, G. (1840). Recherches relatives à la cristallisation, considérée sous les rapports physiques et mathématiques; 1ère Partie: sur la structure des cristaux, et sur les phénomènes physiques qui en dépendent. *C. R. Acad. Sci. Paris* **11**, 394–400.

Delafosse, G. (1843). Recherches sur la cristallisation, considérée sous les rapports physiques et mathématiques. *Mémoires Acad. Sci. Paris* **8**, 641–690.

Delafosse, G. (1852). Mémoire sur une relation importante qui se manifeste, en certains cas, entre la composition atomique et la forme cristalline, et sur une nouvelle appréciation du rôle que joue la silice dans les combinaisons minérales. *Mémoires Acad. Sci. Paris* **13**, 542–579.

Delafosse, G. (1867). *Rapport sur les progrès de la minéralogie.* Imprimerie Impériale, Paris.

Démeste, J. (1779). *Lettres du Docteur Démeste au Docteur Bernnard sur la chimie, la docimasie, la cristallographie, etc.* Didot, Ruault et Clousier, Paris.

Descartes, R. (1637). *Discours de la Méthode plus La Dioptrique et les Météores.* Imprimerie de Ian Maire; Leyde.

Desiraju, G. (1989). *Crystal Engineering. The design of organic solids.* Elsevier Science, Amsterdam.

Desiraju, G. (2010). Crystal engineering: a brief overview. *J. Chem. Sci.* **122**, 667–675.

Dickinson, R. G. (1920). The crystal structure of wulfenite and scheelite. *J. Amer. Chem. Soc.* **42**, 85–93.

Dickinson, R. G. (1923). The crystal structure of tin tetra-iodide. *J. Amer. Chem. Soc.* **45**, 958–962.

Dickinson, R. G. and Raymond, A. L. (1923). The crystal structure of hexamethylene teytramine. *J. Amer. Chem. Soc.* **45**, 22–29.

Dijksterhuis, F. J. (2004). *Lenses and Waves. Christiaan Huygens and the Mathematical Science of Optics in the Seventeenth Century.* Archimedes. New Studies in the History and Philosophy of Science and Technology. Vol. 9. Editor: J. Z. Buchwald. Kluwer Academic Publishers, Dordrecht.

Dirichlet, L. G. (1850). Über die Reduction der positiven quadratischen Formen mit drei unbestimmten ganzen Zahlen. *J. reine angew. Math.* **40**, 209–227.

Dolomieu, D. (1801). *Sur la philosophie minéralogique, et sur l'espèce minéralogique.* Bossange, Masson et Bersson, Paris.

Donnay, J. D. H. and Harker, D. (1937). A new law of crystal morphology extending the law of Bravais. *Am. Miner.* **22**, 446–447.

Dorn, E. (1900). Versuche über Sekundärstrahlen und Radiumstrahlen. *Abhandl. der Naturforsch. Gesellschaft zu Halle.* **22**, 39–43.

Drenth, J. and Looyenga-Vos, A. (1995). Ralph W. G. Wyckoff 1897–1994. *Aeta Crystallogr.* **51**, 649–650.

Duane, W. (1925). The calculation of the X-Ray diffracting power at points in a crystal. *Proc. Nat. Acad. Sci. (Washington)* **11**, 489–493.

Duane, W. and Hunt, F. L. (1915). On X-ray wave-lengths. *Phys. Rev.* **6**, 176–171.

Duane, W. and Patterson, R. A. (1920). On the X-ray spectra of tungsten. *Phys. Rev.* **16**, 526–539.

Duane, W., Palmer, W. H., and Yeh, C.-S. (1921). A remeasurement of the radiation constant, *h*, by means of X-rays. *Phys. Rev.* **18**, 98–99.

Eckert, M. (1999). Mathematics, Experiments, and Theoretical Physics: The Early Days of the Sommerfeld School. *Phys. Persp.* **1**, 238–253.

Eckert, M. (2012*a*). Disputed discovery: the beginnings of X-ray diffraction in crystals in 1912 and its repercussions. *Acta Crystallogr.* **A68**, 30–39.

Eckert, M. (2012*b*). "Max von Laue and the discovery of X-ray diffraction in 1912. *Ann. Phys.* **524**, A83–A85.

Ehrenberg, W. and Mark, H. (1927). Über die natürliche Breite der Röntgenemissionslinien. I. *Z. Phys.* **42**, 807–822.

Ehrenberg, W. and Süsich G. von (1927). Über die natürliche Breite der Röntgenemissionslinien. II. *Z. Phys.* **42**, 823–831.

Ehrenberg, W., Ewald, P. P., and Mark, H. (1927). Untersuchungen zur Kristalloptik der Röntgenstrahlen. *Z. Kristallogr.* **66**, 547–584.

Einstein, A. (1905). Über einen die Erzeugung und Verwandlung des Lichtes betreffenden heuristischen Gesichtspunkt. *Ann. Phys.* **322**, 132–148.

Einstein, A. (1911). Elementare Betrachtungen über die thermische Molekularbewegung in festen Körpern. *Ann. Phys.* **340**, 679–694.

Ellenberger, F. (1988). *Histoire de la géologie.* Chapter IV: *Le dix-septième siècle.* Technique et Documentation. Lavoisier, Paris.

Engel, P. (1986). *Geometric crystallography.* Kluwer Academic Publishers, Dordrecht.

'Espinasse, M. (1956). *Robert Hooke.* William Heinemann Ltd., London.

Ewald, P. P. (1912). *Dispersion und Doppelbrechung von Elektronengittern.* Inaugural-Dissertation zur Erlangung der Doktorwürde, Königl. Ludwigs-Maximilians-Universität zu München. (16 February 1912.)

Ewald, P. P. (1913*a*). Zur Theorie der Interferenzen der Röntgentstrahlen in Kristallen. *Phys. Zeit.* **14**, 465–472. Reproduced in Cruickshank *et al.* (1992).

Ewald, P. P. (1913*b*). Bemerkung zu der Arbeit von M. Laue: die dreizählig-symmetrischen Röntgenstrahlaufnahmen an regulären Kristallen. *Phys. Zeit.* **14**, 1038–1040.

Ewald, P. P. (1913*c*). Bericht über die Tagung der British Association in Birmingham (10 bis 17 September 1913). *Phys. Zeit.* **14**, 1297–1307.

Ewald, P. P. (1914*a*). Die Intensität der Interferenzflecke bei Zinkblende und das Gitter der Zinkblende. *Ann. Phys.* **349**, 257–282.

Ewald, P. P. (1914*b*). Die Berechnung der Kristallstruktur aus Interferenzenaufnahmen mit X-Strahlen. *Phys. Zeit.* **15**, 399–401.

Ewald, P. P. (1914*c*). Interferenzaufnahme eines Graphitkristalls und Ermittlung des Achsenverhältnisses von Graphit. *Sitzungsberich. der Kgl. Bayer. Akad. der Wiss.* 325–327.

Ewald, P. P. (1916*a*). Zur Begründung der Kristalloptik. I.Theorie der Dispersion. *Ann. Phys.* **354**, 1–38.

Ewald, P. P. (1916*b*). Zur Begründung der Kristalloptik. II.Theorie der Reflexion und Brechung. *Ann. Phys.* **354**, 117–173.

Ewald, P. P. (1917). Zur Begründung der Kristalloptik. III. Die Kristalloptik der Röntgenstrahlen. *Ann. Phys.* **359**, 519–556 and 557–597.

Ewald, P. P. (1920). Zum Reflexionsgesetz der Röntgenstrahlen. *Z. Phys.* **2**, 332–342.

Ewald, P. P. (1921). Das 'reziproke' Gitter in der Strukturtheorie. *Z. Kristallogr.* **56**, 129–156.

Ewald, P. P. (1923). *Kristalle und Röntgenstrahlen.* J. Springer, Berlin.

Ewald, P. P. (1924). Über den Brechungsindex für Röntgenstrahlen und die Abweichung von Braggschen Reflexionsgesetz. *Z. Phys.* **30**, 1–13.

Ewald, P. P. (1925). Die Intensitäten der Röntgenreflexe und der Strukturfaktor. *Phys. Zeit.* **26**, 29–32.

Ewald, P. P. (1927). Der Aufbau der festen Materie und seine Erforschung durch Röntgenstrahlen. *Handbuch der Physik.* Berlin: Springer. **24**, 191–369.

Ewald, P. P. (1932). Zur Entdeckung der Röntgeninterferenzen vor Zwanzig Jahren und zu Sir William Bragg's siebzigtem Geburtstag. *Naturwiss.* **20**, 527–530.

Ewald, P. P. (1936). Historisches und Systematisches zum Gebrauch des 'Reziproken Gitters' in der Kristallstrukturlehre. *Z. Kristallogr.* **93**, 396–398.

Ewald, P. P. (1937). Zur Begründung der Kristalloptik. IV. Aufstellung einer allgemeinen Dispersion bedingung, inbesondere für Röntgenfelder. *Z. Kristallogr.* **97**, 1–27.

Ewald, P. P. (1958). Group velocity and phase velocity in X-ray crystal optics. *Acta Crystallogr.* **11**, 888–891.

Ewald, P. P. (1960). Max von Laue, 1879–1960. *Biogr. Mem. Fell. Roy. Soc.* **6**, 135–156.

Ewald, P. P. (1961). The origin of dynamical theory of X-ray diffraction. *J. Phys. Soc. Jap.* **17**, B-II, 48–52.

Ewald, P. P. (1962*a*). *Fifty years of X-ray diffraction.* Published for the International Union of Crystallography. N. V. A. Oosthoek, Utrecht.

Ewald, P. P. (1962*b*). William Henry Bragg and the new crystallography. *Nature* **195**, 320–325.

Ewald, P. P. (1968*a*). Personal reminiscences. *Acta Crystallogr.* **A24**, 1–3.

Ewald, P. P. (1968*b*). Errinerungen an die Anfänge der Münchener Physikalischen Kolloquiums. *Phys. Bl.* **12**, 538–542.

Ewald, P. P. (1969*a*). The myth of myths. Comments on P. Forman's paper on "The discovery of the diffraction of X-rays by crystals". *Archive for History of Exact Science* **6**, 38–71.

Ewald, P. P. (1969*b*). Introduction to the dynamical theory of X-ray diffraction. *Acta Crystallogr.* **25**, 103–108.

Ewald, P. P. (1977). The Early History of the International Union of Crystallography. *Acta Crystallogr.* **A33**, 1–3.

Ewald, P. P. (1979*a*). A review of my papers on crystal optics 1912 to 1968. *Acta Crystallogr.* **A35**, 1–9.

Ewald, P. P. (1979*b*). Max von Laue: Mensch und Werk. *Phys. Bl.* **35**, 337–349.

Ewald, P. P. and Friedrich, W. (1914). Röntgenaufnahmen von kubischen Kristallen, insbesondere Pyrit. *Ann. Phys.* **349**, 1183–1196.

Ewald, P. P. and Hermann, C. (1927). Gilt der Friedelsche Satz über die Symmetrie der Röntgeninterferenzen? *Z. Kristallogr.* **65**, 251–259.

Fajans, K. (1916). Henry G. J. Moseley. *Naturwiss.* **4**, 381–382.

Faraday, M. (1846*a*). Thoughts on ray vibrations. *Phil. Mag.* Series 3. **28**, 345–350.

Faraday, M. (1846*b*). On the magnetic affection of light, and on the distinction between the ferromagnetic and diamagnetic conditions of matter. *Phil. Mag.* Series 3. **29**, 153–156.

Faxèn, H. (1918). Die bei Interferenz von Röntgenstrahlen durch die Wärmebewegung entstehende zerstreute Strahlung. *Ann. Phys.* **359**, 615–620.

Fedorov, E. S. (1885). The elements of the study of figures. *Zap. Miner. Obshch.* (*Trans. Miner. Soc. Saint Petersburg*) **21**, 1–279 (in Russian).

Fedorov, E. S. (1887). Ein Versuch, durch ein kurzes Zeichen die Symbolen aller gleichen Richtungen einer gegebener Abtheilung des Symmetrie Systems auszudrucken. *Zap. Miner. Obshch.* (*Trans. Miner. Soc. Saint Petersburg*) **23**, 99–115 (in Russian).

Fedorov, E. S. (1889). Die Symmetrie der endlichen Figuren. *Zap. Miner. Obshch.* (*Trans. Miner. Soc. Saint Petersburg*) **25**, 1–52 (in Russian).

Fedorov, E. S. (1891*a*). Ueber seine beiden Werke: 1, Die Symmetrie der endlichen Figuren. 2, Die Symmetrie der regelmässigen Systeme der Figuren. *Neues Jahrb. für Mineral.* **1**, 113–116.

Fedorov, E. S. (1891*b*). The symmetry of regular systems of figures. *Zap. Miner. Obshch.* (*Trans. Miner. Soc. Saint Petersburg*) **28**, 1–146 (in Russian). English transl. *Amer. Crystallgr. Ass. Monographs* No 7 (1971).

Fedorov, E. S. (1891*c*). Symmetry in the plane. *Zap. Miner. Obshch.* (*Trans. Miner. Soc. Saint Petersburg*) **28**, 346–390 (in Russian).

Fedorov, E. S. (1892). Zusammenstellung der krystallographischen Resultate des Herrn Schoenflies und der meinigen. *Z. Kristallogr.* **20**, 25–75.

Fedorov, E. S. (1893). Universal-Theodolith-Methode in der Mineralogie und Petrographie. *Z. Kristallogr.* **21**, 574–678.

Fedorov, E. S. (1894). Das Grundgesetz der Krystallographie. *Z. Kristallogr.* **23**, 99–113.

Fedorov, E. S. (1895). Theorie der Kristallstruktur. Einleitung. Regelmässige Punktsysteme (mit übersichtlicher graphischer Darstellung). *Z. Kristallogr.* **24**, 209–252.

Fedorov, E. S. (1896). Theorie der Kristallstructur. II. Teil. Reticuläre Dichtigkeit und erfahrungsgemässe Bestimmung der Krystallstructur. *Z. Kristallogr.* **25**, 113–224.

Fedorov, E. S. (1899). Reguläre Plan- und Raumtheilung. *Abh. Bayer. Akad. Wiss. Naturwiss. Kl.* **20**, 465–588.

Fedorov, E. S. (1902). Theorie der Kristallstructur. I. Teil. Mögliche Structurarten.*Z. Kristallogr.* **36**, 209–233.

Fedorov, E. S. (1912). Die Praxis in der krystallochemischen Analyse und die Abfassung der Tabellen für dieselbe. *Z. Kristallogr.* **50**, 513–576.

Fischer, E. (1962). Christian Samuel Weiss und seine Bedeutung für die Entwicklung der Kristallographie. *Wiss. Z. Humboldt-Univ. Berlin, Math.-Naturwiss.* R. **XI**, 249–254.

FitzGerald, G. (1897). Dissociations of atoms. *The Electrician* **39**, 103–104.

Flack, H. D. (2009). Louis Pasteur's discovery of molecular chirality and spontaneous resolution in 1848, together with a complete review of his crystallographic and chemical work. *Acta Crystallogr.* **A65**, 371–389.

Fomm, L. (1896). Die Wellenlänge der Röntgen-Strahlen. *Ann. Phys.* **295**, 350–353.

Forman, P. (1969). The discovery of the diffraction of X-rays by crystals: a critique of the myths. *Archive for History of Exact Science* **6**, 38–71.

Forster, R. T. (1855). On the molecular constitution of crystals. *Phil. Mag.* Series 4. **10**, 108–115.

Frank, F. C. (1982). Snow crystals. *Contemporary Physics* **23**, 3–22.

Frankenheim, M. L. (1826). Crystallonomische Aufsätze. *Isis* (Jena), **19**, 497–515, 542–565.

Frankenheim, M. L. (1832). Einige Sätze aus der geometrie der geraden Linien. *J. für die reine und angewandte Mathematik* **8**, 178–186.

Frankenheim, M. L. (1835). *Die Lehre von der Cohäsion.* August Schulz, Breslau.

Frankenheim, M. L. (1842). System der Kristalle. *Nova Acta Acad. Naturae Curiosorum.* **19**, No 2, 469–660.

Frankenheim, M. L. (1856). Ueber die Anordnung der Molecule im Kristall. *Ann. Phys.* **173**, 337–382.

Fresnel, A. (1816). Mémoire sur la diffraction de la lumière, où l'on examine particulièrement le phénomène de franges colorées que présentent les ombres des corps éclairés par un point lumineux. *Ann. Chim. Phys. Paris* **1**, 239–280.

Fresnel, A. (1821). Mémoire sur la diffraction de la lumière. *Mém. Acad. Roy. Sci.* **5**, 339–475.

Fresnel, A. (1822). Explication de la réfraction dans le système des ondes. *Ann. Chim. Phys. Paris* **21**, 225–241.

Fresnel, A. (1827). Mémoire sur la double réfraction. *Mém. Acad. Roy. Sci.* **7**, 45–176.

Friauf, J. B. (1927a). The crystal structure of two intermetallic compounds. *J. Am. Chem. Soc.* **49**, 3107–3114.

Friauf, J. B. (1927b). The crystal structure of magnesium di-zincide. *Phys. Rev.* **29**, 34–40.

Fricke, H. (1920). The *K*-characteristic absorption frequencies for the chemical elements magnesium to chromium. *Phys. Rev.* **16**, 202–216.

Friedel, G. (1904). Sur la loi de Bravais et sur l'hypothèse réticulare. *C. R. Acad. Sci. Paris* **139**, 314–315.

Friedel, G. (1907). Etudes sur la loi de Bravais. *Bull. Soc. Fr. Miner.* **30**, 326–455.

Friedel, G. (1913a). Loi générale de la diffraction des rayons Röntgen par les cristaux. *C. R. Acad. Sci. Paris* **156**, 1676–1679.

Friedel, G. (1913b). Sur la diffraction des rayons Röntgen par les cristaux. *Bull. Soc. Fr. Miner.* **36**, 211–252.

Friedel, G. (1913c). Sur les symétries cristallines que peut révéler la diffraction des rayons X. *C. R. Acad. Sci. Paris* **157**, 1533–1536.

Friedel, G. (1926). *Leçons de Cristallographie.* Strasbourg: Berger-Levrault.

Friedrich, W. (1912). Intensitätsverteilung der X-Strahlung, die von einer Platinaantikathode ausgehen. *Annal. Phys.* **344**, 377–430.

Friedrich, W. (1913a). Interferenzerscheinungen bei Röntgenstrahlen und die Raumgitter der Krystalle. *Z. Kristallogr.* **52**, 58–62.

Friedrich, W. (1913b). Röntgenstrahlinterferenzen. *Phys. Zeit.* **14**, 1079–1084.

Friedrich, W. (1922). Die Geschichte der Auffindung der Röntgenstrahlinterferenzen. *Naturwiss.* **10**, 363–366.

Friedrich, W. (1949). Errinerungen an die Entdeckung der Interferenzerscheinungen bei Röntgenstrahlen. *Naturwiss.* **36**, 354–356.

Friedrich, W., Knipping, P., and Laue, M. (1912). Interferenz-Erscheinungen bei Röntgenstrahlen. *Sitzungsberichte der Kgl. Bayer. Akad. der Wiss.* 303–322.

Friedrich, W., Knipping, P., and Laue, M. (1913). Interferenz-Erscheinungen bei Röntgenstrahlen. *Ann. Phys.* (1913) **346**, 971–988.

Gadolin, A. V. (1871). Mémoire sur la déduction d'un seul principe de tous les systèmes cristallogaphiques avec leurs subdivisions. *Acta Societatis Scientiarum Fennicae* **9**, 1–71.

Galiulin, R. V. (2003). To the 150th Anniversary of the Birth of Evgraf Stepanovich Fedorov (1853–1919). Irregularities in the Fate of the Theory of Regularity. *Crystallography Reports* **48**, 899–913.

Gasman, L. D. (1975). Myths and X-rays. *British J. for the Phil. of Sci.* **26**, 51–60.

Gauss, C. F. (1831). Recension der 'Untersuchungen über die Eigenschaften der positiven ternären quadratischen Formen' von Mudwig August Seeber. *Göttingschen gelehrten Anzeigen* **108**, 248–256. Reproduced in *J. reine angew. Math.* (1840). **20**, 312–320.

Geiger, H. and Marsden, E. (1909). On a diffuse reflection of the α-particles. *Proc. Roy. Soc. A* **82**, 495–500.

Gerlach, W. (1963). Münchener Errinerungen. *Phys. Bl.* **19**, 97–102.

Gibbs, J. W. (1881). *Elements of Vector Analysis, arranged for the Use of Students in Physics.* Privately printed, New Haven.

Gibbs, R. E. (1926a). Structure of α-quartz. *Proc. Roy. Soc. A* **110**, 443–445.

Gibbs, R. E. (1926b). The polymorphism of silicon dioxide and the structure of tridymite. *Proc. Roy. Soc. A* **113**, 351–368.

Glasser, O. (1931a). *Wilhelm Connrad Röntgen und die Geschichte der Röntgenstrahlen.* J. Springer, Berlin.

Glasser, O. (1931b). Wilhelm Connrad Roentgen and the discovery of the Roentgen rays. *Am. J. Radiol.* **25**, 437–450. Reproduced in *Am. J. Radiol.* (1995). **165**, 1033–1040.

Glazer, A. M. and Thomson, P. editors (2015). *Crystal clear, the Autobiographies of Sir Lawrence and Lady Bragg.* Oxford University Press, Oxford.

Glocker, E. F. (1831). *Handbuch der Mineralogie.* Johann Leonard Schrag, Nuremberg.

Glocker, R. (1914). Experimenteller Beitrag zur Interferenz der Röntgenstahlen II. *Phys. Zeit.* **15**, 401–405.

Glocker, R. (1921). Über die atomare Streuung von Natrium und Chlor. *Zeit. Phys.* **5**, 389–382.

Goldschmidt, V. M. (1926). *Geochemische Verteilungsgesetze der Elemente. VII: Die Gesetze der Krystallo-chemie (nach Untersuchungen mit T. Barth, G. Lunde und W. Zachariasen).* Skrifter Norske Videnskaps Akad. Oslo, (I) Mat. Natur. Kl., Oslo.

Goldschmidt, V. M. (1929). Crystal structure and chemical constitution. *Trans. Faraday Soc.* **25**, 253–283.

Goldstein, E. (1876). Vorläufige Mittheilungen über elektrische Entadungen in verdünnten Gasen. *Sitzungsber. Akad. König. Preuss. Wissensch. zu Berlin,* 279–295.

Goldstein, E. (1880). Ueber die Entladung der Electricität in verdünnten Gasen. *Ann. Phys.* **247**, 832–856.

Goldstein, E. (1881). Ueber die Entladung der Electricität in verdünnten Gasen. *Ann. Phys.* **248**, 249–279.

Goldstein, E. (1886). Ueber eine noch nicht untersuchte Strahlungsform am der Kathode inducirter Entladungen. *Sitzungsber. Akad. König. Preuss. Wissensch. zu Berlin,* 691–699.

Gonell, H. W. (1924). Röntgenspektrographische Beobachtungen an Cellulose. III. *Zeit. Phys.* **25**, 118–120.

Gonell, H. W. and Mark, H. (1923). Röntgenographische Bestimmung der hexamethylene tetramine and urea. *Zeit. phys. Chem.* **107**, 181.

Goodspeed, A. W. (1896). The Röntgen phenomena. *Science* **3**, 394–396.

Goucher, F. S. (1926). Further studies on the deformation of tungsten single crystals. *Phil. Mag.* Series 7. **2**, 289–309.

Grassmann, J. G. (1829). *Zur physischen Krystallonomie und geometrischen Combinationslehre.* Stettin.

Green, G. (1839a). On the propagation of light in crystallized media. *Trans. Camb. Phil. Soc.* **7**, 1–24; 115–120.

Green, G. (1839b). On the laws of reflexion and refraction of light at the common surface of two non-crystalline media. *Trans. Camb. Phil. Soc.* **7**, 121–140

Grimaldi, F. M. (1665). *Physicomathesis de lumine, coloribus, et iride, aliise annexis; libri duo, In quorum primo afferuntur nova experiment.* Vittorio Bonati, Bologna (posthumous work).

Groth, P. (1888). *Ueber die Molekularbeschaffenheit der Krystalle.* Münchener Akademie der Wissenschaften, Munich.

Groth, P. (1904). *On the structure of crystals and its relation to chemical structure,* in *Report of the Seventy-first Meeting of the British Association for the Advancement of Science.* John Murray, London. pp. 505–509.

Groth, P. (1925). Zur Geschichte der Krystallkunde. *Naturwiss.* **13**, 61–66.

Groth, P. (1926). *Geschichte der mineralogischen Wissenschaften.* Berlin.

Guglielmini, D. (1688). *Riflessioni filosofiche dedotte dalle figure de' sali.* Bologna.

Guglielmini, D. (1705). *De salibus dissertatio epistolaris physico-medico-mechanica.* apud Aloysium Pavinum, Venice.

Guillot, R., Fertey, P., Hansen, N. K., Allé, P., Elkaïm, E., and Lecomte, C. (2004). Diffraction study of the piezoelectric properties of low quartz. *Europ. Phys. J. B (Condensed Matter)* **42**, 373–380.

Guinier, A. (1938a). Un nouveau type de diagrammes de rayons X. *C. R. Acad. Sci. Paris* **206**, 1641–1643.

Guinier, A. (1938b). Structure of age-hardened aluminium-copper alloys. *Nature* **142**, 569–569.

Guinier, A. (1959). L'étude des structures cristallines imparfaites par les rayons X. *Beitr. Phys. Chem. 20 Jahrhunderts.* Braunschweig: Vieweg & Sohn. pp. 178–187.

Guinier, A. (1964). Technische Anwendung der Röntgenanalyse: Métallurgie. *Z. Kristallogr.* **120**, 125–136.

Guyon, E. (2006). Editor. *L'Ecole Normale de l'an III. Leçons de Physique, de Chimie, d'Histoire Naturelle.* Editions de la rue d'Ulm, Paris.

Haga, H. (1907). Über die Polarisation der Röntgenstrahlen und der Sekundärstrahlen. *Ann. Phys.* **328**, 439–444.

Haga, H. and Jaeger, F. M. (1914). Röntgenpatterns of boracite, obtained above and below its inversion temperature. *Proc. Roy. Neth. Acad. (KNAW)* **16**, 792–799.

Haga, H. and Jaeger, F. M. (1916). On the Symmetry of the Röntgen-patterns of Trigonal and Hexagonal Crystals, and on Normal and Abnormal Diffraction-Images of birefringent Crystals in general. *Proc. Roy. Neth. Acad. (KNAW)* **18**, 542–558.

Haga, H. and Wind, C. H. (1899). Die Beugung der Röntgenstrahlen. *Ann. Phys.* **304**, 884–895.

Haga, H. and Wind, C. H. (1903). Die Beugung der Röntgenstrahlen. Zweite Mitheilung. *Ann. Phys.* **315**, 305–312.

Hahn, Th. and Klapper, H. (2003). *Twinning of Crystals*, in *International Tables of Crystallography*, Vol. **D**, 393–448. Ed. A. Authier; Kluwer Academic Publishers, Dordrecht.

Halban, H. von and Preiswerk, P. (1936). Preuve expérimentale de la diffraction des neutrons. *C. R. Acad. Sci. Paris* **203**, 73–75.

Hales, T. C. (2000). Cannonballs and honeycombs. *Notices of the American Mathematical Society.* **47**, 440–449.

Hales, T. C. (2001). The Honeycomb Conjecture. *Disc. Comp. Geom.* **25**, 1–22.

Hales, T. C. (2005). A proof of the Kepler conjecture. *Annals of Mathematics, Second Series* **162**, 1065–1185.

Hall, A. R. (1990). Beyond the fringe: diffraction as seen by Grimaldi, Fabri, Hooke and Newton. *Notes Rec. Roy. Soc.* **44**, 13–23.

Hall, K. T. (2014). *The man in the monkeynut coat, William Asrbury and the Forgotten Road to the Double Helix.* Oxford University Press, Oxford.

Halleux, R. (1975). *L'Etrenne ou la neige sexangulaire.* Paris: Librairie Vrin et Editions du CNRS.

Hallwachs, W. (1888). Ueber den Einfluss des Lichtes auf electrostatisch geladene Körper. *Ann. Phys.* **269**, 301–312.

Hallwachs, W. (1889). Ueber den Zusammenhang des Electricitätsverlustes durch Beleuchtung mit der Lichtabsorption. *Ann. Phys.* **273**, 666–675.

Ham, W. R. (1910). Polarization of Röntgen Rays. *Phys. Rev.* Series 1. **30**, 96–121.

Hardie, C. with essays by Mason, B. J. and Whyte, L. L. (1966). *The Six-sided Snowflake.* Oxford: Oxford University Press.

Hardouin Duparc, O. B. M. (2010). The Preston of the Guinier-Preston zones – Guinier. *Met. and Mat. Trans.* **41B**, 925–934.

Hardouin Duparc, O. B. M. (2011). A review of some elements in the history of grain boundaries, centered on Georges Friedel, the coincident 'site' lattice and the twin index. *J. Mater. Sci.* **46**, 4116–4134.

Hartman, P. (1973). *Structure and Morphology*, in *Crystal growth: an introduction*, P. Hartman Editor, pp. 367–402. North-Holland, Amsterdam.

Hartman, P. and Perdok, W. G. (1955). On the relations between structure and morphology of crystals. *Acta Crystallogr.* **8**, 49–52.

Hartree, D. R. (1925). The atomic structure factor in the intensity of reflexion of X-rays by crystals. *Phil. Mag.* Series 6. **50**, 289–306.

Hartree, D. R. (1928). The Wave Mechanics of an Atom with a Non-Coulomb Central Field. *Proc. Cambridge Phil. Soc.* **24**, 89–132.

Hassel, O. (1926). Die Kristallstruktur einiger Verbindungen von der Zusammensetzung MRO_4. I. Zirkon $ZrSiO_4$. *Z. Kristallogr.* **63**, 247–254.

Hassel, O. and Mark, H. (1924). Über die Kristallstruktur des Graphits. *Zeit. Phys.* **25**, 317–337.

Hatley, C. C. (1924). Index of refraction of calcite for X-rays. *Phys. Rev.* **24**, 486–494.

Haüy, R.-J. (1782*a*). Sur la structure des cristaux de grenat. *Observations sur la physique, sur l'histoire naturelle et sur les arts.* **XIX**, 366–370.

Haüy, R.-J. (1782*b*). Sur la structure des cristaux des spaths calcaires. *Observations sur la physique, sur l'histoire naturelle et sur les arts.* **XX**, 33–39.

Haüy, R.-J. (1784). *Essai d'une théorie sur la structure des cristaux, appliquée à plusieurs genres de substances cristallisées.* Chez Gogué & Née de la Rochelle, Paris.

Haüy, R.-J. (1786). Mémoires sur la structure du cristal de roche. *Mém. Acad. Roy. Sci. Paris.* **89**, 78–94.

Haüy, R.-J. (1792). *Exposition abrégée de la théorie sur la structure des cristaux*. Imprimerie du Cercle Social, Paris.

Haüy, R.-J. (1793). Théorie sur la structure des cristaux. *Ann. Chim.* **XVII**, 225–319.

Haüy, R.-J. (1795). *Leçons de Physique* in *L'Ecole Normale de l'an III. Leçons de Physique, de Chimie, d'Histoire Naturelle* (2006). E. Guyon, Editor. Editions de la rue d'Ulm, Paris.

Haüy, R.-J. (1796). D'une méthode simple et facile pour représenter les différentes formes cristallines par des signes très abrégés, qui expriment les lois de décroissement auxquelles est soumise la structure. *Journal des Mines* **4**, 15–36.

Haüy, R.-J. (1801). *Traité de Minéralogie*. Chez Louis, Paris.

Haüy, R.-J. (1808). Sur l'arragonite. *Journal des Mines* **4**, 241–270.

Haüy, R.-J. (1815). Sur une loi de la cristallisation, appelée loi de symétrie. *Journal des Mines* **23**, 215–235.

Haüy, R.-J. (1822). *Traité de Cristallographie*. Bachelier et Huzard, gendres et successeurs de Mme veuve Courcier, Paris.

Havighurst, R. J. (1925). The distribution of diffracting power in sodium chloride. *Proc. Nat. Acad. Sci. (Washington)* **11**, 502–507.

Havighurst, R. J. (1926). The intensity of reflection of X-rays by powdered crystals, I. Sodium chloride and sodium, lithium and calcium fluorides. *Phys. Rev.* **28**, 869–881.

Havighurst, R. J. (1927). Electron distribution in the atoms of crystals. Sodium chloride and lithium, sodium and calcium fluorides. *Phys. Rev.* **29**, 1–19.

Heilbron, J. L. (1966). The work of H. G. J. Moseley. *Isis* **57**, 336–366.

Heilbron, J. L. (1967). The Kossel–Sommerfeld theory and the ring atom. *Isis* **58**, 450–485.

Heilbron, J. L. (1974). *The life and letters of an English physicist*. University of California Press. Berkeley, CA, USA.

Helmholtz, H. von (1875). Zur Theorie der anomalen Dispersion. *Ann. Phys.* **230**, 582–595.

Hengstenberg, J. and Mark, H. (1928). Über Form und Grösse der Mizelle von Zellulose und Kautschuk. *Z. Kristallogr.* **69**, 271–284.

Hengstenberg, J. and Mark, H. (1929). Ein röntgenographischer Nachweis von Gitterstörungen an Metallen. *Naturwiss.* **17**, 443–443.

Hengstenberg, J. and Mark, H. (1930). Röntgenographische Intensitätsmessungen an gestörten Gittern. *Zeit. Phys.* **61**, 435–453.

Hermite, C. (1850). Sur la théorie des formes quadratiques ternaires. *J. reine und angew. Math.* **40**, 1173–177.

Herschel, J. F. W. (1822). On the rotation impressed by plates of rock salt on the planes of polarization of the rays of light, as connected with certain peculiarities in its crystallization. *Trans. Camb. Phil. Soc.* **1**, 43–52.

Hertz, G. (1920). Über die Absorptionsgrenzen in der *L*-Serie. *Z. Phys.* **3**, 19–25.

Hertz, H. (1883). Versuche über die Glimmentladung. *Ann. Phys.* **255**, 782–816.

Hertz, H. (1887*a*). Ueber sehr schnelle electrische Schwingungen. *Ann. Phys.* **267**, 421–448.

Hertz, H. (1887*b*). Ueber einen Einfluss des ultravioletten Lichtes auf die electrische Entladung. *Ann. Phys.* **267**, 983–1000.

Hertz, H. (1892). Über den Durchgang der Kathodenstrahlen durch dünne Metallschichten. *Ann. Phys.* **281**, 28–32.

Herweg, J. (1909). Über die Polarisation der Röntgenstrahlen. *Ann. Phys.* **334**, 398–400.

Herweg, J. (1913*a*). Über die Beugungserscheinungen der Röntgenstrahlen am Gips. *Phys. Zeit.* **14**, 417–420.

Herweg, J. (1913*b*). Über das Spektrum der Röntgenstrahlen. Erste Mitteilung. *Verhandl. Deutsch. Phys. Gesel.* **15**, 555–556.

Herweg, J. (1914). Über das Spektrum der Röntgenstrahlen. Zweite Mitteilung. Ein Spektrograph für Röntgenstrahlen. Die Linien des Platins und Wolframs. *Verhandl. Deutsch. Phys. Gesel.* **16**, 73–78.

Herzog, R. O. and Jancke, W. (1920). Röntgenspektrographische Beobachtungen an Zellulose. *Z. Phys.* **3**, 196–198.

Herzog, R. O., Jancke, W., and Polanyi, M. (1920). Röntgenspektrographische Beobachtungen an Zellulose. II. *Z. Phys.* **3**, 343–348.

Hessel, J. F. C. (1830). *Krystallometrie oder Krystallonomie und Krystallographie*. In Gehler's Physikalisches Wörterbuch. **8**, 1023–1360. Schwickert, Leipzig.

Hildebrandt, G. (1993). The discovery of the diffraction of X-rays in crystals – A historical review. *Crystal Res. & Technol.* **28**, 747–766.

Hevesy, G. von (1923). Bohrsche Theorie und Radioaktivität. *Naturwiss.* **11**, 604–605.

Hildebrandt, G. (1995). Early experimental proofs of the dynamical theory. *J. Physics D: Applied Physics* **28**, A8–A16.

Hildebrandt, G. (2002). How discoveries are made – or some remarks on the discovery of the Borrmann effect. *Crystal Res. & Technol.* **37**, 777–782.

Hilton, H. (1902). A comparison of various notations employed in theories of crystal-structure, and a revision of the 230 groups of movements. *Phil. Mag.* Series 6. **3**, 203–212.

Hilton, H. (1903). *Mathematical crystallography and the theory of groups of movements*. Clarendon Press, Oxford.

Hittorf, W. (1869). Ueber die Elektricitätsleitung der Gase. *Ann. Phys.* **212**, 1–31, 197–234.

Hjalmar, E. (1920). Präzisionbestimmingen in der *K*-Reihe der Röntgenspektren. Elemente Cu bis Na. *Z. Phys.* **1**, 439–458.

Hjalmar, E. (1923). Röntgenspektroskopische Messungen. Beitrag zur Kenntnis der Röntgenspektren. *Z. Phys.* **15**, 65–109.

Hjalmar, E. and Siegbahn, M. (1925). Anomalous dispersion in the field of X-rays. *Nature* **115**, 85–86.

Höbler, H.-J. (2006). *Christian Samuel Weiss*. In Universität Leipzig, Jubiläen 2006. 131–135. Leipzig.

Hodeau, J.-L. and Guinebretiere, R. (2007). Crystallography: past and present. *Appl. Phys. A* **89**, 813–823.

Hoddeson, L., Braun, E., Teichmann, J., and Weart S. (1992). *Out of the crystal maze*. Oxford University Press, New York.

Hönl, H. (1933). Zur Dispersionstheorie der Röntgenstrahlen. *Z. Phys.* **84**, 1–16.

Hooke, R. (1665). *Micrographia or some Physiological Descriptions of minute bodies, made by magnifying glasses with observations and inquiries thereupon*. Jo. Martyn, and Ja. Allestry, Printers to the Royal Society, London.

Hooykaas, R. (1955). Les débuts de la théorie cristallographique de R. J. Haüy, d'après les documents originaux. *Rev. Hist. Sci.* **8**, 319–337.

Hoppe, G. (1982). Christan Samuel Weiss und das Berliner Mineralogische Museum. *Wissenschaftliche Zeit. Humboldt-Universität zu Berlin, Math.-Nat.* **31**, 245–254.

Hull, A. W. (1916). The maximum frequency of X-rays at constant voltages between 30 000 and 100 000. *Phys. Rev.* **7**, 156–158.

Hull, A. W. (1917*a*). The crystal structure of iron. *Phys. Rev.* **9**, 83–87.

Hull, A. W. (1917*b*). A new method of crystal analysis. *Phys. Rev.* **10**, 661–696.

Hull, A. W. (1919*a*). A new method of chemical analysis. *J. Am. Chem. Soc.* **41**, 1168–1175.

Hull, A. W. (1919*b*). The crystal structure of carborundum. *Phys. Rev.* **13**, 292–295.

Hull, A. W. (1919*c*). The crystal structure of ferro-magnetic metals. *Phys. Rev.* **14**, 540–541.

Hull, A. W. (1921*a*). The crystal structure of calcium. *Phys. Rev.* **17**, 42–44.

Hull, A. W. (1921*b*). X-ray crystal analysis of thirteen common metals. *Phys. Rev.* **17**, 571–588.

Hull, A. W. (1921*c*). Crystal structure of titanium, zirconium, cerium, thorium and osmium. *Phys. Rev.* **18**, 88–89.

Hull, A. W. (1922). Crystal structures of vanadium, germanium and graphite. *Phys. Rev.* **20**, 113–113.

Hull, A. W. and Davey, W. P. (1921). Graphical determination of hexagonal and tetragonal crystal structures from X-ray data. *Phys. Rev.* **17**, 549–570.

Hume Rothery, W. (1926). Researches on the Nature, Properties and Conditions of Formation of Intermetallic Compounds (with special Reference to certain Compounds of Tin). *J. Instit. Metals*. **35**, 295–361.

Hume Rothery, W. (1936). *The Structure of metals and alloys. Institute of Metals Monograph*. **1**, The Institute of Metals, London.

Hume Rothery, W., Mabbott, G. W., and Channel-Evans, K. M. (1934). The freezing points, melting points, and solid solubility limits of the alloys of silver and copper with the elements of the B sub-groups. *Phil. Trans. Roy. Soc. A* **233**, 1–97.

Hunter, G. K. (2004). *Light is a messenger: the life and science of William Lawrence Bragg*. Oxford University Press, Oxford.

Hupka, E. (1913*a*). Über den Durchgang von Röntgentrahlen durch Metalle. *Phys. Zeit.* **14**, 623.

Hupka, E. (1913*b*). Die Streifungen in Interferensbild der Röntgenstrahlen. *Phys. Zeit.* **14**, 623.

Hupka, E. and Steinhaus, W. (1913). Systems of lines obtained by reflection of X-rays. *Nature* **91**, 10.

Hüttig, G. F. (1920). Notiz zur Geometrie der Koordinationszahl. *Zeit. anorg. Chem.* **114**, 24–26.

Huygens, C. (1673). An Extract of a Letter Lately Written by an Ingenious Person from Paris, Containing Some Considerations upon Mr. Newton's Doctrine of Colors, as Also upon the Effects of the Different Refractions of the Rays in Telescopical Glasses. *Phil. Trans. Roy. Soc.* **8**, 6086–6087.

Huygens, C. (1690). *Traité de la lumière*. Aa Van der Pierre, Leyden.

Huygens, C. (1724). *Opera varia*. Aa Van der Janssonios, Leyden.

Imbert, A. and Bertin-Sans, H. (1897). Diffusion des rayons de Röntgen. *C. R. Acad. Sci. Paris* **122**, 524–525.

Innes, P. D. (1907). On the velocity of the cathode particles emitted by various metals under the influence of Röntgen rays, and its bearing on the theory of atomic disintegration. *Proc. Roy. Soc. A* **79**, 442–462.

Ioffe, A. and Kirpitcheva, M. W. (1922). Röntgenograms of strained crystals. *Phil. Mag.* Series 6. **43**, 206–208.

Ito, T., Morimoto, N., and Sadanaga, R. (1951). The crystal structure of boracite. *Acta Crystallogr.* **4**, 310–316.

Jackson, W. W. and West, J. (1928). The structure of muscovite. *Z. Kristallogr.* **76**, 211–227.

Jacquot, J. (1952). Thomas Harriot's Reputation for Impiety. *Notes Res. Roy. Soc. Lond.* **9**, 164–187.

James, R. W. (1925). The influence of temperature on the intensity of reflexion of X-rays from rocksalt. *Phil. Mag.* Series 6. **49**, 585–602.

James, R. W. and Brindley, G. W. (1928). A quantitative study of the reflexion of X-rays by sylvine. *Proc. Roy. Soc. A* **121**, 155–171.

James, R. W. and Firth, E. M. (1927). An X-Ray study of the heat motions of the atoms in a rock-salt crystal. *Proc. Roy. Soc. A* **117**, 62–87.

James, R. W. and Wood, W. A. (1925). The crystal structure of barite, celestite and anglesite. *Proc. Roy. Soc. A* **109**, 598–620.

James, R. W., Waller, I., and Hartree, D. R. (1928). An investigation into the existence of zero-point energy in the rock-salt lattice by an X-Ray diffraction method. *Proc. Roy. Soc. A* **118**, 334–350.

Jameson, R. (1820). *System of Mineralogy in which minerals are arranged according to the Natural History method*. Third Edition, in 3 Volumes. Archibald Constable & Co, Edinburgh.

Jauncey, G. E. M. (1922). The scattering of X-rays by crystals. *Phys. Rev.* **20**, 405–420.

Jauncey, G. E. M. (1945). The birth and early infancy of X-rays. *Am. J. Phys.* **13**, 362–379.

Jenkin, J. (2001). A unique partnership: William and Lawrence Bragg and the 1915 Nobel Prize in Physics. *Minerva* **39**, 373–392.

Jenkin, J. (2008). *William and Lawrence Bragg, father and son*. Oxford University Press, Oxford.

Johansson, C. H. and Linde, J. O. (1925). Röntgenographische Bestimmung der Atomanordnung in den Mischkristallreihen Au-Cu und Pd-Cu. *Ann. Phys.* **78**, 439–460.

Jones, H. C. (1903). *The elements of physical chemistry*. Macmillan, New York.

Jones, H. C. (1913). *A new era in chemistry*. D. Van Nostrand, New York.

Jones, R. and Lodge, O. (1896). The discovery of a bullet lost in a wrist by means of the Roentgen rays. *Lancet* **1**, 476–477.

Jong, W. F. de and Bouman, J. (1938). Das Photographieren von reziproken Kristallnetzen mittels Röntgenstrahlen. *Z. Kristallogr.* **A98**, 456–459.

Jong, W. F. de and Stradner, E. (1956). Zeittafel. Skizze der Entwicklung der geometrischen Kristallographie und der Strukturtheorie. *Miner. and Petr.* **5**, 362–379.

Jordan, C. (1867). Sur les groupes de mouvements. *C. R. Acad. Sci. Paris* **65**, 229–232.

Jordan, C. (1868). Mémoire sur les groupes de mouvements. *Annali di Matematica Pura Appl.* **2**, 167–215, 322–345.

Kaemmel, T. (2006). *Arthur Schoenflies, Mathematiker und Kristallforscher*. Projekte-Verlag.

Kamminga, H. (1989). The International Union of Crystallography: Its Formation and Early Development. *Acta Crystallogr.* **45**, 581–601.

Kargon, R. H. (1966). *Atomism in England from Harriot to Newton*. Oxford University Press, Oxford.

Kasper, J. S. (1964). Metal structures. *Z. Kristallogr.* **120**, 53–70.

Kaufmann, W. (1897). Die magnetische Ablenkbarkeit der Kathodenstrahlen und ihre Abhängigkeit vom Entladungspotential. *Ann. Phys.* **297**, 544–552.

Kato, N. (1958). The flow of X-rays and material waves in an ideally perfect single crystal. *Acta Crystallogr.* **11**, 885–887.

Kato, N. (1960). Dynamical X-ray diffraction theory of spherical waves. *Acta Crystallogr.* **13**, 349–356.

Kato, N. and Lang, A. R. (1959). A Study of Pendellösung fringes in X-ray diffraction. *Acta Crystallogr.* **12**, 787–794.

Katzir, S. (2004). The emergence of the principle of symmetry in physics. *Historical studies in physical and biological sciences* **35**, 35–65.

Kaye, G. W. C. (1909). The emission and transmission of Röntgen rays. *Phil. Trans. Roy. Soc.* **209**, 123–151.

Keene, H. B. (1913a). The reflection of X-rays. *Nature* **91**, 111.

Keene, H. B. (1913b). On the transmission of X-rays through metals. *Nature* **91**, 607.

Keene, H. B. (1913c). Über den Durchgang von Röntgentrahlen durch Metalle. *Phys. Zeit.* **14**, 903–904.

Keller, E. (1914). Diamant-Röntgenbilder. *Annal. Phys.* **351**, 157–175.

Kelvin, Lord (Thomson, W.) (1887). On the division of space with minimum partitional area. *Phil. Mag.* Series 5. **24**, 503–514.

Kelvin, Lord (Thomson, W.) (1893). On the elasticity of a crystal according to Boscovich. *Phil. Mag.* Series 5. **36**, 414–430.

Kelvin, Lord (Thomson, W.) (1894*a*). On homogeneous division of space. *Proc. Roy. Soc.* **55**, 1–16.

Kelvin, Lord (Thomson, W.) (1894*b*). *The molecular tactics of a crystal* (second Boyle lecture). Clarendon Press, Oxford.

Kelvin, Lord (Thomson, W.) (1896). On the generation of longitudinal waves in ether. *Proc. Roy. Soc.* **59**, 270–273.

Kepler, J. (1611). *Strena Seu de Nive Sexangula*. Frankfurt am Main: Gottfried Tampach. English translation: C. Hardie (1966). French translation: R. Halleux (1975). German translations: F. Rossmann, with collaboration of M. Caspar and F. Neuhart (1943). *Neujahrsgabe oder vom sechseckigen Schnee*. Berlin: W. Keiper; H. Strunz and H. Borm (1958). *Über den hexagonalen Schnee*. Regensburg: Bernhard Bosse.

Kirchhoff, G. (1883). Zur Theorie der Lichtstrahlen. *Ann. Phys.* **254**, 663–695.

Knipping, P. (1913). Über den Durchgang von Röntgentrahlen durch Metalle. *Phys. Zeit.* **14**, 996–998.

Koch, P. P. (1912). Über die Messung der Schwärzung photographischer Platten in sehr schmalen Bereichen. Mit Anwendung auf die Messung der Schwärzungsverteilung in einigen mit Röntgenstrahlen aufgenommenen Spaltphotogrammen von Walter und Pohl. *Ann. Phys.* **38**, 507–522.

Kolkmeijer, N. H. (1920). The possible existence of binding rings in diamond; confirmation of Debye and Scherrer's conclusions. *Proc. Roy. Neth. Acad. (KNAW)* **23**, 120–128.

Kossel, W. (1914). Bemerkung zur Absorption homogener Röntgenstrahlen. *Verh. Dtsch. Phys. Gesell.* **16**, I. 898–909; II. 953–963.

Kossel, W. (1916*a*). Über Molekülbildung als Frage des Atombaus. *Ann. Phys.* **354**, 229–362.

Kossel, W. (1916*b*). Bemerkungen zum Seriencharacter der Röntgenspektren. *Verh. Dtsch. Phys. Gesell.* **18**, 339–359.

Kossel, W. (1919). Über die physikalische Natur der Valenzkräfte. *Naturwiss.* **7**, 360–366.

Kossel, W. (1920). Zum Bau der Röntgenspektren. *Z. Phys.* **1**, 119–134.

Kossel, W. (1924). Bemerkung zum scheinbaren selektiven Reflexion von Röntgenstrahlen an Kristallen. *Z. Phys.* **23**, 278–285.

Kossel, W. (1927). Zur Theorie des Krystallwachstums. *Nachrichten Kgl. Gesell. Wiss. Göttingen* **2**, 135–143.

Kossel, W. (1935). Röntgeninterferenzen an der Einkristallantikathode. *Ann. Phys.* **415**, 677–704.

Kossel, W., Loeck, H., and Voges, H. (1935). Die Richtungsverteilung der in einem Kristall entstandenen charakteristischen Röntgenstrahlung. *Z. Phys.* **94**, 139–144.

Kossel, W. and Voges, H. (1935). Röntgeninterferenzen an der Einkristallantikathode. *Ann. Phys.* **415**, 677–704.

Kragh, H. (1987). *The historiography of science*. Cambridge University Press, Cambridge.

Kragh, H. (2011). Resisting the Bohr atom: the early British opposition. *Phys. Perspect.* **13**, 4–35.

Kubbinga, H. (2002). *L'Histoire du concept de "molécule"*. Springer Verlag, Paris.

Kupffer, A. T. (1826). Ueber die Krystallisation des Kupfervitriols, nebst allgemeinen Bemerkungen über das ein- und eingliedrige oder tetartoprismatische System. *Ann. Phys.* **84**, 61–77, 215–230.

Kupffer, A. T. (1831). *Handbuch der rechnenden Krystallonomie*. Buchdrückerei der Kaiserlichen Akademie der Wissenschaften, Saint Petersburg.

Lagrange, J.-L. (1772). Recherches arithmétiques. *Nouveaux Mémoires Acad. Sci. Berlin.* 265–310.

La Hire, P. de (1710). Observations sur une espèce de talc qu'on trouve communément proche de Paris au dessus des bancs de pierre de plâtre. *Mémoires Acad. Roy. Sci.* (1710). 341–352.

Lalena, J. N. (2006). From quartz to quasicrystals: probing nature's geometric patterns in crystalline substances. *Crystallography Reviews* **12**, 125–180.

Lamla, E. (1939). Zur Frage der Umweganregung bei Röntgenstrahlinterferenzen. *Ann. Phys.* **428**, 194–208.

Lang, A. R. (1958). Direct observation of individual dislocations by X-ray diffraction. *J. Appl. Phys.* **29**, 597–598.

Langmuir, I. (1919). The arrangement of electrons in atoms and molecules. *J. Am. Chem. Soc.* **41**, 863–934.

Lannelongue, O.-M., Oudin, P. and Barthélemy, T. (1896). De l'utilité des photographies par les rayons X dans la pathologie humaine. *C. R. Acad. Sci.* **122**, 159–159.

Larmor, J. (1894). A dynamical theory of the electric and luminiferous medium. *Phil. Trans. Roy. Soc.* **185**, 719–822.

Larsson, A. (1927). Precision measurement of the K-series of molybdenum and iron. *Phil. Mag.* Series 7. **3**, 1136–1160.

Larsson, A., Siegbahn, M. and Waller, I. (1924). Die experimentelle Nachweis der Brechung der Röntgenstrahlen. *Naturwiss.* **52**, 1212–1213.

Laub, J. (1908). Über die durch Röntgenstrahlen erzeugten sekundären Kathodenstrahlen. *Ann. Phys.* **331**, 712–726.

Laue, M. von (1912). Eine quantitative Prüfung der Theorie für die Interferenz-Erscheinungen bei Röntgenstrahlen. *Sitzungsberichte der Kgl. Bayer. Akad. der Wiss.* 363–373.

Laue, M. von (1913a). Eine quantitative Prüfung der Theorie für die Interferenz-Erscheinungen bei Röntgenstrahlen. *Ann. Phys.* **346**, 989–1002.

Laue, M. von (1913b). Kritische Bemerkungen zu den Deutungen der Photogramme von Friedrich und Knipping. *Phys. Zeit.* **14**, 421–423.

Laue, M. von (1913c). Die dreizählig-symmetrischen Röntgenstrahlaufnamen an Regulären Kristallen. *Ann. Phys.* **347**, 397–414.

Laue, M. von (1913d). Les phénomènes d'interférence des rayons de Röntgen produits par le réseau tridimensionnel des cristaux. In *Rapport du Conseil de Physique de l'Institut International Solvay*, Bruxelles, 27–31 Octobre 1913. (1921). Gauthier-Villars, Paris. pp. 75–119.

Laue, M. von (1914). Die Interferenzerscheinungen an Röntgenstrahlen, hervorgerufen durch das Raumgitter der Kristalle. *Jahrb. der Radioaktivität und Elektronik (Editor J. Stark).* **11**, 308–345.

Laue, M. von (1915). *Concerning the Detection of X-ray Interferences.* Nobel Lectures, Physics 1901–1921 (1967), Elsevier Publishing Company, Amsterdam.

Laue, M. von (1916). Über die Symmetrie der Kristall-Röntgenogramme. *Ann. Phys.* **355**, 433–446.

Laue, M. von (1918). Röntgenstrahlinterferenz und Mischkristalle. *Ann. Phys.* **361**, 497–506.

Laue, M. von (1926). Lorentz-Faktor und Intensitätsverteilung in Debye-Scherrer Ringen. *Z. Kristallogr.* **64**, 115–142.

Laue, M. von (1931a). Die dynamische Theorie der Röntgenstrahlinterferenzen in neuer Form. *Ergeb. Exakt. Naturwiss.* **10**, 133–158.

Laue, M. von (1931b). Bemerkung zur geschichte der dynamischen Theorie der Röntgenstrahlinterferenzen. *Naturwiss.* **19**, 966.

Laue, M. von (1935). Die Fluoreszenzröntgenstrahlung von Einkristallen (Mit einem Anhang über Elektronenbeugung). *Ann. Phys.* **415**, 705–746.

Laue, M. von (1937). Über die Auffindung der Röntgenstrahlinterferenzen. In *Laue Diagrams, 25 Years of research on X-ray diffraction following Prof. Max von Laue discovery.* pp. 9–10. *Current Science.* Indian Academy of Sciences, Bangalore.

Laue, M. von (1941). *Röntgenstrahl-Interferenzen.* Akademische Verlagsgesellschaft, Beckert & Erler Komandit-Gesellschaft, Leipzig.

Laue, M. von (1943). Zu P. von Groths 100 Geburtstage. *Z. Kristallogr.* **105**, 81–81.

Laue, M. von (1946). Zur Gedächtnis Wilhelm Conrad Röntgens. *Naturwiss.* **33**, 3–7.

Laue, M. von (1948). Address before the first Congress of the International Union of Crystallography at Harvard University, August 1948. In McLachlan and Glusker, 1983.

Laue, M. von (1949). Die Absorption der i Röntgenstrahlen in Kristallen in Interferenzfall. *Acta Crystallogr.* **2**, 106–113.

Laue, M. von (1952a). *Historical Introduction.* pp. 1–5. In *International Tables of Crystallography.* Vol. I. Editor N. F. M. Henry. em Theory of Crystallographic Groups. The Kynoch Press, Birmingham.

Laue, M. von (1952b). *Mein physikalischer Werdegang. Eine Selbstdarstellung.* Bonn: In Schöpfer des neuen Weltbildes (ed. H. Hartmann). Athenäum-V. Translated by Ewald and R. Bethe in Ewald (1962a). Autobiography. pp. 278–307.

Laue, M. von (1952c). Die Energie Strömung bei Röntgenstrahlinterferenzen in Kristallen. *Acta Crystallogr.* **5**, 619–625.

Laue, M. von (1960). *Röntgenstrahl-Interferenzen.* Akademische Verlagsgesellschaft, Frankfurt am Main.

Laue, M. von and Tank, F. (1913). Die Gestalt der Interferenzpunkte bei den Röntgenstrahlinterferenzen. *Ann. Phys.* **346**, 1003–1010.

Laue, M. von and van der Lingen, J. S. (1914a). Experimentelle Untersuchungen über den Debyeeffekt. *Phys. Zeit.* **15**, 75–77.

Laue, M. von and van der Lingen, J. S. (1914b). Beobachtungen über Röntgernstrahlinterferenzen. *Naturwiss.* **2**, 328–329.

Laue, M. von and van der Lingen, J. S. (1914c). Die Temperatureinfluss auf die Röntgernstrahlinterferenzen beim Diamant. *Naturwiss.* **2**, 371–371.

Laves, F. (1932). Zur Klassifikation der Silikate. Geometrische Untersunchungen möglicher Silicium-Sauerstoff-Verbände als Verknupfungsmöglichkeiten regulärer Tetraeder. *Z. Kristallogr.* **82**, 1–14.

Laves, F. (1964). Technische Anwendung der Röntgenanalyse: Anorganische (nichtmetalische) Chemie. *Z. Kristallogr.* **120**, 137–142.

Le Bolloch, D., Livet, F., Bley, F., Shulli, T., M. Veron, and Metzger, T. H. (2002) X-ray diffraction from rectangular slits. *J. Synchrotron Rad.* **9**, 258–265.

Lebesgue, V.-A. (1856). La réduction des formes quadratiques définies positives à coefficients réels quelconques. Démonstration du théorème de Seeber sur les réduites de formes ternaires. *J. Math. Pures et App.* **2**, 401–410.

Legendre, A.-M. (1794). *Eléments de Géométrie*. Firmin Didot, Paris.

Lenard, P. (1894). Ueber Kathodenstrahlen in Gasen von atmosphärischem Druck und im äussersten Vacuum. *Ann. Phys.* **287**, 225–267.

Lenard, P. (1895). Ueber die Absorption der Kathodenstrahlen. *Ann. Phys.* **292**, 255–275.

Lenard, P. (1896). On cathode rays and their probable connection with Röntgen rays. *Report of the Sixty-sixth Meeting of the British Association for the Advancement of Science*. London: John Murray. pp. 709–710.

Lenard, P. (1900). Erzeugung von Kathodenstrahlen durch ultraviolettes Licht. *Ann. Phys.* **307**, 359–375.

Lenard, P. (1902). Ueber die lichtelektrische Wirkung. *Ann. Phys.* **313**, 149–198.

Lewis, G. N. (1916). The atom and the molecule. *J. Am. Chem. Soc.* **38**, 762–785.

Lewis, G. N. (1926). The Conservation of Photons. *Nature* **118**, 874–875.

Lima-de-Faria, J., Editor (1990). *Historical Atlas of Crystallography*. Kluwer Academic Publishers, Dordrecht.

Lindh, A. (1921). Zur Kenntnis des Röntgenabsorptionsspektrums von Chlor. *Z. Phys.* **6**, 606–610.

Lindh, A. (1922). Sur les spectres d'absorption du soufre pour les rayons X. *C. R. Acad. Sci. Paris* **175**, 25–27.

Linnaeus, C. (1735). *Regnum lapideum* in *Systema naturae*. Leiden.

Linton, O. W. (1995). News of X-ray reaches America days after annoucement of Roentgen's discovery. *Am. J. Radiol.* **165**, 471–472.

Lipson, H. (1950). Order-disorder changes in alloys. *Progr. Met. Phys.* **2**, 1–52.

Lipson, H. and Beevers, C. A. (1936). An improved numerical method of two-dimensional Fourier synthesis for crystals. *Proc. Phys. Soc. A* **48**, 772–780.

Lodge, O. (1896*a*). On the rays of Lenard and Röntgen. *The Electrician* **36**, 438–440.

Lodge, O. (1896*b*). On the present hypotheses concerning the nature of Röntgen's rays. *The Electrician* **36**, 471–473.

Lodge, O. (1896*c*). The surviving hypothesis concerning the X-rays. *The Electrician* **37**, 370–373.

Lodge, O. (1912). Becqerel Memorial Lecture. *Journ. Chem. Soc.* **101**, 2005–2042.

Lohne, J. A. (1959). Thomas Harriot: The Tycho Brahe of Optics. *Centaurus* **6**, 113–121.

Lohne, J. A. (1977). *Nova Experimenta Crystalli Islandici Disdiaclastici*. *Centaurus* **21**, 106–148.

Lohr, E. (1924). Kontinuätstheorie der Röntgenstrahlausbreitung in Krystallen. *Sitzgsber. Akad. Wiss. Wien.* **133**, 517–572.

Lomonosov, M. V. (1749). *On the genesis and nature of saltpeter*. Dissertation. Academy of Sciences, St Petersburg.

Lonsdale, K. (1928). The structure of the benzene ring. *Nature* **122**, 810–810.

Lonsdale, K. (1929*a*). The structure of the benzene ring in hexamethylbenzene. *Proc. Roy. Soc. A* **123**, 494–515.

Lonsdale, K. (1929*b*). X-ray evidence on the structure of the benzene nucleus. *Trans. Faraday Soc.* **25**, 351–356.

Lonsdale, K. (1931). An X-ray analysis of the structure of hexachlorobenzene, using the Fourier method. *Proc. Roy. Soc. A* **133**, 536–552.

Lonsdale, K. (1947). Divergent Beam X-ray Photography of Crystals. *Phil. Trans. Roy. Soc. A* **240**, 219–250.

Lorentz, H. A. (1895). *Versuch einer Theorie der elektrischen und optischen Erscheinungen in bewegten Körpern*. Brill, Leiden.

Lorentz, H. A. (1916). *The theory of electrons and its applications to the theory of light and radiant heat*. B. G. Teubner, Leipzig.

Lorenz, L. (1867). Ueber die Identität der Schwingungen des Lichts mit den elektrischen Strömen. *Ann. Phys.* **207**, 243–263.

Love, A. E. H. (1906). *A treatise on the theory of elasticity*. Cambridge University Press.

Machatschki, F. (1928). Zur Frage der Struktur und Konstitution der Feldspäte (Gleichzeitig vorläufige Mitteilung über die Prinzipien des Baues der Silikate). *Zentralblatt Min. A.* 97–100.

Machatschki, F. (1931). Zur Spinellstruktur. *Z. Kristallogr.* **80**, 416–427.

Magnus, A. (1920). Nber Chemische Komplexverbindungen. *Zeit. anorg Chem.* **124**, 289–321.

Maitte, B. (2001). *René-Just Haüy (1743–1822) et la naissance de la cristallographie.* In *Travaux du Comié Français d'Histoire de la Géologie Troisième série.* **XV**.

Malgrange, C. and Authier, A. (1965). Interférences entre les champs d'ondes créés par double réfraction des rayons X. *C. R. Acad. Sci. Paris* **261**, 3774–3777.

Malgrange, C., Velu, E. and Authier, A. (1968). Mise en évidence de la dispersion anormale par mesure de l'indice d'un prisme à l'aide de la double diffraction des rayons X. *J. Appl. Crystallogr.* **1**, 181–184.

Mallard, F.-E. (1879). *Traité de cristallographie géométrique et physique.* Dunod, Paris.

Malmer, I. (1914). The high-frequency spectra of the elements. *Phil. Mag.* Series 6. **28**, 787–794.

Malus, E. L. (1810). *Théorie de la double réfraction de la lumière dans les substances cristallisées.* Garnery, Paris.

Mandelstam, L. von and Rohmann, H. (1913). Reflexion der Röntgenstrahlen. *Phys. Zeit.* **14**, 220–222.

Mark, H. (1925). Über den Aufbau der Krystalle. *Naturwiss.* **13**, 1042–1045.

Mark, H. and Polanyi, M. (1923). Die Gitterstruktur, Gleitrichtungen und Gleitebenen des weien Zinns. *Zeit. Phys.* **18**, 75–96.

Mark, H., Polanyi, M., and Schmid, E. (1923). Vorgänge bei der Dehnung von Zinkkristallen. *Z. Phys.* **12**, 58–77, 78–110, 111–116.

Mark, H. and Weissenberg, K. (1923*a*). Röntgenographische Bestimmung der Struktur gewalzter Metallfolien. *Zeit. Phys.* **14**, 328–341 and **16**, 314–318.

Mark, H. and Weissenberg, K. (1923*b*). Röntgenographische Bestimmung der Struktur des Harnstoffs und des Zinntetrajodids. *Zeit. Phys.* **16**, 1–22.

Marx, E. (1906). Die Geschwindigkeit der Röntgenstrahlen; Experimentaluntersuchung. *Ann. Phys.* **325**, 677–722.

Mason, B. J. (1992). Snow crystals, natural and man made. *Contemporary Physics* **33**, 227–243.

Mauguin, C. (1924*a*). Sur la structure cristalline du corindon et de l'oligiste. *C. R. Acad. Sci. Paris* **178**, 785–787.

Mauguin, C. (1924*b*). Arrangement des atomes dans les cristaux de calomel. *C. R. Acad. Sci. Paris* **178**, 1913–1916.

Mauguin, C. (1924*c*). *La structure des cristaux déterminée au moyen des rayons X.* Société 'Journal de physique'. Paris. Les Presses universitaires de France.

Mauguin, C. (1926*a*). Réseaux polaires et diagrammes de rayons X. *Bull. Soc. Fr. Miner.* **49**, 5–32.

Mauguin, C. (1926*b*). Structure du graphite. *Bull. Soc. Fr. Miner.* **49**, 32–61.

Mauguin, C. (1927). Etude du mica moscovite au moyen des rayons X. *C. R. Acad. Sci. Paris* **185**, 288–291.

Mauguin, C. (1928*a*). Etude des micas non fluorés au moyen des rayons X. *C. R. Acad. Sci. Paris* **186**, 879–881. Etude des micas fluorés au moyen des rayons X. 1131–1133.

Mauguin, C. (1928*b*). Les rayons X ne donnent pas toujours la véritable maille des cristaux. *C. R. Acad. Sci. Paris* **187**, 303–304.

McKeehan, L. W. (1922). The crystal structure of silver-palladium and silver-gold alloys. *Phys. Rev.* **20**, 424–432.

Maxwell, J. C. (1865). A dynamical theory of the electromagnetic field. *Phil. Trans. Roy. Soc.* **155**, 459–512.

McLachlan Jr, D. and Glusker, J. P., Editors (1983). *Crystallography in North America.* American Crystallographic Association.

Meissner, A. (1927). Über piezo-elektrische Kristalle bei Hoch-frequenz. *Zeit. f. techn. Physik* **2**, 74–77.

Menzer, G. (1925). Die Kristallstruktur von Granat. *Zentralbl. Miner. A*, 344–345.

Menzer, G. (1926). Die Kristallstruktur von Granat. *Z. Kristallogr.* **63**, 157–158.

Menzer, G. (1929). Die Kristallstruktur der Granate. *Z. Kristallogr.* **69**, 300–396.

Menzer, G. (1964). Anorganische Strukturen. *Z. Kristallogr.* **120**, 19–31.

Metzger, H. (1918). *La genèse de la science des cristaux.* Ecole Pratique des Hautes Etudes, Paris (Thesis). Reprinted (1969). Librairie scientifique et technique Albert Blanchard, Paris.

Mie, W. (1923). Echte optische Resonanz bei Röntgenstrahlen. *Z. Phys.* **25**, 256–57 and **18**, 105–108.

Miller, W. H. (1839). *A treatise on crystallography.* J. & J. J. Deighton Cambridge.

Millikan, R. A. (1916). A direct photoelectric determination of Planck's *h. Phys. Rev.* **7**, 355–388.

Millikan, R. A. (1917). A new determination of *e*, *N*, and related constants. *Phil. Mag.* Series 6. **34**, 1–30.

Minnigerode, B. (1887). Untersuchungen über die Symmetrieverhältnisse der Krystalle. *Neues Jahrb. f. Mineral.* **5**, 145–166.

Mitscherlich, E. (1819). Über die Kristallisation der Salze, in denen das Metall der Basis mit zwei Proportionen Sauerstoff verbunden ist. *Abh. K. Akad. Wiss. Berlin.* 427–437.

Mitscherlich, E. (1820). Sur la relation qui existe entre les proportions chimiques et la forme cristalline. *Ann. Chim. Phys. Paris* **14**, 172–190.

Mitscherlich, E. (1821). Sur la relation qui existe entre les proportions chimiques et la forme cristalline. II. Mémoires sur les arséniates et les phosphates. *Ann. Chim. Phys. Paris* **19**, 350–419.

Mitscherlich, E. (1823). Sur le rapport qui existe entre la forme cristalline et les proportions chimiques. Sur les corps qui affectent deux formes cristallines différentes. *Ann. Chim. Phys. Paris* **24**, 172–190.

Mitscherlich, E. (1824). Über das Verhältnis der Form der kristallisirten Körper zur Ausdehnung durch die Wärme. *Ann. Phys.* **1**, 125–127.

Mohs, F. (1812). *Versuch einer Elementar-Methode zur naturhistorischen Bestimmung und Erkennung der Fossilien*. Vienna.

Mohs, F. (1820). *Die Charaktere der Klassen, Ordnungen, Geschlechter, und Arten der naturhistorischen Mineral-Systems*. First Edition. In der Arnoldschen Buchhandlung, Dresden. Second Edition, 1821. English translation: *The Characters of the Classes, Orders, Genera, and Species, or the Characteristics of the Natural-History System of Mineralogy*. (1820). Edinburgh.

Mohs, F. (1822). *Grundriss der Mineralogie. Erster Teil. Terminologie, Systematik, Nomenlatur, Charakteristik*. In der Arnoldschen Buchhandlung, Dresden.

Mohs, F. (1823). On the crystallographic discoveries and systems of Weiss and Mohs. In a letter from Frederick Mohs, Esq. Professor of Mineralogy at Freyberg to Professor Jameson, in answer to that of Professor Weiss, in the last number of this Journal. *The Edinburgh Philosophical Journal* **VIII**, 275–290.

Moliere, G. (1939). Quantenmechanische Theorie der Röntgenstrahlinterferenzen in Kristallen. *Ann. Phys.* **35**, 272–313; **36**, 265–274.

Moseley, H. G. J. (1912). The number of β-particles emitted in the transformation of radium. *Proc. roy. Soc. A* **87**, 230–255.

Moseley, H. G. J. (1913). The high frequency spectra of the elements. *Phil. Mag.* Series 6. **26**, 1024–1034.

Moseley, H. G. J. (1914). The high frequency spectra of the elements. II. *Phil. Mag.* Series 6. **27**, 703–713.

Moseley, H. G. J. and Darwin, C. G. (1913*a*). The reflection of X-rays. *Nature* **90**, 594.

Moseley, H. G. J. and Darwin, C. G. (1913*b*). The reflection of X-rays. *Phil. Mag.* Series 6. **26**, 210–232.

Müller, A. (1927). An X-ray investigation of certain long-chain compounds. *Proc. Roy. Soc. A* **114**, 542–561.

Müller, A. and Shearer, G. (1923). Further X-ray measurements of long-chain compounds and a note on their interpretation. *J. Chem. Soc.* **123**, 3156–3164.

Murdock, C. C. (1934). Multiple Laue spots. *Phys. Rev.* **45**, 117–118.

Nagaoka, H. (1904). Kinetics of a system of particles illustrating the line and band spectrum and the phenomena of radioactivity. *Phil. Mag.* Series 6. **7**, 445–455.

Nardroff, R. von (1924). Refraction of X-rays in iron pyrites. *Phys. Rev.* **24**, 143–151.

Naumann, C. F. (1824). Ueber plagiobasische Crystall-Systeme. *Isis* (Jena) **9**, 954–959.

Naumann, C. F. (1826). *Grundriss der Kristallographie*. Verlag von Johann Ambrosius Barth, Leipzig.

Naumann, C. F. (1830). *Lehrbuch der reinen und angewandte Kristallographie*. F. A. Brockhaus, Leipzig.

Naumann, F. E. (1856). *Elemente der theoretischen Krystallographie*. Wilhelm Engelmann, Leipzig.

Navier, H. (1827). Mémoire sur les lois de l'équilibre et d'un mouvement des corps solides élastiques. *Mém. Acad. Sci.* **7**, 375–393.

Neumann, F. E. (1823). *Beiträge zur Krystallonomie*. Ernst Siegfried Mittler, Berlin und Posen.

Neumann, F. E. (1833). Die thermischen, optischen und kristallographischen Axen des Kristallsystems des Gypses. *Ann. Phys.* **27**, 240–274.

Neumann, F. E. (1885). *Vorlesungen über die Theorie der Elastizität der festen Körper und des Lichtäthers*. Editor: O. E. Meyer, B. G. Teubner-Verlag, Leipzig.

Newton, I. (1672*a*). A Letter of Mr. Isaac Newton, Professor of the Mathematicks in the University of Cambridge; Containing His New Theory about Light and Colors. *Phil. Trans. Roy. Soc.* **6**, 3075–3087.

Newton, I. (1672*b*). Answer to Some Considerations upon His Doctrine of Light and Colors; Which Doctrine Was Printed in Numb. 80. of These Tracts. *Phil. Trans. Roy. Soc.* **7**, 5084–5103.

Newton, I. (1673). Mr. Newton's Answer to the Foregoing Letter Further Explaining His Theory of Light and Colors, and Particularly That of Whiteness; together with His Continued Hopes of Perfecting Telescopes by Reflections Rather than Refractions. *Phil. Trans. Roy. Soc.* **8**, 6087–6092.

Newton, I. (1687). *Philosophiæ Naturalis Principia Mathematica*. Sam. Smith, London.

Newton, I. (1704). *Opticks, or a treatise of the reflections, refractions, inflections and colours of light*. William Innys, London.

Niggli, P. (1916). Die Struktur der Kristalle. *Zeit. anorg. Chem.* **94**, 207–216.

Niggli, P. (1919). *Geometrische Kristallographie des Diskontinuums*. Gebrüder Bornträger, Leipzig.

Niggli, P. (1921). Atombau und Atomstruktur. *Z. Kristallogr.* **56**, 167–190.

Niggli, P. (1922). Die Bedeutung des Lauediagrammes für die Kristallographie. *Naturwiss.* **10**, 391–399.

Nishikawa, S. (1914). On the spectrum of X-rays obtained by means of lamellar or fibrous substances. *Proc. Tokyo Math.-Phys. Soc.* **7**, 296–298.

Nishikawa, S. (1915). Structure of some crystals of spinel group. *Proc. Tokyo Math.-Phys. Soc.* **8**, 199–209.

Nishikawa, S. and Matukawa, K. (1928). Hemihedry of zincblende and X-ray reflection. *Proc. Imp. Acad. Japan* **4**, 96–97.

Nishikawa, S. and Ono, S. (1913). Transmission of X-rays through fibrous, lamellar and granular substances. *Proc. Tokyo Math.-Physics Society* **7**, 131–138.

Nix, F. C. and Shockley, W. (1938). Order-disorder transformations in alloys. *Rev. Mod. Phys.* **10**, 1–72.

Okkerse, B. (1963). Consecutive Laue and Bragg reflections in the same perfect crystal. *Philips Res. Repts.* **18**, 413–431.

Ott, H. (1928). Die Kristallstruktur des Graphits. *Ann. Phys.* **390**, 81–109.

Owen, E. A. and Blake, G. G. (1913). X-rays spectra. *Nature* **91**, 135.

Owen, E. A. and Blake, G. G. (1914). X-rays and metallic crystals. *Nature* **92**, 685–686.

Owen, E. A. and Preston, G. D. (1922). Modification of the powder method of determining the structure of metal crystals. *Proc. Phys. Soc.* **3**, 101–108.

Owen, E. A. and Preston, G. D. (1923*a*). X-ray analysis of solid solutions. *Proc. Phys. Soc.* **36**, 14–30.

Owen, E. A. and Preston, G. D. (1923*b*). X-ray analysis of zinc-copper alloys. *Proc. Phys. Soc.* **36**, 49–66.

Owen, E. A. and Preston, G. D. (1923*c*). The atomic structure of two intermetallic compounds. *Proc. Phys. Soc.* **36**, 341–349.

Pardies I. (1972*a*). A Latin Letter Written to the Publisher April 9. 1672 by Ignatius Gaston Pardies Containing Some Animadversions upon Mr. Isaac Newton. *Phil. Trans. Soc.* **7**, 4087–4090.

Pardies I. (1972*b*). A Second Letter of I. Pardies, Written to the Publisher from Paris May 21. 1672. to Mr.Newton's Answer, Made to His First Letter. *Phil. Trans. Soc.* **7**, 5012–5013.

Pasteur, L. (1848*a*). Relation qui peut exister entre la forme cristalline et la composition chimique, et sur la cause de la polarisation rotatoire. *C. R. Acad. Sci. Paris* **26**, 535–538.

Pasteur, L. (1848*b*). Recherches sur le dimorphisme. *Ann. Chim. Phys.* **23**, 267–296.

Pasteur, L. (1848*c*). Relation qui peut exister entre la forme cristalline et la composition chimique, et sur la cause de la polarisation rotatoire. *Ann. Chim. Phys.* **24**, 442–459.

Paterson, M. S. (1952). X-Ray Diffraction by Face-Centered Cubic Crystals with Deformation Faults. *J. Appl. Phys.* **23**, 805–811.

Patterson, A. L. (1928). Über die Messung der Grösse von Kristallteilchen mittels Röntgenstrahlung. *Z. Kristallogr.* **66**, 637–650.

Patterson, A. L. (1934). A Fourier series method for the determination of the components of the interatomic distances in crystals. *Phys. Rev.* **46**, 372–376.

Patterson, A. L. (1935). A direct method for the determination of the components of interatomic distances in crystals. *Z. Kristallogr.* **90**, 517–542.

Patterson, A. L. (1939). The Scherrer formula for X ray particle size determination. *Phys. Rev.* **56**, 978–982.

Paufler, P. (1981). C. S. Weiss in seiner Leipziger Periode. *Wiss. Z. Karl-Marx-Univ. Leipzig, Math.-Naturwiss. R.* **30**, 428–433.

Pauling, L. (1924). The crystal structures of ammonium fluoferrate, fluoaluminate and oxyfluomolybdate. *J. Amer. Chem. Soc.* **46**, 2738–2751.

Pauling, L. (1926). The dynamic model of the chemical bond and its application to the structure of benzene. *J. Amer. Chem. Soc.* **48**, 1132–1143.

Pauling, L. (1927). The sizes of ions and the structure of ionic crystals. *J. Amer. Chem. Soc.* **49**, 765–790.

Pauling, L. (1928*a*). The coordination theory of the structure of ionic crystals. *Festschrift zum 60. Geburtstage Arnold Sommerfelds, Verlag von S. Hirzel, Leipzig.* pp. 1–17.

Pauling, L. (1928*b*). The shared-electron chemical bond. *Proc. Nat. Acad. Sci. USA* **14**, 359–362.

Pauling, L. (1928*c*). The crystal structure of topaz. *Proc. Nat. Acad. Sci. USA* **14**, 603–606.

Pauling, L. (1929). The principles determining the structure of complex ionic crystals. *J. Amer. Chem. Soc.* **51**, 1010–1026.

Pauling, L. (1930*a*). The Structure of the Micas and Related Minerals. *Proc. Nat. Acad. Sci. USA* **16**, 123–129.

Pauling, L. (1930*b*). The structure of some sodium and calcium aluminosilicates. *Proc. Nat. Acad. Sci. USA* **16**, 453–459.

Pauling, L. (1937). The X-ray analysis of crystals. In *Laue Diagrams, 25 Years of research on X-ray diffraction following Prof. Max von Laue discovery.* pp. 20–22. *Current Science.* Indian Academy of Sciences. Bengalore.

Pauling, L. and Sturdivant, J. H. (1928). The crystal structure of brookite. *Z. Kristallogr.* **68**, 239–256.

Pedersen, O. (1991). Steno and the origin of Crystallography. *Stenoniana*, nova series, Copenhagen. **1**, 113–134.

Pepinsky, R. (1955). Crystal engineering: New concepts in Crystallography. *Phys. Rev.* **100**, 971–971.

Perrin, J. (1895). Quelques propriétés des rayons cathodiques. *C. R. Acad. Sci. Paris* **121**, 1130–1134.

Perrin, J. (1896). Quelques propriétés des rayons de Röntgen. *C. R. Acad. Sci. Paris* **122**, 186–188.

Perrin, J. (1897). Rayons cathodiques et rayons de Röntgen. *Ann. Chim. Phys. Paris* **11**, 496–554.

Pertlik, F. (2006). Argumente für die Existenz eines diklinen Krystallsystems in der Fachliteratur des 19, Jahrhunderts, Ein Beitrag zur Geschichte der Kristallographie. *Mitt. Österr. Miner. Ges.* **152**, 17–29.

Perutz, M. F. (1971). Sir Lawrence Bragg. *Nature* **233**, 74–76.

Perutz, M. F. (1990). How Bragg invented X-ray analysis. *Acta Crystallogr.* **46**, 633–643.

Pfeiffer, P. (1915). Die Kristalle als Molekülverbindungen. *Zeit. anorg. Chem.* **92**, 376–380.

Pfeiffer, P. (1920). Die Befruchtung der Chemie durch die Röntgenstrahlenphysik. *Naturwiss.* **8**, 984–989.

Phillips, D. Sir (1979). William Lawrence Bragg. 31 March 1890–1 July 1971. *Bibl. Mems. Fellows Roy. Soc.* **25**, 74–143.

Phragmén, G. (1925). Ueber den Aufbau der Eisen-Silizium Legierungen. *Stahl und Eisen.* **45**, 299.

Pietsch, U., Stahn, J., Davaasambuu, J., and Pucher, A. (2001). Electric field induced charge density variations in partially-ionic compounds. *J. Phys. Chem. Solids* **62**, 2129–2133.

Piper, S. H. (1922). The Fine Structure of Some Sodium Salts of the Fatty Acids in Soap Curds. *Proc. Phys. Soc.* **35**, 269–372.

Piper, S. H. (1929). Some examples of information obtainable from the long spacings of fatty acids. *Trans. Faraday Soc.* **25**, 348–351.

Planck, M. (1937). Zum 25 jährigen Jubiläum der Entdeckung von W. Friedrich, P. Knipping und M. v. Laue. *Verhandl. Dtsche Phys. Ges.* **18**, 77–80.

Poincaré, R. (1896). Les rayons cathodiques et les rayons Röntgen. *Rev. Génér. Sciences.* **7**, 52–59.

Poisson, S. D. (1829). Mémoire sur l'équilibre et le mouvement des corps élastiques. *Mém. Acad. Sci.* **8**, 357–380.

Poisson, S. D. (1831). Mémoire sur les équations générales de l'équilibre et du mouvement des corps solides élastiques et des fluides. *Journ. de l'Ecole Polytechnique.* **13**, 1–174.

Polanyi, M. (1921a). Faserstruktur im Röntgenlichte. *Naturwiss.* **9**, 337–340.

Polanyi, M. (1921b). Das Röntgenfaserdiagramm. *Z. Phys.* **7**, 149–180.

Polanyi, M. (1922). Röntgenographische Bestimmung von Kristallanordnugen. *Naturwiss.* **16**, 411–416.

Polanyi, M. (1934). Über eine Art Gitterstörung, die einen Kristall plastisch machen könnte. *Z. Phys.* **89**, 660–664.

Polanyi, M. and Weissenberg, K. (1922a). Das Röntgenfaserdiagramm. *Z. Phys.* **9**, 123–130.

Polanyi, M. and Weissenberg, K. (1922b). Das Röntgenfaserdiagramm. *Z. Phys.* **10**, 44–53.

Polanyi, M., Schiebold, E., and Weissenberg, K. (1924). Über die Entwicklung des Drehkristallverfahrens. *Z. Phys.* **23**, 337–340.

Posner, E. (1970). Reception of Röntgen's discovery in Britain and the U.S.A. *Brit. Med. Journal* **4**, 357–360.

Poulsen, J. E. and Snorrason, E., Eds. (1986). *Nicolaus Steno (1638–1686). A reconsideration by Danish scientists.* Nordisk Insulinlaboratorium, Gentofte.

Prechtl, J. J. (1808). Theorie der Crystallisation. *J. Chem. Phys. Mineral. Berlin* **5**, 455–504.

Preston, G. D. (1928). The structure of α-manganese. *Phil. Mag.* Series 7. **5**, 1198–1206.

Preston, G. D. (1938a). The diffraction of X-rays by age-hardening aluminium copper alloys. *Proc. Roy. Acad. A* **167**, 526–538.

Preston, G. D. (1938b). Structure of age-hardened aluminium-copper alloys. *Nature.* **142**, 570.

Prins, J. A. (1930). Die Reflexion von Röntgenstrahlen an absorbierenden idealen Kristallen. *Z. Phys.* **63**, 477–493.

Pryce, W. (1778). *Mineralia cornubiensis; A treatise on minerals, mines and mining.* James Phillips, London.

Puluj J. (1896). Über die Entstehung der Röntgenstrahlen und ihre photographische Wirkung. *Wiener Berichte.* **105**, 228–238; 243–245.

Raman, C. V. (1937). Introduction. In *Laue Diagrams, 25 Years of research on X-ray diffraction following Prof. Max von Laue discovery.* pp. 1–2. *Current Science.* Indian Academy of Sciences, Bangalore.

Rashed, R. (1990). A Pioneer in Anaclastics: Ibn Sahl on Burning Mirrors and Lenses. *Isis*, **81**, 464–491.

Rayleigh, Lord (1874). On the manufacture and theory of diffraction-gratings. *Phil. Mag.* Series 4. **47**, 81–93 and 193–205.

Renninger, M. (1955). Messungen zur Röntgenstrahl-Optik des Idealkristalls. I. Bestätigung der Darwin-Ewald-Prins-Kohler-Kurve. *Acta Crystallogr.* **8**, 597–606.

Richards, T. W. (1913). The chemical significance of crystalline form. *J. Am. Chem. Soc.* **35**, 381–396.

Riemann, B. (1867). Ein Beitrag zur Elektrodynamik. *Ann. Phys.* **207**, 237–243.

Rinne, F. (1917). Zur Leptonenkunde als Feinbaulehre der Stoffe. *Naturwiss.* **5**, 49–56.

Rinne, F. (1919). *Einführung in die kristallographische Formenlehre und elementare Anleitung zu kristallographisch-optischen sowie röntgenographischen Untersuchungen.* M. Jänecke, Leipzig.

Rinne, F. (1921). Röntgenographische Feinbaustudien. *Abh d. Sächs. Akad. der Wiss. Math-Phys. Klasse,* **38**, No 3.

Robertson, J. M. (1933a). The crystalline structure of anthracene. A quantitative X-ray investigation. *Proc. Roy. Soc. A* **140**, 79–98.

Robertson, J. M. (1933b). The crystalline structure of naphthalene. A quantitative X-ray investigation. *Proc. Roy. Soc. A* **142**, 674–688.

Robertson, J. M. (1964). The problem and development of organic crystal structure analysis. *Z. Kristallogr.* **120**, 71–89.

Romé de l'Isle, J.-B. L. (1772). *Essai de Cristallographie ou description des figures géométriques propres à différents corps du règne minéral connus vulgairement sous le nom de cristaux.* Didot jeune, Paris.

Romé de l'Isle, J.-B. L. (1783). *Cristallographie, ou Description des Formes Propres à tous les Corps du Règne Minéral, dans L'Etat de Combinaison Saline, Pierreuse ou Métallique.* Didot jeune, Paris.

Röntgen, W. C. (1888). Über die durch Bewegung eines im homogenen elektrischen Felde befindlichen Dielektrikums hervogerufen elektrodynamischen Kraft. *Sitzber. Königl. Preuss. Akad. Wiss. zu Berlin,* 23–28.

Röntgen, W. C. (1895). Über eine neue Art von Strahlen. *Sitzungsber. Der Würzburger Physik-Medic. Gesellsch.* **137**, 132–141. Translated into English by Arthur Stanton (1896). On a new kind of rays. *Nature* **53**, 274–276. (23 January 1896.)

Röntgen, W. C. (1896). Über eine neue Art von Strahlen. II. Mittheilung. *Sitzungsber. Der Würzburger Physik-Medic. Gesellsch.* **138**, 1–11. Translated in English by G. F. Barker, *Röntgen rays* (1899). Harper & Brothers, New York.

Röntgen, W. C. (1897). Weitere Beobachtungen über die Eigenschaften der X-Strahlen. Dritte Mittheilung. *Sitzungsber. König. Preuss. Akad. Wiss. zu Berlin* 392–406. Translated in English by G. F. Barker, *Röntgen rays* (1899). Harper & Brothers Publishers, New York.

Röntgen, W. C. and Ioffe, A. (1913). Über die Elektrizitätsleitung in einigen Kristallen und über den Einfluss der Bestrahlung darauf. *Ann. Ohys.* **346**, 449–498.

Rose, G. (1844). Ueber das Krystallisationssystem des Quarzes. *Annal. Phys.* **62**, 325–337.

Rosenhain, W. (1915). *An introduction to the study of physical metallurgy.* Constable & Co Ltd., London.

Rousseau, G. (1992). On Gauss's proof of Seeber's theorem. *Aequationes Mathematicae* **43**, 145–155.

Runge, C. (1917). Die Bestimmung eines Kristallsystems durch Röntgenstrahlen. *Phys. Zeit.* **18**, 509–515.

Rutherford, E. (1903). The magnetic and electric deviation of the easily absorbed rays from radium. *Phil. Mag.* Series 5. **5**, 177–187.

Rutherford, E. (1911). The scattering of alpha and beta particles by matter and the structure of the atom. *Phil. Mag.* Series 6. **21**, 669–688.

Rutherford, E. (1915a). Henry Gwyn Jeffries Moseley. *Nature* **96**, 33–34.

Rutherford, E. (1915b). H. G. J. Moseley, 1887–1915. *Proc. Roy. Soc. A* **93**, xxii–xxviii.

Rutherford, E. (1925). Moseley's work on X-rays. *Nature* **116**, 316–317.

Rutherford, E. and Andrade, E. N. da C. (1914a). The wave-length of the soft γ-rays from radium B. *Phil. Mag.* Series 6. **27**, 854–868.

Rutherford, E. and Andrade, E. N. da C. (1914b). The spectrum of the penetrating γ-rays from radium B and radium C. *Phil. Mag.* Series 6. **28**, 263–273.

Rutherford, E., Barnes, J., and Richardson, H. (1915). Maximum frequency of the X rays from a Coolidge tube for different voltages. *Phil. Mag.* Series 6. **30**, 339–360.

Sadler, C. A. and Mesham, P. (1915). The Röntgen radiation from substances of low atomic weight. *Phil. Mag.* Series 6. **24**, 138–149.

Sagnac, G. (1896). Sur la diffraction et la polarisation des rayons de M. Röntgen. *C. R. Acad. Sci. Paris* **122**, 783–785.

Sagnac, G. (1897). Sur la transformation des rayons X par les métaux. *C. R. Acad. Sci. Paris* **125**, 230–232 and 942–944.

Sagnac, G. (1901). Rayons secondaires dérivés des rayons de Röntgen. *Ann. Chim. Phys. Paris* **22**, 493–563.

Savchuk, W. (2007). The naturalist I. P. Puljuj and the discovery of X-rays. In M. Kokowski (ed.), *The Global and the Local: The History of Science and the Cultural Integration of Europe*. Proceedings of the 2nd ICESHS (Cracow, Poland, September 6–9, 2006). The Press of the Polish Academy of Arts and Sciences, Cracow.

Scherrer, P. (1918). Bestimmung der Grösse und der inneren Struktur von Kolloidteilchen mittels Röntgenstrahlen. *Nachr. Kng. Ges. Wiss. Göttingen*. 98–100.

Scherrer, P. and Stoll, P. (1922). Bestimmung der von Werner abgeleiteten Struktur anorganischer Verbindungen vermittelst Röntgenstrahlen. *Zeit. anorg. Chem.* **121**, 319–320.

Scherz, G. (1987, 1988). *Niels Stensen; eine Biographie*. Vols. 1 & 2. St Benno Verlag, Leipzig.

Schiebold, E. (1922*a*). Beiträge zur Auswertung der Lauediagramme. *Naturwiss.* **10**, 399–411.

Schiebold, E. (1922*b*). Bemerkungen zur Arbeit: Das Röntgenfaserdiagramm von M. Polanyi. *Z. Phys.* **9**, 180–183.

Schiebold, E. (1924). Über graphische Auswertung von Röntgenphotogrammen. *Z. Phys.* **28**, 355–370.

Schiebold, E. (1929*a*). The fine structure of feldspars. *Trans. Faraday Soc.* **25**, 316–320.

Schiebold, E. (1929*b*). Über den Feinbau der Feldspates. *Forschritte Min. Krist. Petr.* **14**, 62–68.

Schmidt, G. J. M. (1971). Photodimerization in the Solid State. *Pure & Appl. Chem.* **27**, 647–678.

Schneer, C. J. (1960). Kepler's New Year's Gift of a Snowflake. *Isis* **51**, 531–545.

Schneer, C. J. (1971). *Steno: On crystals and the corpuscular hypothesis.* pp. 293–307. In *Dissertations on Steno As Geologist* edited by G. Scherz, Acta Historica Scientiarum Naturalium Et Medicinalium, Edidit Bibliotheca Hauniensis, Vol. **23**. Odense University Press, Copenhagen.

Schoenflies, A. M. (1883). Über die Bewegung eines starren räumlichern Systems. *Zeit. Math. Phys.* **28**, 229–241.

Schoenflies, A. M. (1886). *Geometrie der Bewegung in synthetischer Darstellung*. B. G. Teubner, Leipzig.

Schoenflies, A. M. (1887). Über Gruppen von Bewegungen. *Mathematische Annalen.* **28**, 319–342; **29**, 50–80.

Schoenflies, A. M. (1888*a*). Über reguläre Gebietstheilungen des Raumes. *Nachrichten Kgl. Gesell. Wiss. Göttingen*, 223–237.

Schoenflies, A. M. (1888*b*). Beitrag zur Theorie der Krystallstructur. *Nachrichten Kgl. Gesell. Wiss. Göttingen*, 483–501.

Schoenflies, A. M. (1889). Über Gruppen von Transformationen des Raumes in sich. *Mathematische Annalen.* **34**, 172–203.

Schoenflies, A. M. (1891). *Krystallsysteme und Krystallstruktur*. B. G. Teubner, Leipzig.

Schoenflies, A. M. (1892). Bemerkung zu dem Artikel des Herrn E. von Fedorow, die Zusammentstellung seinr krystallographischen Resultate und der meinigen betreffend. *Z. Kristallogr.* **20**, 25–75.

Schoenflies, A. M. (1915). Über Krystallstruktur. *Z. Kristallogr.* **54**, 545–569; **55**, 321–352.

Scholz, E. (1989). The rise of symmetry concepts in the atomistic and dynamistic schools of crystallography, 1815–1830. *Rev. Hist. Sci.* **42**, 109–122.

Schülke, W. and Brümmer, O. (1962). Vergleichende Untersuchungen von Interferenzen bei kohärenter und inkohärenter Lage der Röntgen-Strahlenquelle zum Kristallgitter. *Z. Naturforsch.* **A17**, 208–216.

Schuster, A. (1884). Bakerian Lecture: The discharge of electricity through gases. (Preliminary Communication) *Proc. Roy. Soc.* **47**, 526–561.

Schuster, A. (1890). The Bakerian Lecture: Experiments on the discharge of electricity through gases. Sketch of a theory. *Proc. Roy. Soc.* **37**, 317–339.

Schwalbe, C. H. (2014). Lars Vegard: key communicator and pioneer crystallographer. *Crystallography Reviews*, **20**, 9–24.

Schwarzschild, M. (1928). Theory of the double X-ray spectrometer. *Phys. Rev.* **32**, 162–171.

Seeber, L. A. (1824). Versuch einer Erklärung des inneren Baues der Festen Körper. *Ann. Phys.* **76**, 229–248, 349–371.

Seeber, L. A. (1831). *Untersuchungen über die Eigenschaften der positiven ternären quadratischen Formen*. Freiburg.

Seemann, H. (1914). Das Röntgenspektrum des Platins. *Phys. Zeit.* **15**, 794–797.

Seemann, H. (1916). Röntgenspektroskopie Methoden ohne Spalt. *Ann. Phys.* **354**, 470–480.

Seemann, H. (1919*a*). Eine fokussierende röntgenspektrometrische Anordnung für Kristallpulver. *Ann. Phys.* **364**, 455–464.

Seemann, H. (1919*b*). Vollständige Spektraldiagramme von Kristallen. *Phys. Zeit.* **20**, 169–175.

Seliger, H. H. (1995). Wilhelm Conrad Röntgen and the glimmer of light. *Physics Today* **48**, 25–31.

Seljakov, N. (1925). Eine röntgenographische Methode zur Messung der absoluten Dimensionen einzelner Kristalle in Körpern von fein-kristallinischem Bau. *Zeit. Phys.* **31**, 439–444.

Selling, E. (1877). Des formes quadratiques binaires et ternaires. *J. Math. Pures Appl.* **3**, 21–60, 153–206.

Senechal, M. (1981). Which tetrahedra fill space. *Mathematics Magazine* **54**, 227–243.

Senechal, M. (1990). *Brief history of geometrical crystallography.* In Lima-de-Faria (1990). pp. 43–59.

Shafranovskii, I. I. (1971). *Die kristallographischen Entdeckungen Niels Stensen.* pp. 244–259. In *Dissertations on Steno As Geologist* edited by G. Scherz, Acta Historica Scientiarum Naturalium Et Medicinalium, Edidit Bibliotheca Hauniensis, Vol. **23**. Odense University Press, Copenhagen.

Shafranovskii, I. I. (1972). *Nicolaus Steno: A Crystallographer, Geologist, Paleontologist, and Anatomist.* Nauka, Leningrad (in Russian).

Shafranovskii, I. I. (1975). Kepler's crystallographic ideas and his tract "The six-cornered snowflake". *Vistas in Astronomy* **18**, 861–876.

Shafranovskii, I. I. (1978). *The History of Crystallography, Vol. I. From Ancient Times to the Beginning of the Nineteenth Century* (in Russian). Nauka, Leningrad.

Shafranovskii, I. I. (1980). *The History of Crystallography, Vol. II. The Nineteenth Century* (in Russian). Nauka, Leningrad.

Shapiro, A. E. (1973). Kinematic optics: A study of the wave theory of light in the seventeenth century. *Arch. Hist. Exact Sci.* **11**, 134–266.

Shirley, J. W. (1951). An early experimental determination of Snell's law. *Am. J. Phys.* **19**, 507–508.

Shirley, J. W. (ed.) (1974). *Thomas Harriot: renaissance scientist.* Oxford University Press, Oxford.

Shirley, J. W. (1983). *Thomas Harriot: a biography.* Oxford University Press, Oxford.

Shull C. G. and Wollan (1948) X-Ray, Electron, and Neutron Diffraction. *Science*, **108**, N° 2795, 69–75.

Siegbahn, M. (1914). Über den Zusammenhang zwischen Absorption und Wellenlänge bei Röntgenstrahlen. *Phys. Zeit.* **15**, 753–756.

Siegbahn, M. (1916a). Relations between the K and L Series of the High-Frequency Spectra. *Nature* **96**, 676–676.

Siegbahn, M. (1916b). Sur l'existence d'un nouveau groupe de lignes (série M) dans les spectres de haute fréquence. *C. R. Acad. Sci. Paris* **162**, 787–788.

Siegbahn, M. (1919a). Röntgenspektroslopische Präzisionsmessungen. *Ann. Phys.* **59**, 56–72.

Siegbahn, M. (1919b). Precision-measurements in the X-ray spectra. *Phil. Mag.* Series 6. I. **37**, 601–612; II. **38**, 639–646.

Siegbahn, M. (1924). *Spektroskopie der Röötgenstrahlen.* Julius Springer, Berlin. Second edition: 1931.

Siegbahn, M. (1925). La réflexion et la réfraction des Rayons X. *J. Phys. Rad.* **6**, 228–231.

Siegbahn, M. (1937). X-ray spectroscopy. In *Laue Diagrams, 25 Years of research on X-ray diffraction following Prof. Max von Laue discovery.* pp. 14–15. *Current Science.* Indian Academy of Sciences. Bengalore.

Siegbahn, M. and Friman, E. (1916). On an X-ray vacuum spectrograph. *Phil. Mag.* Series 6. **32**, 494–496.

Siegbahn, M. and Leide, A. (1919). Precision-measurements in the X-ray spectra. Part III. *Phil. Mag.* Series 6. **38**, 647–651.

Smekal, A. (1927). Über die Grössenordnung der ideal gebauten Gitterbereiche in Realkristallen. *Ann. Phys.* **388**, 1202–1206.

Smekal, A. (1933). *Strukturempfindliche Eigenschaften der Kristalle.* In *Handbuch der Physik*, **XXIV/2**, 2nd edition, Eds. H. Geiger and K. Scheel.

Smekal, A. (1934). Zur Theorie der Realkristalle. *Z. Kristallogr.* **89**, 386–399.

Sohncke, L. (1867). Die Gruppierung der Molecüle in den Krystallen. Eine theoretische Ableitung der Krystallsysteme und ihrer Unterabtheilungen. *Ann. Phys.* **208**, 75–106.

Sohncke, L. (1869). Ueber die Cohäsion des Steinsalzes in krystallographisch verschiedenen Richtungen. *Ann. Phys.* **213**, 177–200.

Sohncke, L. (1874). Die regelmässig ebenen Punkt systeme von unbegrenzter Ausdehnung. *J. reine angew. Math.* **77**, 47–102.

Sohncke, L. (1876). Die unbegränzten regelmässigen Punktsysteme als Grundlage einer Theorie der Krystallstruktur. *Ann. Phys.* **E7**, 337–390.

Sohncke, L. (1879). *Entwickelung einer Theorie der Krystallstruktur.* B.G. Teubner, Leipzig.

Sohncke, L. (1882). Ableitung des Grundgesetzes der Krystallographie aus der Theorie der Kristallstruktur. *Ann. Phys.* **252**, 489–500.

Sohncke, L. (1884). Probable nature of the internal symmetry of crystals. *Nature* **29**, 383–384.

Sohncke, L. (1888). Erweiterung einer Theorie der Kristallstruktur. *Z. Kristallogr.* **14**, 426–446.

Sohncke, L. (1891a). Die Entdeckung des Eintheilungsprincips der Krystalle durch J. F. C. Hessel. *Z. Kristallogr.* **18**, 486–498.

Sohncke, L. (1891b). Die Struktur der optisch drehenden Krystalle. *Z. Kristallogr.* **19**, 486–498.

Sommerfeld, A. (1896). Mathematische Theorie der Diffraction. *Mathematische Annalen*. **47**, 317–374.

Sommerfeld, A. (1899). Theoretisches über die Beugung der Röntgenstrahlen. *Phys. Zeit.* **1**, 105–111.

Sommerfeld, A. (1900). Theoretisches über die Beugung der Röntgenstrahlen. II. Mittheilung. *Phys. Zeit.* **2**, 55–60.

Sommerfeld, A. (1901). Theoretisches über die Beugung der Röntgenstrahlen. *Zeit. f. Mathematik und Physik.* **46**, 11–97.

Sommerfeld, A. (1909). Über die Verteilung der Intensität bei der Emission von Röntgenstrahlen. *Phys. Zeit.* **10**, 969–976.

Sommerfeld, A. (1910). Über die Verteilung der Intensität bei der Emission von Röntgenstrahlen. *Phys. Zeit.* **11**, 99–101.

Sommerfeld, A. (1912). Über die Beugung der Röntgenstrahlen. *Ann. Phys.* **343**, 473–506.

Sommerfeld, A. H. (1913). Sur les photogrammes quaternaires et ternaires de la blende et le spectre du rayonnement de Röntgen. In *Rapport du Conseil de Physique de l'Institut International Solvay*, Bruxelles, 27–31 Octobre 1913. (1921). Gauthier-Villars, Paris. pp. 125–134.

Sommerfeld, A. (1916). Zur Quantentheorie der Spektrallinien. *Ann. Phys.* **356**, I. 1–94; III. Theorie der Röntgenstrahlen. 125–167.

Sommerfeld, A. (1919). *Atombau und Spektrallinien*. Vieweg & Sohn, Braunschweig.

Sommerfeld, A. (1920). Bemerkungen zur Feinstruktur der Röntgenspektren. *Z. Phys.* **1**, 135–146.

Sommerfeld, A. (1937). X-ray spectroscopy and atomic structure. In *Laue Diagrams, 25 Years of research on X-ray diffraction following Prof. Max von Laue discovery*. pp. 16–19. *Current Science*. Indian Academy of Sciences, Bangalore.

Stark, J. (1907). Elementarquantum der Energie. Modell der negativen und der positiven Elektrizität. *Phys. Zeit.* **8**, 881–884.

Stark, J. (1908). The wave-length of Röntgen rays. *Nature* **77**, 320–320.

Stark, J. (1909*a*). Über Röntgenstrahlen und die atomistische Konstitution der Strahlung. *Phys. Zeit.* **10**, 579–586.

Stark, J. (1909*b*). Zur experimentellen Entscheidung zwischen Ätherwellen- und Lichtquantenhypothese. Röntgenstrahlung. *Phys. Zeit.* **10**, 902–913.

Stark, J. (1910). Zur experimentellen Entscheidung zwischen Ätherwellen- und Lichtquantenhypothese. Röntgenstrahlung. *Phys. Zeit.* **11**, 24–31.

Stark, J. (1912). Bemerkung über Zerstreuung und Absorption von β-strahlen und Röntgenstrahlen in Kristallen. *Phys. Zeit.* **13**, 973–977.

Stark, J. and Wendt, G. (1912*a*). Über das Eindringen von Kanalstrahlen in feste Körper. *Ann. Phys.* **343**, 921–940.

Stark, J. and Wendt, G. (1912*b*). Pflanzt sich der Stoss von Kanalstrahlen in einem festen Körper fort? *Ann. Phys.* **343**, 941–957.

Steinmetz, H. and Weber, L. (1939). P. von Groth, Der Gründer der Zeitschrift für Kristallographie. *Z. Kristallogr.* **100**, 5–46.

Steno, N. (1669). *De solido intra solidum naturaliter contento dissertationis prodromus.* Stellae, Florence. reprinted in Stenonis, Nicolai (1910). *Opera Philosophica*, Vilhelm Maar Editor. Vol. II. Copenhagen: Vilhelm Tryde.

Stenström, W. (1919). *Experimentelle Untersuchungen der Röntgenspectra*. Dissertation, Lund.

Stokes, G. G. Sir (1849*a*). On the theories of the internal friction of fluids in motion and of the equilibrium and motion of elastic fluids. *Trans. Cambridge Philos. Soc.* **8**, 287–319.

Stokes, G. G. Sir (1849*b*). On the dynamical theory of diffraction. *Trans. Cambridge Philos. Soc.* **8**, 1–62.

Stokes, G. G. Sir (1896*a*). On the nature of X-rays. *Proc. Camb. Phil. Soc.* **9**, 215–216.

Stokes, G. G. Sir (1896*b*). On the Röntgen rays. *Nature* **9**, 427–430.

Stokes, G. G. Sir (1897). On the nature of the Röntgen rays (the Wilde lecture). *Mem. Proc. Manchester Lit. and Phil. Soc.* **41**, 43–66.

Stokes, A. R. and Wilson, A. J. C. (1942). A method of calculating the integral breadths of Debye–Scherrer-lines. *Proc. Camb. Phys. Soc.* **38**, 313–322.

Stokes, A. R. and Wilson, A. J. C. (1944). The diffraction of X-rays by distorted-crystal aggregates. I. *Proc. Phys. Soc. A.* **56**, 174–181.

Stoney, G. J. (1894). Of the 'electron', or atom of electricity. *Phil. Mag.* Series 5. **38**, 418–420.

Swinton, A. A. C. (1896). Professor Röntgen's discovery. *Nature* **53**, 276–277.

Takagi, S. (1962). Dynamical theory of diffraction applicable to crystals with any kind of small distortion. *Acta Crystallogr.* **15**, 1311–1312.

Takahashi, A. (2008). Nature, formative power and intellect in the natural philosophy of Albert the Great. *Early science and medicine* **13**, 451–481.

Tammann, G. (1918). Über die Verteilung zweier Atomarten in den regulären, Frankenheim-Bravaisschen Raumgittern. *Nachrichten Kgl. Gesell. Wiss. Göttingen*. 190–234.

Tandy, P. (2004). William Barlow (1845–1934): speculative builder, man of leisure and inspired crystallographer. *Proc. Geologists' Association* **115**, 77–84.

Taylor, W. H. (1933). The structure of sanidine and other feldspars. *Z. Kristallogr.* **85**, 425–442.

Taylor, H. F. (1972). W. J. D. Bernal 1901–1971. *Acta Crystallogr.* **A28**, 359–360.

Teichmann, J., Eckert, M., and Wolff, S. (2002). Physicists and Physics in Munich. *Phys. Persp.* **4**, 333–359.

Terada, T. (1913*a*). X-rays and crystals. *Nature* **91**, 135–136; 213.

Terada, T. (1913*b*). On the transmission of X-rays through crystals. *Proc. Tokyo Math.-Physics Society* **7**, 60–70.

Thomas, J. M. (2012). William Lawrence Bragg: The Pioneer of X-ray Crystallography and His Pervasive Influence. *Angewan. Chemie*, **51**, 12946–12958.

Thomas, J. M. and Phillips, D. C. (1990). *Selections and Reflections: the Legacy of Sir Lawrence Bragg*. Science Reviews Ltd., London.

Thomson, G. P. and Reid, A. (1927). Diffraction of cathode rays by a thin film. *Nature* **119**, 890.

Thomson, J. J. (1894). On the velocity of the cathode-rays. *Phil. Mag.* Series 5. **38**, 358–365.

Thomson, J. J. (1896*a*). Longitudinal electric waves and Röntgen's rays. *Proc. Camb. Phil. Soc.* **9**, 49–61.

Thomson, J. J. (1896*b*). *On cathode rays and their probable connection with Röntgen rays. Report of the Sixty-sixth Meeting of the British Association for the Advancement of Science*. London: John Murray. pp. 699–709.

Thomson, J. J. (1897). Cathode rays. *Phil. Mag.* Series 5. **44**, 293–309.

Thomson, J. J. (1898*a*). A theory of the connection between cathode and Röntgen rays. *Phil. Mag.* Series 5. **45**, 172–183.

Thomson, J. J. (1898*b*). On the charge of electricity carried by the ions produced by Röntgen rays. *Phil. Mag.* Series 5. **46**, 528–545.

Thomson, J. J. (1898*c*). On the diffuse reflection of Röntgen rays. *Proc. Camb. Phil. Soc.* **9**, 393–397.

Thomson, J. J. (1899). On the masses of the ions in gases at low pressures. *Phil. Mag.* Series 5. **48**, 547–567.

Thomson, J. J. (1903). *Conduction of electricity through gases*. Cambridge University Press.

Thomson, J. J. (1910). On the scattering of rapidly moving electrified particles. *Proc. Camb. Phil. Soc.* **15**, 465–471.

Thomson, J. J. (1912). Ionization by moving electrified particles. *Phil. Mag.* Series 6. **28**, 449–457.

Townsend, J. S. (1900). Secondary Röntgen rays. *Proc. Camb. Phil. Soc.* **10**, 217–226.

Tu, Y. (1932). A precision comparison of calculated and observed grating constants of crystals. *Phys. Rev.* **40**, 662–675.

Tutton, A. E. H. (1911). *Crystallography and practical crystal measurements*. Macmillan and Co. Ltd., London. Revised Edition (1922).

Tutton, A. E. H. (1912*a*). Crystallo-chemical analysis, a new method of chemical analysis. *Nature* **89**, 503–505.

Tutton, A. E. H. (1912*b*). The crystal space-lattice revealed by Röntgen rays. *Nature* **90**, 306–309.

Tutton, A. E. H. (1917). X-ray analysis and topic axes of the alkali sulphates, and their bearing on the theory of valency volumes. *Proc. Roy. Soc. A* **93**, 72–89.

Underwood, E. A. (1945). Wilhelm Conrad Röntgen (1845–1923) and the early developments of radiology. *Proc. Roy. Soc. Medicine* **38**, 697–606.

van Arkel, A. E. (1925). Über die Verformung des Kristallgitters von Metallen durch mechanische Bearbeitung. *Physica* **5**, 208–212.

van Arkel, A. E. and Burgers, W. G. (1928). Verbreiterung der Debye-Seherrerschen Linien von kaltbearbeitetem Wolframdraht und Wolframband als Funktion der Glühtemperatur und Glühdauer. *Z. Phys.* **48**, 690–702.

Van Besien, L. and Van Besien, Y. (2009). L'étonnant destin de Nicolas Stenon (1638–1686). *Actes. Société Française d'histoire de l'art dentaire*. **14**, 78–81.

van den Broek, A. (1913). Die Radioelemente das periodische System, und die Konstitution der Atome. *Phys. Zeit.* **14**, 32–41.

Vegard, L. (1910). On the polarisation of X-rays compared with their power of exciting high velocity cathode rays. *Proc. Roy. Soc. A.* **83**, 379–393.

Vegard, L. (1916*a*). The structure of silver crystals. *Phil. Mag.* **31**, 83–87.

Vegard, L. (1916*b*). Results of crystal analysis. *Phil. Mag.* **32**, 65–96; 505–518.

Vegard, L. (1921). Die Konstitution der Mischkristalle und die Raumfüllung der Atome. *Z. Phys.* **5**, 17–26.

Vegard, L. (1926). Results of crystal analysis. *Phil. Mag.* Series 6. **1**, 1151–1193.

Vegard, L. (1928). Die Röntgenstrahlen im Dienste der Erforschung der Materie. *Z. Kristallogr.* **67**, 239–258.

Vegard, L. and Schjelderup, H. (1917). Die Konstitution der Mischkristalle. *Phys. Zeit.* **18**, 93–96.

Villard, P. (1900*a*). Sur la réflexion et la réfraction des rayons cathodiques et des rayons déviables du radium. *C. R. Acad. Sci. Paris* **130**, 1010–1012.

Villard, P. (1900*b*). Sur le rayonnement du radium. *C. R. Acad. Sci. Paris* **130**, 1178–1179.

Voigt, W. (1887). Theoretische Studien über die Elasticitätsverhältnisse der Krystalle. I. *Abhandl. König. Gesel. Wiss. Göttingen.* 3–52.

Voigt, W. (1889). Ueber die Beziehung zwischen den beiden Elasticitätsconstanten isotroper Körper. *Ann. Phys.* **274**, 573–587.

Voigt, W. (1910). *Lehrbuch der Kristallphysik.* B. G. Teubner, Leipzig.

Voller, A. and Walter, B. (1897). Mittheilungen über einige Versuche mit Röntgenstrahlen. *Ann. Phys.* **297**, 88–104.

Voronoï, G. (1908). Nouvelles applications des paramètres continus à la théorie des formes quadratiques. Deuxième mémoire. Recherches sur les parallélloèdres primitifs. *J. reine und angew. Math.* **134**, 198–287.

Wagenfeld, H. (1968). Ewald's and von Laue's dynamical theories of X-ray diffraction. *Acta Crystallogr.* **A24**, 170–174.

Wagner, E. (1914). Spektraluntersuchungen an Röntgenstrahlen: nach Versuchen gemeinsam mit Joh. Brentano. *Sitzungsberichte der Kgl. Bayer. Akad. der Wiss.* 329–338.

Wagner, E. (1915*a*). Das Röntgenspektrum des Platins. *Phys. Zeit.* **16**, 30–33.

Wagner, E. (1915*b*). Spektraluntersuchungen an Röntgenstrahlen. *Ann. Phys.* **351**, 868–892.

Wagner, E. (1916). Über vergleichende Raumgittermessungen an Steinsalz und Sylvin mittels homogener Röntgenstrahlen und über deren exakte Wellenlängenbestimmung. *Ann. Phys.* **354**, 625–647.

Wagner, E. (1918). Spektraluntersuchungen an Röntgenstrahlen. Über die Messung der Planckschen Quantenkonstante *h* aus dem zur Erzeugung homogener Bremsstrahlung notwendigen Minimumpotential. *Ann. Phys.* **362**, 401–470.

Wagner, E. and Glocker, R. (1913). Experimenteller Beitrag zur Interferenz der Röntgenstrahlen. *Phys. Zeit.* **14**, 1232–1237.

Wagner, E. and Kulenkampf, H. (1922). Die Intensität der Reflexion von Röntgenstrahlen verschiedener Wellenlänge an Kalkspat und Steinsalz. *Ann. Phys.* **373**, 369–413.

Waller, I. (1923). Zur Frage der Einwirkung der Wärmebewegng auf die Interferenz von Röntgenstrahlen. *Zeit. Phys.* **17**, 398–408.

Waller, I. (1927). Die Einwirkung der Wärmebewegung der Kristallatome auf Intensität, Lage und Schärfe der Röntgenspektrallinien. *Ann. Phys.* **388**, 153–183.

Waller, I. (1928). Über eine verallgemeinerte Streuungsformel. *Zeit. Phys.* **51**, 213–231.

Waller, I. and Hartree, D. R. (1929). On the Intensity of Total Scattering of X-Rays. *Proc. Roy. Soc. A* **124**, 119–142.

Waller, I. and James, R. W. (1927). On the Temperature Factors of X-Ray Reflexion for Sodium and Chlorine in the Rock-Salt Crystal. *Proc. Roy. Soc. A* **117**, 214–223.

Walter, B. (1898). Ueber die Natur der Röntgenstrahlen. *Ann. Phys.* **302**, 74–82.

Walter, B. (1902). Über die Haga und Wind'schen Beugungsversuche mit Röntgenstahlen. *Phys. Zeit.* **3**, 137–140.

Walter, B. and Pohl, R. (1908). Zur Frage der Beugung der Röntgenstrahlen. *Ann. Phys.* **330**, 715–724.

Walter, B. and Pohl, R. (1909). Weitere Versuche über die Beugung der Röntgenstrahlen. *Ann. Phys.* **334**, 331–354.

Wangerin, A. (1896). F. E. Neumann. *Leopoldina*, Halle. **32**, 51–54, 63–66.

Wangerin, A. (1907). *Franz Neumann und sein Wirken als Forscher und Lehrer.* Vieweg, Berlin. Republished by Bibliobazaar (2009).

Warren, B. E. (1929). The crystal structure and the chemical composition of the monoclinic amphibole. *Z. Kristallogr.* **72**, 493–528.

Warren, B. E. (1969). *X-ray Diffraction.* Addison-Wesley, Reading.

Warren, B. E. and Averbach, B. L. (1950). The Effect of Cold-Work Distortion on X-Ray Patterns. *J. Appl. Phys.* **21**, 595–599.

Warren, B. E. and Averbach, B. L. (1952). The Separation of Cold-Work Distortion and Particle Size Broadening in X-Ray Patterns. *J. Appl. Phys.* **23**, 497–497.

Warren, B. E. and Bragg, W. L. (1929). The crystal structure of diopside. *Z. Kristallogr.* **69**, 168–193.

Warren, B. E. and Warekois, E. P. (1955). Stacking faults in cold worked alpha-brass. *Acta Met.* **3**, 473–479.

Wasastjerna, J. A. (1923). On the radii of ions. *Acta Societatis Fennicae Scientiaru.* **1**, 1–25.

Webster, D. L. (1915). The intensities of X-ray spectra. *Phys. Rev.* **5**, 238–243.

Webster, D. L. (1916). Experiments on the emission quanta of characteristic X-rays. *Phys. Rev.* **7**, 599–613.

Weiss, C. S. (1809). *De indagando formarum cristallinarum caractere geometrico principali.* Dissertatio, Leipzig.

Weiss, C. S. (1814–1815). Uebersichtliche Darstellung der verschiedenen naturlichen Abteilungen der Kristallisations-Systeme. *Abh. K. Akad. Wiss. Berlin.* 289–337.

Weiss, C. S. (1816–1817). Ueber eine verbessertre Methode für die Bezeichnung der verschiedenen Flächen eines Kristallisations-Systems. *Abh. K. Akad. Wiss. Berlin.* 286–314.

Weiss, C. S. (1820–1821). Ueber mehrere neubeobachtete Krystallflächen des Feldspathes, und die Theorie seines Krystallsystems im Allgemeinen. *Abh. K. Akad. Wiss. Berlin.* 145–184.

Weiss, C. S. (1823). On the methodical and natural distribution of the different systems of crystallization. In a letter from M. Weiss, Professor of Mineralogy in the University of Berlin, to Dr Brewster. *The Edinburgh Philosophical Journal.* **VIII**, 103–110.

Weiss, C. S. (1829). Über das Dihexaeder, dessen Flächenneigung gegen die Axe gleich ist seinem ebenen Endspitzenwinkel; nebst allgemeineren Betrachtungen über Invertirungskörper. *Abh. K. Akad. Wiss. Berlin.* 3–28.

Weiss, C. S. (1832). Vorbegriffe zu einer Cohäsionlehere. *Abh. K. Akad. Wiss. Berlin.* 57–83.

Weiss, C. S. (1836). Über rechts und links gewundene Bergkrystalle. *Abh. K. Akad. Wiss. Berlin.* 187–205.

Weiss, C. S. (1855). Einige krystallographische Bemerkungen, die sich auf das rhomboëdrische Krystallsystem beziehen. *Abh. K. Akad. Wiss. Berlin.* 7–9; 90–97.

Weissenberg, K. (1924). Ein neues Röntgengoniometer. *Z. Phys.* **23**, 229–238.

Werner, A. G. (1774). *Von der äusserlichen Kennzeichen der Fossilien.* Siegfried Lebrecht Crusius, Leipzig.

Wertheim, G. (1848). Mémoire sur l'équilibre des corps solides homogènes. *Ann. Chim. Phys.* **23**, 52–95.

West J. (1930). A quantitative X-ray analysis of the structure of potassium dihydrogen phosphate (KH_2PO_4). *Z. Kristallogr.* **74**, 306–332.

Westfeld, C. F. G. (1767). *Mineralogische Abhandlungen.* I. E. Dietrich, Göttingen.

Westgren, A. (1932). Zur Chemie der Legierungeng. *Angew. Chemie.* **45**, 33–40.

Westgren, A. and Bradley, A. J. (1928). X-ray analysis of silver aluminium alloys. *Phil. Mag.* Series 7. **7**, 280–288.

Westgren, A. and Lindh, A. E. (1921). Zur Kristallbau des Eisens und Stahl. I. *Z. Phys. Chem.* **98**, 181.

Westgren, A. and Phragmén, G. (1922*a*). Zur Kristallbau des Eisens und Stahl. II. *Z. Phys. Chem.* **103**, 1–25.

Westgren, A. and Phragmén, G. (1922*b*). X-Ray Studies on the Crystal Structure of Steel. *J. Iron & Steel Institute.* **105**, 241–261.

Westgren, A. and Phragmén, G. (1925*a*). X-ray analysis of copper-zinc, silver-zinc and gold-zinc alloys. *Phil. Mag.* Series 6. **50**, 777–788.

Westgren, A. and Phragmén, G. (1925*b*). Zum Kristallaufbau des Mangans. *Zeit. Phys.* **33**, 777–788.

Westgren, A. and Phragmén, G. (1929). X-ray studies on alloys. *Trans. Faraday Soc.* **25**, 379–385.

Wheaton, B. R. (1983). *The tiger and the shark. Empirical roots of the wave-particle dualism.* Cambridge University Press, Cambridge.

Whewell, W. (1825). A general method to calculate the angles made by any planes of crystals, and the laws according to which they are formed. *Phil. Trans. Roy. Soc.* **115**, 87–130.

Whewell, W. (1837). *History of mineralogy,* in *History of the Inductive Sciences,* Third Volume. J. W. Parker, London.

Whiddington, R. (1911). The Production of Characteristic Röntgen Radiations. *Proc. Roy. Soc. A* **85**, 323–332.

Whitaker, M. A. B. (1979). History and quasi-history in physics education. *Phys. Educ.* **14**, 108–112.

Whitrow, G. J. (1938). Robert Hooke. *Phil. of Science* **5**, 403–502.

Wiechert, E. (1896*a*). Die Theorie der Elektrodynamik und die RöntgenSsche Entdeckung. *Phys.-Ökon. Gesel. Königsberg.* **37**, 1–48.

Wiechert, E. (1896*b*). Ueber die Grundlagen der Electrodynamik. *Ann. Phys.* **295**, 283–323.

Wiechert, E. (1897). Ergebniss einer Messung der Geschwindigkeit der Kathodenstrahlen. *Phys.-Ökon. Gesel. Königsberg.* **38**, 3–16.

Wiedemann, G. (1880). Ueber das thermische und optische Verhalten von Gasen unter dem Einflusse electrischer Entladungen. *Ann. Phys.* **246**, 202–257.

Wiederkehr, K. H. (1977). René-Just Haüys Strukturtheorie der Kristalle. *Centaurus* **38**, 278–299.

Wiederkehr, K. H. (1981). Über die Entdeckung der Röntgenstrahlinterferenzen durch Laue und die Bestätitung der Kristallgittertheorie. *Gesnerus* **21**, 351–369.

Wien, W. (1897). Untersuchungen über die electrische Entladung in verdünnten Gasen. *Verh. der physik. Ges. zu Berlin*. Reprinted in *Ann. Phys.* (1898). **301**, 440–452.

Wien, W. (1904). Über die Energie der Röntgenstrahlen. *Phys. Zeit.* **5**, 129–130.

Wien, W. (1907). Über eine Berechnung der Wellenlänge der Röntgenstrahlen aus dem Planckschen Energie Element. *Nachrichten Kgl. Gesell. Wiss. Göttingen.* **5**, 598–601.

Wiener, C. (1863). *Die Grundlage der Weltordnung.* C. F. Winter'sche Verlagshandlung, Leipzig und Heidelberg.

Williams, C. M. (1917). X-ray analysis of the crystal structure of rutile and cassiterite. *Proc. Roy. Soc. A* **93**, 418–427.

Winter, J. G. (1916). *The Prodromus of Nicolaus Steno's dissertation concerning a solid body enclosed by process of nature within a solid.* In *University of Michigan Humanistic Studies*, **XI**, 169–273. New York: The MacMillan Company.

Whitaker, M. A. B. (1979). History and quasi-history in physics education. *Physics Education* **14**, 108–112.

Wollaston, W. H. (1802a). A Method of Examining Refractive and Dispersive Powers, by Prismatic Reflection. *Phil. Trans. Roy. Soc.* **92**, 365–380.

Wollaston, W. H. (1802b). On the Oblique Refraction of Iceland Crystal. *Phil. Trans. Roy. Soc.* **92**, 381–386.

Wollaston, W. H. (1809). The description of a reflective goniometer. *Phil. Trans. Roy. Soc.* **99**, 253–258.

Wollaston, W. H. (1812). On the Primitive Crystals of Carbonate of Lime, Bitter-Spar, and Iron-Spar. *Phil. Trans. Roy. Soc.* **102**, 159–162.

Wollaston, W. H. (1813). On the elementary particles of certain crystals. *Phil. Trans. Roy. Soc.* **103**, 51–63.

Wooster, W. A. and Wooster, N. (1933). A graphical method of interpreting Weissenberg photographs. *Z. Kristallogr.* **84**, 327–331.

Wulff, L. (1887). Ueber die regelmässigen Punksysteme. *Z. Kristallogr.* **13**, 503–566.

Wulff, G. (1913a). Über die krystallographische Bedeutung der Richtungen der durch eine Krystallplatte gebeugten Röntgenstrahlen. *Z. Kristallogr.* **52**, 65–67.

Wulff, G. (1913b). Über die Kristallröntgenogramme. *Phys. Zeit.* **14**, 217–220.

Wulff, G. and Uspenski, N. (1913a). Über die Beschaffenheit der Maxima bei Interferenz der X-Strahlen. *Phys. Zeit.* **14**, 783–785.

Wulff, G. and Uspenski, N. (1913b). Über die Interferenz der Röntgenstrahlen. *Phys. Zeit.* **14**, 785–787.

Wyckoff, R. W. G. (1920a). The crystal structure of cesium dichloro-iodide. *J. Amer. Chem. Soc.* **42**, 1100–1116.

Wyckoff, R. W. G. (1920b). The crystal structure of sodium nitrate. *Phys. Rev.* **16**, 149–157.

Wyckoff, R. W. G. (1920c). The crystal structures of some carbonates of the calcite group. *Am. J. Science.* **50**, 317–360.

Wyckoff, R. W. G. (1921). An outline of the application of the theory of space groups to the study of the structures of crystals. *Am. J. Science* **1**, 127–137.

Wyckoff, R. W. G. (1922). *The Analytical Expression of the Results of the Theory of Space Groups.* Carnegie Institute of Washington, Washington D.C.

Wyckoff, R. W. G. (1923). On the Hypothesis of Constant Atomic Radii. *Proc Nat. Acad. Sci. USA* **9**, 33–38.

Wyckoff, R. W. G. (1924). *The structure of crystals.* American Chemical Society Monograph Series. New York, The Chemical Catalog Company, Inc.

Wyckoff, R. W. G. (1925). Crystal structure of high temperature cristobalite. *Am. J. Science* **9**, 448–459.

Wyckoff, R. W. G. (1927). Die Kristallstruktur von Zirkon und die Kriterien für spezielle Lagen in tetragonalen Raumgruppen. *Z. Kristallogr.* **66**, 73–102.

Wyckoff, R. W. G. and Corey, R. B. (1934). Spectrometric measurements on hexamethylene tetramine and urea. *Z. Kristallogr.* **89**, 462–476.

Wyckoff, R. W. G. and Posnjak, E. W. (1921). The crystal structure of ammonium chloroplatinate. *J. Am. Chem. Soc.* **43**, 2292–2309.

Young, T. (1800). Outlines of Experiments and Inquiries Respecting Sound and Light. *Phil. Trans. Roy. Soc.* **90**, 106–150.

Young, T. (1801). The Bakerian Lecture: On the Mechanism of the Eye. *Phil. Trans. Roy. Soc.* **91**, 23–88.

Young, T. (1802). The Bakerian Lecture: On the Theory of Light and Colours. *Phil. Trans. Roy. Soc.* **92**, 12–48.

Young, T. (1804). The Bakerian Lecture: Experiments and Calculations Relative to Physical Optics. *Phil. Trans. Roy. Soc.* **94**, 1–16.

Young, T. (1807). *A course of lectures on natural philosophy and the mechanical arts*. Joseph Johnson, London. New edition (1845). Ed. Rev. P. Kelland. Taylor and Walton. London.

Zachariasen, W. H. (1929). The crystal structure of potassium chlorate. *Z. Kristallogr.* **71**, 501–506.

Zeeman, P. (1897). On the influence of magnetism on the nature of the light emitted by a substance. *Phil. Mag.* **43**, 226–239.

Zeitz, K. (2006). Max von Laue (1879–1960). Franz Steiner Verlag, Stuttgart.

Ziggelaar, A. (1966). Aux origines de la Théorie des Vibrations Harmoniques: Le Père Ignace Gaston Pardies (1636–1673). Etudes sur Ignace Gaston Pardies I. *Centaurus* **11**, 145–151.

Ziggelaar, A. (1980). How did the wave theory of light take shape in the mind of Christiaan Huygens? *Ann. of Science* **37**, 179–187.

Ziggelaar, A. (1997). *Chaos: Niels Stensen's Chaos manuscript*. In *Acta Historica Scientiarum naturalium et Medicinalium*. Danish National Library of Science and Medicine, Copenhagen. Vol. **44**.

Zimmermann, H. (2004). A coffee break for two–oder der 150. Geburtstag. *Deutsche Gesellschaft für Kristallographie Mitteilungen*. Heft **27**, January issue, 15–21.

Zolotoyabko, E. and Quintana, J. P. (2002). Measurment of the speed of X-rays. *J. Synchrotron Rad.*, **9**, 60–64.

Zwicky, F. (1929). On the imperfections of crystals. *Proc. Nat. Acad. Sci. USA* **15**, 253–259.

AUTHOR INDEX

SUBJECT INDEX